Handbook of Soil Conditioners

BOOKS IN SOILS, PLANTS, AND THE ENVIRONMENT

Soil Biochemistry, Volume 1, edited by A. D. McLaren and G. H. Peterson
Soil Biochemistry, Volume 2, edited by A. D. McLaren and J. Skujiņš
Soil Biochemistry, Volume 3, edited by E. A. Paul and A. D. McLaren
Soil Biochemistry, Volume 4, edited by E. A. Paul and A. D. McLaren
Soil Biochemistry, Volume 5, edited by E. A. Paul and J. N. Ladd
Soil Biochemistry, Volume 6, edited by Jean-Marc Bollag and G. Stotzky
Soil Biochemistry, Volume 7, edited by G. Stotzky and Jean-Marc Bollag
Soil Biochemistry, Volume 8, edited by Jean-Marc Bollag and G. Stotzky
Soil Biochemistry, Volume 9, edited by G. Stotzky and Jean-Marc Bollag

Organic Chemicals in the Soil Environment, Volumes 1 and 2, edited by C. A. I. Goring and J. W. Hamaker
Humic Substances in the Environment, M. Schnitzer and S. U. Khan
Microbial Life in the Soil: An Introduction, T. Hattori
Principles of Soil Chemistry, Kim H. Tan
Soil Analysis: Instrumental Techniques and Related Procedures, edited by Keith A. Smith
Soil Reclamation Processes: Microbiological Analyses and Applications, edited by Robert L. Tate III and Donald A. Klein
Symbiotic Nitrogen Fixation Technology, edited by Gerald H. Elkan
SoilWater Interactions: Mechanisms and Applications, Shingo Iwata and Toshio Tabuchi with Benno P. Warkentin
Soil Analysis: Modern Instrumental Techniques, Second Edition, edited by Keith A. Smith
Soil Analysis: Physical Methods, edited by Keith A. Smith and Chris E. Mullins
Growth and Mineral Nutrition of Field Crops, N. K. Fageria, V. C. Baligar, and Charles Allan Jones
Semiarid Lands and Deserts: Soil Resource and Reclamation, edited by J. Skujiņš
Plant Roots: The Hidden Half, edited by Yoav Waisel, Amram Eshel, and Uzi Kafkafi
Plant Biochemical Regulators, edited by Harold W. Gausman
Maximizing Crop Yields, N. K. Fageria

Transgenic Plants: Fundamentals and Applications, edited by Andrew Hiatt
Soil Microbial Ecology: Applications in Agricultural and Environmental Management, edited by F. Blaine Metting, Jr.
Principles of Soil Chemistry: Second Edition, Kim H. Tan
Water Flow in Soils, edited by Tsuyoshi Miyazaki
Handbook of Plant and Crop Stress, edited by Mohammad Pessarakli
Genetic Improvement of Field Crops, edited by Gustavo A. Slafer
Agricultural Field Experiments: Design and Analysis, Roger G. Petersen
Environmental Soil Science, Kim H. Tan
Mechanisms of Plant Growth and Improved Productivity: Modern Approaches, edited by Amarjit S. Basra
Selenium in the Environment, edited by W. T. Frankenberger, Jr., and Sally Benson
Plant–Environment Interactions, edited by Robert E. Wilkinson
Handbook of Plant and Crop Physiology, edited by Mohammad Pessarakli
Handbook of Phytoalexin Metabolism and Action, edited by M. Daniel and R. P. Purkayastha
Soil–Water Interactions: Mechanisms and Applications, Second Edition, Revised and Expanded, Shingo Iwata, Toshio Tabuchi, and Benno P. Warkentin
Stored-Grain Ecosystems, edited by Digvir S. Jayas, Noel D. G. White, and William E. Muir
Agrochemicals from Natural Products, edited by C. R. A. Godfrey
Seed Development and Germination, edited by Jaime Kigel and Gad Galili
Nitrogen Fertilization in the Environment, edited by Peter Edward Bacon
Phytohormones in Soils: Microbial Production and Function, William T. Frankenberger, Jr., and Muhammad Arshad
Handbook of Weed Management Systems, edited by Albert E. Smith
Soil Sampling, Preparation, and Analysis, Kim H. Tan
Soil Erosion, Conservation, and Rehabilitation, edited by Menachem Agassi
Plant Roots: The Hidden Half, Second Edition, Revised and Expanded, edited by Yoav Waisel, Amram Eshel, and Uzi Kafkafi
Photoassimilate Distribution in Plants and Crops: Source–Sink Relationships, edited by Eli Zamski and Arthur A. Schaffer
Mass Spectrometry of Soils, edited by Thomas W. Boutton and Shinichi Yamasaki
Handbook of Photosynthesis, edited by Mohammad Pessarakli
Chemical and Isotopic Groundwater Hydrology: The Applied Approach, Second Edition, Revised and Expanded, Emanuel Mazor
Fauna in Soil Ecosystems: Recycling Processes, Nutrient Fluxes, and Agricultural Production, edited by Gero Benckiser
Soil and Plant Analysis in Sustainable Agriculture and Environment, edited by Teresa Hood and J. Benton Jones, Jr.
Seeds Handbook: Biology, Production, Processing, and Storage: B. B. Desai, P. M. Kotecha, and D. K. Salunkhe

Modern Soil Microbiology, edited by J. D. van Elsas, J. T. Trevors, and E. M. H. Wellington

Growth and Mineral Nutrition of Field Crops: Second Edition, N. K. Fageria, V. C. Baligar, and Charles Allan Jones

Fungal Pathogenesis in Plants and Crops: Molecular Biology and Host Defense Mechanisms, P. Vidhyasekaran

Plant Pathogen Detection and Disease Diagnosis, P. Narayanasamy

Agricultural Systems Modeling and Simulation, edited by Robert M. Peart and R. Bruce Curry

Agricultural Biotechnology, edited by Arie Altman

Plant–Microbe Interactions and Biological Control, edited by Greg J. Boland and L. David Kuykendall

Handbook of Soil Conditioners: Substances That Enhance the Physical Properties of Soil, edited by Arthur Wallace and Richard E. Terry

Additional Volumes in Preparation

Environmental Chemistry of Selenium, edited by William T. Frankenberger, Jr., and Richard A. Engberg

Sulfur in the Environment, edited by Douglas G. Maynard

Soil–Machine Interactions: A Finite Element Perspective, edited by Jie Shen and Radhey Lal Kushawa

Principles of Soil Chemistry: Third Edition, Revised and Expanded, Kim H. Tan

Handbook of Soil Conditioners

Substances That Enhance the Physical Properties of Soil

Edited by

Arthur Wallace
University of California–Los Angeles, Los Angeles, and Wallace Laboratories, El Segundo, California

Richard E. Terry
Brigham Young University
Provo, Utah

CRC Press is an imprint of the
Taylor & Francis Group, an **informa** business

CRC Press
Taylor & Francis Group
6000 Broken Sound Parkway NW, Suite 300
Boca Raton, FL 33487-2742

First issued in paperback 2019

© 1998 by Taylor & Francis Group, LLC
CRC Press is an imprint of Taylor & Francis Group, an Informa business

No claim to original U.S. Government works

ISBN-13: 978-0-8247-0117-8 (hbk)
ISBN-13: 978-0-367-40068-2 (pbk)

This book contains information obtained from authentic and highly regarded sources. Reasonable efforts have been made to publish reliable data and information, but the author and publisher cannot assume responsibility for the validity of all materials or the consequences of their use. The authors and publishers have attempted to trace the copyright holders of all material reproduced in this publication and apologize to copyright holders if permission to publish in this form has not been obtained. If any copyright material has not been acknowledged please write and let us know so we may rectify in any future reprint.

Except as permitted under U.S. Copyright Law, no part of this book may be reprinted, reproduced, transmitted, or utilized in any form by any electronic, mechanical, or other means, now known or hereafter invented, including photocopying, microfilming, and recording, or in any information storage or retrieval system, without written permission from the publishers.

For permission to photocopy or use material electronically from this work, please access www.copyright.com (http://www.copyright.com/) or contact the Copyright Clearance Center, Inc. (CCC), 222 Rosewood Drive, Danvers, MA 01923, 978-750-8400. CCC is a not-for-profit organization that provides licenses and registration for a variety of users. For organizations that have been granted a photocopy license by the CCC, a separate system of payment has been arranged.

Trademark Notice: Product or corporate names may be trademarks or registered trademarks, and are used only for identification and explanation without intent to infringe.

Visit the Taylor & Francis Web site at
http://www.taylorandfrancis.com

and the CRC Press Web site at
http://www.crcpress.com

Preface

Many books, reviews, and symposia have covered the various materials used in the past to improve the physical properties of soil. Gypsum, lime, compost, biosolids (sewage sludge), and polymers have been the subject of various reports. This volume, however, is the first to integrate several uses and functions of mineral conditioners, organic conditioners, and synthetic conditioners. Its scope is unique. This handbook outlines the usage of many products that can be considered as soil conditioners, and also emphasizes how many of these products interact in beneficial ways.

We define a soil conditioner as a substance—natural or manufactured—that has the ability to improve the physical properties of soil, with emphasis on the soil's ability to grow plants. Soil conditioners enhance the quality of soil for many of its functions.

The definition of soil conditioners presented here is not universally used. Nevertheless, it is important to consolidate the many products that give somewhat similar results when used on soil. Some materials that we classify as soil conditioners also do other things to soil besides changing or improving its physical properties. Appropriate products, then, can also be considered as nutrient sources, as suppliers of organics and microorganisms, as well as soil conditioners.

Any substance added to soil, to benefit either the soil or the crop, is called a soil amendment. Soil conditioners, therefore, are also soil amendments. Technically, fertilizers are also soil amendments, but they are not usually discussed with soil amendments (or conditioners) used to improve aeration, increase water infiltration, decrease erosion, and enhance tilth.

A soil amendment is any material applied to the soil to improve it or the crop it supports. There are three general types of amendments:

1. Fertilizers: Any nutrient or combination of nutrients, consistent with legal requirements, that are applied to the soil
2. Soil conditioners: Any material applied to soil to improve one or more of its physical properties
3. Biological/physiologically active substances: Mostly microorganisms, food for microorganisms, and plant hormones or regulators

The boundaries between these three groups are not strictly defined; there nearly always is overlap. Many materials possess two and sometimes all three of these functions. For example, manure has fertilizer properties; it conditions soil, and it provides microbes, organic carbon, vitamins, and regulators. It is, therefore, an amendment that participates in each of the three subgroups.

Each of the three types of amendments may be subdivided. For purposes of this book, soil conditioners may be classed generally as 1) mined or mineral, 2) organic, 3) synthetic, and 4) waste or by-products. Other classifications are sometimes used.

The major purpose of this book is to discuss the various products added to the soil that improve its physical properties. Some of the products have more value than others, especially for specific intended usages. Some will have more value for particular soils or crops than for others. Precision, extensive soil analysis, and a holistic approach to soil management can help get the most from tools classed as soil conditioners.

Very often, a soil conditioner gives little or no response when it theoretically should. In those cases, there must be reasons. With adequate analysis and a holistic approach, the reasons may be identified so that satisfactory results are ultimately possible. Too often, symptoms of problems are treated without looking for the underlying causes.

Several concerns require increased improvement in the physical properties of soils. Much progress has been made in the past 100 years in enhancing plant nutrition. There is an equal need now to enhance the physical properties of soil. Full response to the high levels of nutrients (fertilizers) applied will seldom be achieved without simultaneous improvement in the physical properties of the soil. Soil conditioners can do this when properly used. As more food will be necessary in the coming decades, because of world population growth, soil improvement will become increasingly important.

We claim that the best management practices are the best management practices only when used with other best management practices. Positive interactions that have been documented among some soil conditioners emphasize this fact. When various inputs are used simultaneously, effects on crops are often greater than the sum of the individual responses. Gypsum, various organics, and water-soluble polymers can interact favorably.

As much as $25 billion is spent in the United States per year for soil care. Worldwide, the value is much larger. The fraction used to improve the physical properties of soil, however, is too small, and much more can be done.

Losses of all kinds caused by soil erosion are reported at $40 billion each year in the United States, and ten times that figure for the whole world. Among techniques for erosion control, in addition to cropping systems and farm layout, are various uses of most of the major soil conditioners including organics, gypsum, lime and water-soluble polymers, especially polyacrylamide. One reason for so much soil erosion, however, is that the techniques for control are not thoroughly known and/or are greatly underutilized.

World population is increasing so rapidly that many people worry that soon the need for food and other resources will not be met. An urgent concern is that the land and production systems are not sustainable. This means that they cannot last indefinitely, especially with current use procedures. Soil conditioners certainly can help. Degraded soils can and must be vastly improved to achieve sustainability and needed production. Levels of soil organic matter can be increased even though most are decreasing throughout the world. Many advantages will result as soil organic levels are increased: scarce water supplies can be used more efficiently, there will be less erosion, and crop yields can be dramatically increased. The simultaneous use of multiple soil conditioners can often make a huge difference in these matters.

In past decades and centuries, lime, manure, gypsum, green manure, and cover crops have been the major conditioners. More recently, a myriad of waste products have also been used as soil conditioners. Most composts are derived from waste products. In recent decades, synthetic polymers have been added as conditioners. Zeolites, pozzolans, and other mined products are also being added. At this time, the scientific basis for most of the newer materials has not yet advanced as much as it has for the older materials. This is reflected in the various chapters of this book.

Water-soluble polymer soil conditioners are relatively new and have much promise for improving physical properties of soils. They are probably most effective when combined with other soil conditioners, especially with organic matter and gypsum. Water-soluble polymers can improve soil aeration and water infiltration, prevent crusting, and protect against compaction. Some results are: less erosion, greater water-use efficiency, higher yields, few soil-borne diseases, and earlier crop maturity.

Much progress has been made in the past ten years in the use of polymer soil conditioners to solve various problems. The products used in the 1990s are far superior to those used in the 1950s.

We firmly believe that even more potent products will become available for future use. Moreover, there is yet much to be gained from further discovery and implementation of beneficial interactions in soil that are possible with polymers and other materials.

The landfill crisis and other environmental concerns have resulted in a surge in recycling, including composting. In the near future, some 50 million megagrams (more than 55 million tons) of compost in the United States will be available for land use. Use of compost on soil is increasing, but for full use of the 50 million MG, much more must be used on farms. In order for this to happen, value-added procedures in which one or more soil conditioners are combined will probably be necessary.

Some new conclusions emerge from this overview of compost soil conditioners. Compost should not be considered to be in competition with other soil conditioners (amendments), but rather as a valuable companion/aid to go with at least some of the others. There are separate benefits for each, and also benefits that are enhanced by interactions with one another. Maximum benefits from a given soil conditioner are obtained under conditions obtained from use in conjunction with other soil conditioners (amendments) that result in synergy or near synergy. Most likely, the only way for compost to achieve its rightful level of use is in combination with other conditioners (amendments). Also, its possible value in plant disease suppression is as yet mostly unexplored.

Soil improvement is important for many reasons. We have to assume that good land must last forever. As soils are used, however, they tend to degrade. Effort and new inputs can reverse this tendency. Not only must soils be protected from misuse and degradation, but they must be so cared for in order for future generations to produce more food on less land. Further, the public also expects environmental benefits from the land. Specifically, it expects soil to be a means for using and detoxifying wastes, as well as a means for recycling substances necessary for life.

Use of soil conditioners is largely in the realm of technology. This book presents a blend of scientific and technological information. Some of the citations used are from trade journals. These sources tend to fine-tune technologies. Many of the relevant findings are reported, often first and exclusively, in those sources.

We have designed this book to address the need of a wide range of concerned groups. It is hoped that the professionals in soil and plant sciences as well as those with practical concerns, will find much of value.

Arthur Wallace
Richard E. Terry

Contents

Part III. Mineral Soil Conditioners

Part IV. Polymer Soil Conditioners

Part V. Example Uses of Soil Conditioners

Contributors

Abdulaziz R. Al-Harbi Department of Plant Production, College of Agriculture, King Saud University, Riyadh, Saudi Arabia

Abdulrasol M. Al-Omran Department of Soil Science, College of Agriculture, King Saud University, Riyadh, Saudi Arabia

M. Ben-Hur Institute of Soils and Water, Agricultural Research Organization, The Volcani Center, Bet Dagan, Israel

D. L. Bouranis Department of Agricultural Biotechnology, Agricultural University of Athens, Athens, Greece

C. E. Clapp Department of Soil, Water, and Climate, Agricultural Research Service, U.S. Department of Agriculture, St. Paul, Minnesota

Alejandro Fierro Center for Horticultural Research, Laval University, Ste-Foy, Quebec, Canada

Gary A. Lehrsch Northwest Irrigation and Soils Research Laboratory, Agricultural Research Services, U.S. Department of Agriculture, Kimberly, Idaho

Guy J. Levy Institute of Soils and Water, Agricultural Research Organization, The Volcani Center, Bet Dagan, Israel

Ruilong Liu Department of Soil, Water, and Climate, University of Minnesota, St. Paul, Minnesota

S. Miyamoto Agricultural Research Center at El Paso, Texas A&M University, El Paso, Texas

Sheldon D. Nelson Department of Agronomy and Horticulture, Brigham Young University, Provo, Utah

Jeffrey Norrie Department of Technology, Acadian Seaplants Limited, Dartmouth, Nova Scotia, Canada

L. Darrell Norton National Soil Erosion Research Laboratory, Agricultural Research Service, U.S. Department of Agriculture, Purdue University, West Lafayette, Indiana

Alan Olness North Central Soil Conservation Research Laboratory, Agricultural Research Service, U.S. Department of Agriculture, Morris, Minnesota

Marcello Pagliai Research Institute for Soil Study and Conservation, Ministry of Agriculture, Florence, Italy

Antonio J. Palazzo Cold Regions Research and Engineering Laboratory, Geochemical Sciences Division, U.S. Army, Hanover, New Hampshire

Michael N. Quigley Cornell Nutrient Analysis Laboratories, Department of Soil, Crop and Atmospheric Sciences, Cornell University, Ithaca, New York

Jack E. Rechcigl Soils and Environmental Quality, Soil and Water Science Department, Range Cattle Research and Education Center, University of Florida, Ona, Florida

Charles W. Robbins Northwest Irrigation & Soils Research Laboratory, Agricultural Research Service, U.S. Department of Agriculture, Kimberly, Idaho

Byron M. Shock Shock Computer Consulting, Ontario, Oregon

Clinton C. Shock Malheur Experiment Station, Oregon State University, Ontario, Oregon

L. L. Somani Department of Agricultural Chemistry and Soil Science, Rajasthan College of Agriculture, Udaipur, Rajasthan, India

Margie Lynn Stratton Soil and Water Science Department, Range Cattle Research and Education Center, University of Florida, Ona, Florida

Malcolm E. Sumner Department of Crop and Soil Sciences, University of Georgia, Athens, Georgia

Donavon H. Taylor Department of Plant and Earth Sciences, University of Wisconsin–River Falls, River Falls, Wisconsin

Richard E. Terry Department of Agronomy and Horticulture, Brigham Young University, Provo, Utah

K. L. Totawat Department of Agricultural Chemistry and Soil Science, Rajasthan College of Agriculture, Udaipur, Rajasthan, India

Nadia Vignozzi Research Institute for Soil Study and Conservation, Ministry of Agriculture, Florence, Italy

Arthur Wallace University of California–Los Angeles, Los Angeles, and Wallace Laboratories, El Segundo, California

Garn A. Wallace Wallace Laboratories, El Segundo, California

C. Frank Williams Department of Agronomy and Horticulture, Brigham Young University, Provo, Utah

X. C. (John) Zhang National Soil Erosion Research Laboratory, Agricultural Research Service, U.S. Department of Agriculture, Purdue University, West Lafayette, Indiana

1
Introduction: Soil Conditioners, Soil Quality and Soil Sustainability

Arthur Wallace *University of California–Los Angeles, Los Angeles, and Wallace Laboratories, El Segundo, California*

Richard E. Terry *Brigham Young University, Provo, Utah*

I. WHY THIS BOOK

Several reasons exist concerning the urgent need for soil improvement. The well being of the world does depend very much upon how the soil is maintained. The present generation is just starting to appreciate this fact. Foremost among the values of soil is its capacity to produce food. But there are other important reasons for giving more priority to the care of soil. Among these are that soil acts as an environmental filter for cleaning air and water, it is a major sink for unwanted or waste gases which if managed properly may even help avert or postpone global climate change, it can decompose and help detoxify organic wastes, and it is a means for recycling of the nutrients needed at all levels of life (Doran et al., 1994). If mismanaged, however, soil can fail to provide these benefits or it may even have some opposite effects. Human health; well-being, and prosperity are all closely related to the quality of soil. Robert (1995) has said that soil is an interface with the lithosphere, the atmosphere, the hydrosphere and the biosphere. This unique role gives soil a central position in the sustainability and quality of the planet. Environmental processes interact with soil.

Soil is not really renewable in the short term and must be cared for not only to maximize output or maximize other values within reason but also to make it possible for future generations to obtain similar benefits from the land. Ownership of land, therefore, is quite different from most other kinds of ownership. Soil has to last into perpetuity. Land ownership therefore is really stewardship. Wise use of soil conditioners is certainly part of the stewardship.

A. Soil Conditioners

For purposes of this book "soil conditioner" is defined as a substance that improves the physical properties of soil. Soil conditioners include both synthetic and natural products. This definition obviously includes substances commonly known as soil amendments; there is then overlap in the meanings of such terms as soil conditioners, amendments, and agricultural minerals. The general use of the term amendment usually implies substances that enhance any and all of the physical, biological, and nutritional properties of soil. Legal definitions in California call all polymers, lignites or humates, wetting agents, and various microbial products used on soil as auxiliary soil and plant substances rather than conditioners or amendments (Chaney et al., 1992). Compost is even defined separately in California. It is "a biologically stable material derived from the composting process. Composting is the biological decomposition of organic matter which inhibits pathogens, viable weed seeds, and odors. Composting may be accomplished by mixing, and piling in a way as to promote aerobic or anaerobic decay, or both."

All of these mentioned substances, regardless of definition, in varying degrees influence the physical properties of soil and have soil conditioning value. Some influence the plant nutrition as well. Many of them are also of considerable importance to the biological properties of soil. A clear-cut separation of products into each of these three categories, i.e., those that influence physical, biological, or nutritional properties, cannot be done because there is considerable overlap.

Perhaps a more modern concept of soil conditioners should state that their use is to enhance soil quality and that they are soil enhancers. That ideally is the case when appropriate soil conditioners are properly used, even though the scientific community is still working on an appropriate definition of soil quality (Doran et al., 1994). Soil conditioners may even help crystallize a useful definition of soil quality.

Soil conditioners as defined here include many kinds of organics, gypsum, lime, various water-soluble polymers, various cross-linked polymers that hold water in soil, zeolites, diatomaceous earth, living plants, microbes, many industrial waste products and others. Of course there is considerable variation in the degree to which these substances improve physical properties of soil.

Soil is usually subject to severe problems that require the use of various soil conditioners to keep it tillable, fertile, and chemically and biologically healthy, and also to prevent its loss to various kinds of erosion. In many ways, intelligent use of soil conditioners can help reach these goals. Conditioners generally do not degrade soil and their value is often long term. Too much can be applied as well as too little.

This book discusses various interactions that do occur among the different conditioners to result in important additive and synergistic effects. A soil conditioner may have limited value when used alone but much more if used with others. This point is of crucial importance. Vast improvement of the soil can result with well-defined holistic management.

The physical and biological properties of soil are undoubtedly as important as the nutritional properties, which are mostly but not entirely in the realms of inorganic chemistry but with involvement of biological cycles. The physical properties of the soil are indirectly related to these factors and have, in the past, received far too little attention. The Western Fertilizer Handbook (1994) devotes less than 10% of its pages to the subject of soil conditioners and amendments. It is quite safe to assume that major advances in improvement of crop yields in the future that are possible through management will come from several different soil conditioners. Near maximum amounts of fertilizers are being added, but average yields are less than half of those theoretically possible. Increases due to fertilizer are getting smaller than in the past. A hypothesis is offered that interactions among different conditioners and amendments is an important key to more efficient crop production. It is therefore imperative that those various disciplines concerned with productivity of the land work more closely together than in the past.

We have come to the understanding (but of course subject to ongoing research) that no one technology including those with soil conditioners really should stand alone. Best management practices are best management practices only if they are combined with other best management practices. Additive and synergistic responses can be obtained, for example, when gypsum, organic matter, and water-soluble polymers are all used according to directions and simultaneously (see Chapters 15, 17, and 19). But even these may not be beneficial for health of the land and plant growth if pronounced limiting factors of different nature have not been first corrected. All must be done together in a holistic approach if the value of each is to be maximized.

The opportunities for increased use of soil conditioners are intriguing and challenging. There is an urgent call for better quality soils that resist erosion. There is a loud cry for more sustainability not only in agriculture but also in everything that we do. There is a pressing movement for a cleaner, safer, and more sensible environment. Some efforts are resulting in recycling of waste materials that, partly at least, end up as potential soil conditioners. More research is needed on how best to use them.

B. Food Scarcity Is Being Forecast

1. Food Scarcity and Its Abundance Begins with the Soil

Soil is the essence of our basic life-support system. In turn the support system for soil includes forests, rivers, swamps, mountains, climate, some mines (gypsum, lime, etc.) and knowledge. This list may be expanded to include other items. The concepts of ecosystems and ecosystem interactions are very important when soil is concerned. Most likely soil can efficiently sustain humanity with food, clothing, and clean environmental maintenance only when it is considered and managed from the holistic and ecosystem points of view. Although the emphasis in this book is on soil conditioners and generally their effect on the physical and other properties of soil, it is recognized that all aspects of the soil must be cared for equally. Only then can the full benefits of land use for society be obtained.

2. Maximizing the Effects of Variables

The returns from land use can be maximized in relationship with different factors. Maximized returns can be on the basis of land, or water, or capital, or time or something else. Some important limitations are discussed here, and all relate to the need for soil conditioners. These variously relate to maximum economic yield (Fageria, 1992), highest possible yield, or highest possible food quality with less regard for total yield. Limitations also involve social, economic, and political values of land use and agrarian societies. More and more, as time goes on, the returns from agriculture must be maximized. This will require many disciplines, many professionals, and many individuals all working together. The effects of limitations must be minimized, sometimes with soil conditioners, if maximum value is to be obtained from the soil.

At least five major constraints to reaching crop yield goals exist globally (Wallace 1994a):

Where Land Is Limited. Many nations strive for food security by producing sufficient of the basic foods to have minimum requirements. Some nations find it extremely difficult to have self-sufficiency because population per unit of arable land is too great. In some developing countries, maximum yields that are possible on the crop land are often needed regardless of cost so long as the cost is in local currency. Maximum economic yield has little meaning to such nations. Nations which have a large population on a limited land base cannot be comfortable with a policy of low-input agriculture, even though such agriculture produces food at low cost. Maximum yield is more important there. When land is limited, use of soil conditioners and other inputs must be maximized if highest possible production is to be obtained.

Where Capital Is Limited. Most farms, even in the USA, are on extremely tight budgets, and operators are prevented from making desired changes for progress because capital is not available. Yield goals must always consider the availability of needed capital. Highest return per dollar available may be most practical, but it may not even be the maximum economic yield if funds are not available to reach maximum economic yield. Soil conditioners do cost money but they can help reach needed goals. Changes in the way capital is obtained and used on farms may be necessary.

Where There Are Environmental Limitations. Environmental limitations are probably universal. Inputs into agriculture generally enter into biological and geochemical cycles. At least some of the effects could be undesirable. Constant study and monitoring are essential, and it is conceivable that yield goals may need to be modified in places for environmental reasons. Soil conditioners do function to decrease certain environmental limitations.

Where There is Expertise Limitation. Information supply with its dissemination and its implementation is still a major hindrance to increased crop production and to obtaining environmental benefits from the land. Information will be of little value if it is not available to users. Also if research does not keep pace with needs, production will decrease and/or the environment will degrade. Proper use of soil conditioners does require expertise.

Where There is Resource Limitation. This is the area of much concern to the environmental community, which is afraid that a decreasing resource base that provides inputs will more and more limit crop production and decrease soil quality with time. Prudent conservation, but within limits, can protect the resource base. The resource base does include all soil conditioners, but the availability of some is no long-term problem. All aspects of the resource base need constant study and evaluation. Water may head the list.

C. Society's New Emphasis on Soil Quality and Sustainability

In 1992 the *American Journal of Alternative Agriculture* published a special issue on soil quality (Youngberg, 1992). Eleven papers were included. Among them were "Coming full circle—the new emphasis on soil quality" (Haberern, 1992), "Soil quality—the key to a sustainable agriculture" (Papendick and Parr, 1992), "New emphasis on soil quality: attributes and relationship to alternative and sustainable agriculture" (Parr et al., 1992), "The need for soil quality index" (Granatstein and Bezdicek, 1992), "Characterization of soil quality" (Arshad and Coen, 1992), and "Soil biological criteria as indicators of soil quality" (Visser and Parkinson, 1992). This was an awesome new beginning for the world's sudden concern about soil quality.

In 1995 the Soil Science Society America (SSSA) made an official statement on soil quality (*Agronomy News*, 1995). The introduction to the statement said that soil quality and soil health have received increased attention since the U.S. National Academy of Sciences published its 1993 report entitled "Soil and Water Quality: An Agenda for Agriculture." This increased recognition that soil resources affect everyone provides a unique opportunity to define soil quality, to suggest its use as a focal point for addressing complex environmental and land-use issues, and to suggest criteria for monitoring and evaluating it. The statement follows (used by permission):

"Soil Quality: A Conceptual Definition.

"Public interest in soil quality is increasing throughout the world as humankind recognizes the fragility of earth's soil, water, and air resources and the need to protect them to sustain civilization. To understand soil quality, however, one must first be aware of the complexity of soil and its intrinsic value.

"Soil is a living system that represents a finite resource vital to life on earth. It forms the thin skin of unconsolidated mineral and organic matter on the earth's surface. It develops slowly from various parent materials and is modified by time, climate, macro- and micro-organisms, vegetation, and topography. Soils are complex mixtures of minerals, organic compounds, and living organisms that interact continuously in response to natural and imposed biological, chemical, and physical forces. Vital functions that soils perform within ecosystems include: (i) sustaining biological activity, diversity, and productivity; (ii) regulating and partitioning water and solute flow; (iii) filtering, buffering, degrading, immobilizing, and detoxifying organic and inorganic materials, including industrial and municipal by-products and atmospheric depositions; (iv) storing and cycling nutrients and other elements within the earth's biosphere; and (v) providing support for socioeconomic structures and protection for archeological treasures associated with human habitation.

"Conceptually, the intrinsic quality or health of a soil can be viewed simply as its 'capacity to function.' More explicitly, SSSA defines soil quality as:

> The capacity of a specific kind of soil to function, within natural or managed ecosystem boundaries, to sustain plant and animal productivity, maintain or enhance water and air quality, and support human health and habitation.

"By encompassing productivity, environmental quality, and health as major functions of soil, this definition requires that values be placed on specific soil functions as they relate to the overall sustainability of alternate land-use decisions. Although unstated, the definition also presumes that soil quality can be expressed by a unique set of characteristics for every kind of soil. It recognizes the diversity among soils, and that a soil that has excellent quality for one function or product can have very poor quality for another. The functions encompassed by soil quality integrate chemical, physical, and biological properties and processes occurring within every soil and are responsive to human use and management decisions. Soil quality in its broadest sense is enhanced by land-use decisions that focus on single functions. Soil quality can be degraded by using inappropriate tillage and cropping practices; through excessive livestock grazing or poor timber harvesting practices; or by misapplication of animal manures, irrigation water, fertilizers, pesticides, and municipal or industrial by-products. To enhance soil quality, everyone must recognize that the soil resource affects the health, functioning, and total productivity of all ecosystems. We must become more aware of potential side effects of soil management and land-use decisions. By addressing the integrative concept of soil quality, SSSA members can provide the leadership for better long-range, landscape-scale planning and land-use, thus helping to protect finite and living soil resources and fragile landscapes.

"To evaluate soil quality for various land-uses, assessments will be required over time and across space. For the evaluations to be meaningful, research must develop soil quality indicators that are sensitive, reliable, reproducible, and capable of detecting changes in soil physical, chemical, and biological properties, processes, and their interactions. The indicators must be qualitative or semiquantitative for immediate use by farmers or land managers, and quantitative for the people responsible for developing long-range, process-based strategies that will improve land-use; enhance plant, animal, and human health; and sustain biological diversity within natural and fabricated ecosystems. Numerous specific soil measurements are being evaluated, but to help identify the attributes that will be most useful for soil quality evaluations, research and education are needed to: (i) identify the vital functions of soil for which assessments need to be made; (ii) determine the appropriate methods, measurements, and criteria for interpreting those assessments; (iii) develop and evaluate accessible databases as repositories for high quality, reliable soils information; and (iv) design improved land use practices that will ensure the economic viability, environmental safety, social acceptability and value that is appropriate for each soil resource. This is a great challenge, but our premise is that determining how to evaluate soil quality provides a unique and important opportunity for the SSSA to demonstrate the utility of our profession and to serve as a focal point for addressing complex environmental and land-use issues."

To promote new research on soil quality, the Natural Resources Conservation Service under the United States Department of Agriculture has recently formed a soil quality institute. This is welcome.

1. Can Soil Quality Be Measured?

When growers have soil analyzed, they usually test how much fertilizer they need—nitrogen, phosphorus, potassium and other nutrients. But there's another factor: tilth, or overall quality of the soil. Can it be measured?

Nutrients are just one part of the picture of a healthy soil. There are other important physical, chemical and biological characteristics that may be as important as nutrients to get optimal yields over a long period of time. There has been no organized way to test or interpret overall soil health. That is being changed with new research, however. For example, Oregon State University researchers in a new project will study the biological, chemical, and physical properties of soils to determine which properties are most sensitive to changes in soil quality as influenced by soil management and cover crops (Carpenter, 1995).

Carpenter says that interest is increasing with growers in the understanding and measurement of soil quality. In some cases their yields are slowly dropping over time, but it is not obvious why; there is no disease problem. They would also like to know as fast as possible, for example, whether a change in soil management such as a soil conditioner or adoption of a winter cover crop system will improve their soil. New tests for measurement of soil quality may help.

Measurements for soil quality such as microbial biomass, carbon fractions, or soil enzyme activity may be able to indicate changes years sooner than traditional methods of measuring changes in total organic material in the soil. Aeration and tilth should not be overlooked. Traditional tests may take five to ten years for a given practice in the field before researchers can measure detectable changes, whereas some new soil quality tests might detect differences after only one or two years. This is the hope. Because of the inherent complexity of soils, considerable research is likely needed before definable soil quality standards can be established.

2. Sustainability

Wallace (1994a) compared and contrasted ten different definitions of sustainability as related to agriculture. The one liked best was that of Fretz (1992):

> Sustainable agriculture has been defined as an overarching, interconnected framework of technologies, practices, and systems developed in response to environmental, social, economic, political, agronomic, and horticultural issues. Sustainability achieves maximum efficiency, while enhancing or maintaining environmental quality. Sustainable agricultural systems are resource-conserving. Sustainable practices maximize nutrient recycling. Sustainable systems protect groundwater and surface water resources and reduce soil erosion to a minimum. Sustainable systems must view the economics of production not only as short term, but also as long term. Sustainable must be defined as forever.

This definition was further expanded (Fretz et al., 1993).

3. Conventional Agriculture vs. Sustainability

Some questions raised by the critics of conventional agriculture regarding sustainability include (Wallace, 1994a):

1. Can agriculture be done in a way that does not result in degradation or erosion of the soil?

2. Can damage to ecosystems and natural resources resulting from use of some agricultural inputs be avoided?
3. How long can nonrenewable natural resources be dissipated without irreversible damage to agricultural productivity? It has been appropriately said that no soil can sustain the constant depletion of critical elements contained in plant parts harvested and moved to other locations (Buol, 1995). Unless compensated for, the soil will not be sustainable.
4. Can all the phases of the agriculture, including its support systems, be in harmony with nature's cycles of carbon, nitrogen, sulfur, phosphorus, potassium, water, and others so that no necessary components someday fail to function?
5. Can the output be sufficient to justify the energy required for the inputs?
6. Can the practices in use all be socially acceptable? Do any of the practices make the produce unusable, for either real or perceived reasons, for intended use?
7. Is the land that is being used sufficiently free from near-or-far-away industrial and urban pollution that could with time render agriculture on that land impossible?
8. Are any aspects of the agriculture impossible to maintain on a permanent basis or for at least fifty or more years when new breakthroughs could come for sources of energy? In other words, is the agriculture in acceptable equilibrium with the resources required to keep it functional (Fageria, 1992)?
9. Is the agriculture economically viable? This question may never be answerable as long as supplies and prices are manipulated by politicians.
10. Will increased urban and industrial uses of agricultural land and available water in time severely hinder the capacity of the world to produce adequate and quality food?

The questions are "terrible" mostly because the answers are generally not yet known.

Paine and Harrison (1993) say that healthy, sustainable agroecosystems must resemble, to some extent, natural ecosystems. In a natural ecosystem, energy and nutrient flow are relatively stable and self-regulating. Agriculture cannot be wholly like a natural ecosystem because of the removal from the farm of much of the produce with little hope of equivalent return. The produce may even be shipped to other nations. In most cases inputs will nearly always be needed to replace nutrient and other losses.

Carpenter (1995) has commented on the alternative sustainable agriculture program of the United States Department of Agriculture (USDA). In 1988 the USDA set aside funds for studies concerning alternative agriculture. Now known as Sustainable Agriculture and Research Education (SARE), the program has four regional centers. SARE is committed to expanding knowledge and adoption of sustainable farming and ranching practices that are environmentally sound, economically viable, and socially acceptable. SARE research aims to maintain an abundant and safe food supply for generations to come, and to preserve natural resources and vital rural communities. These aims are noble but are extremely challenging. SARE scientists across the country are researching numerous topics with applications to alternative agriculture, including nutrient cycling, nitrogen fixation, the effects of grazing on watersheds, manure composting, plastic mulches, optimum timing of fertilizer application, reduced tillage, bacterial sprays, and biological control of insects and weeds and others. SARE has been changed somewhat by new legislation in 1996.

Some people call striving for sustainable agriculture an "alternative" or "holistic" approach. Others call it an ecological approach. Agriculturists prefer to call it a "systems approach" which adds and integrates knowledge from all disciplines including biotechnology. Biotechnology can help fine-tune food production efficiency and develop pest-resistant plants and animals and even totally new foodstuffs.

The sustainable agriculture approach according to Carpenter (1995) is not construed to be a hippie approach, or a new-age approach, or a totally organic approach, or a step back. Rather its goal is to use all available good practices as possible. We cannot go back, because the world population has increased over five-fold in the last 100 years. The earth has stayed the same size, however. Populations are still increasing to add to the challenge. Any system to be used must consider the total picture. But the good practices have to be researched and identified.

We do live in an era of rapidly diminishing resources. Most major rivers have been dammed. Aquifers have shrunk. We live with declining oil reserves which allow tremendous energy input into agricultural systems. Productionists and environmentalists increasingly are concerned with the same resources: land, water, forests, wildlife. Competition is becoming severe. The sustainable agricultural movement funded by the USDA could help ease tensions among diverse interests and help plot the way to healthier soils, cleaner waters, and more pleasing rural landscapes (Carpenter, 1995). But feeding more people needs to be in the equation also.

4. Some Requirements for Sustainable Agriculture

A tentative list of some ingredients or philosophies variously necessary for a sustainable agriculture include (Wallace, 1994a):

1. Nitrogen from native soil organic matter should not be considered as a major source of nitrogen for growing crops; soil organic matter should be protected as soil organic matter or more should be added to the land to maintain a steady state to preserve soil quality.
2. Soil organic matter levels will be maintained at near their original levels or whatever is needed to have optimum levels. All useful organics will be applied to the land even if farmers need financial help to do so.

3. Inputs into agriculture generally should not exceed, in energy equivalents for cost, what can be returned from the inputs.
4. Inputs should not be classified by consideration of whether or not they are chemically derived or biologically derived. Other more important criteria should be used, such as their sustainability and any contaminants in them.
5. Pesticides in use are to be regulated by type and residue levels that give acceptable risk. Alternatives will continue to be sought and used.
6. All soils should be monitored for heavy-metal balance, and levels will be established above which new additions to availability will not be tolerated. Soils having levels of available heavy metals above given standards should not be used for certain crops. It is important to prevent soils from reaching excess levels because removal or inactivation are very difficult (McBride, 1995).
7. Procedures are to be used that prevent groundwater contamination or other types of contamination in any conceivable form.
8. Plant breeding and genetic engineering are to be used as much as possible to minimize potential toxicity problems, but new plant cultivars will undergo tests for natural toxins before they are used.
9. Inputs will be made to save the land as a higher priority than maximizing production and independently of narrow definitions of sources of inputs.
10. Land should be taken from production if it cannot reasonably meet criteria for safety. Such land may be used for conservation purposes, however.
11. Since consumers will have to pay, directly or indirectly, any additional costs for sustainability, they are entitled to the quality and quantities of produce they want and will pay for. The full cost of what they want must be paid and not necessarily by the farmer.
12. Steps will be taken to minimize wind and water erosion of soil, even if they require changes in farming practices and use of organic water-soluble polyacrylamides with gypsum or lime or other soil conditioners to stabilize soil.
13. Reasonable risk analysis based on sound science and without unusual assumptions will be used in the decision process concerning what is not safe. All new risks are to be compared with background risks for significance.
14. Conservation methods of irrigation are to be promoted. Only in extremely urgent circumstances will farmers be required to let salt levels increase in soil because of insufficient water supplies. Reclamation procedures will be promoted where needed.
15. Tillage methods in use are those that conserve soil organic matter.
16. Organic carbon will be added to soils systematically and even with inorganic nitrogen applications when such can conserve soil organic matter. Crop residues are part of the addition. Anoxic conditions must not develop, however, unless they are specifically needed.

17. Use of nitrate nitrogen fertilizer will be minimized if and where it can adversely affect soil organic matter or groundwater. Alternate nitrogen management procedures are to be developed and promoted.
18. It is recognized that soil is not a renewable resource within the time frame in which we must work (Friend, 1992). The soil must be conserved. Relevant soil conditioners will help.
19. Mining of soils for nutrients and erosion of them must not exceed their regeneration or replacement rates unless appropriate inputs can be used.
20. None of the practices used in agriculture can irreversibly pollute the soil or its surroundings or render the crop produce unsuitable for intended use, either real or perceived.
21. The land cannot be maintained in a way that is not permanently possible at least within an acceptable time frame.
22. The agriculture must be profitable to all parties concerned.
23. The agriculture must have no adverse effects on adjacent or far away environments or people.

All these ingredients and philosophies require continuing research. Soil conditioners directly or indirectly relate to most of them.

The 1991 Yearbook of Agriculture of the USDA was devoted entirely to environmental issues related to agriculture (Smith, 1991). It nicely presented many topics of current concern including soil conservation, water quality, clear air, landfills and recycling, integrated pest management, food safety, sustainable agriculture, and others. Soil conditioners as tools to solve the various problems were not addressed, however. The role of soil conditioners on soil quality and sustainability has not been appropriately appreciated as yet.

As the USA has moved from LISA (Low Input Sustainable Agriculture) in 1985 to SARE (Sustainable Agriculture Research and Education) in 1990 to FAIR (Federal Agricultural Improvement and Reform) in 1996, concepts of sustainability and what is needed have evolved considerably (Bird, 1996; Goulart, 1996). Brumfield (1996) says that sustainable agriculture is not a break with modern agriculture, or another name for organic gardening, or only for small farms, or only with livestock, or a step backwards or the like. The current approach to sustainability by the USDA is intended to provide a full range of options and will not impose difficult or impossible limitations on growers. The approach is expected to result in blending of ecological system approaches, prudent recycling, and existing technologies in best management combinations. In a way it says to do the best that is possible.

5. Different Types of Sustainability

Since sustainability appears to have multiple meanings or different meanings according to various interest groups it may be best to provide an assortment of definitions all of which may be used simultaneously. Some separate concepts of sustainability were proposed by Doxan (1996):

1. Environmental sustainability. This includes ability to assimilate pollutants from human activities. It is associated with little or no risk from inputs like pesticides. The quality of groundwater and other water is protected. Protection against decreases in biodiversity is arranged in this type of sustainability. It includes the total environment.
2. Resource sustainability. This implies that agriculture is done in a way that there will never in the future be a shortage of any nutrient or other needed input. Needed good soil, energy, water, any other inputs, and manpower will always be available.
3. Social sustainability. Future generations will never be jeopardized because of loss of soil quality. Land will always be available with both suitable quantity and quality of produce. Harmonious relationships must exist between urban and farm interests.
4. Economic sustainability. Agriculture must be profitable to all parties. Actually some mechanism must be involved to pay for maintenance and improvement of soil quality with the view to needs of future generations. All this becomes more and more difficult as agricultural trade becomes global. Extreme competition can curtail caring for the land. This is a challenge which must be solved.

6. Attitudes About Sustainability

Elmore (1996) has given a new perspective on the attitudes concerning sustainability. What one considers as sustainable depends very much upon how that person considers himself or herself in relationship with nature. Elmore divides people into four groups according to their personal philosophies.

The first group extols nature over the individual. Conservation of resources has highest priority and the person is willing to be removed for the good of the land. Although this appears to be unselfish, Elmore says it is pointless if people are removed from the world.

The second group considers the individual and the rest of nature as equals. If a man considers himself to be just another animal, that person generally tends to think even less of the animals. But if this impersonal attitude prevails, is food production important? Is quality of human life important?

The third group values nature and ecosystems for their economic value or how much good they are for the individual. People are placed above ecosystems, but the reason for protecting an ecosystem is because it has tangible values to present and future generations. Elmore says that this leads to an egotistical attitude in which nature is considered as a machine to serve man's needs. Self-interest reigns. Nature tends to be defied and much of it is destroyed. He says that this view cannot promote a sustainable agriculture or a sustainable world.

The fourth group considers that man has a mandate, divine or otherwise, to be a steward of the ecosystems of nature. This group, Elmore says, extols deity over all creation, other persons over self, and views humans as faithful, humble caretakers of ecosystems. This view he says will most likely result in a sustainable agriculture.

One conclusion then is to ask ourselves how our beliefs influence long-term stability and sustainability of everything on the earth. Our beliefs or philosophy rather than economics may determine how diligent we are in using many different kinds of soil conditioners and management procedures to improve soil quality.

Population may not be the largest environmental problem of the world. Instead, it may be consumerism, which destroys and exhausts many of the natural resources of the earth (Wallace, 1994f). In 1776, a Scottish moral philosopher, Adam Smith, wrote a book on *Wealth of Nations* (Robison, 1992). Three major points of his that explain why nations prosper are specialization, trade, and freedom of choice. The economic system of free enterprise advocated by Adam Smith has become known as capitalism. Even though capitalism is also known as the theory of self-interest and selfishness, Adam Smith did outline a procedure that would avoid this serious problem (Robison, 1992). Men and women, he said, have a natural need to sympathize one with another. Without a mutual empathy for others, society would degenerate to a war of "all against all." Material consumerism with its devastating effects on natural resources is a form of capitalism not envisioned by Adam Smith.

People must care for one another for the free enterprise system to succeed morally. The ingredient needed is charity, which essentially is working hard to bless the lives of others. The opposite is selfish craving for material things. Included in all this must be an intense desire to care for the soil with future generations in mind.

Proper care of the soil does cost money including that for needed soil conditioners. Among the options to finance better care of the soil are paying more for food, tax-subsidy programs, taking less profit at the farm level, decreasing the level of consumerism, and various combinations of the options. Use of revenues to care for the soil properly needs to be guided by good research, which is also paid for by one or more of the options.

D. Society's Emphasis on Waste Management

Society wants wastes handled in the most environmentally sound way that is possible but consistent with economics. Between 1960 and 1990, the waste stream in the USA increased by 122%, with a total of 178×10^6 MG generated in 1990 (*Statistical Abstracts of the USA*, 1994). During the same period, landfill space continued to decline. In 1988, there were 7924 landfills remaining in the USA, and by 1994 the number had dropped to 3558 (Steuteville, 1995). This combination of events increased the need for recycling and reuse to reduce the waste stream going into limited landfill space (Heckman and Kluchinski, 1996). In 1990, the waste stream included 32 × 10^6 MG of leaves plus yard trimmings (*Statistical Abstracts of the USA*, 1994) all of which can be recycled. See Chapter 2.

Landfill problems, objections to ocean dumping of sewage sludge, and clean air laws all combined to create changes in approaches to waste management in the late 1980s and early 1990s. A central hope in all of these concerns is that wastes can become useful. Recycling is gradually becoming a way of life. Many organizations and journals and magazines have arisen in support of environmentally acceptable waste management. Examples are the Composting Council, MSW (municipal solid waste) Management, and others. One result of these matters is that many wastes are being transformed into soil conditioners. Some examples are given below.

1. Municipal Solid Waste

The waste stream of municipalities traditionally has been placed in landfills in recent decades, but the environmental movement and the perceived shortage of landfill space have caused the various jurisdictions to create laws mandating the removal of much of the MSW from landfills. By the year 2000, many cities must recycle at least 50% of the waste stream. Much of this is organic material that can be composted for use as a soil conditioner (see Chapter 2). A new MSW management journal that features recycling of multiple wastes appeared in 1991.

2. Animal and Crop Production Residues and Sewage Sludge (Biosolids)

This category includes manures, bedding litter, blood and bone meal, spent mushroom compost, straw and other crop residues, wastewater biosolids, and others. Some are transformed into useful soil conditioners by composting processes (see Chapters 2 to 5). Others are used as soil conditioners without composting.

The value of these wastes is their ability to improve the soil both by providing plant nutrients and by improving soil structure and other characteristics. Decomposition of these wastes in the soil enhances the availability of nutrients for crop growth, and a portion of the organics is incorporated into the soil organic matter. This enhances the stability of soil aggregates and increases the moisture and nutrient holding characteristics of the soil.

Vegetative growth rates of several types of plant communities have been shown to increase following treatment of the soil with biosolids, including grasses and legumes (Seaker and Sopper, 1988), desert shrubs (Sabey et al., 1990), and trees and shrubs in loblolly pine forests (McLeod et al., 1986).

Composting alone is supposed to add 50 million MG of soil conditioners to the annual supply in the USA alone in the year 2000 (Hyatt, 1993). The potential value of this to the farm lands cannot be questioned, but there are economic problems (see Chapter 2) and a very urgent need for production of value-added composts (Wallace, 1995b).

Physical properties of soils treated with composts have been shown to improve. Infiltration and retention of water is enhanced and soil temperatures are favorably modified by compost applications (Dick and McCoy, 1993).

3. Industrial By-Products

Gypsum, lime, sulfur, and pyrite are mined products, but in various combinations they are also industrial by-products. An ASA monograph has discussed the possible use on land and problems of such as fly ash, coal ash, flue-gas desulfurization lime-gypsum, and phosphogypsum (Karlen et al., 1995). Chapters 7, 9, and 11 in this book add more information.

4. Other Products Originating From Wastes

Some examples are given in Chapter 18. They include wood ashes, plasterboard gypsum, recycled plastics, and other things. Huge quantities of other gypsum and lime waste are also available (Chapters 2, 8, 9, and 11). Shredded rubber tires are used in athletic fields as a means for creating less compacted turf and as a bulking agent in composts (see Chapter 18).

II. THE ENVIRONMENTAL MOVEMENT AND SOIL CONDITIONERS

Reference to the environment was made in the previous section; industrial by-products are used as soil conditioners in efforts to improve the environment. Soil conditioners produced as by-products of environmental programs can also be used to help solve industrial problems. It works both ways. (See Chapters 3, 5, 6, 7, 8, and 9.)

A. Pollution

1. Pesticides

Degradation of at least some pesticides in soil can be enhanced when the microbial potential of soil is enhanced by addition of some soil conditioners (Block and Clark, 1990). Useful conditioners include organics and soil porosity enhancing water-soluble polyacrylamides (See Chapters 2 and 15). In addition, the use of some plant nutrients can enhance the remediation process. Singh et al. (1996) and Agassi et al. (1995) have shown that removal of pesticides from farms by runoff water can be decreased with water-soluble polyacrylamide (see Chapter 17).

Cover crops may also facilitate the degradation of agricultural pesticides. A number of researchers have described an increase in pesticide decomposition in the rhizospheres of a variety of crop species. Sandmann et al. (1984) found an increased population of 2,4-D-degrading microbes in the rhizosphere of sugarcane. Seibert et al. (1981) reported an increase in production of atrazine metabolites by microorganism in the corn rhizosphere, especially in the presence of decomposing roots. Anderson et al. (1994) found greater degradation of atrazine, metolachlor, and trifluralin in the rhizosphere soil of herbicide resistant *Kochia* sp., compared with nonvegetated soil. See Chapter 21.

2. Heavy Metals

Heavy metals have contaminated land in many parts of the world; they are largely the result of the industrial age (Adriano, 1991). The problem and its control have been the topic of a myriad of studies. An example is a large number of papers in a typical symposium (Adriano, 1991). Biosolids are a source of heavy metals on land. Even the agronomic rate (see Chapter 5) to supply annual needs for the nitrogen requirement of a crop when used for several consecutive years results in a large accumulation of metals. The USEPA, however, after consideration of relevant literature concerning field experiments, has defined a clean sludge in terms of heavy metals (mg kg^{-1}: Zn 2800, Cu 1200, Ni 420, Cd 39, Pb 300, As 41, Hg 17) where unlimited amounts could be added to land if all these metals were below the limit (USEPA, 1993). The conclusion is based on the facts that in combination in organic matter or as poorly soluble salts, the metals are sufficiently bound or inactivated that they do not become available enough to plants to be of concern especially to consumers of the crops (Clapp et al., 1994). National limits on heavy metal loading in soil do vary (MacLean et al., 1987).

Although there may be some problems associated with the USEPA conclusions involved (McBride, 1995), the regulations do suggest a means for decreasing heavy-metal problems on land. Application of large quantities of organic soil conditioners should help decrease metal availability because of inactivating reactions. See Chapter 5.

3. Industrial Pollution

Acid rain has been a widely recognized and publicized means of soil pollution. It was thought to be largely a plant (tree) problem and a lake problem, but recent studies indicate the tremendous damage that has been done with decades of acid rain on the cation balance in forest soils (Likens et al., 1996; Kaiser, 1996). Half of the calcium in those soils has been lost by leaching due to the acid rain. A remedy that has been suggested is application of lime to the soil. Although this idea has merit it may be prohibitively expensive, if regular agricultural lime is used for this purpose. It may, however, be a place where large amounts of industrial by-product lime-gypsum and various kinds of ashes from burning wood products may be applied to the land. See Chapter 11.

4. The Soil as a Remediation Machine

Many studies have been conducted on remediation of contaminated land (see Chapter 21). Some of the procedures drastically change the soils, but others let the biological potential of the soil itself do the remediation. This is one form of bioremediation (Atlas, 1995). It is possible to remove much of the contamination associated with oil and oil-derived products from land with this procedure (Block and Clark, 1990). Sometimes the soil is cultivated with the addition of fertilizer nitrogen, but the biological potential can be increased even more with use of soil conditioners.

In situ bioremediation involves the use of indigenous microbes to transform or decompose chemical contaminants. The simultaneous use of soil conditioners to enhance soil physical, chemical, and microbial conditions could enhance and speed up bioremediation.

In land farming, the degradation of petroleum wastes or spills is enhanced by supplementing the soil with nutrients and oxygen. This is done by fertilizing and cultivating the soil. The use of soil conditioners to enhance physical properties, including infiltration, aeration, and water holding capacity, and to enhance conditions for microbial growth, such as aeration, nutrients, and available organic carbon, could greatly enhance bioremediation (Bollag and Bollag, 1995).

Soil microbes involved in bioremediation often cannot grow and thrive on the carbon of the chemical pollutant alone. It may then be necessary to supplement the microbes with readily available carbon sources, such as crop residues, animal manures, wastewater effluent or biosolids, or plant root exudates provided by cover crops.

Establishment of selected plant species on soils at bioremediation sites could enhance microbial degradation of organic chemical contaminants. The root zone, especially the soil immediately surrounding the root, known as the rhizosphere, sustains microbial populations an order of magnitude or more above those in nonvegetated soil by providing a niche suitable to diverse microbial species and by facilitating co-metabolic transformations. Co-metabolism allows microorganisms to decompose an organic chemical pollutant while most of the microbe's energy and nutrient needs are supplied by plant residues. Current and potential uses of soil conditioners in remediation of contaminated soils are reviewed in Chapter 21.

B. Erosion

Erosion by both wind and water are world-wide problems (Glanz, 1995; Pimentel et al., 1995). The problems have been going on for ages. Dust storms are a troublesome problem in some regions, notably in the western part of the

United States. During the dry year of 1934 a dense cloud developed over a large part of Colorado, Kansas, New Mexico, Oklahoma, and Texas. It was 2400 km long, and it darkened the sky as it moved eastward. Its load of dust was dropped on the land as it advanced, some of it falling on New York and Boston and far out over the Atlantic Ocean (Bear et al., 1986).

So severe was this dust storm that the United States almost immediately established the Soil Conservation Service in 1935. Bear et al. (1986) say that this organization, with Hugh Bennett as its first chief and later known as the father of soil conservation, was the most important move the world has ever made to protect the natural environment. Bennett's book (1939) is still a standard for soil conservation.

In the 1938 Yearbook of Agriculture, *Soils and Men*, six different chapters were devoted to the topic of soil erosion (Bennett and Lowdermilk, 1938; Munns et al., 1938; Enlow and Musgrave, 1938; Kell, 1938; Nichols and Chambers, 1938; Utz, 1938). All these papers resulted in substantial progress in understanding both wind and water erosion. Strip and contour farming grew out of those studies. Better management of natural ecosystems grew out of those studies. The need for coordinated and multiple approaches to erosion control was recognized and adopted. But none of these chapters mentioned use of any of the more prominent soil conditioners other than those like cover crops. See Chapter 18.

Of soil erosion, Bennett and Lowdermilk in 1938 said that soil erosion has been going on ever since man's tilling of the earth has bared the soil to rain and wind. It is not to be confused with geologic erosion. This kind of erosion is part of the complex process of rock weathering and is essential to the formation of soil. Soil erosion, on the other hand, is a vastly accelerated process brought about mostly by human interference with the normal equilibrium between soil building and soil removal. Where the land surface is left bare, soil is exposed directly to the abrasive action of the elements of weather. Processes of an extremely rapid order are set in motion, and soil is displaced much faster than it can be formed. Unless adequate measures are taken to guard against this highly accelerated phenomenon of soil removal, it becomes the most potent single factor contributing to the deterioration of productive land.

Soil erosion has exerted a profound influence on civilization. History is largely a record of human struggle to wrest land from nature. Too frequently, however, man's conquest of the land has been disastrous. Over extensive areas, his use of the earth has resulted in complete or almost complete destruction of the soil resource on which he is dependent. So profound is the relationship among erosion, the productivity of the land, and the prosperity of people, that the history of mankind may be interpreted in terms of the soil and what has happened to it (Glanz, 1995; Pimentel et al., 1995).

Recent studies suggest that man, rather than climatic change, caused once rich and populous regions to be reduced to poverty or abandonment. Abuse or neglect of the land is believed to have played a major part in the decline of civilizations now extinct.

With 60 years of progress since 1935 in controlling soil erosion, hill and gully erosion are almost a thing of the past, but the most devastating form of erosion is still occurring (Glanz, 1995). Wind and water can strip away soil in extensive sheets and at the same time leave little visible evidence of damage. It is this form of soil erosion now in need of major attention.

Pimentel et al. (1995) say that globally, 75 billion MG per year of soil are removed from land, mostly agricultural land. The rate of loss they say has been intensified in recent years with much of the loss in developing countries. They further say that soil erosion is perhaps the major environmental threat to the sustainable and productive capacity of agriculture.

Pimentel et al. (1995) place the economic value of soil erosion in the USA at $44 billion per year. This is for all categories of loss including crop production. Others claim that this figure is too high. There are several off-site costs associated with soil erosion. Pimentel et al. say that the erosion of soil per ha per year in the USA is 17 MG but can be reduced to 1 MG with proper control procedures including internal ridge-plantings, no-till cultivation, crop rotations, living mulches, agroforests, terracing, contour planting, cover crops, and wind breaks (Carter, 1994). See Chapters 2–5, 7, and 15–19 for these and other methods of control.

C. Dwindling Resources

Sustainability of agriculture requires that resources needed for it be always available. This is of much concern for the environmental community, which in general thinks that sustainability requires little input from the outside and much reliance on recycling, legumes, and crop rotations. There are several problems related to the resources needed for farming (Gardner, 1996):

1. Water

Even if water is not considered as a soil conditioner, it has much to do with conditioners. It is an agricultural input. The shortage of clean water is in many areas of the world the greatest resource constraint to agriculture (CAST, 1996). Despite the increasing stresses on water for agriculture, most world-wide irrigation systems waste large amounts of water. Some farmers and homeowners do, too.

Several techniques are improving the efficiency of water use in irrigation (Postel, 1993). Possible savings range from 10 to 50% or more. Drought, for example, has forced California farmers to improve efficiency of water use (McMullin, 1993), but additional efficiency is expected. Drip irrigation, including underground drip; low energy precision application; surge flow irrigation; and monitors with computer controls, mostly following Israel's lead, have all increased efficiency 25% or more (Postel, 1993). There is additional potential for savings of water with soil improvement by use of water-soluble polyacrylamide soil conditioners (Wallace, 1991) (see Chapter 19). Better care of water delivery systems in many parts of the world can greatly improve water-use efficiency (Postel, 1993).

Typical environmental points of view about pricing of irrigation water are expressed by Postel (1989). She calls the general world procedure for providing irrigation water a free ride. "Much of the pervasive overuse and mismanagement of water in agriculture stems from the near-universal failure to price it properly." Irrigation systems throughout the world are often built, operated, and maintained by public agencies that charge too little for these services. Farmers' fees in Pakistan, for example, cover only 13 per cent of the government's costs. In most of the Third World, governmental revenues from irrigation average no more than 10–20% of the full cost of delivering the water. Such undercharging deprives agencies of the funds needed to maintain canals and other irrigation works. It leads to gross inefficiency (Postel, 1993).

Pricing reforms for water are equally needed in the United States, according to Postel (1989). The Federal Bureau of Reclamation typically supplies water to a quarter of the irrigated land in the western USA—more than 4 million hectares—under long-term contracts (typically 40 years) at greatly subsidized prices. Farmers benefiting from the Central Valley Project in California have repaid only 5 per cent of the project's costs over the last 40 years: $50 million out of $931 million (Postel, 1989). Largely because of this underpricing, one-third of the bureau's water irrigates "hay, pasture, and other low-value activities." Industry claims an average of 60 times more value for its use of water compared with farms. This results in severe competition for water and indicates the false value placed on food by many; food at least at the farm level is underpriced, which explains some of the underpricing for water.

Populations are increasing in many parts of the world. As water demands continue to rise and water-supply projects get more difficult to build, water budgets are becoming badly imbalanced. Shrinking groundwater reserves, falling water tables, and demands that far exceed available supplies are important signals of water stress. But perhaps the most significant sign of trouble comes from examining the health of aquatic environments. The damming, diverting, and pollution of watercourses with little regard for the environmental services they provide and the species they support have created havoc on the world's wetlands, deltas, lakes, and rivers, according to Postel (1989).

Several groups or needs compete for water—people, industry, natural ecosystems, and farms. Conflict exists concerning water for greater agricultural productivity, industrial expansion, and urban growth, and as a key life-support for all species and natural communities. More water devoted to human needs means less for ecosystems, and in many areas nature seems to be losing out (Postel, 1993).

Urban waste water free from industrial contamination can supply an important portion of irrigation needs for agriculture and at the same time supply nutrients for considerable savings of fertilizer cost (Kasperson and Kasperson, 1977; Postel, 1993). It can enhance crop production near cities, even in rainfed areas; it is close to a complete hydroponic nutrient solution (Wallace et al., 1978). It is expected that reclaimed waste water will supply 16% of Israel's water needs in the near future. California will soon be able to reach 5 to 10%, which will be very beneficial. This can happen most everywhere and can be increased.

Thorne and Anderson (1942) some time ago pointed out that over the history of the world irrigated agriculture is not permanent. The history of the ancient Tigris–Euphrates Valley is relevant (Rush, 1987). Over a period of about 4700 years, a civilization rose, prospered, and then died due to loss of productivity of the land because of salt accumulation and climate change caused by loss of vegetation. Tax records over several centuries in the cuneiform writings revealed that wheat with little salt tolerance gradually gave way to barley with considerable salt tolerance; then it disappeared. In California, with a much shorter history of irrigation, severe salt problems are developing on the land. Both agriculture and the environment could eventually be losers. In Egypt, with 30 years of the Aswan High Dam, some soils have become more salinized in that short period than with 5000 years of previous irrigation. Good science and more use of soil conditioners are needed to make irrigated agriculture more sustainable. The efficiency of use of other inputs will then increase simultaneously.

A recent Gallup poll (Pacific Coast Nurserymen, 1993) indicated that farmers in the USA consider water quality to be the most serious environmental problem associated with agriculture. Most concern about water quality is with pesticide use, and the trend is to fewer and safer pesticides. Regulation of nitrates in groundwater now requires precise management of both fertilizers and soil organic matter. The nitrate problem will probably decrease both use of certain fertilizers and irrigation water but with increased efficiency of use of both (Zwingle, 1993). Relationships with soil conditions need more consideration.

2. Land

There is virtually no new land in the world that can be added to agriculture, so the need to maintain the land resources of the world is the first and foremost requirement for a sustainable agriculture. More food will be required in the future, and from all indications it must be obtained from less land. Crop land, especially productive crop land, will be an increasingly precious resource. In the past decade, 200 new cities world wide have developed mostly at the expense of agricultural land (Gardner, 1996). Urban expansion and new highways usually take the best agricultural land. When aquifers are overdrawn, agriculture land usually goes out of production. Global warming could result in increasing higher water levels in oceans and may result in loss of agricultural land (Trumbore et al., 1996). Severe erosion has taken land out of cultivation.

Great loss of land will require more and more attention to the integrity of the land that is left. Soil conditioners with associated research and development will become increasingly important for this purpose (see Chapters 2–7, 11, 13, and 15–18).

3. Oil and Natural Gas

Agriculture is currently dependent upon oil and natural gas for energy to operate machines, for manufacture of some inorganic fertilizers and some pesticides, and for synthesis of some polymer soil conditioners. Just how long there will be supplies of these materials is unknown. For some time it has been said to be a matter of only a few decades.

The supply of coal will last much longer than that of oil, but making all the needed products from coal instead of from oil may be a problem. Then, too, the global warming problem is always associated with fossil fuels and their use. The world generally wants and needs less use of fossil fuels in order to decrease the amount of carbon dioxide in the atmosphere (Trumbore et al., 1996). These are unsolved problems.

4. Forests

The close relationship between forests and other related natural resources for agriculture may not be obvious, but healthy forests are essential for the global hydrological cycle. Healthy forests help provide the biological diversity that keeps pests in balance. Healthy forests help keep the waterways clean to make irrigation water more available. Healthy forests help to sequester carbon dioxide, which can decrease the problems of global warming. Healthy forests help to remove potentially harmful industrial waste products from the atmosphere and thus give less potential for harming crops.

Some types of soil conditioners are obtained from forests or as by-products of forests (see Chapters 2 and 3), but these are probably quite secondary to the indirect conditioner values that can be provided by healthy forests. The growth and productivity of forest vegetation can be enhanced by the application of wastewater and biosolids (Henry and Cole, 1994) and other soil conditioners that provide nutrients and improve soil structural characteristics.

5. Raw Materials

The world is not likely soon to deplete its lime, gypsum, sulfur, lignite, pyrite, zeolite, or pozzolan resources. These are important soil conditioners. Also, waste forms of most of these are stacked unused, and in places are environmental problems. More use of them on land would be beneficial to all concerned.

Resource supplies of peat are in some jeopardy, especially since there is environmental pressure for conservation of it (see Chapter 18).

The status of raw materials for phosphorous and potassium fertilizers is more crucial than those for most soil conditioners. Phosphorous is the most critical, as the supplies of rock phosphate are finite (Western Fertilizer Handbook, 1994). If low-grade potassium sources are useable, supplies of it may last indefinitely.

D. Soil Degradation

Some soils, especially those containing moderate to high amounts of clay, refuse to grow plants if soil organic matter levels are very low. Just how much organic matter is really needed depends upon many factors, but for a soil to be healthy the level should approach one that is as high as is reasonably possible. Research is needed on what is an optimum level of soil organic matter. Kinds of soil organic matter may be more important than the total.

Soil organic matter levels in general have decreased considerably with years of cultivation compared with native soil conditions. This has resulted in many soil problems such as erosion, compaction, and loss of fertility. Agriculture can never be considered sustainable as long as those soil degrading processes related to the supply of soil organic matter continue; there will be erosion, compaction, and decreasing fertility (Wallace, 1994b; 1994c). This really is one aspect of sustainability that cannot be contested. Levels of soil organic matter may not need to be maintained at previous high levels, but they should be maintained at levels considered as ideal.

There are over 20 reasons why soil organic matter is important (see Chapters 2–5):

1. It slows down soil erosion from both wind and water.
2. It is a source of nutrients of plants and regulates availability of some nutrients in a reasonably orderly fashion.
3. It sustains a healthy diverse soil microbial population to make the soil into a functional ecosystem (Turco et al., 1996). Nutrient recycling then better takes place.
4. It is a buffer against pH and other changes.
5. It holds water.
6. It increases the cation-exchange capacity of the soil, which results in protection against loss of some nutritional elements.
7. It decreases compaction of the soils and also decreases crusting or sealing of soil following rain or irrigation events.
8. It stores some nutrients from season to season.
9. It makes soil warmer in the spring so that the cropping season may be started earlier.

10. It makes soil easier to till, especially when slightly too wet. Where non-tillage is practiced it helps add permeability to water, air, and nutrients.
11. It makes other inputs more valuable because of additive and synergistic effects.
12. It protects against some soil-borne plant diseases.
13. It gives better aerated, more permeable soil.
14. It protects plants against heavy-metal and salt toxicities.
15. It helps to detoxify pesticides and prevents their leaching to groundwaters.
16. It is a storage mechanism for excess atmospheric carbon dioxide. The greenhouse effect can be decreased as larger and larger amounts of organic matter are stored in the soil (Wallace, 1994d).
17. It is a means for obtaining higher crop yields especially when combined with other best management practices.
18. It promotes breakdown of toxic substances in addition to pesticides.
19. It promotes soil formation from parent minerals.
20. It serves as an energy source for microbes and other organisms such as earthworms that improve soil.
21. It has economic benefits.
22. It is the essence of soil quality.

To obtain and enhance these values of soil organic matter and keep soil from degrading, it is necessary to add or return to the soil a critical amount of organic matter each year. This is usually around 10 MG ha^{-1} if the soil is tilled and will vary according to the temperature, climate, and some other factors (Wallace, 1994c). Crop residues count, but additional sources from off the farm or produced on the farm are usually needed for true equilibrium or to increase levels of soil organic matter.

E. The Old Argument—Organic vs. Inorganic

Organic growers have sometimes been accused of doing the right thing for the wrong reason. The fact remains that much of what they do is right for good reasons. There really is no point in contesting this argument. Both organic and inorganic types of inputs are variously necessary. Many inorganic products are called organic merely because they occur naturally, but this doesn't mean that they are different from processed products. When the organic–inorganic argument comes down to a matter of soil degradation, choice will or should always favor protection of the soil.

The Rodale Institute in Pennsylvania (USA) has been conducting an experiment since 1981 in which organic and conventional production systems are being compared (Fioina et al., 1996). The authors reported that organic corn yields were significantly lower than the conventional system during the first four years of the study but were comparable over the next ten years. In the last five years the organic production exceeded an average of eight MG per ha. This study, which will continue, indicates the positive effects that can result from improving the physical properties of the soil. This is one value associated with the organic system.

Two different points of view from the recent farm press (*Farm Chemicals*, 1994) illustrate that there is no consensus yet concerning the relative merits of organic and conventional farming. One reader said that "Organic Farming" has nothing to do with people starving or sacrificing wildlife, because "he understands the dynamics of a living soil." Organic farming can save the soil; he has this understanding because he has been to the field and looked at the issue from both sides for 20 years. He further says that the journal should have a commitment to do the same for the future of the agricultural dealers that read the magazine, for the farmers they serve, and possibly even for the future of the magazine. "Organic Farming" is indeed a threat to continued traditional technological, industrial, and economic development, but not to the land, he claims.

Another reader said that the agricultural-dealer industry is unable, but not unwilling, to mount an effective program to counter the misinformation and regulatory misadventure being taught in the organic movement. This is a result of agriculture becoming so efficient that is has become an insignificant minority in a population otherwise freed from thinking about sources of food. According to this reader the critical missing link is the food processing and retail grocery industry who gain the lion's share of the agri-food dollars spent in the developed world, and who mostly deny their inseparable link to the farm in favor of "pandering to the latest consumer fads." The conventional farmer is too busy, there is too little money, and there are too few of them to enter into the debate. The argument continues mostly unresolved. Both sides have valid concepts, however.

The bottom line of all of this perhaps is that in order for farmers to use as much technology as possible, with full use of soil conditioners and even some organic methods as needed, they need greater returns from their land. Then if greater returns can be translated to more care of the land, good soil will be available for future generations. One known way for obtaining more money per unit of land is the organic movement, which has been able to command higher produce prices in the market than conventionally produced products (Klinkenborg, 1995).

III. SOIL CONDITIONER OVERVIEW

There are many kinds of soil conditioners and several ways to classify them.

A. Organic–Inorganic

One classification of soil conditioners is as organic or inorganic. Separation on this basis is easy if only the chemistry is considered. Actually there are a number of organic products that can be used as soil conditioners, but perhaps the most popular are the products described as composts. But also included are biosolids (sewage sludges), straws, sawdusts, manures, feathers, litter, peat mosses, bark, lignites or humates, paper, all crop residues, seaweeds, animal products, and more (see Chapters 2, 5, and 18). This grouping is different from the organics approved by the "Organic Movement."

In contrast to organic soil conditioners are the inorganic types. Many of these are mined, but some are by-products of manufacturing and some are both. Inorganic soil conditioners include gypsum, lime, sulfur, pyrite, fly ash, phosphogypsum, flue-gas lime-gypsum, zeolites, diatomaceous earth, pozzolan, clays. See Chapters 7–11, and 18.

B. Synthetic–Nonsynthetic

Most soil conditioners are nonsynthetic or are derived from nonsynthetic sources. A few are synthetic, but these are among the most powerful. These include the water-soluble polymers used to stabilize clay (see Chapters 14–17) and the hydrogel polymers (see Chapters 13 and 18). Some industrial by-products are also considered synthetic because of the way they are produced. Phosphogypsum is an example.

C. Clay–Sand Soils

This classification is based on uses of soil conditioners according to soil type. Soils that are predominantly clay differ vastly in their needs from soils that are predominantly sand. Even different type of clay dictates what can be best be done. Organic soil conditioners can benefit both clayey soils and sandy soils, but for different reasons (see Chapter 2). Clay soils need to be made more tillable, better aerated, less erodable, more permeable to water, and less compactable (see Chapter 15). Sandy soils need higher water-holding capacity, higher cation-exchange capacity, and even the ability to hold anions from excessive leaching. This can be achieved (see Chapters 13 and 18). Types of clays are important in the amount of lime that can be added (see Chapter 11).

D. Tropic–Temperate Climates

This possible classification implies that use of soil conditioners will vary according to climate. Use will vary as soils vary and will also vary as economics vary. Most developed countries are in temperate climates where soil conditioners are more commonly used. But organic amendments are much more rapidly oxidized in tropical climates than in temperate climates (Jenny, 1930; 1933; Trumbore et al., 1996). Climate influences how much liming can be done (Espinosa, 1996). Effective soil conditioners for tropical countries require different approaches compared with temperate areas (see Chapter 4).

E. Farms–Landscapes

Soil conditioner needs vary considerably between farm lands and landscape areas. The differences are largely matters of economics. When a single tree in a landscape can be worth tens of thousands of dollars or even more, the approach to soil conditioning must reflect concern for that much value (see Chapter 19). Those who plow, plant, and care for such expensive plant materials (any landscape is expensive to a degree) often underestimate the importance of proper soil conditioning with soil conditioners, which can modify soil areas as needed for a given situation. When need for soil conditioners is underestimated or ignored, landscape plantings can and do variously fail with high financial losses.

Soil conditioning on farms is often crucial and can be a matter between outstanding success and utter failure (see Chapters 2 and 15). But often it is possible to modify the degree of soil conditioning to match the value of a crop. See Chapter 17. Soil conditioning for farms may concern only one year, but for landscaping the conditioning may need to last for 10 to 20 or more years. What is done then differs tremendously.

IV. INTERACTIONS INVOLVING SOIL CONDITIONERS

A. Failure to Get Additional Response—The Plateau Question

Brown (1994) and Brown and Kane (1994) are among the workers who are endeavoring to calculate or forecast how well the world can feed itself in coming decades. Actually they are attempting to sound an alarm in that they believe that yield increases world-wide, which were spectacular during the years of the Green Revolution (1960s to 1980s), have now slowed down to almost a halt. They collected data and published charts indicating that in the 1990s and worldwide there were little or no yield increases. They expect the trend to continue on the supposition that new technologies including the promise of biotechnology have been exhausted or are not available.

One basis on which these conclusions can be challenged is that the average yields in the charts presented are perhaps only half of what they could be and far less than that of the genetic potential (Wittwer, 1975). Brown and Kane base their arguments on worldwide response to fertilizer since about 1985. Farmers have added more fertilizer at least in some of the years since then without appreciable yield increase. By fertilizer, Brown and Kane mean nitrogen, phosphorous, and potassium, the three major or primary plant nutrients. They say that many nations have raised yields to the point that there is little or no response to further addition of these plant nutrients. A plateau has been reached (Figure 1).

When yields "plateau" at levels much less than the potential for the climate and the genetic capacity for the crop, it is because some other factor is more limiting than the one varied (Wallace, 1990b; Wallace and Wallace, 1993). If an experiment were conducted in which different increments of one variable such as nitrogen are applied, a plateau will always be obtained if the treatment levels go high enough. The effect is the Law of Diminishing Returns (Fig. 1). What it really means is that there is less and less response to nitrogen as it is increased or even no further response because some other factors eventually become more limiting than nitrogen.

Brown and Kane (1994) have been describing this very situation on a worldwide basis but without explanation. They have assumed that the yield is mostly a function of fertilizer usage, which is only partially correct. It is now time and prudent to describe plant yield trends also in terms of use of soil conditioners as well as for NPK fertilizers. When physical conditions of soil are improved, and all other factors are favorable, the plateau levels can be increased. The plateau can even be made to disappear. If all limiting factors are improved simultaneously, yields should increase on a straight line relationship right up to the genetic potential with no plateau (Fig. 1). Yields therefore can yet be increased considerably as a worldwide average. The challenge of today is to overcome the present plateau barrier, whatever the cause or causes. Soil conditioners will be important tools in the process.

B. Law of the Maximum

Yields of crops can be most improved by combinations of best management practices. When combined, each technology can add to the value of the others. This is particularly important for uses of more than one soil conditioner–soil amendment simultaneously. For economic and environmental reasons, there is urgent need for increasing the efficiency of inputs used in production of crops. Most always this can be accomplished with precision management. Inputs and management practices interact with each other, and favorable interactions can be manipulated. Both Liebig's Law of the Minimum (describing severe stresses) and Mitscherlich's Law of the Minimum (describing less severe stresses) are real, but they are different; responses to inputs differ for each (Wallace, 1990c; Wallace and Wallace, 1993; Wallace, 1995a). Differential responses to the different types of stresses are described in Fig. 2 and Table 1. All the factors that limit crop yields can be classed as being one or the other of these two types, but there is overlap.

Each factor that limits crop growth must be identified and classed as to which stress type and degree of stress must be obtained by diagnostic procedures in order to practice efficient crop management. A Sufficiency Value of between 0.0 to 1.0 needs to be assigned to each stress factor (Fig. 2 and Table 1). Liebig-type stresses permit little or no response to correction of other stresses

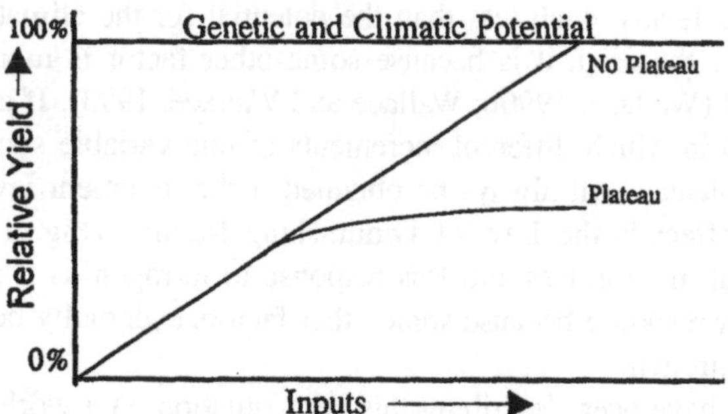

Figure 1 Plateau representing the Law of Diminishing Returns vs. production if all limiting factors are overcome simultaneously.

until those of that type are first corrected, since those of that type limit in a severe manner. When there are no remaining Liebig stresses, inputs to correct Mitscherlich-type limiting factors can then give responses in direct proportion to the amount of input until the need for that input is satisfied as explained in Figure 2 and Table 1. It is with this type of limiting factors (zone C in Figure 2) that the Multiple Action Yield Fraction (MAYF), which is obtained from multiplying all Sufficiency Values together, indicates the expected yield. Order of input for this Mitscherlich type of stress in not important, although logic dictates the order of some of them. High-crop yields are possible only if there are no remaining Liebig-type limiting factors (zone A in Fig. 2), but some of the Mitscherlich type (zone C in Fig. 2) may remain, especially if they are too expensive to correct. Those inputs that are free or inexpensive should be chosen first. Increases in efficiency and also larger absolute responses for individual inputs are obtained as yields become higher.

These phenomena are extrapolated to a concept of the Law of the Maximum; the greatest response for any one factor is obtained when no others are limiting and input factors interact to multiply the value of the others so that the MAYF is near the top of C in Figure 2 (Wallace and Wallace, 1993, 1994). The higher C is in Fig. 2, the less plateau effect will exist (Fig. 1). Computerization and prediction of responses to inputs and their economics are possible after Liebig-type limiting factors are corrected. Limiting factors are multiple, and their interactions, which are graphed in MAYF plots, (Wallace 1994e, Wallace 1995a), determine final yields (Fig. 2 and Table 1). The Multiple Action Yield Fraction may be used to maximize specific aspects of crop production, such as per unit of land or per unit of irrigation water. Management skill is essential for obtaining high yields, and the skill must come from a grower, a consultant, an extension agent, or other. In essence, the Law of the Maximum is getting the most possible out of the inputs. Response to an input can be several times as much

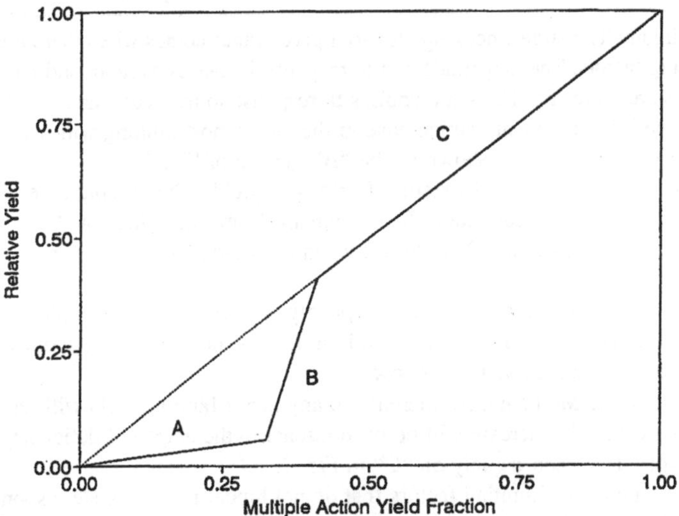

Figure 2 All sufficiency values, which range from 0.00 to 1.00, before or after use of any inputs (inputs increase the sufficiency values) are multiplied together to give the Multiple Action Yield Fractions (MAYF). In the A zone, inputs result in little or no yield response; decreased yield is even possible. This is the result of the severe Liebig Law of the Minimum where at least one sufficiency value is closer to zero than to 1.0. When inputs for corrections are sufficient, some synergism (B zone) is possible as the yields return to the C zone where the Mitscherlich Law of the Minimum operates and sufficiency values are close to 1.0. Inputs that further increase yields above zone B result in sequential additivity until 1.0 is reached. Approaching 1.0 where limiting factors are minimal is defined as the Law of the Maximum. Responses to a given input are greater the higher up the C zone the fraction goes. This is an important consequence of the Law of the Maximum. It means that each input is more valuable when used with a number of other best-management practices. See Table 1 for further definition of terms.

if the yield base (MAYF) is near the top of zone C compared with near the bottom of zone C.

1. Some Salient Points

Some points that need consideration for achieving high yields or maximum economic yields include (Wallace, 1990a; Wallace and Wallace, 1993):

1. Yield improvement is multidisciplinary.
2. Plant breeding will be a continuing need in yield improvement programs, as it interfaces with every old and new cultural development. For some crops, especially perennial ones like trees, the genetic potential is as yet essentially untouched.

Table 1 Definitions of Terms Related to Stress Interactions in Crop Production

Law of the Maximum: The largest net response to a given input comes when there are no other limiting factors. The magnitude of the response increases as more and more limiting factors are corrected. The same applies to response to new cultivars.

Liebig's Law of the Minimum: Only an increase in the factor most limiting will result in a significant increase in yield. Shown as the "A" portion of Fig. 2.

Liebig synergism: The synergism is the result of prior low yields when Liebig-type limitation is present and then corrected. Yields increase from somewhere on "A" to somewhere on "B" or even to "C" in Fig. 2 when Liebig-type limitations are corrected.

Liebig-type limitations: Limiting factors that completely prevent or nearly completely prevent response to correction of other less limiting factors until those severe ones are first corrected. Sufficiency values are low.

Mitscherlich's Law of the Minimum: An increase in any factor limiting yield will result in a yield increase, and the increase will be proportional to the degree of deficiency of all such factors. This happens only on "C" in Fig. 2.

Mitscherlich-type limitations: Limiting factors that do not hinder proportionate response to correction of other limiting factors. Sufficiency values generally are near or not far from 1.0.

Multiple Action Yield Fraction (MAYF): The fraction (0.00 to 1.00) obtained when all sufficiency values for a given crop are multiplied together. The result may or may not correspond with the yield, depending on types of limiting factors. When any sufficiency values are 0.0 or near 0.0, Liebig type MAYF cannot correspond much with yield changes. Physiological interactions in plants under these conditions interfere with the MAYF. If they did not, the line C in Fig. 2 would continue down to the origin.

Sequential additivity: If the per cent effect of multiple inputs when applied together is the same as when they are applied singly and then the individual responses are multiplied together, the interaction is sequentially additive. This fits Mitscherlich-type limitations only where sufficiency values are all high approaching 1.0.

Sufficiency value: The numerical value between 0.00 (total limitation) and 1.00 (no limitation) that gives the degree of limitation for any individual factor involved in crop production. Laboratories with field help should be able to determine the values.

Synergism: Responses to inputs that are greater than those predicted from the effects of the individual inputs. Most likely only for the B portion of Fig. 2.

3. In diagnostic programs, complete analysis of essential and nonessential elements and soil physical properties is important. All possible limiting factors must be quantitized (Wallace and Wallace, 1994).
4. As plant breeding is used to increase the nutrient efficiency status of plants by manipulating the genetics of susceptibility to nutrient deficiencies, it is important that complete packages of nutrient efficiency be developed in the genetic or breeding process.

5. A worthy goal for plant breeding and biotechnology would be to develop plants resistant to any Liebig-type factors (A and B portions on a Multiple Action Yield Fraction plot, Wallace 1994e; 1995a). Salinity, drought, and temperature extremes are examples.
6. With more new cultivars, including some from genetic engineering and plant breeding and with higher yields, it will be increasingly important for knowledge of how to and ability to manipulate unfavorable interactions that relate to the environment to minimize their impact. Management must also provide an environment involving nutrient, biological and physical properties of soil suitable to maximum potential of a crop.
7. Planning for yield goals must consider capital limitations, land limitation, environmental limitations, resource limitations, and supply and demand limitations.
8. Forty or more controllable factors limit the yield of a crop (Wallace, 1990a, 1990c). Economics, farm and orchard management ability, and lack of interest and/or ignorance are the major barriers for overcoming these limiting factors.
9. A 25 to 50% or even greater gain in yield of many crops can be expected in the next 20 years with use of principles related to the Law of the Maximum and also on soil conditioners (Wallace, 1984). More production on a smaller land base could even release considerable marginal land that could be returned to conservation purposes.

Maximum yields of crops obtained through research are often three or more times those of average yields. Maximum yield is not necessarily a major goal for most growers; it may be too costly per unit of return, but considerably higher yields are possible in many cases and can be cost effective. Intense management and multidisciplinary approaches are needed if this is to be accomplished. Many farmers are already very successful in this best-management practice approach to cropping, but even they can do better.

2. Examples of Limiting Factors

Examples of limiting factors (Wallace, 1990a; Wallace, 1990c; and Wallace and Wallace, 1993) include

1. Lack of Water. Many crops are rain fed and are subject to hazards of weather. Development of capacity for supplemental irrigation, when and if needed, can take away much of the risk for this variable, but water supplies for irrigation may not always be plentiful, so they must be managed carefully with considerable emphasis on water-use efficiency. Where irrigation is practiced intensively, there are techniques for improvement of efficiency, which include pulse irrigation, drip and deep-drip irrigation, high-frequency low-pressure irrigation, and others. It is

possible that plant breeders can improve the drought resistance of crops for the future. Soil conditioners can considerably increase water-use efficiency. New methodologies for improvement of water relations in soils need consideration.

2. Low Fertility. There are many components of fertility. There are some 14 essential mineral elements for all or most plants, each of which could be limiting singly or in combination. Some plants require other elements. Ratio of nutrients to each other is usually important. Farmers, scientists, consultants, advisers, and industry have worked together for almost 150 years to diagnose and correct plant nutrient problems. Now plant breeders with the help of many disciplines are making it possible with some crops to use the diversity of germ plasm to develop new cultivars with increased efficiency for uptake and use of virtually all of the essential nutrients. Additionally, information from plant physiologists can be used to help increase efficiency in uptake and use of nutrients.
3. Poor Stand. A poor stand can greatly limit production and result in waste of resources. It is essential that the magnitude of the stand be considered in decisions concerning additional inputs. Many reasons exist for poor stands, including inadequate seedbed preparation and methods of seeding, lack of control of soil-borne diseases, soil crusting, which can be controlled in most cases with soil conditioners, and other factors.
4. Poor Soil Structure. Improvement in soil structure has not coincided with improvement in soil fertility during the past 100 years. Soil structure often remains as a Liebig-type limiting factor that hinders the possibility for yield improvement with inputs to overcome other limiting factors. In many cases, significant yield improvement is possible only if serious attention is given to improvement of soil structure. Often salinity and sodicity need correction (see Chapters 7–9). Water-soluble polymer soil conditioners used with organic matter and gypsum can help (see Chapters 15, 17, and 18). Modified organic farming may be necessary. Crop rotations on a full scientific basis could be more useful than in the past. All techniques could be combined.
5. Weeds, Diseases, Insects. All these pests need full consideration and evaluation on whether their effects are of the Liebig type or the Mitscherlich type so that management decisions can be factored into operations. They even relate to physical properties of soil.

C. Precision Technology

1. Laboratory Diagnosis

Sustainable, environmentally safe, and profitable agriculture of the future really requires a new generation of diagnostic laboratories capable of extreme precision in a very complicated regulated world (Wallace and Wallace, 1994; and Wallace, 1995a). Analyses are only part of what is needed; the meanings of the numbers must be determined in terms of categories of disciplines and philosophies, and then prudent decisions need to be made, and they concern much more than fertilizer recommendations.

What is diagnosis? It is more than analysis, although analysis is crucial. It is more than sampling, although without proper sampling, many soil and plant samples are hardly worth analyzing. An effective diagnostic laboratory includes a wide spectrum of services to provide the capacity to recognize and quantify all of the variables related to crop production that concern each user. It is further required that the information made available be integrated into useful action plans. A diagnostic laboratory therefore really does not market analysis. Rather it markets programs, goals, and systems such as water conservation, sustainability, environmental compliance, precision and efficiency in agriculture, maximum economic yields, integrated pest management, or other specialties. Laboratories need then to consider needs for all possible soil conditioners. Need for various soil conditioners is emphasized in various aspects of soil analysis. See Chapter 19.

A more intense precision agriculture can minimize unwanted accumulation or migration of input residues, can save money, and can avoid interactions that would decrease yields (Wallace, 1994e). Precision has meanings beyond the quantity of an input. It includes correct placement in the soil or on or in the plant, it includes timing of the input, it includes differential fertilization of fields according to needs, it includes relationships with other inputs to create proper balances, it includes correctly leveling, draining, and contouring the land, and much more. Use of precision directly and indirectly has or can have beneficial effects on the environment. Precision agriculture can make it possible to produce necessary food and fiber on much less land than is currently being used. Increasing precision is a procedure to give maximum economic yields.

The trend towards so-called high-precision application, where different parts of fields are differentially fertilized according to analyses and global positioning systems, is encountering some difficulty because of interactions of the test values with other parameters (Fixen, 1996). Virtually all of the parameters involved that correlate positively or negatively with the nutrient analyses can be modified by various soil conditioners. These factors are clay content of soil, soil acidity, lime in soil, soil aeration, and others. The laboratory diagnosis must be expanded over time to include the possibility of such interactions. A further advantage in the use of greater precision in the application of soil conditioners and amendments on crop lands is that the general public will more and more consider growers to be true environmental stewards (Parker, 1995). Farmers will then be more appreciated.

V. CONCLUSIONS

There is hope that declining soil quality (or health) worldwide can be reversed, which will require wide use of soil conditioners in addition to conservation methods in farming. Much recycling of organic matter is necessary, and farms need to be considered as closely as possible as natural ecosystems that function well only as other related ecosystems also function. Soil conditioners will be useful as they can make the soil more functional as an ecosystem and more efficient as a support for crops.

As more and more attention must be given to the land, it seems that there will always be a role for small farms where individual attention can be given to each problem spot of land. Computers are helping to do it on large farms. They may have to do it on small farms also.

The very high yields that approach the maximum attainable have largely been obtained on small experimental plots where a knowledgeable grower can do everything right and where all problems are properly addressed. Knowledge and its use are extremely important regardless of size of farm.

High-acreage corporate farming does sacrifice a bit of individual attention, but perhaps modern computers can correct some of that, otherwise small farms may be better but the prices have to be right to make more small farms possible.

Globally, in the long run, as some natural resources are diminishing and populations are increasing, reliance may need to shift more to family farms. This may not happen everywhere, because the trend to industrialization of farms is equally strong. But even so the need for training to improve skills will continue to increase in importance. Just what is compatible with society must eventually be decided by society itself. Finances are involved.

Higher yields are expected and will be needed in the future. Recognition of the importance of the Law of the Maximum in being able to get the most possible from the inputs into agriculture is the essence of plans for greater crop production. There are ways to lessen the effects of the Law of Diminishing Returns.

Yaalon (1996) has pointed out recently in the Soil Issues section of the journal *Soil Science* that a new era exists for soil science. Improvement in soil productivity is not necessarily the major focus (or only focus) of research in soil science. There is now a shift towards long-term oriented environmental and regional stability aspects. The present 50,000 soil scientists in the world are not nearly enough. And the number must be increased if systems of soil and land use management are to be developed to maintain and enhance soil usefulness and quality. Although much of this book concerns technology, behind each technology is the urgent need for more basic research.

REFERENCES

Adriano, D. C., ed. (1991). Metals in soils, water, plants, and animals. *Water, Air, & Soil Poll. J. 57–58*:3–930.

Agassi, M., Letey, J., Farmer, W. J., and Clark, P. (1995). Soil erosion contribution to pesticide transport by furrow irrigation. *J. Environ. Qual. 24*:892.

Agronomy News. (1995). SSSA Statement on Soil Quality. June, p. 7.

Anderson, T. A., Kruger, E. L., and Coats, J. R. (1994). Enhanced degradation of a mixture of three herbicides in the rhizosphere of a herbicide-tolerant plant. *Chemosphere 28*:1551.

Arshad, M. A., and Coen, G. M. (1992). Characterization of soil quality:—physical and chemical criteria. *Amer. J. Altern. Agri. 7*:25.

Atlas, R. M. (1995). Bioremediation. *Chem. Eng. News 73* (14):32.

Bear, F. E., Pritchard, H. W., and Akin, W. E.. 2nd ed. (1986) *Earth: The Stuff of Life*. University of Oklahoma Press, Norman and London.

Bennett, H. H., (1939). *Soil Conservation*. McGraw-Hill, New York.

Bennett, H. H., and Lowdermilk, W. C. (1938). General aspects of the soil-erosion problem. In: *Soils and Man Yearbook of Agriculture 1938*, U.S. Govt. Printing Office (G. Hambridge, ed.), pp. 581–608.

Bird, G. W. (1996). Sustainable agricultural research and education (SARE) program. *Hort Tech. 6*:359–362.

Block, N., and Clark, T. P. (1990). Biological treatment of soils contaminated by petroleum products. In: *Petroleum Contaminated Soils* (P.T. Kostecki, and E.J. Calabrese, eds.). Lewis Publishers, Chelsea, ML, USA, pp. 167–175.

Bollag, J. M., and Bollag, W. B. (1995). Soil contamination and the feasibility of biological remediation. In: *Bioremediation Science and Application* (H.D. Skipper and R.F. Turco, eds.). Am. Soc. Agron. Special Pub. No. 35. Madison, WI.

Brown, L. R. (1994). *Who Will Feed China? Wake-up Call for a Small Planet*. The Worldwatch Environmental Alert Series. W. W. Norton, New York and London, p.163.

Brown, L. R., and Kane, H. (1994). *Full House—Reassessing the Earth's Population Carrying Capacity*. The Worldwatch Environmental Alert Series. W.W. Norton, New York and London, p. 261.

Brumfield, R. G. (1996). Sustainable horticulture: an overview. *Hort Tech. 6*:352.

Buol, S. W. (1995). Sustainability of soil use, Ann. Rev. Ecology Syst. *26*:25.

Carpenter, J. (1995). From the team and wagon to the space age. *Utah State Univ. 1*(4)(Winter):15–19.

Carter, M. R., ed. (1994). *Conservation Tillage in Temperature Agroecosystems*. Lewis, Boca Raton, Florida.

CAST (1996). Future of irrigated agriculture. Council for Agricultural Science and Technology, Task Force Report No. 127.

Chaney, D. E., Drinkwater, L. E., and Pettygrove, G. S. (1992). Organic soil amendments and fertilizers. UC Sustainable Agriculture Research and Education Program, University of California, Division of Agriculture and Natural Resources, Pub. 21505.

Clapp, C. E., Larson, W. E., and Dowdy, R. H., eds. (1994). *Sewage Sludge: Land Utilization and the Environment*. American Society of Agronomy, Madison, WI.

Dick, W. A., and McCoy, E. L. (1993). Enhancing soil fertility by addition of compost. In: H.A.J. Hoitink and H.M. Kenner, eds., *Science and Engineering of Composting: Design, Environmental, Microbiological and Utilization Aspects*. Renaissance, Worthington, OH, p. 622–644.

Doran, J. W., Coleman, D. C., Bezdicek, D. F., and Stewart, B. A. (1994). Defining soil quality for a sustainable environment. *SSSA Special Publication Number 35*. Soil Society of America, American Society of Agronomy, Madison, WI, USA.

Doxan, L. E. (1996). Landscape sustainability: environmental, human, and financial factors. *Hort Tech 6*:362.

Elmore, R. M. (1996). Our relationship with the ecosystem and its impact on sustainable agriculture. *J. Prod. Agric. 9*:42.

Enlow, C. R., and Musgrave, G. W. (1938). Grass and other thick-growing vegetation in erosion control. *Soils and Man, Yearbook of Agriculture 1938*. U. S. Govt. Printing Office, 615–633.

Espinosa, J. (1996). Liming tropical soils—management challenge. *Better Crops 80*(1):28.

Fageria, N. K. (1992). *Maximizing Crop Yields*. Marcel Dekker, New York.

Farm Chemicals (1994). Letters to the Editor: Organic Farming Revisited. *Farm Chemicals 157*(1):33.

Fiorina, L. J., Peters, S. E., Wagoner, P. and Drinkwater, L. E. (1996). A comparison of corn/soybean cropping systems under conventional and organic management for fifteen years. 1996 Agronomy Abstracts, p. 59.

Fixen, P. (1996). Will yield monitors erode soil test credibility? *Custom Applicator 26*(1):39.

Fretz, T. A. (1992). Sustainable agriculture: our role as horticulturists. *ASHS Newsletter 8*(5):3–4.

Fretz, T. A., Keeney, D. R. and Saterrett, S. B. (1993). Sustainability: defining the new paradigm. *HortTech. 3*:118–126.

Friend, J. A. (1992). Achieving soil sustainability. *J. Soil Water Cons. 47*:156.

Gardner, G. (1996). *Preserving Agricultural Resources*. State of the World 1996, a Worldwatch Institute Report on Progress Toward a Sustainable Society. W.W. Norton, New York and London, pp. 78–94.

Glanz, J. (1995). Erosion study finds high price for forgotten menace. *Science 267*: 1088.

Goulart, B. L. (1996). Welcome to reality: an overview of a low-input sustainable agriculture (LISA) project in small fruit. *Hort Tech. 6*:354.

Granatstein, D., and Bezdicek, D. F. (1992). The need for a soil quality index: local and regional perspectives. *Ameri. J. Altern. Agri. 7*:11.

Haberern, J. (1992). Opinion: coming full circle—the new emphasis on soil quality. *Ame. J. Altern. Agri. 7*:3.

Heckman, J. R., and Kluchinski, D. (1996). Waste management—chemical composition of municipal leaf waste and hand-collected urban leaf litter. *J. Environ. Qual. 25*:355.

Henry, C. L., and Cole, D. W. (1994). Biosolids utilization in forest lands. In: *Sewage Sludge: Land Utilization and the Environment* (C. E. Clapp, W. E. Larson, and R. H. Dowdy, eds.). Soil Sci. Soc. Am. Misc. Pub. Madison, WI, pp. 89–94.

Hyatt, G. W. (1993). U.S. potential for compost markets (1993). 1993 Agron. Abs., p. 22. Also, Executive summary: U.S. potential for compost markets, 14 pp. Amer. Soc. Agron. Ann. Meet. Cincinnati OH, Nov. 7–12.

Jenny, H. (1930). A study of the influence of climate upon the nitrogen and organic matter content of the soil. *Mo. Agr. Expt. Sta. Res. Bul. No. 152.*

Jenny, H. (1933). Soil fertility losses under Missouri conditions. *Mo. Agr. Expt. Sta. Res. Bul. No. 324.*

Kaiser, J. (1996). Acid rain's dirty business: stealing minerals from soil. *Science 272*:198.

Karlen, D. L., Wright, R. J., and Kemper, W. O. eds. (1995). *Agriculture Utilization of Urban and Industrial By-products.* ASA Special Publication Number 58. Madison, WI.

Kasperson, R. E., and Kasperson, J. X., eds (1977). *Water Re-use and the Cities.* University Press of New England, Hanover, NH.

Kell, W. V. (1938). Strip cropping. In: *Soils and Man, Yearbook of Agriculture 1938.* U.S. Govt. Printing Office, p. 634–645.

Klinkenborg, V. (1995). Review: A farming revolution: sustainable agriculture. *National Geographic 188*(6 Dec.):60–89.

Likens, G. E., Driscoll, C. T., and Buso, D.C. (1996). Long-term effects of acid rain: response and recovery of a forest ecosystem. *Science 272*:244.

MacLean, K. S., Robinson, A. R., and MacConnell, H. J. (1987). The effect of sewage-sludge on the heavy metal content of soils and plant tissue. *Comm. Soil Sci. Plant Anal. 18*:1303.

McBride, M. B. (1995). Toxic metal accumulation from agricultural use of sludge: are USEPA regulations protective? *J. Environ. Qual. 24*:5–18.

McLeod, K. W., Davis, C. E., Sherrod, K. C., and Wells, C. G. (1986). Understory response to sewage sludge fertilization of loblolly pine plantations. The forest alternative for treatment and utilization of municipal and industrial wastes (D. W. Cole, ed.). Univ. of Washington Press, Seattle, pp. 308–323.

McMullin, E. (1993). Hard lessons: the silver lining in the drought has been the valuable experience growers have gained in managing precious water resources. *Calif. Farmer 276*(7):6–7.

Munns, E. N., Preston, J. F., and Sims, I. H. (1938). Forests for erosion control. *Soils and Men, Yearbook of Agriculture 1938.* U.S. Govt. Printing Office, pp. 609–614.

Nichols, M. L., and Chambers, T. B. (1938). Mechanical measures of erosion control. *Soils and Men, Yearbook of Agriculture 1938.* U.S. Govt. Printing Office, pp. 646–665.

Pacific Coast Nurseryman (1993). Gallup poll of growers sees water quality as most serious environmental issue facing them. *Pacific Coast Nurs. 52*(4):12.

Paine, L. K., and Harrison, H. (1993). The historical roots of living mulch and related practices. *HortTech. 3*:137.

Papendick, R. I., and Parr, J. F. (1992). Soil quality—the key to a sustainable agriculture, *Ame. J. Altern. Agri. 7*:2.

Parker, D. R. (1995). Retailers and agriculture's emerging technologies. *Agricultural Retailer. 39*(9):6.

Parr, J. F., Papendick, R. I., Hornick, S. B., and Meyer, R. E. (1992). Soil quality: attributes and relationship to alternative and sustainable agriculture. *Amer. J. Altern. Agri. 7*:5.

Pimentel, D., Harvey, C., Resosudarmo, P., Sinclair, K., Kurz, D., McNair, M., Crist, S., Shpritz, L., Fitton, L., Saffouri, R., and Blair, R. (1995). Environmental and economic costs of soil erosion and conservation benefits. *Science 267*:1117.

Postel, S. (1989). Water for agriculture: facing the limits. *Worldwatch Paper 93*, December, Worldwatch Institute, Washington, D.C.

Postel, S. (1993). Facing water scarcity. *State of the World* (L. Starke, ed.), Worldwatch Institute, Washington, D.C., pp. 22–41.

Robert, M. (1995). Soil, an environmental interface. *C.R. Acad. Agri. Fr. 81*(8):93–112.

Robison, J. J. (1994). Economic insights from the Book of Mormon. *J. Book of Mormon Studies 1*:35–53.

Rush, D. W. (1987). Salinity: agriculture's first environmental toxicant. *Solution Sheet 3*(3):1,4. UNOCAL.

Sabey, B. R., Pendleton, R.L., and Webb, B.L. (1990). Effect of municipal sewage sludge application on growth of two reclamation shrub species in copper mine spoils. *J. Environ. Qual. 19*:580.

Sandmann, E. R., Loos, I. C., and Loos, M. A. (1984). Enumeration of 2,4-D-degrading microorganisms in soils and crop plant rhizospheres using indicator media: high populations associated with sugarcane (*Saccharum officinarum*). *Chemosphere 13*:1073.

Seaker, E. M., and Sopper, W. E. (1988). Municipal sludge for minespoil reclamation. II. Effects on organic matter. *J. Environ. Qual. 17*:598.

Seibert, K., Fuehr, F., and Cheng, H. H. (1981). Experiments on the degradation of atrazine in the maize rhizosphere. *Proceedings of the Theory and Practical Use of Soil Applied Herbicides Symposium.* European Weed Resource Society, Paris, France, pp. 137–146.

Singh, G., Letey, J., Hanson, P., Osterli, P., and Spencer, W. F. (1996). Soil erosion and pesticide transport from an irrigated field. *J. Environ. Sci. Health B31*(1):25–41.

Smith, D. T., ed. (1991). *Agriculture and the Environment, The 1991 Yearbook of Agriculture.* U.S. Govt. Printing Office, Washington, D.C.

Statistical Abstract of the United States (1994). 114th ed. U.S. Bureau of the Census, Washington, D.C.

Steutevelle, R. (1995). The state of garbage in America, Part 2. *BioCycle J. Compost. Recycl. 34*(4):54–63.

Thorne, D. W., and Anderson, D. A. (1942). Irrigation and permanent agriculture. *Yearbook, Association Pacific Coast Geographer 8*:8.

Trumbore, S. E., Chadwick, O. A., and Amundson, R. (1996). Rapid exchange between soil carbon and atmospheric carbon dioxide driven by temperature change. *Science 272*:393–396.

Turco, R. F., Kennedy, A. C., and Jawson, M. D. (1996). Microbial indicators of soil quality. *SSSA Special Publication Number 35.* Soil Science Society of America, American Society of Agronomy. Madison, WI, pp. 75–90.

US EPA. (1993). Standards for disposal of sewage sludge; final rules: Part II. *Federal Register, Feb. 19, 1993*, pp. 9247–9415.

Utz, E. J. (1938). The coordinated approach to soil-erosion control. *Soils and Men. Yearbook of Agriculture 1938.* U.S. Govt. Printing Office, pp. 666–678.

Visser, S., and Parkinson, D. (1992). Soil biological criteria as indicators of soil quality: soil microorganisms. *Ameri. J. Altern. Agri. 7*:33–37.

Wallace, A. (1984). The next agricultural revolution. *Comm. Soil Sci. Plant Anal.* *15*:191–197.

Wallace, A. (1990a). Crop management through multidisciplinary approaches to different types of stresses—law of the maximum. *J. Plant Nutr.* *13*:313–325.

Wallace, A. (1990b). Moisture levels, nitrogen levels—clue to the next limiting factor on crop production. *J. Plant Nutr.* *13*:451–457.

Wallace, A. (1990c). The interacting nature of limiting factors including micronutrients on crop production: getting high yields from present cultivars. Proc. Int. Plant physiology held Feb. 15–20, 1988 in New Delhi, India (S. K. Sinha, P. V. Sane, S. C. Bhargava, and P. K. Agrawal, eds). InPrinte Exclusives Pub., New Delhi, India, 2:1164–1168.

Wallace, A. (1991). Possible roles for water-soluble polymer to alleviate drought conditions. *1991 Agron. Abs.*, p. 344.

Wallace, A. (1994a). Sense with sustainable agriculture. *Commun. Soil Sci. Plant Anal.* *25*:5.

Wallace, A. (1994b). Soil organic matter is essential to solving soil and environmental problems. *Commun. Soil Sci. Plant Anal.* *25*:15.

Wallace, A. (1994c). Soil organic matter must be restored to near original levels. *Commun. Soil Sci. Plant Anal.* *25*:29.

Wallace, A. (1994d). Strategies to avoid global greenhouse warming—stashing carbon away in soil is one of best. *Commun. Soil Sci. Plant Anal* *25*:37.

Wallace, A. (1994e). High-precision agriculture is an excellent tool for conservation of natural resources. *Comm. Soil Sci. Plant Anal.* *25*:45.

Wallace, A. (1994f). Consumerism, perhaps our biggest environmental problem. *Commun. Soil Sci. Plant Anal.* *25*:159.

Wallace, A. (1995a). Agronomic and horticultural aspects of iron and the law of the maximum. *Iron Nutrition in Soils and Plants* (J. Abadia, ed). Kluwer, Dordrecht, pp. 207–216.

Wallace, A. (1995b). The economics of value-added composts. In: *Soil Conditioner and Amendment Technologies*, Vol. 1, Wallace Labs, El Segundo, CA, pp. 207–210.

Wallace, A., and Wallace, G. A. (1993). Chapter 10, Limiting factors, high yields, and law of the maximum. *Horticultural Reviews* *15*:409.

Wallace, A., and Wallace, G. A. (1994). Role of soil and plant analyses in safe, sustainable agriculture. *Comm. Soil Sci. Plant Anal.* *25*:55.

Wallace A., Patel, P. M., Berry, W. L., and Lunt, O. R. (1978). Reclaimed sewage water: a hydroponic growth medium for plants. *Resource Recovery Conservation* *3*:191.

Western Fertilizer Handbook (1994). Interstate Printers and Publishers, Danville, Illinois, 8th ed. Calif. Fertilizer Assoc.

Wittwer, S. W. (1975). Food production: technology and the resource base. *Science* *188*:579.

Yaalon, D. H. (1996). Soil science in transition: soil awareness and soil care research strategies. *Soil Science* *161*:3.

Youngberg, G. (1992). Special issue on soil quality, *Ame. J. Altern. Agri.* *7*:2–68.

Zwingle, E. (1993). Ogallala Aquifer, wellspring of the High Plains. *National Geographic* *183*(3):80.

2

Organic Mulches, Wood Products, and Composts as Soil Amendments and Conditioners

Margie Lynn Stratton and Jack E. Rechcigl *Range Cattle Research and Education Center, University of Florida, Ona, Florida*

I. INTRODUCTION

Public interest in organic soil amendments often is accompanied by confusion between mulches and composts. Strictly defined, mulches are materials applied across the soil surface, whereas composts are humus-like products of an engineered process (Stratton et al., 1995). Under this definition, mulches can include many inorganic materials such as stone, gravel, sand, clay aggregates, perlite, vermiculite, plastics, glass, landscape geotextiles, and others (Borland and Weinstein, 1989; Locascio et al., 1985; Wolfe et al., 1989; Sweeney et al., 1987). Organic mulches are carbon-containing materials derived from living matter, such as sawdust, woodchips, hay, straw, leaves, and also compost. Carbon-containing materials that are not organic but that are used as mulches include polyethylene plastic materials (Manrique and Meyer, 1984; Midmore, 1983; Midmore, 1984; Phene and Sanders, 1976), photodegradable plastic film (Taber, 1991; Wolfe, 1989), and layers of reflective chalk (Midmore, 1983; 1984).

Soil-amending mulches and composts are derived from organic (carbon-containing) materials that are more readily decomposable than plastics, and that when partially decomposed or stabilized impart humus to the soil. Humus is the relatively resistant, somewhat decomposed, dark-brown to black substance of complex chemical and physical nature that results from partial decomposition of organic materials in soils (Stratton et al., 1995). Composts, some wood products, and mulches discussed here decompose into humus and impart to the soil all the beneficial qualities derived from traditional organic matter additions to the soil whether from crop residues, leaf litter, or intentional soil amending procedures.

Organic matter in soil is important for the physical and chemical properties that it imparts (Allison, 1973; Black and Siddoway, 1979; McCalla and Army, 1961; Stevenson, 1982; Arshad and Coen, 1992; Wallace, 1994). Historically, farmers have relied on the return of organic matter to the soil to sustain long-term fertility and crop production (Follett and Peterson, 1988; Black, 1973; Oveson, 1966; Rasmussen et al., 1980). Organic amendments such as spent mushroom substrate increased water-holding capacity, decreased bulk density, and subsequently increased crop productivity (Wang et al., 1984). Where mulch has been used in addition to incorporation of compost to the soil, yields of tomatoes have been increased by the additive or synergistic effects of mulches and soil-incorporated organic matter (Steffen et al., 1993). These are just two examples among many in the following sections reporting the beneficial effects of wood products, mulches, and composts as soil amendments.

II. WOOD PRODUCTS

Wood products include organic carbon-containing plant-derived materials such as lumber, sawdust, paper, logyard residues, construction waste lumber, after-the-season Christmas trees, land-clearing residues, sawmill by-products, and wood ash (Schuman 1995; Folk and Campbell, 1990; Mackay et al., 1989; Clapham and Zibilske, 1992; Ohno, 1992; Erich, 1991). These products may be burned, acid-treated, alkaline-treated, or otherwise chemically treated. These products may be whole, chopped, shredded, chipped, or as finely divided as sawdust. Larger particles take longer to decompose (Maas and Adamson, 1972). The decomposition of oak (Quercus spp.) logs has been evaluated with the conclusions that sapwood decayed more slowly than heartwood, which lost 50% of its mass the first year and none in the second year of the study, and inner bark had the highest plant nutrient quality (Schowalter, 1992). Wood products such as preservative-treated lumber have been disposed on land or used in building and while not considered soil amendments have caused concern about contamination of the soil. For example, railway and utility lumber uses have contaminated nearby soils with copper naphthenate and chlorophenols, particularly pentachlorophenol (Harp and Grove, 1994; Wan, 1992; Lamar and Dietrich, 1990). Such products are undesirable as soil amendments or conditioners.

Technically, the term wood products also refers to wood ash, the residue from burned wood. Wood ash is not an organic soil amendment as it does not increase soil organic matter directly. Wood ash is an alkaline mineral soil amendment that is primarily a source of K, Ca, and Mg for improving soil fertility and plant nutrition (Ohno, 1992; Erich, 1991) and a liming amendment (Clapham and Zibilske, 1992). Wood ash is sometimes used in potting media (Butler and Bearce, 1995).

Generally, wood products are used as mulches, particularly the stable, coarsely divided materials (Norden, 1990). However, sawdust and bark have also been used as soil amendments (Cronan et al., 1992; Nus and Brauen, 1991). Bark has been used successfully as an amendment in potting mixes and as a soil conditioner (Bilderback, et al., 1982; Brown and Pokorny, 1975; Carvalho, et al., 1984; Chong, et al., 1991; Dunn and Latimer, 1956; Ellis, et al., 1986; Grzeszkiewicz, 1978; Harder and Baker, 1971; Howard, 1973; Howard, 1970; Odneal and Kaps, 1990; Pokorny, 1969, 1983; Saini and Hughes, 1975; Sarles, 1973). Even salty bark has been studied as a soil amendment (Bollen, 1971). The degradation of bark and its contribution as a soil conditioner is the subject of a review by Cappaert et al. (1976). One effect in the soil due to amendment with bark is attributed to moisture retention and subsequent moisture release by hardwood bark (Spomer, 1975). Wood residue amendments have increased water infiltration and increased vegetation establishment in revegetation of mine spoils (Schuman et al., 1994). In addition to positive effects on soil moisture, wood amendments may have other effects such as the reduction of nematode (*Meloidogyne* spp.) infestations by sawdust or hardwood bark amendments (Siddiqui and Alam, 1990; Kushwaha et al., 1983; Malek and Gartner, 1975) or suppression of *Fusarium* wilt (Pera and Calvet, 1989). Sawdust may be one of the most commonly used wood products for direct incorporation as a soil amendment. One study compares the resistance of sawdusts, peats, and bark to decomposition with the conclusion that organic wood products incorporated into the soil decompose more quickly if the product is finely divided, as with sawdust (Maas and Adamson, 1972).

Sawdust is also used as a mulch (Shutak and Christopher, 1952; Roberts and Mellenthin, 1959; Posey and May, 1954). As mulch on blueberries (*Vaccinium ashei* Reade cv. 'Climax') sawdust has increased fruit productivity as well as the radial spread of roots, which is reported to improve growing conditions for this fruit (Lareau, 1989; Patten et al., 1988). Weathered sawdust was used as a soil amendment in plots where rabbiteye blueberries (*Vaccinium ashei* Reade cv. Tifblue) were grown and reportedly produced greener plants with greater linear growth (Cummings et al., 1981).

Since wood products have a high C:N ratio (Poincelot, 1978), their decay can lead to a depletion of available soil N by N immobilization (Allison, 1981). One management decision if crops are to be grown immediately after sawdust incorporation would be the addition of supplemental N. To alleviate the problem with N immobilization some researchers add sewage sludge as a source of N (Sabey, 1975). Plants can be grown successfully directly in sawdust if adequate fertilization is supplied as in hydroponically grown greenhouse crops or in potting media (Cheng, 1987; Allen V. Barker, personal communication). Pine bark has also been amended with micronutrient fertilizers (Niemiera, 1992) or arcillite (Warren and Bilderback, 1992) prior to use in growing plants. Pine bark has been amended with composted hardwood bark in potting mixes, and the Ca, Mg, P, and micronutrient supply has been increased by the addition of the composted material (Svenson and Witte, 1992; Saini, 1974). Wood products, if not contaminated, and if finely divided and incorporated into soil with careful attention to plant nutrition, may be good sources of organic matter for amending soils.

III. MULCHES

Mulches have many functions in soil applications (Table 1). Conservation of moisture by suppressing evaporation from the soil and increasing infiltration of water into the soil are principal functions of mulches. Weed control, temperature regulation, erosion protection, decoration, and to smaller extents insect control, disease control, and plant nutrition are also important functions of mulches. Composts as mulches can serve all of those functions, but usually composts are incorporated into soil for improving soil fertility, principally through additions of plant nutrients and organic matter. Ordinarily, organic materials that are not suitable for direct incorporation into soil should be composted before application to soils.

A. Kinds of Mulches

Kinds or types of mulches refer to the materials from which the mulches are made and sometimes to the functions of the mulches. Organic mulches function in each of the capacities listed in Table 1 and receive primary attention in this chapter. Mulching and minimum tillage practices are similar in their effects on soil properties, and these practices will be compared in this section.

1. Organic Mulches

The first uses of mulches are those of materials derived from plants, including all kinds of crop residues and materials collected for land application (Table 2). Many organic substances other than those listed have been used and evaluated as mulches (Borland and Weinstein, 1989; Casale et al., 1995; Gallargo-Laro and Nogales, 1987; Lanini et al., 1988; Opitz, 1974; Stephensen and Schuster, 1945; McCalla and Duley, 1943; Jacks et al., 1955; Sances and Ingham, 1995). Organic matter for mulches should be a stable material that will remain in place at least through the growing season of an annual crop and longer for perennial crops. Highly carbonaceous, coarse materials have long-lasting properties. These kinds of materials should remain on the soil surface and not be incorporated into soil, unless considerable decomposition (composting) has occurred. In general, soil-incorporated wood and carbonaceous plant products result in mulches that can immobilize N (Borland and Weinstein, 1989; Gouin, 1992). Gouin (1992) states that the problem with wood chip mulches is N starvation

Table 1 Functions and Actions of Mulches

Function	Action
Moisture conservation	Evaporation is suppressed, and infiltration of water is enhanced by layer of mulch. (Robinson, 1988; Bruce et al., 1995.)
Soil temperature	Heat transfer from solar radiation is reduced in summer; roots may be protected from cold air temperatures in winter. (Ashworth and Harrison, 1983; Skroch et al., 1992.)
Weed control	Layer of mulch is too thick or too impenetrable for weed seedlings to emerge. (Robinson, 1988; Skroch et al., 1992.)
Erosion protection	Movement of water across rough surface of mulch is slowed. Mulch acts as a barrier to movement of water. Infiltration of water is increased. Khatibu et al., 1984; Meyer and Mannering, 1963.)
Disease control	Mulch is a protective layer between soil and plants, preventing disease transmission to plants, keeping produce off soil, or suppressing disease growth in soil by competition by mulch-borne microorganisms. (Hoitink et al., 1993; Borland and Weinstein, 1989.)
Insect control	Mulch layer acts as a barrier to insects keeping adults from entering soil or laying eggs on or near base of plants. (Zehnder and Hough-Goldstein, 1987.) Light colored mulches may repel insects.
Plant nutrition	Plant nutrients from thick organic mulches are leached into the soil. (Himelick and Watson, 1990.)
Decoration	Various kinds of mulches are more ornamental than bare soil. (Skroch et al., 1992.)

of the plants due to a high C:N ratio of between 700:1 and 800:1 for ground wood, and as much as 200:1 to 300:1 for recycled yard waste. Gouin (1992) suggests composting of these materials to achieve a C:N ratio of 60:1 before use as a mulch. As a rule, a moderately thin mulch of about two inches (5 cm) thick will not present much of a problem with immobilization of nutrients as long as the mulch remains on the soil surface. Reactivity between soil and mulch is restricted to the interface. Plants growing in thickly mulched soil often show nutrient deficiencies because the root mass of the crop enters into the moist layer of mulch rather than penetrating into the soil (Allen V. Barker, personal communication). Microbial activity in the mulch depletes available nutrients, and the plants are nutritionally deprived if a major portion of their root volume is in the mulch, rather than in the soil.

Table 2 Some Examples of Plant-Derived Materials Used as Mulches

Alfalfa. (Castle et al., 1995.)
Oat stems. (Guenzi and McCalla, 1966.)
Setaria grass. (Devaux and Haverkort, 1987.)
Mature maize leaves and dried grass. (Midmore, 1983.)
Barley straw. (Manrique and Meyer, 1984.)
Rice straw or hulls. (Midmore, 1983.)
Rice hulls and rice. (Borland and Weinstein, 1989.)
Corn cobs, corn stalks. (Freire, 1984; Rice, 1971, 1974.)
Evergreen boughs. (Borland and Weinstein, 1989.)
Grass clippings. (Castle et al., 1995.)
Hay. (Borland and Weinstein, 1989.)
Cocoa or cottonseed hulls. (Borland and Weinstein, 1989.)
Pecan, rice, or peanut hulls. (Castle et al., 1995.)
Leaf mold. (Borland and Weinstein, 1989.)
Uncomposted leaves. (Borland and Weinstein, 1989.)
Manure. (Castle et al., 1995.)
Peat. (Borland and Weinstein, 1989.)
Pine cone chips. (Borland and Weinstein, 1989.)
Pine needle straw. (Borland and Weinstein, 1989.)
Waste paper including third-class mail and newspaper. (Warmund, et al., 1995; Pellett and Heleba, 1995.)
Wood chips and wood fibers, bark, sawdust, wood shavings. (Borland and Weinstein, 1989; Castle et al., 1995.)

Concerns other than C:N ratios and nutrient content regarding choice of mulch materials include, but are not limited to, particle size, hindrance of air and water exchange, difficulty in wetting, or an increase in pH or temperature (Borland and Weinstein, 1989). Materials must allow permeability of water and gases; texture and color should be uniform; material should be easy to apply; material must be resistant to fire and to wind and water erosion, must not be toxic to plants, and should be resistant to decay. Plants that are mulched should remain mulched throughout their lives. Roots of mulched plants are more lateral and shallow in distribution, with many of the roots being right at the soil–mulch interface or into the mulch. Decomposition of the mulch can expose the roots to drying and other unfavorable atmospheric conditions.

Organic mulch materials differ in their durability (McCalla and Duley, 1943). High-lignin straw mulches were more resistant to decomposition than were alfalfa (*Medicago saliva* L.) residues (McCalla and Duley, 1943). Materials such as paper, wood chips, and sawdust with high lignin contents are recalcitrant to decay (Allison, 1973, 1981). The more durable organic matters provide longer-lasting benefits as mulches than readily decomposable plant residues. If mulch is incorporated into the soil and allowed to decay, then the benefits from the mulch are similar to the benefits from compost. The decaying mulch adds humus to the soil from the surface downward (Allison, 1973, 1981). Rapid decay of the incorporated product is considered to be a favorable trait. Factors influencing the rapidity of residue decomposition include the nature of the material (C:N, cellulose, lignin content), soil and climatic conditions, and types and populations of soil organisms (McCalla and Duley, 1943; Spaulding and Eisenmenger, 1938; Tenney and Waksman, 1929; Waksman and Gerretsen, 1931). The decay rate of surface-applied plant residues has been studied with reports that rye (*Secale cereale* L.) and wheat (*Triticum aestivum* L.) due to their higher lignin and cellulose contents produce a more weather-resistant mulch than did legumes (Triplett, 1986).

Yields of soybeans (*Glycine max* L.) were increased by 74%, corn (*Zea mays* L.) by 61%, and wheat by 36% by surface-applied plant residues from each crop (Ghidey et al., 1985). These yield enhancements are related to the rates of decomposition of each residue. The rates of decomposition reflect the capacity of mulches to add nutrients to the soil as well as their benefit imparted through moisture conservation and weed suppression. Surface-applied cornstalks decomposed by 50% of mass in 56 days, whereas incorporated corn residues decomposed 65% in 35 days (Parker, 1962). Surface-applied straw decomposed 45% by mass in 78 days, whereas that which was incorporated lost 82% of mass (McCalla and Duley, 1943). Alfalfa residues on top or incorporated decomposed 72% during the same 78-day period (McCalla and Duley, 1943). McCalla and Duley suggest high C:N ratio as a factor in the decomposition rate of residues. C:N ratios varied among the residues, as wheat straw has less than 1% total N; corn stalk residues have about 1 to 2% total N, whereas the alfalfa residues have about 2 to 4% total N (McCalla and Duley, 1943; A.V. Barker, personal communication; Buchanan-Smith et al., 1996). These differences in N concentrations govern the C:N ratios and rates of decompositions of the residues.

Production and storage of mulch are important in end quality of product (Bilderback and Pokorny, 1987; Svenson and Witte, 1989). In bark production, particles smaller than 1 cm are used for container media, whereas larger particles are used for mulch (Hoitink, 1979; Pokorny and Delaney, 1975). Too fine particles become compressed or compacted and possibly can impede air and water movement on soil surfaces. Many recommendations of storage include conditions of moisture and dimensions of storage piles, such as storage at moisture of 40% of total mass (Hoitink, 1979) and in windrows 2 m high by 4 m wide (Bell, 1973; Golueke, 1972; Pokorny and Delany, 1975). Storage in these conditions leads to some composting of the material and to improvements in physical and chemical properties, such as division of coarse particles and narrowing of C:N ratios. It is debatable whether fresh materials should be used as container media or as mulches. Generally, fresh materials should be applied as mulches, whereas composted or microbially resistant materials can be used successfully as soil amendments or as mulches.

2. No-Tillage Agriculture and Mulches

Some of the benefits of mulching may be solely due to the effects of not tilling the soil. Air and water infiltration rates are significantly higher under no-till agriculture than with moldboard plowing (Radcliffe et al., 1988). Mulches impart similar conditions to those of not tilling the land. Mulches can prevent formation of impermeable crust and can slow runoff in manners similar to those of no-tillage systems (Langdale et al., 1983). Growing crops in no-till culture results in a different soil environment from conventional tillage (Doran, 1980; Fenster and Peterson, 1979). Crop rooting patterns are affected (Drew and Saker, 1980; Ellis et al., 1977; Hodgson et al., 1977; Wilhelm et al., 1982), giving patterns similar to those in mulched soil. Increased soil phosphatase activity occurs under no-till culture (Doran, 1980) along with changes in soil P dynamics, in general allowing greater availability of P to crops (Kunishi et al., 1982; Westerman and Edlund, 1985).

Conventional tillage practices can allow losses of as much as 60% of the soil organic carbon over a period of 60 years (Jenny, 1941; Haas et al., 1957; Tiessen and Stewart, 1983). Long-term mulching through surface additions of organic matter can lead to enhancement in soil organic matter and aid in reversing these losses. Also, mulching of land lessens the needs for tillage for seedbed preparation and for weed control and consequently leaves the land less disturbed mechanically than conventionally tilled or unmulched land.

3. Stubble Mulch

Stubble mulch is the residue of shoots and intact roots left as a covering across a field and in the soil after the harvest of a crop. The residue consists of the roots and bases of stems of the crop (collectively termed stubble) and any above ground parts of the plant that remain after the harvest. The term stubble mulch usually refers to field conditions after harvest grain crops, wheat, oats (*Avena sativa* L.), barley (*Hordeum vulgare* L.), rye, corn, and soybeans but sometimes to conditions after harvest of hay. Stubble mulch planting or stubble mulch tillage is a method of reduced tillage agriculture in which seeds or seedlings of the next crop are planted into the stubble mulch. Stubble mulch tillage is less disturbing to soil than conventional moldboard plowing or disking, but more disturbing than no-till culture. Benefits from stubble mulch tillage are intermediate between those of no-tillage systems and moldboard plowing with respect to a number of soil physical and chemical properties.

Dry matter and grain yield, N concentration and N uptake, and fertilizer recovery were lowest in no-till, highest in conventional plowing, and intermediate in stubble mulch (Varvel et al., 1989). Stubble mulch tillage has been reported to promote N and C retention (Black and Siddoway, 1979; Lamb et al., 1985; Unger, 1968). In a 44-year-long stubble mulch tillage experiment in Oregon, organic N and C increased in the top 75 cm by 26 and 32%, respectively, over moldboard plowing (Rasmussen and Rohde, 1989). Below 75 cm the organic N and C were the same for the two treatments in this long term experiment (Rasmussen and Rohde, 1989). Under stubble mulch tillage soil had higher C content than under conventional tillage (Christensen et al., 1994). Stubble mulch tillage with wheat and sorghum (*Sorghum bicolor* (L.) Moench) resulted in less leaching of nitrate below the root zone (Eck and Jones, 1992), higher grain yields of winter wheat (Smika, 1990), and increased soil microaggregates (Cambardella and Elliott, 1993). Stubble mulch tillage may lower soil pH (Unger, 1991) and under stubble mulch tillage the acidifying effects of ammonium sulfate application were concentrated in the upper 7 cm of the soil profile as opposed to 22 cm for moldboard plowing (Rasmussen and Rohde, 1989).

A technique similar to stubble mulch tillage called no-burn mulching has been used on sugarcane (*Saccharum* spp.) plantings in Brazil (Ball-Coelho et al., 1993). By refraining from burning of sugarcane plantings in preparation for the harvest and keeping the residue as mulch, from 2600 to 4800 kg C ha^{-1} were retained on the field (Ball-Coelho, et al., 1993). The mulch layer did not reduce tillering nor reduce N and P accumulation in sugarcane (Ball-Coelho et al., 1993). The harvestable yield of the first ratoon sugarcane crop from the mulched fields was 17 Mg ha^{-1} (wet mass) greater than that from the burn-harvested fields. Table 3 lists benefits of no-tillage, stubble mulch and organic mulches.

4. Living Mulches

A living mulch is a covering of living plant material that is planted or allowed to remain living on the land after another crop has been planted. Living mulches impart some of the same benefits as surface-applied organic mulches, such as increased soil carbon, increased rain infiltration, and reduced soil erosion (Bruce et al., 1995). Living mulches reduce erosion and suppress weeds (Hargrove et al., 1982; Sweet, 1982). Management of living mulches can be difficult, for the mulch can be competitive with the principal crops. Often living mulches are a cover crop turned into the soil as an organic amendment that will impart the same properties to the soil as humus.

B. Effects on Soil Physical Properties

Mulch is beneficial to crop production through soil and water conservation, weed control, improved soil structure, higher infiltration and retention of water, stabilization of temperatures, and enhanced biological activity (Table 1) (Ekern,

Table 3 Benefits of No-Tillage Systems, Stubble Mulches, and Organic Mulches

Moisture retention. (Fenster and Peterson, 1979; Smika, 1990.)
Higher surface water contents, decreased freeze–thaw cycling and drying by sublimation, and decreased fluctuations in water content. (Layton et al., 1993.)
Increased available nutrients. (Follett and Peterson, 1988; Tracy et al., 1990.)
Increased nutrient cycling. (Follett and Schimel, 1989.)
Increased microbial biomass, respiration, and total N. (Follett and Schimel, 1989.)
Increased yield and P uptake in winter wheat. (Stecker et al., 1988.)

1967; Jacks et al., 1955; Lal et al., 1980; Merwin, et al., 1994; Tumulhairwe and Gumbs, 1983). The benefits can occur with land-applied mulches, with stubble mulch, or with living mulches. Most of the following discussion applies to land-applied, organic mulches.

1. Soil Structure

In general, under mulched conditions, bulk density of the soil will be lower than with unmulched conditions (Tindall et al., 1991; Himelick and Watson, 1990; Lemon and Erickson, 1952). This result was noted with straw mulch (Tindall et al., 1991) and with rice hulls and rice straw mulches (Lemon and Erickson, 1952). However, soil strength increased under no-till compared to conventional tillage (Tollner et al., 1984). Bulk density with no-till culture may be higher in the layer directly under the no-till surface, but density is lower at 15 to 30 cm depths than that of tilled culture (Tollner et al., 1984). Much of the effect of mulches on soil structure may be imparted to the soil as the mulch decomposes and becomes more humus-like.

2. Soil Erosion

Mulching reduces erosion (Khatibu et al., 1984; Meyer and Mannering, 1963; McGregor et al., 1988). This benefit may be one of the most important uses for mulch in some geographical areas. Applied mulches, no-tillage mulches, stubble mulches, and living mulches are effective in prevention of erosion. The effects of mulch on erosion control are due to the many factors that lessen the impact of a rain event on a soil surface. Covering the soil surface with mulch decreases the amount and the velocity of runoff and decreases the impact of falling raindrops by dissipating the energy from the raindrops (Mannering and Meyer, 1963; Wischmeier and Mannering 1965; Foster and Meyer, 1972; Hudson, 1957; Kramer et al., 1941). Resistance to penetration of soil was lowered under rice hull and grass mulched plots (Manrique, 1995). Leaving crop residues after harvest gives mulches that are effective in prevention of erosion, even more effective than planting of a cover crop, which will take time to become established. Removal of corn stover allowed 40% more runoff than did incorporating the corn residues (Wischmeier, 1973).

Surface residues can reduce soil erosion by increasing the size and stability of wet or dry soil aggregates (Siddoway, 1963; Black and Power, 1965; Woodruff et al., 1972; Black, 1973; Dickey et al., 1984; Layton et al., 1993). The placement of plant residues is reported to affect soil erosion control (Tanaka, 1986). Shallow incorporation improves erosion-reducing capabilities of surface-applied plant residues compared to moldboard plowing-under of the residues (Wischmeier, 1973). With incorporated corn residues (15% cover), soil loss was 10.5 Mg ha^{-1}, whereas with wheat residues, where 79% remained on surface, losses were only 1.3 Mg ha^{-1}. With bare land without cover, losses were 16.6 Mg ha^{-1} (McGregor et al., 1990). Benefits may not appear until a year or more after applications of mulches. A single application of corn stover or straw mulch did not increase infiltration or decrease runoff as much in the first year as in the second year after application (Wischmeier and Mannering, 1965).

3. Soil Moisture

Soil moisture under mulch is increased through the processes of minimizing soil surface evaporation (Himelick and Watson, 1990; Lal, 1975; Tumulhairwe and Gumbs, 1983; Kolb et al., 1985; Spiers, 1983; Bond and Willis, 1969), increasing water retention with increasing organic matter (Firth et al., 1994), and increasing infiltration rate (Hassan, 1985; Oster et al., 1986; Lal et al., 1980; Hill and Blevins, 1973; Blevins et al., 1983; Mannering and Fenster, 1983). Infiltration rate is also increased by the practice of no-till culture (Tollner et al., 1984). Soil moisture increased under mulch in experiments with 7 species of shade trees (Watson, 1988). Straw mulch increased infiltration rate and decreased evaporation (Tindall et al., 1991).

Mulching increased water use efficiency in plots that were irrigated every 8 days (Manrique, 1995). Under straw mulch, more favorable moisture conditions for potato (*Solanum tuberosum* L.) growth were maintained (Zehnder and Hough-Goldstein, 1987). In other experiments, a mulch cover of 2.5 cm thickness of corn residue reduced evaporation loss and stabilized soil moisture content (Chung and Horton, 1987). One reported advantage from mulching is the increased water use efficiency and water availability that may allow the farmer to adjust planting dates advantageously, as was reported with planting of potatoes during the end of the rainy season in Rwanda so that plants mature during the dry season, resulting in increased tuber yields and reduced late blight incidence (Devaux and Haverkort, 1987).

Soil moisture under mulch is increased through minimizing soil surface evaporation (Himelick and Watson, 1990; Lal, 1975; Tumulhairwe and Gumbs, 1983; Kolb et al., 1985; Spiers, 1983; Bond and Willis, 1969). Mulch also increases water retention as a result of increasing organic matter (Firth et al., 1994). A mulch of kleingrass (*Panicum coloratum* L.) increased water storage on minespoils (Hauser and Chichester, 1989). Under mulched macadamia (*Macadamia integrifolia*) trees, organic matter was increased in the top 15 cm, allowing increased field capacity, higher permanent wilting point, and increased available moisture (Firth et al., 1994). Soil moisture under mulch was higher during dry times than under unmulched conditions (Himelick and Watson, 1990). Mulch has been reported to decrease moisture stress, especially under droughty conditions (Spiers, 1983).

Mulch can be a barrier to water penetration in light rain (Midmore, 1983) but may increase infiltration rate during heavy rains (Hassan, 1985; Oster et al., 1986; Lal et al., 1980). Many people believe that a mulch should be absorbent of water to be effective in moisture conservation. To the contrary, mulches should not be absorbent, for water must pass through the mulch and into the soil. Water held in the mulch will be lost largely by evaporation. In other experiments, removing mulch sharply decreased the infiltration rate, whereas applying mulch in conventional tillage culture increased infiltration rate (Radcliffe et al., 1988). Soil moisture in one report was affected by the age of the rye straw mulch as it was tilled in with the effect that the aged material retained more moisture than the fresher material (Munawar et al., 1990).

Some problems may be encountered with wet conditions maintained under a mulch, where soil is not permitted to dry. If soil is too wet under mulch, increases in anaerobiosis and denitrifying bacteria may cause loss of N (Doran et al., 1984). In areas of high rainfall, thick mulching may lead to waterlogged conditions that provide an ideal environment for diseases (Manrique, 1995).

4. Soil Temperature

Mulches significantly affect soil temperature, largely through stabilization by insulation (Cooper, 1973; Fraedrich and Ham, 1982; Ashworth and Harrison, 1983, Unger, 1978). Organic mulches are the most effective ones for preventing soil warming; for example, temperatures under wheat straw mulches in Kansas in summer were consistently lower than under bare soil or gravel or black polyethylene mulches (Hanks et al., 1961). Soil temperatures were also reported lower under straw mulch (Tindall et al., 1991). In warm regions, use of organic mulches results in cooler soil temperatures, which can be especially favorable in tropical and subtropical zones (Manrique and Meyer, 1984; Devaux and Haverkort, 1987; Midmore, 1983; Lal, 1974, 1978). In the South of the United States, where high temperatures can be limiting to pepper *(Capsicum annuum)*, applications of straw mulch increased pepper yields (Roberts and Anderson, 1994). Researchers report that straw mulching resulted in favorable soil temperature conditions for growth of a potato crop in Virginia (Zehnder and Hough-Goldstein, 1987).

Crop residue mulches decrease the average daily soil temperature by lowering the maximum temperature achieved (Van Doren and Allmaras, 1978; Fontin and Pierce, 1990). Mulch also reduces fluctuations in soil temperature (Unger, 1978). A mulch cover of 2.5 cm thickness of corn residue reduced the daily variation in soil temperatures (Chung and Horton, 1987). The mulch intercepts some solar energy, thereby reducing the heat available for warming the soil and preventing rapid heat loss from warmed soil (Manrique, 1988). Loosely packed mulches, such as hay, absorb radiation but transmit little energy downward by conduction because of still air entrapped below or within the mulch (Waggoner et al., 1960). This same action serves to insulate soils against cooling.

Lower soil temperature that results from crop residues on the soil surface (Griffith et al., 1973; Al-Darby and Lowery, 1987; Gupta et al., 1988) may adversely affect seedling emergence and growth in areas where excessive soil heat is not a concern or where soils warm slowly in the spring (Rickerl et al., 1992). Temperatures under an oat (*Avena sativa* L.) straw and an inert poplar (*Populus* spp.) excelsior mulch were 2.2°C lower than bare soil and delayed corn development in a study in Ontario, Canada (Fortin and Pierce, 1991). Crop residues left on soil caused significant suppressions in growing season soil temperatures (Van Doren and Allmaras, 1978; Gupta et al., 1983), which were accompanied by developmental delays in corn growth compared to growth with conventional tillage (Mock and Erbach 1977; Al-Darby and Lowery 1987; Swan et al., 1987; Fontin and Pierce, 1990). One indirect problem associated with cool soil temperatures may be disease promotion such as *Rhizoctonium solani* (Gupta et al., 1983).

C. Effects on Soil Chemistry

Mulches affect soil chemistry by inhibition of air exchange, retention of moisture, and influences of organic acids produced by the decay of the mulch (Waggoner et al., 1960; Blevins et al., 1977). Oxygen deprivation to roots, stimulation of plant diseases, and denitrification may be expected in mulched soil. Effects on soil pH or on nutrient solubilities may develop slowly but may be the most important effects of mulches on soil chemistry.

1. Soil Aeration

The overall effects of mulching on soil oxygen are not predicted easily, since they are site specific, depending on climate, soil physical properties, and mulches. Mulches are reported to immobilize oxygen near the soil–mulch interface (Waggoner et al., 1960). However, the soil oxygen diffusion rate, a parameter assessing oxygen availability for root growth, was similar under mulch, bare soil, or grass with seven species of shade trees (Watson, 1988). The oxygen diffusion rate was significantly higher in plots mulched with rice (*Oryza sativa* L.) hulls and rice straw than in unmulched plots (Lemon and Erickson, 1952). In areas of high rainfall, thick mulching can lead to waterlogging and anoxic conditions, which can provide an environment for plant diseases (Manrique, 1995) and denitrifying bacteria (Doran et al., 1984). Anoxic conditions affect reduction–oxidation processes, which in turn affect solubilities of multivalent elements (Bohn et al., 1979).

2. Soil Fertility, Plant Nutrition, and Effects on Crops

Soil acidity affects nutrient availability (Bohn et al., 1979). The pH of the soil in no-till systems or under mulch decreases as a result of addition or retention of organic matter, with organic acids produced from decomposition of plant-derived materials accumulating or leaching into the soil (Blevins et al., 1977; Tindall et al., 1991; Himelick and Watson, 1990). The effects of mulch on soil pH ultimately depend upon the mulching material and the initial soil pH and buffering capacity (Blevins et al., 1977; Tindall et al 1991; Himelick and Watson, 1990). Lower pH under mulched trees increased available Mn, for example. (Himelick and Watson, 1990). Since pH influences availability of several plant nutrients, the effects of mulch on nutrient availability should be included in future research.

Soil N recovery and fertilizer N recovery were lowest in no-till, highest in conventional plowing, and intermediate in stubble mulch culture (Varvel et al., 1989). Broadcast P accumulated in the upper 1 cm of soil under no-till culture (Eckert and Johnson, 1985; Ehlers et al., 1972; Fink and Wesley, 1974; Moschler et al., 1972). No-till culture increased extractable P and exchangeable Ca, Mg, and K near the soil surface (Lal, 1976: Eckert, 1985; Blevins et al., 1977). No-till culture increased extractable Ca, Mg, P, Mn, and Zn in the top 7.5 cm but decreased K compared to soils under conventional tillage (Hargrove et al., 1982). Under no-till culture, Fe and Mn are reported to be converted into more available exchangeable or organic complexes (Shuman and Hargrove, 1985).

Through decomposition, mulch contributes plant nutrients to the soil. An outstanding example of this action was with legume mulches in a tropical climate (Schroth et al., 1992). The legume mulches were rapidly leached of N and P (about 50%) and K (75 to 80%) during the first 11 days after application and of 90% of most nutrients by 42 days after application (Schroth et al., 1992). In another example, from an 8-year tropical experiment, mulch contibuted 10% of the corn crop N (10 kg ha^{-1}) within 60 days after planting (Haggar et al., 1993). To maintain a mulch in the tropics, frequent reapplication is practiced (Manrique and Meyer, 1984), potentially contributing much N, P, and K to the soil. With macadamia mulch significantly increased soil K, Ca, and P (Firth et al., 1994). Tropical climates permit rapid decomposition of mulch, and such a great extent of plant nutrients contributed to the soil would not be expected in cooler climates.

Mulching can result in higher concentrations of nutrients in foliage (Himelick and Watson, 1990; Merwin and Stiles, 1994). In an established apple (*Malus domestica* Borkh.) orchard, soil under hay straw mulch had higher concentrations of P, K, and B than soil under living crown vetch (*Coronilla varia* L.) mulch, sod, or traditional herbicide treatments (Merwin and Stiles, 1994). Geiger et al. (1992) reported that 5 years of pearl millet (*Pennisetum glaucum* R. Br.) residue mulch resulted in slightly higher available P, higher exchangeable bases, and lower A1 saturation than in unmulched soil.

More Mn and N and less P were reported in foliage of mulched trees than in unmulched plantings (Himelick and Watson, 1990). In another experiment, leaf K was higher in apple leaves grown in soil with hay straw mulch than in soils with living crown vetch mulch, sod, or traditional herbicide treatments (Merwin and Stiles, 1994). The mulch as it decomposes contributes organic acids to the underlying soil, resulting in increased bioavailability of plant nutrients. Alternatively, under sodic conditions in an irrigated regime, K, Na, Mg, Cl, and Na absorption by blueberries (*Vaccinium* sp. L) were decreased by mulch (Patten et al., 1988).

Several researchers have reported restricted yields under no-till systems relative to conventional tillage (Gallaher, 1984; Hargrove and Hardcastle, 1984; Thurlow et al., 1984; Touchton and Johnson, 1982). Lower yields in no-till culture may likely be due to factors involved in decomposition of crop residues, primarily immobilization of N. Nelson et al. (1977) and Hargrove (1985) reported higher yields under no-till culture than under conventional tillage and attributed the increased yields to increased water infiltration and decreased evaporation. Increased crop yields have also been explained as attributable to increased soil moisture resulting in greater root growth and greater P accumulation (Stecker et al., 1988). No differences in yield were reported in no-till corn (Singh et al., 1966; Triplett and Van Doren, 1969; Moeschler et al., 1972; Fink and Wesley, 1974). Stubble mulch tillage resulted in higher yields of winter wheat (Smika, 1990).

Mulch has increased yields of vegetables, grasses, trees, and fruits (Manrique, 1995; Tindall et al., 1991; Sances and Ingham, 1995; Lal, 1974; Maurya and Lal, 1981; Mbagwu, 1990; Hauser and Chichester, 1989; Firth et al., 1994; Merwin and Stiles, 1994; Fraedrich and Ham, 1982; Spiers, 1983; Christopher and Shutak, 1947; Shutak et al., 1949). As mulch was increased, potato tuber yields increased in a study in the tropics (Manrique, 1995). Tomato yields were higher under straw mulch than in bare soil (Tindall et al., 1991). The use of mulch increased yields of bell pepper (Sances and Ingham, 1995). Increased yields with the use of mulch have been attributed, in part, to improved moisture conditions (Zehnder and Hough-Goldstein, 1987). For example, mulching increased potato tuber yields in mulched plots in the tropics and increased water-use efficiency in plots that were irrigated every 8 days (Manrique, 1995). However, in one set of experiments, no yield differences were reported for mulched root crops or cassava (*Manihot* sp. Adans.) (Ghuman and Lal, 1983).

Mulch increased the yields of cereal and legume crops (Lal, 1974; Maurya and Lal, 1981). Five years of pearl millet residue mulch increased yields of millet biomass (Geiger et al., 1992). Increased yields of corn (Mbagwu, 1990), especially during drier seasons, was attributed at least in part to increased N-use efficiency under mulch. A mulch of kleingrass (*Panicum coloratum* L.) (13.5 Mg dry matter ha^{-1}, the first year and 6.8 Mg dry matter ha^{-1} the second year) increased forage yields of kleingrass on minespoils (Hauser and Chichester, 1989).

Mulched macadamia trees showed significant yield increases in healthy trees and in those with symptoms of decline after mulch was applied (Firth et al., 1994). Hay straw mulch and living crown vetch mulch were compared in an established apple orchard (Merwin and Stiles, 1994). Trunk cross-sectional area and fruit yields were higher under straw mulch (Merwin and Stiles, 1994). Wood chip mulch increased top growth of maples (*Acer* spp. L.) (Fraedrich and Ham, 1982). Blueberry yields were improved by mulching (Spiers, 1983). Mulch increased yields of highbush blueberry (*Vaccinium corymbosum* L.) (Christopher and Shutak, 1947; Shutak et al., 1949). However, in another study with blueberries, no differences were reported (Townsend, 1973).

Mulch, if applied too thickly, may hinder seedling emergence by forming an impenetrable barrier (Manrique and Meyer, 1984, Midmore, 1984). Summer bermudagrass (*Cynodon dactylon* Pers.) establishment was reduced under barley straw mulch due to restricted rooting from stolon nodes into the underlying soil (Sowers and Welterlen, 1988). Alternatively, high soil crust strength may impede seedling establishment, and crust strength may be decreased by mulch, which decreases bulk density and increases soil moisture, thereby promoting seedling emergence (Khan and Rafey, 1985; Bowen and Coble, 1967). Once emergence was complete, mulched potato plants grew taller, with longer stems, profuse branching, and quick canopy cover, and tuber initiation was hastened under mulch and resulted in rapid tuber enlargement and high tuber yields (Manrique and Meyer, 1984) and improved tuber size and quality relative to unmulched plots (Manrique, 1995).

Mulch affects root development (Watson, 1988; Spiers, 1983; Himelick and Watson, 1990; Atkinson, 1980; Gough, 1980; Kramer et al., 1941; Spiers, 1983). Mulch increased the root densities and root surface areas in seven species of shade trees (Watson, 1988). Blueberry root growth and plant establishment was improved by mulch (Spiers, 1983). Mulch increased the development of fine roots and accompanying mycorrhizae (Himelick and Watson, 1990). Mulching promoted uniform horizontal root proliferation in the upper 15 cm of the soil (Atkinson, 1980; Gough, 1980; Kramer et al., 1941; Spiers, 1983). However, sawdust mulch increased the radial spread of roots but did not promote rooting in the top 15 cm of soil (Patten et al., 1988). Some researchers have suggested that the effects of mulch on rooting proliferation are due to the ameliorating effects of mulch on soil temperatures (Ashworth and Harrison, 1983; Skroch et al., 1992). Cooper (1973) showed temperatures for maximum root extension and branching were 25 to 32°C, compared to 8 to 15°C for maximum root thickness. Mulch may permit maintenance of soil temperatures within these ranges.

Researchers have reported decreased light reflection off straw mulch compared to light-colored painted mulches which reportedly resulted in 15% less yield of plants grown with straw mulch (Matheny et al., 1992). In the tropics a reduction in the light absorbed by the soil surface is considered advantageous in restricting heating of soil and decreasing light penetration thereby preventing sunburn in potato tubers and increasing marketable potato tuber yields (Devaux and Haverkort, 1987).

In selection of mulches, consideration must be given to allelopathic effects of plant residues. Plant residues from several species inhibit the growth of other plants and soil organisms (Rice, 1974). Juglone, an allelopathic agent, has been isolated from black walnut foliage (*Juglans nigra* L.) (Thompson, 1971; Lee and Campbell, 1969). Juglone has been reported to inhibit the growth of crimson clover (*Trifolium incarnatum* L.), crown vetch (*Coronilla varia* L.), hairy vetch (*Vicia villosa* Roth.), and *Lespedeza* spp. (Rietveld, 1983). Juglone also inhibits the growth of the N_2-fixing organisms *Frankia* spp. and *Bradyrhizobium japonicum* (Dawson and Seymore, 1983). Plant extracts of *Populus balsamifera* L. inhibited nodulation, nitrogenase activity, and growth of *Alnus* spp. B. Ehrh. (Jobidon and Thibault, 1982). Extracts of leaves of red oak (*Quercus rubra* L.), white oak (*Quercus alba* L.), sugar maple (*Acer sacchrum* Marsch.), and sycamore (*Plantanus occidentalis* L.) have allelopathic effects on plants or soil microorganisms (Lodhi, 1978).

Numerous phytotoxic substances have been associated with various crop residues (McCalla and Nordstadt, 1974). Crop residues have been shown to limit growth of several species reportedly through exudation, leaching or microbial production of allelopathic chemicals (Lodhi et al., 1987). Corn residues and other plant residues had an adverse effect on the nodulation of *Phaseolus vulgaris* L. (Freire, 1984; Rice, 1971, 1974). Guenzi and McCalla (1966) suggest phenolic acids released during corn stover decomposition as the allelopathic agent in their study. Extracts of rye or corn plant residues inhibited seedling growth of many crop species (Rice, 1974). Milled peanut (*Arachis hypogaea* L.) hulls, milled almond (*Prunus* sp. L.) hulls, chicken manure, a horse and cow manure mixture, cow manure, and alfalfa hay were inhibitory to at least one growth parameter of citrus (*Citrus* sp. L.) or avocado (*Persea americana Mill*), but three mulches tested—yard waste of wood chips, grass, and leaves; rice hulls; and rice hulls and paper—were not harmful to growth (Castle et al., 1995). In these experiments, the mulches that were injurious released as much as 1000 mg NH_3 kg^{-1} (Castle et al., 1995).

D. Effects on Soil Organisms and Pests

Mulches have varied effects on soil microflora. The material used for mulch has a significant effect on whether the mulch increases or decreases the incidence or severity of plant diseases (Davis 1994; Elmer and Ferrandino, 1991). *Trichoderma*, in addition to being a strong colonizer of compostable materials, has potential to control plant diseases (Hoitink and Fahd, 1986; Hoitink et al., 1993; Papavizas, 1985). Potting media containing high levels of *Trichoderma* have been reported to suppress *Pythium* and *Rhizoctonia* (Papavizas, 1985) Some *Trichoderma* strains (TH1 and 8MF2) have biocontrol potential and have been reported to promote growth if added to compost after autoclaving (Jackson et al., 1991; Lynch, 1987). *Trichoderma* sp. also enhanced growth and development of vesicular–arbuscular mycorrhizal fungal (VAMF) mycelium (Calvet et al., 1993). Mycorrhizae can enhance nutrient uptake, promote plant uniformity, and reduce transplanting injury (Biermann and Linderman, 1983). Microflora can be competitive against pathogenic organisms, reducing the incidences of plant diseases (Spring et al., 1980).

Bacterial soft rot was higher when wheat straw mulch rather than hardwood mulch or pine (*Pinus* sp. L.) bark mulch were applied on plots of basil (*Ocimum basilicum* L.) (Davis, 1994). Late blossom end rot was reduced by paper mulch, whereas early blossom end rot was increased by black polyethylene mulch on tomatoes (*Lycopersicon esculentum* Mill.) (Elmer and Ferrandino, 1991). Phytophthora root rot was a serious problem in hay straw mulch in an established apple orchard (Merwin and Stiles, 1994). Plant residues alter populations of *Rhizoctonium solani* (Frank and Murphy, 1977). Two primary effects of mulching are suggested; the innoculum potential of *Rhizoctonium solani* may be affected by altered soil temperature (Gupta et al., 1983); and plant residues provide substrates for growth of *Rhizoctonia* (Papavizas, 1970).

Under sawdust mulch, populations of bacterial- and fungal-feeding nematodes (*Nygolaimus, Mononchidae, Aporcelaimidae*) were increased (Yeates et al., 1993). Field plots of potato mulched with wheat straw showed an increase in population of predators of Colorado potato beetle (*Leptinotarsa decemlineata*) within 2 to 3 weeks after application (Brust, 1994). Numbers of overwintered adult beetles, egg masses, and larvae were lower in mulched plots than in unmulched plots (Zehnder and Hough-Goldstein, 1987). Carrot psyllid (*Trioza apicalis*) injury was reduced by fresh spruce and pine sawdust alongside the plant rows (Nehlin et al., 1994). However, meadow voles were a serious problem in hay straw mulch in an established apple orchard (Merwin and Stiles, 1994).

E. Effects on Weeds

An important benefit of mulches is weed control (Billeaud and Zajicek, 1989). Broadleaf and annual grass weeds were reduced 78% under straw mulch compared to unmulched plantings (Sowers and Welterlen, 1988). Weed control with straw mulch on potato fields was comparable to that with conventional herbicides (Manrique, 1995). Municipal solid waste as a mulch applied at 224 Mg ha^{-1} was as effective as glyphosate [isopropylamine salt of N-(phosphonomethyl)glycine] in control of weeds in vegetable crop alleys in Florida (Roe et al., 1993). The distillation wastes of citronella java (*Cymbopogon winterianus* Jowitt) as a mulch applied at 3 Mg ha^{-1} resulted in yields of citronella java, lemongrass (*Citronella flexuosus* (Stapf) and palmarosa (*Citronella martinii* Stapf var. motia) equivalent to the weed-free plots and superior to herbicidal weed control plots (Singh et al., 1991).

Effectiveness of weed control is dependent on mulch material (Manrique, 1995). Black plastic mulches are reported as more effective for weed control than organic mulches in thin layers (Skroch et al., 1992). Pine bark was effective and durable (Manrique, 1995) because it decomposes slowly and requires minimal reapplication (30% replenishment after 630 days) (Manrique, 1995; Skroch et al., 1992).

F. Management Issues

Indirect negative effects of mulching include prevention of sidedress fertilization and limited fertilizer application at planting if the mulches are applied prior to fertilization (Manrique, 1995). In shallow potato plantings if mulch is improperly managed mulching may lead to a greater percentage of stolons becoming aerial shoots and more tubers becoming sunburned (Manrique, 1995). Thickness of mulch is an important management issue. Effects of mulch are dependent on depth and extent of coverage (Colvin et al., 1981; Blevins, 1981). In a Nigerian study 2.0 Mg ha^{-1} mulch were optimum for maximum water infiltration and retention, for the formation of water-stable aggregates, and for minimizing bulk density (Mbagwu, 1991). Corn and cowpea (*Vigna unguiculata* L. Walp) yields were highest at 4.0 Mg ha^{-1} mulch in the same study (Mbagwu, 1991). In areas of high rainfall, thick mulching may lead to waterlogging conditions and provide an environment for disease (Manrique, 1995). If soil is too wet under mulch, increases in denitrifying bacteria may cause loss of N (Doran et al., 1984). Gouin (1983) discusses the possibility of the use of too much mulch. For example, mulch, if applied too thickly, may hinder seedling emergence through the deep impenetrable barrier (Manrique and Meyer, 1984; Midmore, 1984). Manrique (1995) suggested using a thin layer at planting so as not to impede emergence and to follow with thicker application once the seedlings are established (Manrique, 1995). Manrique (1995) also suggested that in areas with high winds and rapid biodegrading conditions (high sun and rain) no mulch cover is preferred over intermittent or only partial coverage because of potential damage to exposed shallow roots. With potatoes it is suggested that the mulch layer be removed once full canopy is achieved, so that disease and pests are not encouraged (Manrique, 1995).

Careful processing and storage of mulch is recommended to prevent what some authors term mulch toxicity (Svenson and Wittc, 1989). Mulch toxicity is most likely similar to immature compost, which is addressed below. Storage of mulch in large piles may also be a fire hazard.

IV. COMPOSTS

A. Process and Compost Maturity

Composting is defined in general terms as the practice of employing biological reduction of organic wastes to humus or humus-like substances. Traditionally, composting transforms biodegradable organic wastes into a soil amendment or fertilizer, sometimes referred to as artificial manures (Allison, 1973). Composting provides an on-farm means of utilizing plant residues, such as straw, that if incorporated into soil directly, without additional fertilizer, produce nutrient deficiencies in crops. Composting is used also as a means of converting objectionable wastes, such as biosolids (sewage sludge), garbage, organic trash, food processing wastes, and farm manures, into materials suitable for application to land. The benefits of additions of compost-generated humus to soils are the same as the benefits imparted by rich natural humus levels in soils. Compost enriches soils with plant nutrients and improves physical features of soil, such as tilth and water-holding capacity.

The most commonly composted materials include those with which most people are familiar, kitchen vegetable scraps, yard clippings, and green wastes from landscaping and farming activities. Most of these materials, with the exception of the most recalcitrant woods and leaves, are easily compostable in the personal backyard compost pile. On an industrial level, many communities are investigating roadside pickup, chipping of materials, and large-scale composting of green wastes in a municipal setting (Carra and Cossu, 1990; Christopher and Asher, 1994). Composting of biosolids or animal manures is a means of reduction or stabilization of wastes prior to utilization on land or disposal.

1. C:N ratio

Many raw materials for composting are carbonaceous materials, rich in C and low in N. Paper, twigs, wood chips, dead leaves, and residues of dead plants have C:N ratios exceeding 200:1 (Golueke, 1972, 1977). Before these materials can be added to agricultural land the C:N ratio must be narrowed to about 35:1 (Alexander, 1977). Addition of materials with wide C:N ratios to soil induces microbial consumption of soil-borne nutrients (immobilization).

Composting accomplishes a narrowing of C:N ratios (Wiles and Stone, 1975). Carbonaceous materials alone, such as wood chips, decay slowly in composting; therefore, N-rich materials such as grass clippings, biosolids, and farm manures which have C:N ratios below 35:1 may be mixed with carbonaceous materials to accelerate rates of composting (Poincelot, 1978; Golueke, 1972, 1977). Mixing coarse carbonaceous materials with nitrogenous materials also imparts porosity in the mass (bulking). Finished composts have C:N ratios of 15:1 to 30:1 (Wiles and Stone, 1975).

The narrowing of the C:N ratio during composting is accomplished by microorganisms, which use the carbonaceous materials as their source of C, use the nitrogenous materials as their source of N, and consume other mineral nutrients, such as P, S, Ca, Mg, and K. During composting, C is lost to the atmosphere mainly as CO_2, and N is lost mainly as NH_3 gas; however, the loss of C as CO_2 exceeds the loss of N as NH_3. The end result is a narrowing of the C:N ratio. The C:N ratio of the bodies of microorganisms ranges from 5:1 to 15:1 (Alexander, 1977). During composting, a series of microorganisms grow and die contributing additional N-rich material to the mass, narrowing the C:N ratio.

The C:N ratio is not the only factor governing the rate of decomposition of composting materials. Materials such as paper, wood chips, and sawdust with high lignin contents are recalcitrant to decay (Allison, 1973). Even though lignin may have a wide C:N ratio, it does not contribute to immobilization of N or other nutrients. Peatmoss is a lignin-rich material that can be added to potting media with little or no possibility of nutrient immobilization; yet, the benefits of the organic matter in the media are realized. Compost is suggested as a substitute for peatmoss in soil-based or soilless media (Smith, 1992; Verdonck et al., 1985). Compost for this purpose must be at a stable C:N ratio so that the compost will have no effect on nutrient availability and will be long lasting as an amendment.

2. Compost Maturity

Curing, stabilization, or maturation of compost is required if the product is malodorous, is only partially decomposed, or is derived from an anaerobic composting process. Immature compost has been reported to contain many phytotoxic compounds such as ethylene, ammonia, phenolic acids, acetic acid, and other volatile (short-chain) fatty acids. (Zucconi et al., 1982a; Zucconi et al., 1982b; Golueke, 1977; Harper and Lynch, 1982; Lynch, 1978; Wong and Chu, 1985). Haug suggests some guidelines for measuring the degree of stabilization: 1) decline in temperature at the end of batch composting, 2) a low level of self-heating in the final product, 3) analysis of organic content yielding a desirable C:N ratio, 4) O_2 uptake rate of end product 1/30 that of substrate, 5) presence of NO_3^- with concurrent absence of NH_4^+ and starch, 6) lack of insect attraction or insect larvae, 7) characteristic lessening of obnoxious odor during composting and absence of odor upon rewetting of the end product, 8) rise in redox potential, and 9) experience of the operator (Haug, 1980). Poincelot (1978) suggests that compost is stabilized when decomposition no longer uses N and the C:N ratio is 10 to 12. Zucconi and De Bertoldi (1987) suggest using a germination bioassay to assess maturity of compost. Brinton and Droffner (1994) define stability in composting as low CO_2 respiration and lack of continued self-heating. Plant nutrient availability, cation exchange capacity, electrical conductivity, C:N ratio, NO_3-N content, and respiration have been used to assess compost maturity (Chanyasak and Kubota, 1981; Garcia et al., 1991; Garcia et al., 1992; Katayama, 1985; Ianotti et al., 1994). Some researchers report that total C:N ratio is not usually a good indicator of maturity for organic wastes (Garcia et al., 1991; Garcia et al., 1992; Katayama, 1985). Several methods of assessing compost maturity have been reviewed, with the authors suggesting that no method is adequate to assess the maturity of composts from all the different types of substrates (Jiminez and Garcia, 1989).

B. Effects on Soil Physical Properties

Physical properties such as bulk density, water-holding capacity, porosity, and aggregate stability are soil properties that may be influenced by compost applications (Doran and Parkin, 1994; He et al., 1992). Several researchers have studied the effects of composts on soil properties (de Bertoldi, 1987; Cook et al., 1994; Elliott and Stevenson, 1977; Hernando et al., 1989; Mays, et al., 1973; Hortenstein and Rothwell, 1972; Scanlon et al., 1973). Reported changes in soil physical properties occurred with compost applications as low as 13.6 Mg ha^{-1} (Hernando et al., 1989) but in other cases effects were not detected until the application was 149 Mg ha^{-1} (Mays et al., 1973). The most dramatic soil physical property responses to the application of compost to land have been on marginal soils with poor soil structure and low levels of soil organic matter and plant nutrients (Hortenstein and Rothwell, 1972; Scanlon et al., 1973).

Repeated applications of MSW (municipal solid waste) composts to agricultural lands can result in significant improvements in physical, microbiological, and chemical properties of soils similar to the effects imparted to soils after sewage biosolids or other types of organic matter are applied (Epstein, 1975; Khaleel et al., 1981; Page et al., 1987). The effects of compost applications to European soils were reviewed by Gallardo-Lara and Nogales (1987) and Guidi and Petruzzelli (1989) who indicated that the positive effects well outweighed the negative effects. Changes in soil physical properties generally are attributed to an increase in organic matter from the addition of composts.

1. Soil Structure

A significant beneficial effect of compost applications to soil is improvement of soil structure due to the increased integrity of aggregates stabilized by the interaction of microorganisms and the mineral fractions of the soil (Gallardo-Lara and Nogales, 1987; Guidi and Petruzzelli, 1989). Little improvement in soil structure was detected if organic residues were applied to sterilized soil which indicates the importance of microorganisms (Chesters et al., 1957). Adding organic matter increases the growth of microorganisms, with the growth also being influenced by the seasons (Guidi et al., 1988; Lynch, 1981; Tisdall et al., 1978). With stabilization of aggregates, bulk density is decreased and porosity is increased (Heitkamp and Cerniglia, 1989; Mays et al., 1973).

The high content of organic matter in compost and the resultant effects of the organic matter on the humic fractions and nutrients in soil increase microbial populations, activity, and enzyme production, which in turn increase aggregate stability (Tisdale and Oades, 1982; Dong et al., 1983; Haynes and Swift, 1990; Perucci, 1990). Soil microbial biomass and enzyme activity are considered important indicators of soil improvement due to the addition of organic matter (Perucci, 1990).

2. Bulk Density

One beneficial effect of organic matter or compost additions to soil is the decrease in soil bulk density (De Smet et al., 1991; Soane, 1990). In a study in Washington, D.C., composted sewage biosolids or composted MSW were surface-applied in a restoration project in a heavily trafficked and compacted parkland (De Smet et al., 1991). Improvements to the soil after compost application included an increase in water infiltration rate, a decrease in bulk density, and an increase in pore volume (Cook et al., 1979). Recent research also indicates that soil treated with MSW compost is less compactable than untreated soil (Spugnoli et al., 1993). In studies on sandy soils of Saudi Arabia, additions of MSW compost decreased soil bulk density and penetration resistance of the soil (Sabrah et al., 1995).

3. Soil Erosion

Organic matter content of soil is significantly correlated with erosion prevention (Young and Onstad, 1978). Several investigators stressed the importance of organic matter in water-stable aggregates through the formation of organo-mineral complexes (Tisdale and Oades, 1982; Dong et al., 1983; Haynes and Swift, 1990). The extent to which a soil erodes depends on the strength of soil aggregates to withstand raindrop impact and surface flow (Meyer, 1981). Humic and fulvic acids and humins are important as persistent binding agents in mineral-organic complexes, and 52 to 92% of soil organic matter may be involved in these complexes (Edwards and Bremmer, 1967; Hamblin, 1977). Fungal hyphae are associated with temporary binding agents which may impart a degree of short-term stability to aggregates (Tisdale and Oades, 1982).

Compost applications to prevent erosion can be an important measure in soil conservation. Through increases in soil organic matter, compost applications to soil are effective in reducing soil erosion and water runoff (Nearing et al., 1990). Digested papermill sludge amended with N and P reduced erosion on a steeply sloping strip-mine reclamation site (Watson and Hoitink, 1985).

4. Soil Moisture

Incorporating MSW compost into sandy plots where wheat was grown in experiments in Saudi Arabia resulted in significant increases in water retention at field capacity, amount of plant-available water in the soil, and water-holding pores, while significantly decreasing water movement under saturated conditions (Sabrah et al., 1995). In another experiment, saturated hydraulic conductivity was reduced by incorporating organic matter into the soil (Sabrah, 1993). Additions of MSW compost increased the water infiltration rate and pore volume of the soil (Cook et al., 1979). Water-holding capacity is also increased by applications of compost to soil (Hernando et al., 1989; Bengston and Cornette, 1973).

C. Effects on Soil Chemistry

1. Soil Fertility, Plant Nutrition, and Effects on Crops

Compost has traditionally been used as a soil amendment for garden plants and as a substrate for mushroom culture (*Agaricus bisporus*) (Kitto, 1979; Minnich and Hunt, 1988; Bisht and Harsh, 1985; Weigant et al., 1992). Recently, compost use has been suggested for pasture and orchard improvement, commercial vegetable production, containerized nursery crop operations, and turfgrass production (Bevacqua and Mellano, 1993; Bugbee et al., 1991; Chong et al., 1994; Cisar, 1994; Flanagan et al., 1993; Hornick, 1988; Korcak, 1986; Lohr and Coffey, 1987; Madan and Vasudevan, 1989; Mandelbaum et al., 1988; Marchesini et al., 1988; Maynard and Hill, 1994; Mays and Giordano, 1989; Ozores-Hampton et al., 1994; Petruzzelli et al., 1989; Purman and Gouin, 1992; Smith, 1992; Wang and Blessington, 1990; Wong and Chu, 1985).

Addition of organic matter to soil may increase cation exchange capacity (CEC) significantly. For each percent of humus in the soil, CEC is increased 2 meq $100g^{-1}$ (Sopher and Baird, 1978). Cation exchange capacity of soil is increased by compost applications (Bengston and Cornette, 1973). Cation exchange capacity is important in plant nutrition and fertility management, as CEC effectively constitutes temporary storage for cations and therefore is considered an indicator of the nutrient-holding capacity of a soil. A low CEC may allow for greater leaching of cations. Thus additions of compost may improve cation retention in the root zone.

Amending the soil with compost may alter pH (Stratton et al., 1995). The pH of soils affects ion availability to, and absorption by, plants (Woodbury, 1992; Heckman et al., 1987; Hue, 1988; Street et al., 1977; Tadesse et al., 1991). An increase in pH can bring about strong adsorption on soil particles or, in some cases, precipitation of Cd, Mn, Pb, and Zn, among other metals, which in turn allows for lower accumulation of these metals in plant tissue. Decreases in accumulation of Cd by plants with increasing soil pH is well documented

(Heckman et al., 1987; Hue, 1988; Tadesse et al., 1991; Williams and David, 1976). Availability of Zn can be decreased by organic matter additions to soil by formation of insoluble Zn organic complexes with humic acids, thereby lessening risk of Zn toxicity of plants (Stevenson and Ardakani, 1972). Minor essential elements may be supplied by compost amendments to deficient croplands. Increases in Zn concentrations in soil were noted after the addition of compost in one study (Giusquiani et al., 1988).

Depending on soil conditions, compost may increase exchangeable and water-soluble Mn (Tisdale et al., 1985). Although Zn and Mn can be toxic to plants, their introduction into aerobic soil from compost is not usually considered a problem (Leeper, 1978), as high concentrations of Fe in composts inhibit Zn uptake by plants, and Mn availability to plants is reduced by the oxidizing conditions in most soils (Mengel and Kirkby, 1987). Copper is bound very tightly to inorganic exchange sites in soils and is not readily available to plants (Grimme, 1983). Iron, Cu, Mn and Zn are essential elements that may be supplied by biosolids compost additions (Tisdale et al., 1985).

Improvements in the physical, chemical, and microbiological characteristics of cultivated soil have resulted in increased crop production from additions of compost to soil (Selby et al., 1989; Darmody et al., 1983; Steffen, 1979; Pera et al., 1983). Some of these soil improvements are long term and can have significant effects on yield. Composted animal and plant residues continued to improve sunflower yields after 6 years of cropping (Allievi et al., 1993). Water hyacinth (*Eichhornia crassipes*) compost applied to a rice-based cropping system has been reported to increase rice yield (Sharma and Mittra, 1991). Yields of wheat and gram (*Cicer arietinum*) grown after the rice harvest were increased significantly without any further additions of inorganic fertilizer in the same study (Sharma and Mittra, 1991).

Land-applied, poultry manure–based compost has been reported to produce a gradual, long-term response in crop yield (Altierii et al., 1991; Astier, 1990). Seedlings of onion (*Allium cepa*), lettuce (*Lactuca sativa*), or snapdragon (*Antirrhinum majus*) transplanted to plots amended with composted sewage biosolids had more vigorous growth and produced higher yields than those transplanted to unamended plots (Bevacqua and Mellano, 1993). In these experiments, compost-amended plots also produced increased growth of turfgrass (*Festuca arundinacea*) relative to turfgrass grown without compost amendment (Bevacqua and Mellano, 1993). Recently, Astier et al. (1994) reported increased yields of broccoli (*Brassica oleracea*) after woolypod vetch (*Vicia dasycarpa* var. lana) was added to composted poultry manure prior to land application. Composted poultry manure alone or woolypod vetch alone also increased yield of broccoli in this study. Astier et al. (1994) amending calcareous soils with MSW compost increased the growth and yield of tomato (*Lycopersicon esculentum*) and squash (*Curcurbita maxima*) (Ozores-

Hampton et al., 1994). Mays and Giordano (1989) landspread MSW compost annually and grew corn for 14 years with yield increases of 55% to 153% over unamended soils. Composted plant material added to soil increased the yield of sweet potato (*Ipomoea batatas*) (Waddell, 1972; Wohlt, 1986a; Wohlt, 1986b; Floyd et al., 1985; Floyd et al., 1988; Preston, 1990) and corn (Hue et al., 1994).

Much of the effect of application of compost on crop yield and production is derived from the plant nutrients, particularly N, in composts (Ozores-Hampton et al., 1994; Woodbury, 1992; Maynard, 1993). Relatively high applications of composted manure (19–30 Mg ha^{-1}) are required if all of the necessary N is to be supplied solely from compost in the short term. Plant analyses have indicated that adequate N and K were supplied by refuse compost and that adequate N, P, and K were supplied by composted sewage sludge (Smith, 1992; Maynard, 1993; Ozores-Hampton et al., 1994). Additions of MSW compost decreased incidence of nitrate leaching after heavy rains in a Connecticut field experiment on a fine sandy loam (Maynard, 1994).

Other researchers report an increase in plant nutrients from an increase in organic matter due to amendment with MSW compost (He et al., 1992; Shiralipour et al., 1992). In a laboratory experiment, a sandy loam or a clayey silt was incubated for 12 months with or without MSW compost (Giusquiani et al., 1988). Total N, soluble P, and exchangeable K increased in the MSW-amended soil compared to the unamended soil (Giusquiani et al., 1988). Koma Alimu et al. (1977) reported that yield responses from MSW additions to soil in their experiments were due to increases in K. Composted pig (*Sus scrofa*) slurry applied to wheat in field experiments significantly increased leaf K (Gonzales et al., 1991). Increased plant nutrition from composted animal manure applied to soil also increased the protein content of potato (*Solanum tuberosum*) (Srikumar and Ockerman, 1990).

Plant analyses have indicated that adequate N and K were supplied by refuse compost and that adequate N, P, and K were supplied by composted sewage sludge (Maynard, 1993; Ozores-Hampton, et al., 1994; Smith, 1992). Sulfur availability to plants may be enhanced by applications of composts to agricultural land (Gallardo-Lara et al., 1990). Although more research is needed in this area and some reports are inconclusive or conflicting, a review by Gallardo-Lara and Nogales (1987) has addressed some of the effects of MSW compost on N, P, and S in soils.

Earthworm casts are reported to be higher in available N, P, Ca, and Mg than the substrates upon which they feed (de Vleeschauwer and Lal, 1981; Satchell, 1983). The activity and numbers of microorganisms producing acid phosphatases is increased near earthworm casts in the soil (Satchell and Martin, 1984; Stewart and Chaney, 1975; Satchell et al., 1984). Combined together, these specific effects appear to raise P availability in soil amended with vermicompost (Buchanan and Gliessman, 1990).

Several studies suggest that organic sources of P are more effective for plant absorption than inorganic sources (Sample et al., 1980; Meek et al., 1982; Swaider and Morse, 1984; Mishra and Bangar, 1986; Singh et al., 1987). This result may arise from increases in soluble P (Azvedo and Stout, 1974), increases in the microbial P pool (Coleman et al., 1983), or increased hydrolysis of organic P by microbial activity (Mishra and Bangar, 1986). The increase in availability of soluble P from additions of compost is an effect that is described as resulting from phosphohumic complexes that minimize immobilization processes, anion replacement of phosphate by humate ions, and coating of sesquioxide particles by humus to form a cover that reduces the phosphate-fixing capacity of the soil (Sample et al., 1980; Swaider and Morse, 1984; Tisdale et al., 1985). As an example, in a field study with broccoli, P use efficiency was increased with compost amendments (Buchanan and Gliessman, 1990).

D. Effects on Soil Organisms and Pests

Additions of compost or compost extracts have increased microbial populations with various effects (Jansen and McGill, 1995; Jansen et al., 1995). Suppression of soil-borne plant pathogens after applications of organic and especially composted organic materials has been reported (Cook and Baker, 1983; Hoitink and Kuter, 1984; Hoitink et al., 1982; Spring et al., 1980). Applications of composted organic residues suppress *Phytophthora cinnamoni*, *Rhizoctonium solani*, *Fusarium oxysporum*, and *Pythium aphanidermatum* with many crops (Ellis et al., 1986; Heckman et al., 1987; Hoitink and Fahd, 1986; Mandelbaum et al., 1988; Nelson and Hoitink, 1982; Pera and Filippi, 1987; Spring et al., 1980).

Among other disease-suppressing organisms are mesophilic strains of bacteria (*Bacillus* sp.) (Hardy and Sivasithamparam, 1991; Hoitink et al., 1993; Yohalem et al., 1994). A species of bacteria commonly found in compost, *Bacillus subtilis*, has been shown to produce antifungal volatiles (Fiddaman and Rossal, 1993). Wood waste compost was at least as effective as fungicides in controlling *Phytophthora* root rots (Hoitink et al., 1993). Topdressing with composts made with plant refuse or animal wastes suppressed dollar spot (*Sclerotinia homoeocarpa*) of turfgrass as effectively as a commercial fungicide (Nelson and Craft, 1991). Extract of spent mushroom compost inhibited in vitro *Venturia inaequalis*, the causal agent of apple scab (Yohalem et al., 1994). Suppression of *Rhizoctonia* in compost media has been reported by several researchers (Nelson and Hoitink, 1982; Chen et al., 1988; Kuter et al., 1983; Kwok et al., 1987; Tunlid et al., 1989). Composted manure suppressed damping-off by *Rhizoctonium solani* in potting media experiments with radish seedlings (*Raphanus sativus*) (Voland and Epstein, 1994). Composted grape (*Vitis vinifera*) marc and composted cattle manure suppressed diseases caused by *Rhizoctonium solani* and *Sclerotium rolfsii* (Mandelbaum et al., 1985; Gorodecki and Hadar, 1990).

Trichoderma, in addition to being a strong colonizer of compostable materials, has potential to control plant diseases (Hoitink and Fahd, 1986; Hoitink et al., 1993; Papavizas, 1985). Potting media containing high levels of *Trichoderma* have been reported to suppress *Pythium* and *Rhizoctonia* (Papavizas, 1985). Some *Trichoderma* strains (TH1 and 8MF2) have biocontrol potential and have been reported to promote growth if added to compost after autoclaving (Jackson et al., 1991; Lynch, 1987). *Trichoderma* sp. also enhanced growth and development of vesicular-arbuscular mycorrhizal fungal (VAMF) mycelium (Calvet et al., 1993). Mycorrhizae can enhance nutrient uptake, promote plant uniformity, and reduce transplanting injury (Biermann and Linderman, 1983). Other microorganisms, such as thermophilic fungi (*Scytalidium thermophilum*) promoted growth of the edible mushroom (*Agaricus bisporus*) (Weigant et al., 1992). The response was attributed to increased CO_2 production (Weigant et al., 1992).

Soil amendments have been reviewed for their efficacy in suppressing nematode populations (Muller and Gooch, 1982). Organic amendments such as sawdust, green manure, poultry manure, and compost have reduced nematode populations (Muller and Gooch, 1982). Land applications of MSW compost decreased juvenile populations of *Meloidogyne incognita* in field squash in southern Florida but did not affect the plant-parasitic stages of nematode populations (Muller and Gooch, 1982). Products containing chitin, such as blue crab (*Callinectes sapidus*) scrap, have been shown to suppress nematode populations (Rodriguez-Kabala et al., 1984; Rodriguez-Kabala et al., 1989). A compost of blue crab scrap and cypress (*Taxodium distichum*) sawdust suppressed populations of nematodes (*Meloidogyne javanica*) in container-grown tomato (Rich and Hodge, 1993). Thus compost may have many beneficial effects on plant diseases and pests.

1. Effects on Worms

Vermicomposting of many organic wastes has been studied (Edwards, 1983; Huhta and Haimi, 1988; Loehr et al., 1984; Appelhof, 1988). Aerobic sewage sludge and animal manure have been shown to be good substrates for earthworm growth (Hartenstein et al., 1979; Neuhauser et al., 1980; Tomlin and Miller, 1980: Mitchell et al., 1980). Cattle (*Bos taurus*) biosolids, brewery waste, spent mushroom compost, or potato wastes may be used for earthworm culture without prior composting, but pig wastes need to be composted for 2 weeks prior to worm culture (Edwards, 1988). Poultry biosolids need to be leached of salts, and NH_3 must be allowed to volatilize until acceptable levels are reached (Edwards, 1988). Culture of earthworms in animal or vegetable wastes requires temperatures of 4 to 30°C (15–20°C optimum), moisture content 60 to 90% (80–90% optimum), aerobic conditions, NH_3 below 0.5 mg g^{-1}, salt content below 0.5%, and pH between 5 and 9 (Edwards, 1988). Vermicomposted cattle solids have been analyzed before and after vermicomposting with the results that K and NO_3^- increased, pH stayed about the same (7.4–8.6), and NH_4^+ decreased significantly (Edwards, 1988). Earthworm culture is reported to increase the overall rate of decomposition (Mitchell et al., 1980; Brown and Mitchell, 1981), decrease the proportion of anaerobic to aerobic decomposition resulting in a decrease of NH_4^+ production and volatile S compounds (Mitchell et al., 1980), and decrease the incidence of pathogenic bacteria (Brown and Mitchell, 1981).

E. Effects on Weeds

Compost used as a thick layer of mulch would be as effective as other mulches for weed control (addressed in the earlier sections of this chapter). Weed seeds are seldom a problem in properly processed compost (Stratton et al., 1995).

F. Management Issues

Compost with a high C:N ratio could cause plants and microorganisms to compete for N. Microorganisms often compete effectively, and the N is assimilated into the microorganism cells (immobilization). Thus composts with a high C:N ratio may leave little N available for plant development. Immature composts may have high C:N ratios (greater than 40:1), which initially immobilize N after application of compost to soil (Cisar, 1994). Additions of N fertilizer may overcome the problem, as was the case in research in which St. Augustine turfgrass sod was produced in 5 months on 100% MSW with additional N fertilizer compost (Cisar, 1994).

Most organic contaminants that might occur in composting decomposed during 6 months of composting, so mature composts usually are no cause for concern (He et al., 1992; Lemmon and Pylypiw, 1965). Organic contaminants in MSW composts that are land-applied are not expected to leach, as vertical leaching becomes a problem only at extremely low soil organic content (Hsu et al., 1993). Repeated applications of composts to soils has led to concerns of accumulation of metals in soils, but recent regulatory procedures are allowing the processing of composts virtually devoid of heavy metal contaminants (Petruzzelli et al., 1989; Gallardo-Lara and Nogales, 1987). Increased soluble salts in soils can occur from amendment with composted sewage sludge, MSW, and animal manures (Gallardo-Lara and Nogales, 1987; Bevacqua and Mellano, 1994). The potential for high salinity in mixes has limited the use of some composts as soil amendments for potted crops (Chong et al., 1991a). However, leaching of salts was rapid in experiments growing woody ornamental plants in pots with trickle irrigation (Chong et al., 1991a; Chong et al., 1991b).

Pathogenicity is one of the major concerns in the use of MSW and sewage biosolids in agriculture. The objective of regulations from the U.S. Environmental Protection Agency and other agencies for treatment of biosolids is to kill parasitic worm (*helminth*) eggs and pathogenic viruses and bacteria and to reduce vector (flies and rodents) attraction. Biosolids that meet the standards of pathogen reduction set by the US EPA are called Class A. Composting can be used to eliminate pathogens from MSW and biosolids for land application. Composting should be carried out at relatively high temperatures (>55°C) for 3 to 15 days, depending on the process of composting (Farrell, 1993). Pretreatment of biosolids by liming, anaerobic digestion, irradiation, or oxyozonation substantially reduces pathogens, relative to those present in raw biosolids; so considerable advantages in safety and destruction of pathogens might be obtained by use of pretreated sewage biosolids.

Composted MSW may have a high pH and contain large amounts of $CaCO_3$ (Chong et al., 1991a; Rich and Hodge, 1993). Therefore applications of MSW compost to acid soils may be beneficial. Compost amendment increased pH in a soil with low buffering capacity (Buchanan and Gliessman, 1990). Consideration must also be given to applications of composts with high pH to high pH soils. Optimization of composting parameters such as moisture content, temperature, aeration, and duration has significant effects on compost disinfection, pathogen destruction, and odor minimization (Richard, 1990). Maturation and curing is important to prevent N immobilization and NH_3 or NH^{+}_{4} toxicity to plants (He et al., 1992; Inbar, 1990). Several methods for assessing compost maturity have been reviewed by He et al. (1992) and Inbar et al. (1990).

The occurrence and fate of organic contaminants, heavy metals in contaminated composts, high salt composts, pathogenocity of composts, and immature composts used in land application is considered highly dependent upon management of composting processes and applications (Rechcigl and Rechcigl, 1997; Rechcigl, 1995a; Rechcigl, 1995b).

V. SUMMARY

Soil properties may be improved through the effects of mulch, wood products, or compost application. Organic soil amendments may have advantageous effects on bulk density, water-holding capacity, porosity, aggregate stability, erosion, and bioremediation of previously contaminated soils. Water quality may be improved through the binding of leachable compounds or the decomposition of potentially hazardous contaminants.

Beneficial effects of mulch, wood products, or compost applications to crops are many and varied. Most are due to improvement of soil physical and chemical properties and nutrient enhancement, and result in increases in crop quality and yield. Compost applications have significantly positive effects on plant disease suppression, especially soil-borne root pathogens.

Adverse effects from use of mulches, wood products, or compost applications to land are almost always attributable to poor management or poor control of factors such as particle size, contaminants, maturity of materials, placement of materials, supplemental fertilization, irrigation, or poor sitting or design of land application efforts.

ACKNOWLEDGMENTS

The authors would like to acknowledge gratefully the continuing support of Lisa Roberts, Christina Markham, Allen V. Barker, Wayne Smith, Aziz Shiralipour, and the Center for Biomass Programs at the University of Florida.

REFERENCES

Al-Darby, A. M., and Lowery, B. 1987. Seed zone soil temperature and early-growth of corn with three conservation tillage systems. *Soil Sci. Soc. Am. J. 51*:768.

Alexander, M. 1977. *Introduction to Soil Microbiology*. 2d ed. Wiley, New York.

Allievi, L., Marchesini, A., Salardi, A., Piano, C., and Ferrari, V. 1993. Plant quality and soil residual fertility six years after a compost treatment. *Bioresource Technology 43*.

Allison, F. E. 1973. *Soil Organic Matter and Its Role in Crop Production*. Elsevier, Amsterdam.

Allison, F. E. 1981. Decomposition of wood and bark sawdusts in soil, nitrogen requirements, and effects on plants. Washington DC: Agricultural Research Service, US Dept. of Agriculture 58.

Altierii, M. A., Trujillo, J., Astier, M., Gersper, P. P, and Bakx, W. 1991. Low-input technology proves viable for limited-resource farmers in Salinas Valley. *Calif. Agric. 45*(2):20.

Appelhof, M. 1988. Domestic vermicomposting systems. In: *Earthworms in Waste and Environmental Management* (Edwards, C. A., and Neuhauser, E. F., eds.) SPB Academic Publishing, The Hague, The Netherlands, p. 157.

Arshad, M. A., and Coen, G. M. 1992. Characterization of soil quality: physical and chemical criteria. *Am. J. Alt. Agr. 7*:25–31.

Ashworth, S., and Harrison, H. 1983. Evaluation of mulches for their use in the home garden. *Hort. Science 18*:180.

Astier, M. 1990. Developing low-input energy saving vegetable cropping systems for small farmers in Salinas Valley. Internal Report, Division of Biological Control and Department of Soil Science, U.C. Berkeley, ACBE, Washington. D.C.

Astier, M., Gersper, P. L., and Buchanan, M. 1994. Combining legumes and compost: a viable alternative for farmers in conversion to organic agriculture. *Compost Sci. Util. 2*(1):80.

Atkinson, D. 1980. The distribution and effectiveness of roots of tree crops. *Hort. Rev. 2*:424.

Azvedo, J., and Stout, P. R. 1974. Farm animal manures: an overview of their role in agricultural environment. *Calif. Agric. Expt. Sta. Serv. Manual #44*, University of California, 108.

Ball-Coelho, B., Tiessen, H., Stewart, J. W. B., Salcedo, I. H., and Sampaio, E. V. S.B. 1993. Residue management effects on sugarcane yield and soil properties in northeastern Brazil. *Agron. J. 85*(5):1004.

Bell, R. G. 1973. The role of composts and composting in modern agriculture. *Compost Sci. 14*:12.

Bengston, G. W., and Cornette, J. J. 1973. Disposal of composted municipal waste in a plantation of young slash pine: effects on soil and trees. *J. Environ. Qual. 2*:441.

Bevacqua, R. F., and Mellano, V. J. 1993. Sewage sludge compost's cumulative effects on crop growth and soil properties. *Compost Sci. Util. 1*(3):34.

Bevacqua, R. F., and Mellano, V. J. 1994. Cumulative effects of sludge compost on crop yields and soil properties. *Communications Soil Sci. Plant Anal. 25*(3/4):395.

Biermann, B. J., and Linderman, R. G. 1983. Increased geranium growth using pretransplanted inoculation with mycorrhizal fungus. *J. Am. Soc. Hort. Sci. 108*:972.

Bilderback, T. E., and Pokorny, F. A. 1987. Problems associated with high stacking and compaction of pine bark during storage. *Proc. South. Nurs. Assn. Res. Cr. 32*:45.

Billeaud, L. A., and Zajicek, J. M. 1989. Influence of mulches on weed control, soil pH, soil nitrogen content, and growth of *Ligustrum japonicum*. *J. Env. Hort. 7*(4):155.

Bisht, N. S., and Harsh, N. S. K. 1985. Biodegradation of Lantana camara and wastepaper to cultivate *Agaricus bisporus* (Lange) singer. *Agric. Wastes 12*(3):167.

Black, A. L. 1973. Soil property changes associated with crop residue management in a wheat-fallow rotation. *Soil Sci. Soc. Am. Proc. 37*:943.

Black, A.L., and Power, J. F. 1965. Effect of chemical and mechanical fallow methods on moisture storage, wheat yields, and soil erodibility. *Soil Sci. Soc. Am. Proc. 29*:465.

Black, A. L., and Siddoway, F. H. 1979. Influence of tillage and wheat straw residue management on soil properties in the Great Plains. *J. Soil Watern Conserve. 34*:220.

Blevins R. L. 1981. Cover crops and crop residues. In *Soil Science News and Views*. Dept. of Agron. Univ. of Kentucky, Lexington KY.

Blevins, R. L., Thomas, G. W., and Cornelius, P. L. 1977. Influence of no-tillage and nitrogen fertilization of certain soil properties after 5 years of continuous corn. *Agron. J. 69*:383.

Blevins, R. L., Thomas G. W., Smith, M. S., Frye, W. W., and Cornelius, P. L. 1983. Changes in soil properties after 10 years continuous no-tilled and conventionally tilled corn. *Soil Tillage Res. 3*:123.

Bohn, H. L., McNeal, B. L., and O'Connor, G. A. 1979. *Soil Chemistry*. John Wiley, New York.

Bollen, Walter B. 1971. Salty bark as a soil amendment. *US Pacific Northwest Forest and Range Experiment Station. USDA Forest Service Research Paper PNW-128.*

Bond, J. J., and Willis, W. O. 1969. Soil water evaporation: surface residue rate and placement effect. *Proc. Soil Sci. Soc. Am. 33*:445.

Borland, J., and Weinstein, G. 1989. Mulch: is it always beneficial? *Grounds Maintenance. 24*(2):10 and 120.

Bowen, H. D., and Coble, C. G. 1967. Environmental requirements for germination and emergence. *Proc. Conference on Tillage for Greater Crop Production. ASAE Pub. Proc. 168*:10.

Brinton, W. F., and Droffner, M. W. 1994. Microbial approaches to characterization of composting processes. *Compost Sci. Util. 2*(3):12.

Brown, B. A., and Mitchell, M. J. 1981. Role of the earthworm, *Eisenia foetida*, in affecting survival of *Salmonella enteritidis* ser. typhimurium, *Pedobiologia* 22:434.

Brown, E. F., and Pokorny, F. A. 1975. Physical and chemical properties of media composed of milled pine bark and sand. *J. Am. Soc. Hort. Sci. 100*:119.

Bruce, R. R., Langdale, G. W., West, L. T., Miller, W. P. 1995. Surface soil degradation and soil productivity restoration and maintenance. *Soil Sci. Soc. Am. J. 59*(3):654.

Brust, G. E. 1994. Natural enemies in straw-mulch reduce Colorado potato beetle populations and damage in potato. *Biological Control: Theory. Appl. Pest Management 4*(2):163.

Buchanan, R. A., and Gliessman, S. R. 1990. The influence of conventional and compost fertilization on phosporus use efficiency by broccoli in a phosphorus deficient soil. *Am J. Alt. Agric. 5*:38.

Buchanan-Smith, J. G., and Beitz, D. C. 1996. *Nutrient Requirements of Beef Cattle.* National Academy Press, Washington, DC.

Bugbee, G. J., Frink, C. R., and Migneault, D. 1991. Growth of perennials and leaching of heavy metals in media amended with municipal leaf, sewage sludge and street sand compost. *J. Environ. Hort. 9*(1):47.

Butler, S. H., and Bearce, B.C. 1995. Greenhouse rose production in media containing coal bottom ash. *J. Environ. Hort. 13*(4):160.

Calvet, C., Barea, J. M., and Pera, J. 1993. In vitro interactions between the vescular-arbuscular mycorrhizal fungus *Glomus mosseae* and some saprophytic fungi isolated from organic substances. *Soil Biol. Biochem. 24*(8):775.

Cambardella, C. A., and Elliott, E. T. 1993. Carbon and nitrogen distribution in aggregates from cultivated and native grassland soils. *Soil Sci. Soc. Am. J. 57*(4):1071.

Cappaert, I., Verdonck, O., and De Boodt, M. 1976. Composting of bark from pulp mills and the use of bark compost as a substrate for plant growth. Part 2. The effect of physical parameters on the composting rate of bark. Growth experiments with bark compost. *Compost Sci. 17*:18.

Carra, J. S., and Cossu, R. 1990. *International Perspectives on Municipal Solid Wastes and Sanitary Landfilling*. Academic Press, San Diego.

Carvalho, G., Beca, R. A. G., Sampaio, M. N., Neves, O., and Pereira, M. C. 1984. Use of pine bark for preparation of activated carbon and as a soil conditioner. *Agricultural Wastes 9*:231.

Castle, W. L., Minassian, V., Menge, J. A., Lovatt, C. J., Pond, E., Johnson, E., Guillement, F. 1995. Urban and agricultural wastes for use as mulches on avocado and citrus and for delivery of microbial biocontrol agents. *J. Hort. Sci. 70(2)*:315.

Chanyasak, V., and Kubota, H. 1981. Carbon/organic nitrogen ratio in water extract as measure of composting degradation. *J. Ferment. Technol 59*(3):215.

Chen, W., Hoitink, H. A. J., and Schmitthenner, A. F. 1988. Factors affecting the suppression of Pythium damping-off in container media amended with composts. *Phytopathology 77*:755.

Cheng, B. T. 1987. Sawdust as a greenhouse growing medium. *J. Pl. Nut. 10*(9/10):1437.

Chesters, G., Attoe, O. J., and Allen, O. N. 1957. Soil aggregation in relation to various soil constituents. *Soil Sci. Soc. Am. Proc. 21*:272.

Chong, C., Cline, D. L., Rinker, D. L., and Hamersma, B. 1991a. An overview of reutilization of spent mushroom compost in nursery container culture. *Landscape Trades 13*(11):14.

Chong, C., Cline, D. L., Rinker, D. L., and Allen, O. B. 1991b. Growth and mineral nutrient status of containerized woody species in media amended with spent mushroom compost. *J. Am. Soc. Hort. Sci. 116*(2):242.

Chong, C., Cline, R. A., and Rinker, D. L. 1994. Bark- and peat-amended spent mushroom compost for containerized culture of shrubs. *Landscape Tradesman 29*(7):781.

Christensen, N. B., Lindemann, W. C., Salazar-Sosa, E., and Gill, L. R. 1994. Nitrogen and carbon dynamics in no-till and stubble mulch tillage systems. *Agron. J. 8*(2):298.

Christopher, T., and Asher, M. 1994. *Compost This Book*. Sierra Club Books, San Francisco.

Christopher, E. P., and Shutak, V. G. 1947. Influence of several soil management practices upon the yield of cultivated blueberries. *J. Am. Soc. Hort. Sci. 49*:211.

Chung, S. O., and Horton, R. 1987. Soil heat and water flow with a partial surface mulch. *Water Resources Research 23*(12):2175.

Cisar, J. L. 1994. Municipal solid waste compost offers a new soil amendment source for turf. *Grounds Maintenance 29*(3):52.

Clapham, W. M., and Zibilske, L. M. 1992. Wood ash as a liming amendment. *Commun. Soil Sci. Pl. Anal. 23*(11/12):1209.

Coleman, D. C., Reid, C. P. P., and Cole, C. V. 1983. Biological strategies of nutrient cycling in soil systems. *Adv. Ecology 13*:1.

Colvin, T. S., Laflen, J. J., and Erbach, D. C. 1981. *A review of residue reduction by individual tillage implements*. In J. C. Siemens, ed, Crop production with conservation in the 80's. ASAE Publ. 7–81, St. Joseph, MI, pp. 102–110.

Cook, R. J., and Baker, K. F. 1983. *The Nature and Practice of Biological Control of Plant Pathogens*. Am Phyto Soc. St. Paul, MN.

Cook, R. N., Patterson, J. C., and Short, J. R. 1979. Compost saves money in parkland restoration. *Compost Sci. Util. 20*(2):43.

Cook, B. D., Halbach, T. R., Rosen, C. J., and Moncrief, J. R. 1994. Effect of a waste stream component on the agronomic properties of municipal solid waste compost. *Compost Sci. Util.* 2(2):75.

Cooper, A. J. 1973. *Root Temperature and Plant Growth–A Review*. Commonwealth Bureau of Hort. and Plantation Crops. East Malling, England.

Cronan, C. S., Lakshman, S., and Patterson, H. H. 1992. Effects of disturbance and soil amendments on dissolved organic carbon and organic acidity in red pine forest floors. *J. Env. Qual. 21*(3):457.

Cummings, G. A., Mainland, C. M., and Lilly, P. J. 1981. Influence of soil pH, sulfur, and sawdust on rabbiteye blueberry survival, growth and yield. *J. Am Soc. Hort. Sci. 106*(6):783.

Darmody, R. G., Foss, J. E., McIntosh, M., and Wolf, D. C. 1983. Municipal sewage sludge compost-amended soils: some spatiotemporal treatment effects. *J. Environ. Qual 12*:231.

Davis, J. M. 1994. Comparison of mulches for fresh-market basil production. *HortScience 29*:267.

Dawson, J. E., and Seymore, P. E. 1983. Effects of juglone concentration on growth in vitro of *Frankia* Ar13 and *Rhizobium japonicum* strain 71. *J. Chem. Ecol. 9*:295.

de Bertoldi, M., Ferranti, M. P., L'Hermite, P., and Zucconi, F. 1987. *Compost: Production, Quality and Use*. Elsevier Applied Science, London.

De Smet, J., Wontroba, J., De Bood, M., and Hartmann, R. 1991. Effect of application of pig slurry on soil penetration resistance and sugar beet emergence. *Soil Tillage Res. 19*:297.

Devaux, A., and Haverkort, A. J. 1987. The effects of shifting planting dates and mulching on late blight (*Phytophthora infestans*) and drought stress of potato crops grown under tropical highland conditions. *Expl. Agric. 23*:325.

De Vleeschauwer, D. D., and Lal, R. 1981. Properties of worm casts under secondary tropical forest regrowth. *Soil Sci. 132*(2):175.

Dickey, E. C., Fenster, C. R., Mickelson, R. H., and Laflen, J. M. 1984. Tillage and erosion in a wheat-fallow rotation. In *Conservation Tillage*. Great Plain Agri. Council Pub 110:183.

Dong, A., Chester, G., and Simsiman, G. V. 1983. Soil dispersibility. *J. Soil Sci. 136*:208.

Doran, J. W. 1980. Soil microbial and biochemical changes associated with reduced tillage. *Soil Sci. Soc. Am. J. 44*:765.

Doran, J. W., and Parkin, T. B. 1994. Defining and assessing soil quality. In *Defining Soil Quality for a Sustainable Environment* (Doran, J. W., Coleman, D. C., Bezdicek, D. F., and Steward, B. A., eds). SSSA Special Publication Number 35, Madison, WI, p. 3–21.

Doran, J. W., Wilhelm, W. W., and Power, J. F. 1984. Crop residue removal and soil productivity with no-till corn, sorghum, and soybean. *Soil Sci. Soc. Am. J. 48*:640.

Drew, M. C., and Saker, L. R. 1980. Direct drilling and ploughing: their effects on the distribution of extractable phosphorus and potassium, and of roots, in the upper horizons of two clay soils under winter and wheat and spring barley. *J. Agric. Sci. 94*:411.

Dunn, S., and Latimer, L. P. 1956. *The Influence of Waste Bark on Plant Growth*. New Hampshire Agricultural Experiment Station. The University of New Hampshire, Durham, N.H.

Eck, H. V., and Jones, O. R. 1992. Soil nitrogen status as affected by tillage, crops, and crop sequences. *Agron. J. 84*(4):660.

Eckert, D. J. 1985. Effect of reduced tillage on the distribution of soil pH and nutrients in soil profiles. *J. Fert. Issues 2*:86.

Eckert, D. J., and Johnson, J. W. 1985. Phosphorus fertilization in no-till corn production. *Agron. J. 77*:789.

Edwards, C. A. 1983. Earthworms, organic waste and food. *Span. Shell Chem. Co. 26*(3):106.

Edwards, C. A. 1988. Breakdown of animal, vegetable and industrial organic wastes by earthworms. In: *Earthworms in Waste and Environmental Management*. (Edwards, C. A., and Neuhauser, E. F., eds.) SPB Academic Publishing, The Hague, The Netherlands, 1988, 21.

Edwards, A. P., and Bremner, J. M. 1967. Microaggregates in soils. *J. Soil Sci. 18*:64.

Ehlers, W., Pape, G., and Bohm, W. 1972. Changes in the content of calcium lactate soluble potassium and phosphorus in tilled and zero tilled soils during a growing season. *Z. Pflansenernähr. Bodenkd. 133*:24.

Ekern, P. C. 1967. Soil moisture and soil temperature changes with the use of black vapor-barrier mulch and their influence on pineapple *(Ananas comosus* (L.) Merr.) growth in Hawaii. *Soil Sci. Soc. Am. Proc. 31*:270.

Elliott, L. F., and Stevenson, F. J. 1977. *Soils for Management of Organic Wastes and Waste Waters*. Soil Science Society of America, Madison, WI.

Ellis, F. B., Elliott, J. G., Barnes, B. T., and Horse, K. R. 1977. Comparison of direct drilling, reduced cultivation and ploughing on the growth of cereals. 2. Spring barley on a sandy loam: soil physical conditions and root growth. *J. Agric. Sci. 89*:631.

Ellis, M. A., Ferree, D. C., and Madden, L. V. 1986. Evaluation of metalaxyl and captafol soil drenches, composted hardwood bark soil amendments, and graft union placement on control of apple collar rot. *Plant Disease 70*:24.

Elmer W. H., and Ferrandino, F. J. 1991. Early and late-season bloom-end rot of tomato following mulching. *Hort. Sci. 26*(9):1154.

Epstein, E. 1975. Effect of sewage sludge on some soil physical properties. *J. Environ. Qual. 4*(1):139.

Erich, M. S. 1991. Agronomic effectiveness of wood ash as a source of phosphorus and potassium. *J. Env. Qual. 20*(3):576.

Farrell, J. B. 1993. Fecal pathogen control during composting. In: *Science and Engineering of Compost Design Environmental, Microbiological and Utilization Aspects* (Hoitink, H. A. J., and Keener, H. M., eds.). Renaissance Publications, Worthington, Ohio.

Fenster, C. R., and Peterson, G. A. 1979. Effects of no-tillage fallow as compared to conventional tillage in a wheat-fallow system. *Nebr. Agric. Exp. Stn. Res. Bull. 289*.

Fiddaman, P. J., and Rossal, S. 1993. Effect of substrate on the production of antifungal volatiles from Bacillus subtilis. *J. Appl. Bact. 76*(4):395.

Fink, R. J., and Wesley, D. 1974. Corn yield as affected by fertilization and tillage system. *Agron. J. 66*:637.

Firth, D. J., Lobel, J. R., and Johns, G. G. 1994. Effect of mulch, Ca, and Mg on growth, yield, and decline of macadamia. *Tropical Agriculture* 7(3):170.

Flanagan, M. S., Schmidt, R. E., and Reneau, R. B., Jr. 1993. Municipal solid waste heavy fraction for production of turfgrass sod. *Hort. Sci. 28*(9):914.

Floyd, C. N., Lefroy, R. D. B., and D'Souza, E. J. 1985. Composting and crop production of volcanic ash soils in the Southern Highlands of Papua New Guinea. *AFTSEMU Technical Report 12*.

Floyd, C. N., D'Souza, E. J., and Lefroy, R. D. B. 1988. Soil fertility and sweet potato production on volcanic ash soils in the highlands of Papua New Guinea. *Field Crops Res. 19*:1.

Folk, R. L, and Campbell, A. G. 1990. Physical and chemical properties of classified logyard trash: a mill study. *For. Prod. J. 40*(4):22.

Follett, R. F., and Peterson, G. A. 1988. Surface soil nutrient distribution as affected by wheat fallow tillage systems. *Soil Sci. Soc. Am. J. 52*:141.

Follett, R. F., and Schimel, D. S. 1989. Effect of tillage practices on microbial biomass dynamics. *Soil Sci. Soc. Am. J. 53*:1091.

Fortin, M. C., and Pierce, F. J. 1990. Developmental and growth effects of crop residues on corn. *Agron J. 82*:710.

Foster, G. L., and Meyer, L. D. 1972. Erosion mechanics of mulches. *A.S.A.E. Paper no. 72*:754.

Fraedrich, S. W., and Ham, D. L. 1982. Wood chip mulching around maples: effect on tree growth and soil characteristics. *J. Arboric. 8*:85.

Frank, J. A., and Murphy, H. J. 1977. The effect of crop rotations on the Rhizoctonia disease of potatoes. *Am. Potato J. 54*:315.

Freire, J. R. 1984. Important limiting factors in soil for the *Rhizobium*-legume symbiosis. In: *Biological Nitrogen Fixation*. Alexander, M., ed. Plenum. New York.

Gabriels, D. 1988. Use of organic waste materials for soil structurization and crop production initial field experiment. *Soil Technology 1*(1):89.

Gallaher, R. N. 1984. Soybean root as affected by tillage in old tillage studies *Proc. No tillage System Conf. 7th.* (*In*: T. Touchton and R. E. Stevenson, ed.) Headland, AL. Auburn Univ., Auburn, AL.

Gallardo-Laro, F., and Nogales, R. 1987. Effect of the application of town refuse compost on the soil-plant system: a review. *Biological Wastes 19*:35.

Gallardo-Lara, F., Navarra, A., and Nogales, R. 1990. Extractable sulphate in two soils of contrasting pH affected by applied town refuse compost and agricultural wastes. *Biol. Wastes 33*(1):39.

Garcia, C., Hernandez, T., and Costa, F. 1991. Changes in carbon fractions during composting and maturation of organic wastes. *Environ. Manag. 15*:433.

Garcia, C., Hernandez, T., Costa, F., and Ayuso, M. 1992. Evaluation of the maturity of municipal solid waste compost using simple chemical parameters. *Commun. Soil Sci. Plant Anal. 23*:1501.

Geiger, S. C., Manu, A., and Bationo, A. 1992. Changes in a sandy Sahelian soil following crop residue and fertilizer additions. *Soil Sci. Soc. Am. J. 5*(1):172.

Ghidey, F., Gregory, J. M., McCarty, T. R., and Alberts, E. E. 1985. Residue decay evaluation and prediction. *Transactions ASAE 28*(1):102.

Ghuman, B. S., and Lal, R. 1983. Mulch and irrigation effects on plant-water relations and performance of cassava and sweet potato. *Field Crops Res. 7*:13.

Giusquiani, P. L., Marucchini, C., and Businelli, M. 1988. Chemical properties of soils amended with compost of urban waste. *Plant Soil 109*:73.

Golueke, C. G. 1972. *Composting Study of the Process and Its Principles*. Rodale Press.

Golueke, C. G. 1977. *Biological Reclamation of Solid Wastes*. Rodale Press, Emmaus, PA.

Gonzalez, J. L., Benitez, I. C., Perez, M. I., and Median, M. 1991. Pig-slurry composts as wheat fertilizers. *Biores. Tech. 40*(2):125.

Gorodecki, B., and Hadar, Y. 1990. Suppression of Rhizoctonia solani and Sclerotium rolfsii disease in container media containing composted separated cattle manure and composted grape marc. *Crop Protection 9*:271.

Gough, R. E. 1980. Root distribution of 'Coville' and 'Lateblue' highbush blueberry under sawdust mulch. *J. Amer. Soc. Hort. Sci. 105*:576.

Gouin, F. R. 1983. Over-mulching: a national plague. *Weeds, Trees and Turf. 22*(9):22,24.

Gouin, F. R. 1992. Mulch mania. *American Nurseryman 176*(7):97.

Griffith, D. R., Mannering, J. V., Galloway, H. M., Parsons, S. D., and Rickey, C. B. 1973. Effect of eight tillage planting systems on soil temperature, percent stands, plant growth, and yields of corn live Indiana soils. *Agron. J. 65*:321.

Grimme H. 1983. Aluminum induced magnesium deficiency in oats. *Z. Pflanzenernähr. Bodenk. 146*:666.

Grzeszkiewicz, H. 1978. Pine bark as a soil improver in gladiolus culture. *Acta Hort. 82*:31.

Guenzi, W. D., and McCalla, T. M. 1966. Phenolic acids in oats, wheat, sorghum, and corn residues and their phytotoxity. *Agron. J. 58*:303.

Guidi, G., Pera, A., Giovannetti, M., Poggio, G., and de Bertoldi, M. 1988. Variations of soil structure and microbial population in a compost amended soil. *Plant Soil 106*(1):113.

Guidi, G., and Petruzzelli, G. 1989. Effect of compost on chemical and physical properties of soil. In: Compost Production and Use: Technology, Management, Application and Legislation. *Proc. Int. Symp. on Compost*, S. Michele all'Adige, Italy, June 20–23, 1989, p. 53.

Gupta, S. C., Larson, W. E., and Linden, D. R. 1983. Tillage and surface residue effects on soil upper boundary temperatures. *Soil Soc. Soc. Am. J. 47*:1212.

Gupta, S. C., Schneider, E. C., and Swan, W. B. 1988. Planting depth and tillage interactions on corn emergence. *Soil Sci. Soc. Am. J. 52*:1122.

Haas, H. J., Evans, C. E., and Miles, E. F. 1957. Nitrogen and carbon changes in Great Plains soils as influenced by cropping and soil treatments. *USDA Tech. Bull. 1164*.

Haggar, J. P., Tanner, E. V. J., Beer, J. W., and Kass, D. C. L. 1993. Nitrogen dynamics of tropical agroforestry and annual cropping systems. *Soil Biol. Biochem. 25*:1363.

Hamblin, A. P. 1977. Structural features of aggregates in some East Anglian silt soils. *J. Soil Sci. 28*:23.

Hanks, R. J., Bowers, S. A., and Bark, L. D. 1961. Influence of soil surface conditions on net radiation, soil temperature, and evaporation. *Soil Sci. 91*:233.

Harder, R., and Baker, G. O. 1971. Utilization of waste bark as a soil conditioner. *Compost Sci. 12*(2):6.

Hardy, G. E., and Sivasithamparam, K. 1991. Suppression of Phytophthora root rot by a composted Eucalyptus bark mix. *Aust. J. Bot. 39*:154.

Hargrove, W. L. 1985. Influence of tillage on nutrient uptake and yield of corn. *Agron. J. 77*:763.

Hargrove, W. L., and Hardcastle, W. S. 1984. Conservation tillage practices for winter wheat production in the Appalachian Piedmont. *J. Soil Water Conserv. 39*:324.

Hargrove, W. L., Reid, J. T., Touchton, J. T., and Gallaher, A. N. 1982. Influence of tillage practices on the fertility status of an acid soil double-cropped to wheat and soybeans. *Agron. J. 74*:684.

Harp, K. L., and Grove, S. L. 1994. Evaluation of wood and soil samples from copper naphthenate-treated utility poles in service. *Proceedings of the Annual Meeting of the American Wood-Preservers' Association 89*:167.

Harper, S. H. T., and Lynch, J. M. 1982. The role of water-soluble components in phytotoxicity from decomposing straw. *Plant Soil 65*:11.

Hartenstein R., Neuhauser, E. F., and Kaplan, D. L. 1979. Reproductive potential of the earthworm Eisenia foetida. *Oecologia* (Berlin) *43*:329.

Hassan, F. A. 1985. Drip irrigation and crop production in arid regions. *Proc. 3rd Intl. Drip/Trickle Irr. Congr. ASAE.* St. Joseph. MI, p. 150.

Haug, T. H. 1980. *Compost Engineering, Principles and Practice.* Ann Arbor Science Publishers, Ann Arbor, MI.

Hauser, V. L., and Chichester, F. W. 1989. Water relationships of claypan and constructed soil profiles. *Soil Sci. Soc. Am. J. 53*(4):1129.

Haynes, R. J., and Swift, R. S. 1990. Stability of soil aggregates in relation to organic constituents and soil water content. *J. Soil Sci. 41*:73.

He, X. T., Traina, S. J., and Logan, T. J. 1992. Chemical properties of municipal solid waste composts. *J. Environ. Qual. 21*:318.

Heckman, J. R., Angle, J. S., and Chaney, R. L. 1987. Residual effects of sewage sludge on soybean: A. Accumulation of heavy metals. *J. Environ. Qual. 16*:113.

Heitkamp, M. A., and Cerniglia, C. E. 1989. Polyclinic aromatic hydrocarbon degredation by a Myobacterium sp. in microcosms containing sediment and water from a pristine ecosystem. *Appl. Environ. Microbiol. 55*(8):1968.

Hernando, S., Lobo, M. C., and Polo, A. 1989. Effect of application of a municipal refuse compost on the physical and chemical properties of a soil. *Sci. Total Environ. 82*:589.

Hill, J. D., and Blevins, R. L. 1973. Quantitative soil moisture use in corn grown under conventional and no-tillage methods. *Agron. J. 65*:945.

Himelick, E. B., and Watson, G. W. 1990. Reduction of oak chlorosis with wood chip mulch treatments. *J. Arboriculture 16*:275.

Hoitink, H. A. J. 1979. Mass production of composted tree barks container media. *Ohio Florist Assn. Bulletin 599*:3.

Hoitink, H. A. J., and Fahd, P. C. 1986. Basis for the control of soil borne plant pathogens with composts. *Ann. Rev. Phytopath. 24*:93.

Hoitink, H. A. J., and Kuter, J. A. 1984. Role of composts in suppression of soil-borne plant pathogens of ornamental plants. *BioCycle 25*:40.

Hoitink, H. A. J., Nelson, E. B, and Gordon, D. T. 1982. Composted bark controls soil pathogens of plants. *Ohio Report on Research and Development in Agriculture, Home Economics, and Natural Resources* (Ohio Agricultural Research and Development Center) *67*(1):7.

Hoitink, H. A. J., Inbar, R., and Boehm, M. J. 1993. Compost can suppress soil-borne diseases in container media. *Am. Nurserym. 178*(16):91.

Hornick, S. B. 1988. Use of organic amendments to increase the productivity of sand and gravel spoils: effect on yield and composition of sweet corn. *Am. J. Alternative Agric. 3*(4):156.

Hortenstein, C. C., and Rothwell, D. F. 1972. Use of municipal compost in reclamation of phosphate-mining sand tailings. *J. Environ. Qual. 1*:415.

Howard, E. J. 1970. A survey of the utilization of bark as fertilizer and soil conditioner. *Pulp. Pan. Mag. Can. 71*(23/24):53.

Howard, E. J. 1973. The utilization of bark as soil conditioner and fertilizer. *Svensk Papperstidn. 76*(1):33.

Hsu, S. M., Schnoor, J. L., Licht, L. A., St. Clair, M. A., and Fannin, S. A. 1993. Fate and transport of organic compounds in municipal solid waste compost. *Compost Sci. Util. 1*(4):36.

Hudson, N. W. 1957. Erosion control research progress report on experiments at Henderson Research Station. *Rhodesia Agric. J. 54*:297.

Hue, N. V. 1988. A possible mechanism for manganese phytotoxity in Hawaii soils amended with a low-manganese sewage sludge. *J. Environ. Qual. 17*:473.

Hue, N. V., Ikawa, H., and Silva, J. A. 1994. Increasing plant-available phosphorus in an ultisol with a yard-waste compost. *Commun. Soil Sci. Plant Anal. 25*(19/20):3291.

Huhta, V., and Haimi, J. 1988. Reproduction and biomass of Eisenia foetida in domestic waste. In: *Earthworms in Waste and Environmental Management* (Edwards, C. A., and Neuhauser, E. F., eds.). SPB, Academic Publishing, The Hague, The Netherlands.

Iannotti, M. E., Grebus, D. A., Toth, B. L., Madden, L. V., and Hoitink, H. A. J. 1994. Oxygen respirometry to assess stability and maturity of composed municipal solid waste. *J. Environ. Qual. 23*:1177.

Inbar, Y., Chen, Y., Hadar, Y., and Hoitink, H. A. J. 1990. New approaches to compost maturity. *BioCycle 31*(12):64.

Jacks, G. V., Brind, W. D., and Smith, R. 1955. Mulching. *Commonwealth Bureau of Soil Science, Tech. Comm. No. 49*.

Jackson, A. M., Whipps, J. M., and Lynch, J. M. 1991. In vitro screening for the identification of potential biocontrol agents of *Allium* white rot. *Mycol. Res. 95*(4):430.

Janzen, R. A., and McGill, W. B. 1995. Community-level interactions control proliferation of *Azospirillum brasilense* Cd in microcosms. *Soil Biology Biochemistry 27*(2):189.

Jansen, R. A., Cook, F. D., McGill, W. B. 1995. Compost extract added to microcosms may simulate community-level controls on soil microorganisms involved in element cycling. *Soil Biol. Biochem 27*(2):18.

Jenny, H. 1941. *Factors of Soil Formation*. McGraw-Hill, New York.

Jimenez, E. I., and Garcia, V. P. 1989. Evaluation of city refuse compost maturity: a review. *Biol. Wastes 27*:115.

Jobidon, R., and Thibault, J. R. 1982. Allelopathic growth inhibition of nodulated and unnodulated *Alnus crispa* seedlings by *Populus balsamifera. Am. J. Bot 69*:1213.

Katayama, A. 1985. *Application of Gel Chromatography for Monitoring Decomposition of Organic Wastes in Soil*. Dissertation, the Graduate School of Tokyo Institute of Technology, Department of Environmental Chemistry and Engineering, 1985, 113.

Khaleel, R., Reddy, K. R., and Overcash, M. R. 1981. Changes in soil physical properties due to organic waste applications: a review. *J. Environ. Qual. 10*:133.

Khan, A. R., and Rafey, A. 1985. Effect of mulch harrowing and irrigation on physical properties of soil and peanut seedling emergence. *Zeitschrift fur für Acker und Pflanzenbau* (Journal of agronomy and crop science) 155 (4):227.

Khatibu, A. I., Lal, R., and Jana, R. K. 1984. Effects of tillage methods and mulching on erosion and physical properties of a sandy clay loam in an equatorial warm humid region. *Field Crops Res. 8*:239.

Kitto, D. 1979. *Composting the Organic Natural Way*. Thorsons Publishing Group, Wellingborough, Northamptonshire, England.

Kolb, W., Schwarz, T., and Trunk, R. 1985. Effect of mulching on rooting, cost of maintenance, and growth of low perennials and shrubs. *Rasen-Grünflächen-Begrünungen 16*:120.

Koma Alimu, F. X., Angie, I. E. S., and Janssen, B. H. 1977. Evaluation of municipal refuse from Dahomey (Benin) as an organic manure. *Proc. Soil Organic Matter Sym. 1976. Volume 2*: 277–287, Braunschweig, Germany.

Korcak, R. F. 1986. Renovation of a pear orchard site with sludge compost. *Commun. Soil Sci. Plant Anal. 17*(11):1159.

Kramer, A., Evinger, E. L., and Schrader, A. L. 1941. Effect of mulch and fertilizers on yield and survival of dryland and highbush blueberries. *Proc. Amer. Soc. Hort. Sci. 38*:455.

Kunishi, H. M., Bandel, V. A., and Mulford, F. R. 1982. Measurement of available soil phosphorus under conventional and no-till management. *Commun. Soil Sci. Plant Anal. 13*:607.

Kushwaha, J. S., Prasad, D., and Vimal, O. P. 1983. Effect of urea and urine alone and in combination with sawdust on tomato growth and nematode development in an alluvial soil. *Indian J. Ent. 45*(4):479.

Kuter, G. A., Nelson, E. B., Hoitink, H. A. J., and Madden, L. V. 1983. Fungal populations in container media amended with composted hardwood bark suppressive and conducive to Rhizoctonia damping-off. *Phytopathol. 73*:1450.

Kwok, O. C. H., Fahd, P. C., Hoitink, H. A. J., and Kuter, G. A. 1987. Interactions between bacteria and *Trichoderma hamatum* in suppression of Rhizoctonia damping-off in bark compost media. *Phytopathol. 77*:1206.

Lal, R. 1974. Soil temperature, soil moisture and maize yield from mulched and unmulched tropical soil. *Plant and Soil 40*:128.

Lal, R. 1975. Role of mulching techniques in tropical soil and water management. International Institute of Tropical Agriculture, Nigeria. Technical Bulletin No. 1:28.

Lal, R. 1976. No-tillage effects on soil properties under different crops in Western Nigeria. *Soil Sci. Soc. Am. J. 40*:762.

Lal, R. 1978. Influence of within and between row mulching on soil temperature, soil moisture, root development and yield of maize (*Zea mays*) in a tropical soil. *Field Crops Res. 1*:127.

Lal, R., De Vleeschauwer, D., and Malafa Nganje, R. 1980. Changes in properties of a newly cleared tropical alfisol as affected by mulching. *Soil Sci. Soc. Am. J. 44*:827.

Lamar, R. T., and Dietrich, D. M. 1990. In situ depletion of pentachlorophenol from contaminated soil by *Phanerochaete* spp. *Appl. Env. Microbiol. 56*(10):3093.

Lamb, J. A., Peterson, G. A., and Fenster, C. R. 1985. Fallow nitrate accumulation in a wheat-fallow rotation as affected by tillage systems. *Soil Sci. Soc. Am. J. 49*:1441.

Langdale, G. W., Perkins, H. F., Barnett, A. P., Reardon, J. L., and Wilson, R. L., Jr. 1983. Soil and nutrient losses with in-row, chisel plant soybeans. *J. Soil Water Conserve. 38*:297.

Lanini, W. T., Shribbs, J. M., and Elmore, C. E. 1988. Orchard floor mulching trials in the U.S.A. *Le Fruit Belgique 56*:228.

Lareau, M. J. 1989. Growth and productivity of highbush blueberries as affected by soil amendments, nitrogen fertilization and irrigation. *Acta Hort. 241*:12.

Layton, J. B., Skidmore, E. L., and Thompson, C. A. 1993. Winter-associated changes in dry-soil aggregation as influenced by management. *Soil Sci. Soc. Amer. J. 57*(6):1568.

Lee, K. C., and Campbell, R. W. 1969. Nature and occurrence of juglone in *Juglone nigra* L. *Hort. Sci. 4*:297.

Leeper, G. W. 1978. *Managing the Heavy Metals on the Land.* Marcel Dekker, New York.

Lemmon, C. R., and Pylypiw, H. M. 1992. Degradation of diazinon, chlorpyrifos, isofenphos, and pendimethalin in grass and compost. *Bull. Environ. Contam. Toxicol. 48*:409.

Lemon, E. R., and Erickson, A. E. 1952. The measurement of oxygen diffusion in the soil with a platinum electrode. *Soil Sci. Soc. Amer. Proc. 16*:160.

Locascio, S. J., Fiskell, J. G. A., Graetz, D. A., and Hauck, R. D. 1985. Nitrogen accumulation by pepper as influenced by mulch and time of fertilizer application. *J. Amer. Soc. Hort. Sci. 110*(3):325.

Lodhi, M. A. K. 1978. Allelopathic effects of decaying litter of dominant tress and their associated soil in a lowland forest community. *Am. J. Bot 65*:340.

Lodhi, M. A. K., Bilal, R., and Malik, K. A. 1987. Allelopathy in agroecosystems: wheat toxicity and its possible roles in crop rotation. *J. Chem. Ecol. 13*:1881.

Loehr, R. C., Martin, J. H., Neuhauser, E. F., and Malecki, M. R. 1984. *Waste Management Using Earthworms—Engineering and Scientific Relationships*, PB84-193218, NTIS, Springfield, VA.

Loehr, R. C., Martin, J. H., Jr., and Neuhauser, E. F. 1988. Stabilization of liquid municipal sludge using earthworms. In: *Earthworms in Waste and Environmental Management.* (Edwards, C. A., and Neuhauser, E. F., eds.). SPB, Academic Publishing, The Hague, The Netherlands.

Lohr, V. I., and Coffey, D. L. 1987. Growth responses of seedlings to varying rates of fresh and aged spent mushroom compost. *Hort Science. 22*(5):913.

Lynch, J. M. 1978. Production and phytotoxicity of acetic acid in anaerobic soils containing plant residues. *Soil Biol. Biochem. 10*:131.

Lynch, J. M. 1981. Promotion and inhibition of soil aggregate stabilization by soil microorganisms. *J. Gen. Microbiol. 126*:371.

Lynch, J. M. 1987. In vitro identification of Trichoderma harzianum as potential antagonist of plant pathogens. *Curr. Microbiol. 16*:49.

Maas, E. F., and Adamson, R. M. 1972. Resistance of sawdusts, peats, and bark to decomposition in the presence of soil and nutrient solution. *Soil Sci. Soc. Amer. Proc. 36*(5):767.

Mackay, D. C., Carefoot, J. M., and Sommerfeldt, T. G. 1989. Nitrogen fertilizer requirements for barley when applied with cattle manure containing wood shavings as a soil amendment. *Can. J. Soil Sci. 9*(3):515.

Madan, M., and Vasudevan, P. 1989. Silkworm litter: use as nitrogen replacement for vegetable crop cultivation and substrate for mushroom cultivation. *Biol. Wastes 27*(3):209.

Malek, R. B., and Gartner, J. B. 1975. Hardwood bark as a soil amendment for suppression of plant parasitic nematodes on container-grown [tomato] plants. *HortScience 10*(1):33.

Mandelbaum, R., Gorodecki, B., and Hadar, Y. 1985. The use of composts for disease suppressive container media. *Phytoparasitica 13*:158.

Mandelbaum, R., Hadar, Y., and Chen, Y. 1988. Composting for agricultural wastes for their use as container media: effect of heat treatments on suppression of Pythium aphanidermatum and microbial activities in substrates containing compost. *Biol. Wastes 26*:261.

Mankau, R., and Das, S. 1974. Effect of organic materials on nematode bionomics in citrus and root-knot nematode infested soil. *J. Nematology 4*:138.

Mannering, J. V., and Fenster, C. R. 1983. What is conservation tillage? *J. Soil Water Conserve. 38*:141.

Mannering, J. V., and Meyer, L. D. 1963. The effect of various rates of surface mulch on infiltration and erosion. *Soil Sci. Soc. Amer. Proc. 27*:84.

Manrique, J. V. 1995. Mulching in potato systems in the tropics. *J. Pl. Nutr. 18*(4):593.

Manrique, L. A., and Meyer, R. E. 1984. Effects of soil mulches on soil temperature, plant growth, and potato yields in an aridic isothermic environment in Peru. *Turrialba 34*:413.

Marchesini, A., Allievi, L., Comotti, E., and Ferrari, A. 1988. Long-term effects of quality-compost treatment on soil. *Plant Soil 106*:253.

Matheny, T. A., Hunt, P. G., and Kasperbauer., M. J. 1992. Potato tuber production in response to reflected light from different colored mulches. *Crop Sci. 32*:1021.

Maurya, P. R., and Lal, R. 1981. Effects of different mulch materials on soil properties and on the root growth and yield of maize (*Zea mays*) and cowpea (*Vigna unguiculata*). *Field Crops Res. 4*:33.

Maynard, A. 1993. Evaluating the suitability of MSW compost as a soil amendment in field grown tomatoes. Part A: yield of tomatoes. *Compost Sci. Util. 1*(2):34.

Maynard, A. A. 1994. Effect of annual amendments of compost on nitrate leaching nursery stock. *Compost Science and Utilization* 2(3):54.

Maynard, A. A., and Hill, D. E. 1994. Impact of compost on vegetable yields. *BioCycle 35*(3):66.

Mays, D.A., and Giordano, P. M. 1989. Landspreading municipal waste compost. *BioCycle 30*(3):37.

Mays, D. A., Terman, G. L., and Duggan, J. C. 1973. Municipal composts: effects on crop yields and soil properties. *J. Environ. Qual.* 2:89.

Mbagwu, J. S. C. 1990. Maize (Zea mays) response to nitrogen fertilizer on an ultisol in southern Nigeria under two tillage and mulch treatments. *J. Sci. Food Agric. 52*(3):35.

McCalla, T. M., and Army, T. J. 1961. Stubble mulch farming. *Adv. Agron. 13*:125.

McCalla, T. M., and Duley, F. L. 1943. Disintegration of crop residues as influenced by subtillage and plowing. *J. Amer. Soc. Agron. 35*:291–315.

McCalla, T. M., and Norstadt, F. A. 1974. Toxicity problems in mulch tillage. *Agric Env. 1*:175.

McGregor, K. C., Bengtson, R. L., and Mutchler, C. K. 1988. Effects of surface straw on interrill runoff and erosion of Grenada silt loam soil. *Trans ASAE 31*(1):111.

McGregor, K. C., Mutchler, C. K., and Romkens, M. J. M. 1990. Effects of tillage with different crop residues on runoff and soil loss. *Trans ASAE 33*(5):1551.

Meek, B. D., Graham, L., and Donovan, T. 1982. Long-term effects of manure on soil nitrogen, phosphorus, potassium, sodium, organic matter, and water infiltration rates. *Soil Sci. Soc. Am. J. 46*:1014.

Mengel, K., and Kirkby, E. A. 1987. *Prinicples of Plant Nutrition.* 4th ed. International Potash Institute, Worblaufen-Bern, Switzerland, 513.

Merwin, I. A., and Stiles, W. C. 1994. Orchard groundcover management impacts on apple tree growth and yield, and nutrient availability and uptake. *J. Am. Soc. Hort. Sci. 119*(2):209.

Merwin, I. A., Stiles, W. C., and van Es, H. M. 1994. Orchard groundcover management system impacts on soil physical properties. *J. Amer. Soc. Hort. Sci. 119*:216.

Meyer, L. D. 1981. How rain intensity affects interrill erosion. *Trans. ASAE 24*:1472.

Meyer, L. D., and Mannering, J. V. 1963. Crop residues as surface mulches for controlling erosion on sloping land under intensive cropping. *Trans ASAE 6*(4):322.

Midmore, D. J. 1983. The use of mulch for potatoes in the hot tropics. *CIP Cirulare* (March), International Potato Center Lima Peru.

Midmore, D. J. 1984. Potato (*Salanum spp.*) in the hot tropics. I. Soil temperature effects on emergence, plant development and yield. *Field Crops Res. 8*:255.

Millner, P. D., Lumsden, R. D., and Lewis, J. A. 1982. Controlling plant disease with sludge compost. *BioCycle 23*:50.

Minnich, J., and Hunt, M. 1979. *The Rodale Guide to Composting*. Rodale Press, Emmaus, PA.

Mishra, M. M., and Bangar, K. C. 1986. Rock phosphate composting: transformation of phosphorus forms and mechanisms of solubilization. *Biol. Agric. Hort. 3*:331.

Mitchell, M. J., Hornor, S. G., and Abrams, B. L. 1980. Decomposition of sewage sludge in drying beds and the potential role of the earthworm, Eisenia foetida. *J. Environ. Qual. 9*:373.

Mock, J. J., and Erbach, D. C. 1977. Influence of conservation-tillage environments on growth and productivity of corn. *Agron. J. 69*:337.

Moschler, W. W., Shear, G. M., Martens, D. C. Jones, G. D., and Wilmouth, R. R. 1972. Comparative yield and efficiency of no-till and conventionally tilled corn. *Agron. J. 64*:229.

Muller, R., and Gooch, P. S. 1982. Organic amendments in nematode control. An examination of the literature. *Nematropica 12*:319.

Munawar, A., Blevins, R. L., Frye, W. W., and Saul, M. R. 1990. Tillage and cover crop management for soil water conservation. *Agron. J. 82*:773.

Nearing, M. A., Deer-Ascough, L., and Laflen, J. M. 1990. Sensitivity analysis of the WEPP hillslope profile erosion model. *Trans. ASAE 33*:839.

Nehlin, G., Valterova, I. and Borg-Karlson, A. K. 1994. Use of conifer volatiles to reduce injury caused by carrot Psyllid, *Trioza apicalis*, forster (Homoptera, Psylloidea). *Journal of Chem. Ecol. 20*(3):771.

Nelson, E. B., and Craft, C. M. 1991. Suppression of dollar spot on bentgrass and annual bluegrass turf with compost-amended topdressings. *Plant Dis. 76*(9):954.

Nelson, E. B., and Hoitink, H. A. J. 1982. Factors affecting suppression of *Rhizoctonium solani* in container media. *Phytopathol. 72*:275.

Nelson, L. R., Gallaher, R. N., Bruce, R. R., and Holmes, M. R. 1977. Production of corn and sorghum grain in double-cropping systems. *Agron. J. 69*:41.

Neuhauser, E. F., Hartenstein, R., and Kaplan, D. L. 1980. Growth of the earthworm *Eisenia foetida* in relation to population density and food rationing. *Oikos 35*:93.

Niemiera, A. X. 1992. Micronutrient supply from pine bark and micronutrient fertilizers. *HortScience 27*(3):272.

Norden, D. E. 1990. Comparison of pine bark mulch and polypropylene fabric ground cover in blueberries. *Proceedings of the Annual Meeting of the Florida State Horticulture Society 102*:206.

Nus, J. L., and Brauen, S. E. 1991. Clinoptilolitic zeolite as an amendment for establishment of creeping bentgrass on sandy media. *HortScience* 2(2):117.

Odneal, M. B, and Kaps, M. L. 1990. Fresh and aged pine bark as soil amendments for establishment of highbush blueberry. *HortScience* 25(10):1228.

Ohno, T. 1992. Neutralization of soil acidity and release of phosphorus and potassium by wood ash. *J. Env. Qual.* *21*(3):433.

Opitz, K. 1974. Mulching citrus and other subtropical tree crops. *Cooperative Extension One Sheet Answers No. 271.*

Oster, J. D., Hoffman, G. J., and Robinson, F. E.. 1986. Dealing with salinity: Management alternatives: crop, water, and soil. *Calif. Agr.* *38*:29.

Oveson, M. M. 1966. Conservation of soil nitrogen in a wheat summer fallow farm practice. *Agron. J.* *58*:444.

Ozores-Hampton, M., Schaffer, B., Bryan, H. H., and Hanlon, E. A. 1994. Nutrient concentrations, growth and yield of tomato and squash in municipal solid-waste-amended soil. *HortScience* 29(7):785.

Page, A. L., Logan, T. J., and Ryan, J. A. 1987. *Land Application of Sludge: Food Chain Implications.* Lewis Publishers, Chelsea, MI.

Papavizas, G. C. 1985. Trichoderma and gliocladium: biology, ecology, and potential for biocontrol. *Ann. Rev. Phytopathol.* *23*:23.

Papavizas, G. C. 1970. *Colonization and growth of Rhizoctonia solani in soil.* In: *Biology and Pathology of* Rhizoctonia solani (J. R. Parmeter, Jr., ed.). University of California Press, Berkeley.

Parker, D. T. 1962. Decomposition in the field of buried and surface-applied cornstalk residue. *Soil Sci. Soc. Am. Proc.* *26*(6):559.

Patten, K. D., Neuendorf, E. W., and Peters, S. C. 1988. Root distribution of 'Climax' blueberry as affected by mulch and irrigation geometry. *J. Am. Soc. Hort. Sci.* *113*(5):657.

Pellet, N. E., and Heleba, D. A. 1995. Chopped newspaper for weed control in nursery crops. *J. Env. Hort.* *13*(4):77.

Pera, J., and Calvet, C. 1989. Suppression of fusarium wilt of carnation in a composted pine bark and composted olive pumice. *Plant Disease* *73*:699.

Pera, A., and Filippi, C. 1987. Controlling of fusarium wilt in carnation with bark compost. *Biol. Wastes* *22*:218.

Pera, A., Valini, G., Sireno, I., Bianchin, M.L., and de Bertoldi, M. 1983. Effect of organic matter on rhizosphere microorganisms and root development of sorghum plants in two different soils. *Plant Soil* *74*:3.

Perucci, P. 1990. Effect of the addition of municipal solid-waste compost on microbial biomass and enzyme activities in soil. *Biol. Fertil. Soils* *10*(3):221.

Pertruzzelli, G., Lubrano, L., and Guidi, G. 1989. Uptake by corn and chemically extractability of heavy metals from a four year compost treated soil. *Plant Soil* *116*:23.

Phene, C. J., and Sanders, D. C. 1976. High frequency trickle irrigation and row spacing effects on yield and quality of potatoes. *Agron. J.* *68*:602.

Poincelot, R. P. 1978. The biochemistry of composting, in 1977 National Conference on Composting of Municipal Residues and Sludges, August 23–25, 1977. *Information Transfer, Inc. 33.*

Pokorny, F. A. 1969. Pine bark in the soil improves turf quality. *Ga. Agr. Res.* *11*(1):5.

Pokorny, F. A. 1983. Pine bark as a soil amendment [Chemical properties, advantages, use in forest tree nurseries]. *Technical Publication R8-TP-USDA Forest Service Southern Region 4*:131.

Pokorny, F. A., and Delany, S. 1975. Synthesizing a pine bark potting substrate from component particles. *Proc. Southern Nurs. Assn. Res. Conf. 20*:24.

Posey, H. G. and May, J. T. 1954. Some effects of sawdust mulching of pine seedlings. *Agricultural Experiment Station of the Alabama Polytechnic Institute*.

Preston, S. R. 1990. Investigation of compost x fertilizer interactions in sweet potato grown on volcanic ash soils in the highlands of Papua New Guinea. Trop. *Agric. 67*(3):239.

Purman, J. R., and Gouin, F. R. 1992. Influence of compost again and fertilizer regimes on the growth of bedding plants, transplants and poinsettia. *J. Environ. Hort. 10*(1):522.

Radcliffe, D. E., Tollner, E. W., Hargrove, W. L., Clark, R. L., and Golabi, M. H. 1988. Effect of tillage practices on infiltration and soil strength of a typic hapludult soil after ten years. *Soil. Sci. Sco. Am. J. 52*:798.

Rasmussen, P. E., and Rohde, C. R. 1989. Soil acidification from ammonium–nitrogen fertilization in mold board plow and stubble-mulch-wheat-fallow tillage. *Soil Sci. Soc. Am. J. 53*:119.

Rasmussen, P. E., Allmaras, R. R., Rohde, C. R., and Roager, N. C., Jr. 1980. Crop residue influences on soil carbon and nitrogen in a wheat-fallow system. *Soil Sci. Soc. Am. J. 44*:596.

Rechcigl, J. E., ed. 1995a. *Soil Amendments and Environmental Quality*. Lewis Publishers, Boca Raton, FL.

Rechcigl, J. E., ed. 1995b. *Soil Amendments Impacts on Biotic Systems*. Lewis Publishers, Boca Raton, FL.

Rechcigl, N. A., and Rechcigl, J. E., eds. 1997. *Environmentally Safe Approaches to Crop Disease Control*. Lewis Publishers, Boca Raton, FL.

Rice, E. L. 1971. Inhibition of nodulation of inoculated legumes by leaf leachates from pioneer plant species from abandoned fields. *Am J. Bot. 58*:368.

Rice, E. L. 1974. *Allelopathy*. Academic Press, New York.

Rich, J. R., and Hodge, C. H. 1993. Utilization of blue crab scrap compost to suppress *Meloidogyne javanica* on tomato. *Nematropica 23*(1):1.

Richard, T. 1990. Clean compost production. *BioCycle 3*(2):46.

Rickerl, D. H., Curl, E. A., Touchton, J. T., and Gordon, W. B 1992. Crop mulch effects on *Rhizoctonia* soil infestation and disease severity in conservation-tilled cotton. *Soil Biology and Biochemistry 24*(6):553.

Rietveld, W. J. 1983. Allelopathic effects of juglone on germination and growth of several herbaceous and woody species. *J. Chem Col. 9*:295.

Roberts, B. W., and Anderson, J. A. 1994. Canopy shade and soil mulch affect yield and solar injury of bell pepper. *Hort Sci. 29*(4):258.

Roberts, A. N., and Mellenthin, W. M. 1959. Effects of sawdust mulches. II. Horicultural crops. Agricultural Experiment Station, Oregon State College.

Robinson, D. W. 1988. Mulches and herbicides in ornamental plantings. *Hort. Sci. 23*:547.

Rodrigues-Kabana, R., Morgan-Jones, G., and Ownley Ginitis, B. 1984. Effects of chitin amendments to soil on heterodera glycines, nocrobial populations and colonization of cysts by fungi. *Nematropica 14*:10.

Rodrigues-Kabana, R., Boube, D., and Young, R. W. 1989. Chitinous materials from blue crab for control of root-knot nematode. I. Effect of urea and enzymatic studies. *Nematropica 19*:53.

Roe, N. E., Stoffella, P. J., and Bryan, H. H. 1993. Municipal solid waste compost supresses weeds in vegetable crop alleys. *Hort. Sci. 28*(12):1171.

Sabey, B. R. 1975. Effect of 50–50 mixture of wood bark and sludge on soil properties and plant growth. Fort Collins Agriculture Experiment Station, Colorado State University.

Sabrah, R. E. A. 1993. Field and laboratory measurements of saturated hydraulic conductivity of a sandy soil incorporated with organic manure. *Egyptian J. App. Sci. 8*:171.

Sabrah, R. E. A., Abdel Magid, H. H., Abdel-Aal, S. I., and Rabie, R. K. 1995. Optimizing physical properties of a study soil for higher productivity using town refuse compost in Saudi Arabia. *J. Air Environments 29*(2):253.

Saini, G. R. 1974. Effects of soil compaction and shredded tree bark on phosphorus "A" values of two New Brunswick potato soils. *Can. J. Soil Sci. 54*(4):501.

Saini, G. R., and Hughes, D. A. 1975. Shredded tree bark as a soil conditioner in potato soils of New Brunswick, Canada. *SSSA Spec. Publ. Soil Sci. Soc. Am. 7*:139.

Sample, E. C., Soper, R. J., and Racz, G. J. 1980. Reactions of phosphate fertilizers in soils. In *The Role of Phosphorus in Agriculture* (Khasawneh, F. E., Sample, E. C., and Kamprath, E. J., eds.) ASA-CSSA-SSSA Madison, Wisconsin, 263.

Sances, F. V., and Ingham, E. L. 1995. Suitability of organic compost and broccoli mulch soil treatments for commercial strawberry production on the California central coast. *1995 Annual International Research Conference on Methyl Bromide Alternatives and Emissions Reductions*, p. 19.

Sarles, R. L. 1973. Bark—a Cinderella story (bark mulches and soil conditioners: new products for the wood industry). *Logger Timber Processor 21*(9):14.

Satchell, J. E. 1983. Earthworm microbiology. In *Earthworm Ecology*, J. E. Satchell, ed. Chapman and Hall, 495.

Satchell, J. E., Martin, K., 1984. Phosphatase activity in earthworm species. *Soil Biol. Biochem. 16*(2):191.

Satchell J. E., Martin, K., and Krishnamoorthy, R. V. 1984. Stimulation of microbial phosphatase production by earthworm activity. *Soil Biol. Biochem. 16*:195.

Scanlon, D. H., Duggan, C., and Bean, S. D. 1973. Evaluation of municipal compost for strip mine reclamation. *Compost Sci. Util. 14*:4.

Schowalter, T. D. 1992. Heterogeneity of decomposition and nutrient dynamics of oak (Quercus) logs during the first 2 years of decomposition. *Canad. J. Forest Res. 22*(2):161.

Schroth, G., Zech, W., and Heimann, G. 1992. Mulch decomposition under agroforestry conditions in a sub-humid tropical savanna process and influence of perennial plants. *Plant and Soil 147*:1.

Schuman, G. E. 1995. Revegetating bentonite mine spoils with sawmill by-products and gypsum. In: Karlen, D. L., Wright, R. J., and Kemper, W. O. eds., *Agriculture Utilization of Urban and Industrial By-products*, p. 261.

Schuman, G. E., Depuit, E. J., and Roadifer, K. M. 1994. Plant responses to gypsum amendment of sodic bentonite mine spoil. *J. Range Management 47*(3):206.

Selby, M., Carruth, J., and Golob, B. 1989. End use markets for MSW compost. *Bio-Cycle 30*(11):56.

Sharma, A. R., and Mittra, B. N. 1991. Effect of different rates of application of organic and nitrogen fertilizers in a rice-based cropping system. *J. Agric. Sci.*117(3):313.

Shiralipour, A., McConnell, D. B., and Smith, W. H. 1992. Uses and benefits of MSW compost: a review and assessment. *Biomass Bioenergy 3*:297.

Shuman, L. M., and Hargrove, W. L. 1985. Effect of tillage on the distribution of manganese, copper, iron, and zinc in soil fractions. *Soil Sci. Soc. Am. Proc. 49*:1117.

Shutak, V. G., and Christopher, E. P. 1952. Sawdust mulch for blueberries. Agricultural Experiment Station Univ. of Rhode Island.

Shutak, V. G., Christopher, E. P., and McElroy, L. 1949. The effect of soil management on the yield of cultivated blueberries. *J. Amer. Soc. Hort. Sci. 53*:253.

Siddiqui, M. A., and Alam, M. M. 1990. Sawdusts as soil amendments for control of nematodes infesting some vegetables. *Biol. Wastes 33*(2):123.

Siddoway, F. H. 1963. Effects of cropping and tillage methods on dry aggregate soil structure. *Soil Sci. Soc. Am. Proc. 27*:452.

Singh, T. A., Thomas, G. W., Moschler, W. W., and Martens, D. C. 1966. Phophorus uptake by corn (*Zea mays* L.) under no-till and conventional practices. *Agron. J. 58*:157.

Singh, C. P., Singh, Y. P., and Singh, M. 1987. Effect of different carbonaceous compounds on the transformation of soil nutrients. II. Immobilization and mineralization of phosphorus. *Biol. Agric. Hort. 4*:301.

Singh, A., Singh, K. and Singh, D. V. 1991. Suitability of organic mulch (distillation waste) and herbicides for weed management of perennial aromatic grasses. *Tropical Pest Management 37*(2):162.

Skroch, W. A., Powell, M. A., Bilderback, T. E., and Henry, P. H. 1992. Mulches, durability, aesthetic value, weed control, and temperature. *J. Environ. Hort. 10*:43.

Smika, D. E. 1990. Fallow management practices for wheat production in the central great plains. *Agron. J. 82*(2):319.

Smith, S. R. 1992. Sewage sludge and refuse composts as peat alternatives for conditioning impoverished soils: effects on the growth response and mineral status of *Petunia grandiflora. J. Hort. Sci. 67*(5):703.

Soane, B. D. 1990. The role of the organic matter in soil compactability: a review of some practical aspects. *Soil Tillage Res. 16*:179.

Sopher, C. D., and Baird, J. V. 1978. *Soils and Soil Management.* Reston Publishing Co., Restan, VA.

Sowers, R. S., and Welterlen, M. S. 1988. Seasonal establishment of bermudagrass using plastic and straw mulches. *Agron. J. 80*(1):144.

Spaulding, M. F., and Eisenmenger, W. S. 1938. Factors influencing the rate of decomposition of different types of plant tissue in soil and the effect of the products on plant growth. *Soil Sci. 45*:427.

Spiers, J. M. 1983. Irrigation and peat moss for the establishment of rabbiteye blueberries. *HortScience 18*:936.

Spomer, L. A. 1975. Availability of water absorbed by hardwood bark soil amendment. *Agron. J. 69*(4):589.

Spring, D. E., Ellis, M. A., Spotts, R. A., Hoitink, H. A. J., and Schmitthenner, A. F. 1980. Suppression of the apple collar rot pathogen in composted hardwood bark. *Phytopathology 70*:1209.

Spugnoli, P., Partent, A., and Baldi, F. 1993. Compaction of soil treated with municipal solid waste compost using low-pressure and traditional tyres. *J. Agric. Eng. Res. 56*:189.

Srikumar, T. S., and Ockerman, P. A. 1990. The effects of fertilization and manuring on the content of some nutrients in potato (var. provita). *Food Chem. 37*:47.

Stecker, J. A., Sander, D. H., Anderson, F. N., and Peterson, G. A. 1988. Phosphorus fertilizer placement and tillage in a wheat–fallow cropping sequence. *Soil Sci. Soc. Am. J. 52*(4):1063.

Steffen, R. 1979. The value of composted organic matter in building soil fertility. *Compost. Sci. Util. 20*:34.

Steffen, K. L., Dann, M. S., Fager, K., Fleischer, S. J., and Harper, J. K. 1993. Short-term and long-term impact of an initial large scale SMS soil amendment on vegetable crop productivity and resource use efficiency. *Compost Sci. Util. 2*(4):75.

Stephensen, R. E., and Schuster, C. E. 1945. Effect of mulches on soil properties. *Soil Sci. 59*:219.

Stevenson, F. J., and Ardakani, M. S. 1972. Organic matter reactions involving micronutrients in soils. In: *Micronutrients in Agriculture*. SSSA, Madison WI, p. 79.

Stevenson, F. J. 1982. *Humus Chemistry: Genesis, Composition, Reactions*. John Wiley, New York.

Stewart, B. A., and Chaney, R. L. 1975. Wastes: use or discard. *Proc. Soil Conserv. Soc. Am. 30*:160.

Stratton, M. L., Barker, A. V., and Rechcigl, J. E., 1995. Compost. In: *Soil Amendments and Environmental Quality* (J. E. Rechcigl, ed.). Lewis Publishers, Boca Raton, FL.

Street, J. J., Lindsay, W. L., and Sabey, B. R. 1977. Solubility and plant uptake of cadmium in soils amended with cadmium and sewage sludge. *J. Environ. Qual. 6*:72.

Svenson, S. E., and Witte, W. T. 1989. Mulch toxicity: prevent plant damage by carefully processing and storing organic mulch. *Am. Nurseryman 169*(2):45.

Svenson, S. E., and Witte, W. T. 1992. Ca, Mg, and micronutrient nutrition and growth of *Pelargonium* in pine bark amended with composted hardwood bark. *J. Environ. Hort. 10*(3):125.

Swaider, J. M., and Morse, R. D. 1984. Influence of organic amendments on solution phosphorus requirements for vegetables in mine-spoil. *J. Am. Soc. Hort. Sci. 109*:150.

Swan, J. B., Schneider, E. C., Moncrief, J. F., Paulson, W. H., and Peterson, A. E. 1987. Estimating corn growth, yield, and grain moisture from air growing degree days and residue cover. *Agron. J. 79*:53.

Sweeney, D., Graetz, W., Bottcher, A. B., Locascio, S. J., and Campbell, K. L. 1987. Tomato yield and nitrogen recovery as influenced by irrigation method, nitrogen source, and mulch. *Hort. Sci. 22*(1):27.

Sweet, R. D. 1982. Observations on the uses and effects of cover crops in agriculture. In J. C. Miller and S. M. Bell, eds. *Workshop Proceedings, Crop Production Using Cover Crops and Sods as Living Mulches*. IPPC Doc. 45-A-82:7.

Taber, H. G. 1991. Photodegradable plastic film strength loss as affected by UV radiation and vegetable cropping systems. *ASP 23*:293.

Taber, H. G., and Cox, D. F. 1993. Degradation of plastic *compost* bags used for yard *waste. Am. Soc. Hort. Sci. 3*:59.

Tadesse, W., Shuford, J. W., Taylor, R. W., Adriano, D. C., and Sajwan, K. S. 1991. Comparative availability to wheat of metals from sewage sludge and inorganic salts. *Water Air Soil Pollution 55*:397.

Tanaka, D. L. 1986. Wheat residue loss for chemical and stubble mulch fallow. *Soil Sci. Soc. Am. J. 50*(2):434.

Tenney, F. G., and Waksman, S. A. 1929. Composition of natural organic materials and their decomposition in the soil IV. The nature and rapidity of decomposition of the various organic complexes in different plant materials under aerobic conditions. *Soil Sci. 28*:55.

Thompson, R. H. 1971. *Naturally Occurring Quinones*. 2d ed. Academic Press, New York.

Thurlow, D. L., Elkins, C. B. and Hiltbold, A. E. 1984. Effect of in row chisel at planting on yield and growth of full season soybean. In: J. T. Touchton and R. E. Stevenson, ed., Proc. Southcast No-tillage Conf. 7th Headland, AL. 10 July, Auburn Univ., Auburn, AL.

Tiessen, H., and Stewart, J. W. B. 1983. Particle-size fractions and their use in studies of soil organic matter II. Cultivation effects on organic matter composition in size fractions. *Soil Sci. Soc. Am. J. 47*:509.

Tindall, J. A., Beverly, R. B., and Radcliffe, D. E. 1991. Mulch effect on soil properties and tomato growth using micro-irrigation. *Agron. J. 83*(6):1028.

Tisdale, J. M., and Oades, J. M. 1982. Organic matter and water-stable aggregates in soil. *J. Soil Sci. 33*:141.

Tisdale, S. L., Nelson, W. L., and Beaton, J. D. 1985. *Soil Fertility and Fertilizers*. Macmillan, New York.

Tisdall, J. M., Cockroft, B., and Uren, N. C. 1978. The stability of soil aggregates as affected by organic materials, microbial activity and physical disruption. *Aust. J. Soil Res. 16*:9.

Tollner, E. W., Hargrove, W. L., and Langdale, G. W. 1984. Influence of conventional and no-tillage practices on soil physical properties in the Southern Piedmont. *J. Soil Water Conserv. 39*:73.

Tomlin, A. D., and Miller, J. J. 1980. Development and fecundity of the manure worm, *Eisenia foetida* (Annelida: Lumbricidae) under laboratory conditions. In: *Proceedings of VII Internat. Colloq. Soil Zoology* (Dindal, D., ed.) p. 673.

Touchton, J. T., and Johnson, J. W. 1982. Soybean tillage and planting method effects on yield of double-cropped wheat and soybeans. *Agron. J. 74*:57.

Townsend, L. R. 1973. Effect of soil amendments on the growth and productivity of highbush blueberry. *Can. J. Plant Sci. 53*:571.

Tracy, P. W., Westfall, D. F., Elliott, E. T., and Peterson, G. A. 1990. Tillage influence on soil sulfur characteristics in winter wheat–summer fallow systems. *Soil Sci. Soc. Am. J. 54*:1630.

Triplett, G. B., 1986. Crop management practices for surface tillage systems. In: M. A. Sprague and G. B. Triplett, eds. *No-Tillage and Surface-Tillage Agriculture: The Tillage Revolution*. John Wiley, New York, p. 149.

Triplett, G. B., Jr., and Van Doren, D. M., Jr. 1969. Nitrogen, phosphorus, and potassium fertilization of non-tilled maize. *Agron. J. 61*:637.

Tumulhairwe, J. K., and Gumbs, F. A. 1983. Effects of mulching and irrigation on the production of cabbage (*Brassica oleracea*) in the dry season. *Trop. Agric. (Trinidad) 60*(2):122.

Tunlid, A., Hoitink, H. A. J., Low, C., and White, D. C. 1989. Characterization of bacteria that suppress *Rhizoctonia* damping-off in bark compost media by analysis of fatty acid biomarkers. *Appl. Environ. Microbiol. 55*(6):1368.

Unger, P. W. 1968. Soil organic matter and nitrogen changes during 24 years of dryland wheat tillage and cropping practices. *Soil Sci. Soc. Am. Proc. 32*:427.

Unger, P. 1978. Straw mulch effects on soil temperature and sorghum germination and growth. *Agron. J. 709*:858.

Unger, P. W. 1991. Organic matter, nutrient, and pH distribution in no- and conventional-tillage semiarid soils. *Agronomy J. 83*(1):186.

Van Doren, D. M., Jr., and Allmaras, R. R. 1978. Effect of residue management practices on the soil physical environment, micro-climate, and plant growth. In: W. R. Oschwald, ed., *Crop Residue Management Systems*. ASA Spec. Publ. 31:49.

Varvel, G. E., Havlin, J. L., and Peterson, T. A. 1989. Nitrogen placement evaluation for winter wheat in three fallow tillage systems. *Soil Sci. Soc. Am. J. 53*:288.

Verdonck, O., Boodt, M., de Stradiot, P., and Penninick, R. 1985. The use of tree bark and tobacco waste in agriculture and horticulture. In: *Composting of Agricultural and Other Wastes*. (J. K. R. Gasser, ed.) Elsevier Applied Science Publishers, p. 203.

Voland, R. P., and Epstein, A. H. 1994. Development of suppressiveness to diseases caused by *Rhizoctonia solani* in soils amended with composted and noncomposted manure. *Plant Dis. 78*(5):461.

Waddell, E. 1972. The mound builders: agricultural practices, environment, and society in the Central Highlands of New Guinea. *Am. Ethnol. Soc. Monograph 53*:253.

Waggoner, P. E., Miller, P. M., and DeRoo, H. C. 1960. Plastic mulching—principles and benefits. *Bull. No. 634 Conn. Agric. Exp. Stn., New Haven.*

Waksman, S. A., and Gerretsen, F. C. 1931. Influence of temperature and moisture upon the nature and extent of decomposition of plant residues by microorganisms. *Ecology 12*:33.

Wallace, A. 1994. Soil organic matter is essential to solving soil and environmental problems. *Commun. Soil Sci. Plant Anal. 25*(1–2):15.

Wan, M. T. 1992. Utility and right-of-way contaminants in British Colombia: chlorophenols. *J. Env. Qual. 21*(2):225.

Wang, Y., and Blessington, T. M. 1990. Growth and interior performance of poinsettia in media containing composted cotton burrs. *HortScience 25*(4):407.

Wang, S. H., Lohr, V. I., and Coffey, D. L. 1984. Spent mushroom compost as a soil amendment for vegetables. *J. Am. Soc. Hort. Sci. 109*:698.

Warmund, M. R., Starbuck, C. J., and Finn, C. E. 1995. Micropropagated 'Redwing' raspberry plants mulched with recycled newspaper produce greater yields than those grown with black polyethylene. *J. Small Fruit and Vitic. 3*(1):63.

Warren, S. L., and Bilderback, T. E. 1992. Arcillite: effect of chemical and physical properties of pine bark substrate and plant growth. *J. Environ. Hort. 10*(2):63.

Watson, G. W. 1988. Organic mulch and grass competition influence tree root development. *J. Arboriculture 14*(8):200.

Watson, M. R., and Hoitink, H. A. J. 1985. Long term effects of papermill sludge in stripmine reclamation. *Ohio. Rep. Res. Dev. Agric. Home Econ. Nat. Res. Ohio Agric. Res. Dev. Cent. 70*(2):19.

Weigant, W. M., Wery, J., Buitenhuis, E. T., and de Bont, J. A. M. 1992. Growth-promoting effect of thermophilic fungi on the mycelem of the edible mushroom *Agaricus bisporus. Appl. Environ. Microbiol. 58*(8):2654.

Westerman, R. L., and Edlund, M. G. 1985. Deep placement effects of nitrogen and phosphorus on grain yield, nutrient uptake and forage quality of winter wheat. *Agron. J. 77*:803.

Wiles, C. C., and Stone, G. E. 1975. Composting at Johnson City, Final report on joint USEPA-TVA Project, SW-31r.2. *U.S. Environmental Protection Agency.*

Wilhelm, W. W., Mielke, L. N., and Fenster, C. R. 1982. Root development of winter wheat as related to tillage practice in western Nebraska. *Agron. J. 74*:85.

Williams, C. H., and David, D. J. 1976. The accumulation in soil of cadmium residues form phosphate fertilizers and their effect on the cadmium content of plants. *Soil Sci. 121*:86.

Wischmeier, W. H. 1973. Conservation tillage to control water erosion. In *Proc. of the National Conservation Tillage Conference.* Ankeny, JA: Soil Conservation Soc. of America, p. 133.

Wischmeier, W. H., and Mannering, J. V. 1965. Effect of organic matter content of the soil on infiltration. *J. Soil Water Conserv. 20*(4):150.

Wohlt, P. B. 1986a. Kandep: challenge for development. In Technical Bulletin No. 2. Division of Primary Industry, Department of Enga Province, PNG.

Wohlt, P. B. 1986b. Subsistence systems of Enga Province. In Technical Bulletin No. 3. Division of Primary Industry, Department of Enga Province. *PNG.*

Wolfe, D. W. 1989. Effects of environment on degradation rates of photodegradable plastic mulches. *Proc. Natl. Agr. Plastics Congr. 21*:53.

Wolfe, D. W., Albright, L. D., and Wyland, J. 1989. Modeling row cover effects on microclimate and yield I. Growth response of tomato and cucumber. *J. Am. Soc. Hort. Sci. 114*(4):562.

Wong, M. H., and Chu, L. M. 1985. The responses of edible crops treated with extracts of refuse compost of different ages. *Agric. Wastes 14*(1):63.

Woodbury, P. B. 1992. Trace elements in municipal solid waste composts: a review of potential detrimental effects on plants, soil biota, and water quality. *Biomass and Bioenergy 3*:239.

Woodruff, N. P., Lykes, L., Siddoway, F. H., and Fryrear, D. W. 1972. How to control wind erosion. USDA Agric. Information Bull. 354.

Yeates, G. W., Wardle, D. A., and Watson, R. N. 1993. Relationships between nematodes, soil microbial biomass and weed-management strategies in maize and asparagus cropping systems. *Soil Biol. Biochem 25*(7):869.

Yohalem, D. S., Harris, R. F., and Andres, J. H. 1994. Aqueous extracts of spent mushroom substrate for foliar disease control. *Compost Sci. Util. 2*(4)67.

Young, R. A., and Onstad, C. A. 1978. Characterization of rill and interrill eroded soil. *Trans. ASAE 21*:1126.

Zehnder, G. W., and Hough-Goldstein, J. 1990. Colorado potato beetle (*Coleoptera: Chrysomlidae*) population development and effects on yield of potatoes with and without straw mulch. *J. Econ. Entomology 83*:1982.

Zucconi, F., and de Bertoldi, M. 1987. Compost specifications for the production and characterization of compost from municipal solid waste. In *Compost: Production, Quality and Use* (de Bertoldi, M., ed). Elsevier, London, p. 30.

Zucconi, F., Forte, M., Monaco, A., and de Bertoldi, M. 1982a. Biological evaluation of compost maturity. In *Composting Theory and Practice for City, Industry and Farm.* (Staff of Compost Science/Land Utilization, eds). JG Press, Emmaus, PA, p. 341.

Zucconi, F., Pera, A., Forte, M., and de Bertoldi, M. 1982b. Evaluating toxicity of immature compost. In *Composting Theory and Practice for City, Industry and Farm* (Staff of Compost Science/Land, Utilization, eds). JG Press, Emmaus, PA, 61.

3
Paper Sludges as Soil Conditioners

Jeffrey Norrie *Acadian Seaplants Limited, Dartmouth, Nova Scotia, Canada*

Alejandro Fierro *Center for Horticultural Research, Laval University, Ste-Foy, Quebec, Canada*

I. INTRODUCTION

The production of pulp for paper manufacture, either from virgin wood or recycled paper, generates large amounts of solid waste. Paper sludges represent approximately 87% of such wastes. A conservative estimate of average sludge generation from papermills is about 50 dry kg per ton of finished paper (NCASI, 1992). Most of the paper sludges produced from mills are landfilled or lagooned, about 10–20% is incinerated, and only a small proportion is used as soil conditioner (Latua-Somppi et al., 1994; NCASI, 1992).

In the 1950s, land application of papermill waste was considered as a treatment system method to facilitate filtration and microbial decomposition of the waste residues (NCASI, 1959). Today, an opposite point of view is often taken, whereby sludges are applied specifically to condition the soil. Indeed over the past several years, papermill sludges have received increasing attention as potential organic amendments and conditioners for various types of soils (Barclay, 1991; NCASI, 1984, 1991). Increasing landfill costs and mounting pressure to reduce incineration of organic residuals has resulted in research to evaluate disposal alternatives for agricultural and sylvicultural land applications (NCASI, 1992), as a container culture medium (Chong and Cline, 1993), and for mine soil reclamation (Feagley et al., 1994a,b). In support of this research, paper sludge applications to recondition agricultural and urban soils have been shown to be successful alternatives to other disposal methods (Enzor, 1988; Hatch and Pepin, 1985). Market surveys have shown that fresh (Enzor, 1988) and composted sludges are gaining popularity with landscape contractors, nurseries, greenhouses, and municipal agencies (Smyser, 1982). Moreover, applications to agricultural fields are common in several states with up to 100% of available paper sludges being applied during months of high demand.

Paper sludges may be considered as organic amendments as well as a lime amendment (for particular deinking sludges). As organic amendments, they quickly raise the soil organic matter content and boost soil quality by improving biological, chemical, and physical properties of the soil, especially on deteriorated or degraded soils. As a lime amendment, deinking and primary sludges increase and buffer soil pH when incorporated at high doses. Moreover, nutrients in certain sludges may be beneficial for plant growth.

This chapter presents an overview of the use of papermill sludges as soil conditioners and amendments for agricultural, horticultural, and sylvicultural applications. In addition, pertinent research will be examined in order to understand the advantages and disadvantages associated with paper sludges as alternative sources of organic matter. To evaluate objectively the potential for papermill sludge use as a soil conditioners, we must first examine the origin, composition, qualities, and characteristics associated with the various sludges arising from paper production, processing, and recycling.

II. ORIGIN OF PAPERMILL RESIDUES OR SLUDGES

Three different methods are commercially used to break down wood fiber (comprised principally of cellulosic compounds) mechanically or chemically to create wood pulp. This process is called pulping. Mechanical mills use stone discs (with or without added heat) to grind the fibers. Kraft mills use sodium hydroxide and sodium sulphide to break down the fibers, while sulphite mills use calcium sulfite, magnesium sulfite, ammonium sulfite, or sodium sulphite (NCASI, 1984). Many mills use chlorine to bleach the paper, while other mills use hydrogen peroxide (NCASI, 1991; Ferguson, 1992).

Solid wastes arising from these processes, commonly referred to as sludges or pulps, are available in two different classes. Primary or clarifier sludges are composed principally of wood fibers that have physically settled out of the wood/water slurry, or wastewater, during initial virgin fiber treatment, and are unsuitable for further processing. Secondary sludges arise from microbial decomposition used to remove suspended solids that were not removed during primary clarification of wastewater. Secondary solids contain higher concentrations of nitrogen and phosphorus, added to the slurry mixtures to aid microbial activity, increase flocculation, and expedite the subsequent removal of suspended particles in the wastewater. An additional class of papermill sludges are the tertiary or deinking sludges, which are similar to primary sludges in fiber composition but also contain wastes filtered from the fiber stream during screening and deinking processes resulting from newsprint, magazine, and used-paper recycling.

Sludge dewatering to approximately 40% solids prior to disposal is necessary in order to reduce volume and to facilitate handling. Most mills with facilities for secondary wastewater treatment combine secondary and primary sludges prior to dewatering. This alleviates problems associated with dewatering secondary sludges alone (NCASI, 1991), which normally have less than 5% solids. Common methods of dewatering include belt and screw presses as well as centrifugation.

A. Pulping

Cellulose is a fibrous substance used in all pulp and paper products and, along with hemicellulose and lignin, is predominant in woody plants. Whereas cellulose and, to a lesser degree, hemicelluloses are desirable for paper manufacturing, lignin is usually removed during processing. However, lignin can add to the desired qualities in some papers (e.g., newsprint).

1. Mechanical Pulping

Wood chips are either physically or chemically mashed and treated to produce pulp. Water is required in abundance for both processes. Mechanical or groundwood pulp is produced by either a stone process or a refiner process. In the former process, a grindstone is used to tear wood fibers away from short logs. In the refiner process, wood chips are ground between parallel discs and, in some cases, have been heat pretreated to soften the wood chips before grinding. In both processes, wood-fiber-containing water is filtered and cleaned and held for further treatment (NCASI, 1984).

2. Chemical Pulping

There are several methods of chemical pulping which involve cooking wood chips in the presence of several chemicals. These chemicals break down the lignin network within the wood fibers while not affecting the cellulose or hemicellulose constituents. In the Kraft process, wood chips are cooked in a sodium hydroxide and sodium sulfide solution to remove lignin effectively. Another chemical process called sulfite pulping cooks wood chips as in the Kraft process but uses a solution of calcium sulfite, magnesium sulfite, ammonium sulfite, or sodium sulfite to acheive the same pulp slurry. Recycled paper is also prepared into a slurry by mixing with water and cooked in an alkali, detergent, solvent soup to remove coating agents (clay, $CaCO_3$, TiO_2, adhesives, ink, and fillers). The slurry is filtered and washed to produce reclaimed cellulose fibers for further papermaking (NCASI, 1984).

B. Bleaching

Once the pulping process is complete, the brown or deeply colored wood pulp is bleached to produce different shades of the final paper product. Various lignins and resins, produced naturally from wood, cause variations in pulp color. Hydrosulfites and peroxides used under high temperature and pH are the most common bleaching agents for mechanically ground pulp. However, Kraft pulps, which are usually darker in color, require several stages of bleaching. Chemical agents used include chlorine, calcium or sodium hypochlorite, and chlorine dioxide (NCASI, 1984).

C. Sludge Production

Wastewater treatment produces sludges. Sludges are complex mixtures of solid material. Primary and deinking sludges arise from sedimentation processes utilizing both physical and chemical means to remove coarse, suspended and colloidal solids from raw wastewater. Owing to the above processes, these sludge solids include wood fiber, lime, clay fillers, coating agents, and chemicals used in the papermaking process. Chemical characteristics of paper sludges can be either organic or inorganic in nature and can include either natural or man-made components. The natural or organic constituents have been described above, while other organic and inorganic chemicals are contributed through the addition of pulping chemicals. The use of chlorine bleaching materials has caused concern from an environmental standpoint. Chlorinated products resulting from the reaction between lignins, resins, and chlorine are of primary concern. Other compounds of specific interest include dioxines and furans, both potential carcinogens. Other polycyclic aromatic hydrocarbons (PAH) have been identified in very small amounts in paper sludges.

Secondary treatment is a biological process followed by sedimentation to remove remaining suspended solids from wastewaters already treated in the clarification (primary) process. Microorganisms are used in this process (flocculation) to increase the microbial mass, which is subsequently removed after settling out. In secondary treatment, wastewaters are oxygenated and enriched with nutrients (nitrogen and phosphorus) in the presence of microorganisms. The result is a decrease in BOD (biological oxygen demand), in COD (chemical oxygen demand), and in resinous and fatty acid compounds (CQVB and MICT, 1993). The resulting sludges contain mainly settled organic solids and microbial biomass and are rather rich in nitrogen and phosphorus. In many instances, the nutrient rich secondary sludges are mixed with nutrient poor primary sludges. This will become an important feature later in this chapter when discussing the importance of low C/N ratios for sludge amendments.

All three paper sludges can be used as soil conditioners. As rich sources of organic matter, paper sludges can bind and stabilize soil particles against erosion while increasing soil porosity and facilitating water and air movement

(Diehn and Zuercher, 1990; Diehn, 1991). They can also improve the environment for soil micro-organisms which in turn liberate bound nutrients for plants (Elkins et al., 1984).

III. CURRENT SLUDGE PRODUCTION AND DISPOSAL METHODS

An average papermill produces around 100 dry MG/day (1 MG=1 metric tonne) of paper sludges (NCASI, 1991), or about 50 dry kg per tonne of paper produced (NCASI, 1992). Papermills in the United States produce approximately 4–6 million metric MG of dry sludge per year (NCASI, 1992). In Canada, Ontario mills produce around 2000 MG of sludge per day, amounting to over 750,000 MG per year (Chong and Cline, 1991). In Quebec, where our research was carried out, about 160,000 MG of dry sludge is produced annually from provincial mills (Trépanier et al., 1996), while the Daishowa Inc. plant in Quebec City produces about 20,000 MG per year.

Traditionally, papermill sludges have been disposed of by landfilling (or lagooning, when not dewatered), by incineration, and to a significantly lesser degree by land application. In 1989, about 70% was landfilled or lagooned, 21% incinerated, 8% landspread, and 1% recycled or reused (NCASI, 1992). When incinerated, ash disposal is still a significant concern, since the volume produced is still about 25% of the original dewatered sludge.

Given large increases in tipping or dumping costs at landfill sites all over North America, where rates of $100–150 per metric tonne are commonplace, and mounting pressure to recycle and reuse discarded materials, research is urgently needed to evaluate alternative uses for paper sludge residues. To this end, research on several raw and composted paper sludge mixtures for use in container culture has been under way for several years in Ontario. This research has focused mainly on the use of primary or secondary sludges, or mixtures of the two (Chong and Cline, 1993). Research on sludge applications to field crops has been limited. However, research on sludge application for the revegetation or reclamation of mine soils has resulted in the identification of effective sludge application rates (Feagley et al., 1994 a,b) in addition to many plant species capable of establishment in amended soils (Fierro et al., 1997). Many of these species will be identified later in the chapter.

A. Other Alternatives for Paper Sludge Disposal

There are several alternatives to landfilling, burning, or land application of sludges. A very effective method of sludge utilization is through composting. Although municipal sludges can contain various constituents that render it unsuitable for composting, paper sludge composting has met with commercial success. Monitoring is still necessary during the process to ensure a mature compost (decomposed and stabilized organic material, low C/N ratio).

Other methods of sludge disposal include brick fabrication, vermi-composting, use as a high fiber feed supplement, and use as a component of degradable fiber gardening pots (NCASI, 1991). It has also been suggested that paper sludges may be ideal for some forms of mushroom production. The success of these alternatives uses depends on the composition and characteristics of each sludge, whether chemical, physical, or both.

IV. CHARACTERIZATION OF SLUDGES

Sludge composition may vary considerably depending on the papermaking process; for deinking sludges it is based primarily on the paper sources used for recycling (i.e., newsprint, glossy magazines, etc.) and the chemicals involved in the deinking process. The ratios between organic and inorganic fractions, moisture content, and the concentration of undesirable contaminants also varies with sludge origin (Chong, 1993). Calcium is usually present in large concentrations owing to Kraft processing, while nitrogen and phosphorus are prevalent in secondary sludges. Differences in storage and spatial application can also play a role in applying accurate amendment rates (Kraske and Fernandez, 1990).

Paper sludges possess many properties that can make them effective soil conditioners. Characteristics that should be considered and examined before application as soil conditioners include the physical properties, pH, heavy metal and organic compound concentration, organic carbon fractions (carbon–nitrogen [C/N] ratio), decomposition patterns and rates, plant nutrient availability (especially nitrogen concentration), salinity, and sodicity.

In general, primary sludge composition is fairly consistent from one mill to the next, while deinking sludge composition varies according to the process and the origin of the waste paper. Primary and deinking sludges have a water content somewhere between 40 and 70% (Table 1). Their organic fraction consists mainly of wood fiber. The main difference between the sludges is in their inorganic or mineral fraction (ash), which is considerably more important in deinking sludges. The inorganic fraction of deinking sludges varies with the kind of waste paper used for recycling. Along with the inks, deinking removes other noncellulosic materials from the fiber, including coatings, adhesives, dyes, and fillers such as calcium carbonate and clay, as well as obvious trash such as paper clips, bailing wire, and staples (Ferguson, 1992). For example, waste paper with high amounts of filler, such as clay-coated magazines, will produce a sludge higher in inorganic matter than old newspaper. Secondary or biological sludges are usually mixed with primary or deinking sludges prior to dewatering and disposal, resulting in what is called combined sludge.

Table 1 Paper Sludge Characteristics and Elemental Content

Variable or element[1]	Ranges: Primary	Secondary	Deinking	Combined	References[b]
pH (water)	6.4–7.6	6.6–7.9	7.8–9.1	6.5–7.3	2,3,4,5,6,7,11, 12
EC (dS m^{-1})	0.19–0.7	3.9	0.09–0.12	0.29	3,5,6,11
Dry density (g cm^{-3})			0.11–0.3		10,11
Moisture			34–69		1,6
Water content @ 33kPa (g g^{-1})			0.65		12
Ash (%)	4.2	9.9	18–40.2	6.8	1,3
Organic C (%)	38–44	42–43.1	32.8–44.1	34.6–44.7	1,2,3,4,6,7,11
N (%)	0.08–0.4	4.3–7.7	0.15–0.3	1.2–1.4	1,2,3,4,6,7,11, 12
C/N	111–478	5–10	126–344	28–32	1,2,3,4,5,6,7, 11
P (mg kg^{-1})	580–1000	7700–12000	60–100	2468	3,6,7,11
K	120–800	2000–6000	154	712	3,4,6,7
Ca	2100–8100	6300	8312	3712	3,4,6
Mg	610–3150	2300	623	1594	3,4,6
Fe			428		3,4,6
S	1500	7000	786	2557	6
Mo			<2		3,6
B	1.1	1.8	<5	1.3	6
Na	440	1300	974	521	3,6
Pb	4–9.2	11–82	<20	1–59	6,8,9,12
Ni	5–24	6–36	<100	6–34	3,6,8,9,12
Cu	7–58	23–519	46–200	1–270	3,6,8,9,12
Cd	0.3–1.6	0.4–1.6	0.09–1	<1–2	3,6,8,9
Zn	30–94	88–144	39–180	2–200	3,6,8,9,12
Ext. Ca[a]	45–1780	3800	7600	2280	3,5,11
Ext. Mg[a]	13–280	1215	311	510	3,5
Ext. K[a]	0.4		59		5,11
Ext. P[a]	228		38		5,11
Cr	3.6–12	10–32	4.7–30	1–30	3,6,8,9,12
Mn			19.6		6

[1]Total elemental concentrations are presented, unless otherwise specified. Combined: primary + secondary or deinking + secondary sludge.

[a]Extractible elements.

[b]References: 1, Latua-Somppi et al., 1994; 2, Honeycutt et al., 1988; 3, Zhang et al., 1993; 4, Zibilske, 1987; 5, Feagley et al., 1994a; 6, Trépanier et al., 1996; 7, Henry, 1991; 8, Hatch and Pepin, 1985; 9, Barclay, 1991; 10, NCASI, 1991; 11, Fierro et al., 1996; 12, Pridham, N. F., and R. A. Cline, 1988.

A. Physical Properties

Raw paper sludge is usually a light gray color, deinking sludge being a more pronounced gray, and the smell is that of moist newspaper. The physical appearance resembles paper-maché and is composed of aggregates of more or less compressed short wood fiber with a particle size distribution ranging from 1 to 20 mm, with intermediate particles being most abundant. The initial water content is usually 40–70% (it is indeed a very absorbant material), and water retention is about 0.65 kg per kg (or 65%) at a matric potential of –0.033 MPa (Table 1). The dry bulk density of deinking sludge is about 0.11 to 0.30 g ml^{-1}. Primary sludge bulk density can be even lower because it contains less inorganic material.

B. pH

The pH of papermill sludges is usually quite basic, especially in deinking sludges (Table 1). This is caused by the use of pulping and deinking chemicals such as sodium hydroxide and sodium silicate (Na_3SiO_3) as well as calcium carbonate, which is used as paper filler (Ferguson, 1992). The pH of interstitial water and leachate from paper sludges has been found to range from near neutral to over 12 (NCASI, 1979). Moreover, the pH of sludge-containing wastewaters range from 6 to 9 (Thacker, 1986).

C. Heavy Metals and Organic Compounds

Perhaps the most commonly asked questions about papermill sludges relate to their heavy metal content. In the past, the greatest proportion of hazardous materials in deinking sludges came from inks and pigments found in waste paper. Recently, however, ink manufacturers have virtually eliminated most heavy metals from commercial printing inks. Copper and barium still remain common constituents of printing inks (NCASI, 1991). It should be noted that heavy metals often originate from sources other than inks, e.g., from the wood and chemicals used in papermaking, among other materials (Barclay, 1991).

Extractability and distribution patterns between specific forms of heavy metals have been found to vary widely for each metal within sludges. This may be due to characteristics and solubility associated with individual metals in the extraction media (Canarutto et al., 1990). However, heavy metal content is generally much lower than that found in municipal sludges already permitted for application to agricultural lands (Chong, 1993; Thacker, 1986). In fact, Barclay (1991) suggests that paper industry sludges are more suitable than municipal sludges as soil amendments. In our research, we found heavy metal concentrations to be higher in unamended topsoils in urban areas than in paper sludges applied to test plots.

Polychlorinated biphenyls (PCBs) can occasionally be found in deinking sludges, but their source is almost exclusively old file stock, which contains carbonless paper. However, carbonless paper manufacturers stopped using

PCBs about 25 years ago, so the sporadic incidence of this organic compound will diminish over time (RPN, 1992)

The National Coalition for Air and Stream Improvement (NCASI) has been reporting on the state of the industry and research in paper manufacturing, including sludge research, since the 1950s (NCASI, 1959). In 1984, the NCASI examined a number of organic compounds in paper by-products. The study found that recycling and deinking of newspapers contributed little to the presence of organochlorines, resins, fatty acids, or phenolics (NCASI, 1984). In virgin wood processing, very low concentrations are present in paper sludges (Bellamy and Chong, 1995). Furthermore, many mills are now using peroxides in the bleaching process in place of the environmentally unfriendly chlorine bleaches. The levels of organochlorines are much lower in paper sludges if peroxides are used in the bleaching process (RPN, 1992). An examination of potential risks associated with exposure to dioxins in papermill sludges clearly indicated little or no significant threat due to exposure for a family living their entire lives (including hunting and farming) on sludge amended land (Keenan, 1989). However, the significance of trace organic compounds that may accumulate during the land application of sludges has been given little study.

D. Decay Patterns and Nitrogen Availability

Detailed research is currently underway at Laval University on the dynamics of carbon and nitrogen behavior in sludges incorporated into the soil. Decomposition rates for sludges in soil are usually intermediary between faster rates, such as those reported for straw (Chesire and Chapman, 1996), and slower rates reported for wood residues and temperate hardwood leaf litter (Schuman and Belden, 1991; Melillo et al., 1982). Initial carbon fractions (cellulose, lignin, hemicellulose, and solubles) of papermill sludges are similar to those of wood residue (LRN, 1993), but they decompose faster. This is probably due to cellulose fiber in sludge, fiber that is short and partially broken down, thereby presenting a greater exposed surface area to microbial attack.

When papermill sludges are used as soil conditioners, plant nitrogen availability is among the most important considerations in determining application rates and nitrogen supplements for a given soil. When incorporated into soil, paper sludges can act either as a net sink or a net source for plant available nitrogen. When mixed with secondary sludge, primary and deinking sludges are usually a source, but when applied alone, they act as a nitrogen (and phosphorus) sink for some time (weeks to months) after incorporation. This is usually a response of the available carbon (excluding lignin) to total nitrogen ratio of a particular sludge. In other words, when using C_{total}/N_{total} ratio as an indicator of whether or not nitrogen will be immobilized, we should consider how much of that carbon is readily available to microorganisms. If lignin-carbon is substracted from this total, we obtain a better indicator of the net immobilization potential. In the short term (~110 weeks), plant growth was closely related to nitrogen application in spite of wide variations in adjusted C/N and C/P ratios in paper sludges (Fierro et al., 1997).

The initial C/N ratio of primary and deinking papermill residues generally ranges from 100:1 to 400 (Table 1). In comparison, secondary sludges can have ratios as low as 6–7 due to the addition of N and P to microbially treated wastewaters (Watson and Hoitink, 1985). Municipal sludges, on the other hand, have ratios typically around 10:1 (Watson and Hoitink, 1985). Many fertilizer amended sludges have ideal C/N ratios around 20–30 (Bockheim et al., 1988). Sludge amendments can have a more dramatic effect on plant growth when given supplemental fertilizers to enhance sludge fertility (Feagley et al., 1994 a,b). Plant growth in sludge amended soil is usually related to the level of available nitrogen found in the sludges (Chong et al., 1987) or to those amounts supplied in chemical fertilizer. As we will see later, in experiments examining turf establishment and color on sludge amended soil having a range of C/N ratios, the best results were found in plots receiving a split application of urea to adjust C/N ratios to around 20.

E. Availability of Other Plant Nutrients

The application of a soil conditioner may alter the availability of one or more essential plant nutrients other than nitrogen. Whether the amendment modifies soil properties that control the solubility of nutrients (i.e. pH, CEC etc.) or simply dilutes or adds to them, it is important to be aware of potential increases or decreases in availability and, if possible, the magnitude of change. In the case of paper sludges, one can at least check the elemental analysis results as well as other chemical properties of both the sludge and the receiving soil. Afterwards, depending on the application rate, one can have a fairly good idea of the potential effect on nutrient availability following incorporation of a sludge to a particular soil.

Available phosphorus is usually very low in primary and deinking sludges, while in secondary or combined sludges rates can be rather high, considering that these latter sludges are enriched in phosphorus. When using deinking or primary sludges, phosphorus fertilization is essential even if sludges are incorporated into a fertile soil (Fierro et al., 1997). Some microbial phosphorus immobilization can be expected. Sulfur is also likely to be immobilized, but this is of little concern in plant nutrition, since paper sludges are rich in sulfur (Table 1).

Exchangeable potassium, calcium, and magnesium are usually unbalanced in paper sludges from a plant nutrition standpoint (Table 1). A balanced saturation of calcium, potassium, and magnesium requires the addition of the latter two elements depending on content in the receiving soil.

Micronutrients in paper sludges usually fall between reasonable limits for plant nutrition (Table 1). Exceptions may be molybdenum and boron, which are often found in extremely low concentrations. However, availability of native soil molybdenum may increase following sludge incorporation as a result of increased pH. Copper may be found in rather high concentrations mainly in deinking sludges but is below toxic levels.

F. Salinity and Sodicity

Salinity and sodium buildup are two of the concerns in the more widespread use of paper sludges as soil amendments. Indeed, calcium chloride is added in the deinking process to enhance the dissociation of ink, coatings, and fillers from the pulp fiber (NCASI, 1991). However, sludge electrical conductivity (EC) is usually low (Feagley et al., 1994a; Trépanier et al., 1996). In some cases, EC can be high for some combined sludges but excess salts are leached away within a few days (Chong and Cline, 1993).

Sodium hydroxide, sodium bisulfite, and sodium silicate are commonly used in paper manufacturing processes and represent important sources of sodium. Sodicity is an obvious concern when incorporating paper sludges. Nevertheless, the very high calcium (as Ca^{2+}) concentrations commonly found in sludges effectively offset any risk of sodium toxicity since the sodium adsorption ratio is quite low.

V. PAPERMILL SLUDGES AS SOIL CONDITIONERS

Papermill sludges have been applied to soil at rates of 55 MG/hectare for strip mine reclamation (Watson and Hoitink, 1985) as well as 32, 63, and 94 dry MG/hectare in a red pine plantation (Bockheim et al., 1988). Other applications of paper sludges in sylviculture were beneficial at application rates of around 45 MG/hectare but typically range from between 9 and 224 dry MG/hectare (Thacker, 1986).

An annual application of 12 dry MG/ha of deinking sludge improves many soil physical properties and organic matter content of soils under potato cropping. A further supplement of 90 kg N/ha is however required to avoid yield decreases associated with microbial nitrogen immobilization (Trépanier et al., 1996). Research is underway for strawberry and cabbage with similar results. One-time, heavy applications (>100 dry MG/ha) of deinking sludges supplemented with N and P dramatically improve sustained ground cover and plant dry matter production of a revegetated sand pit (Fierro et al., unpublished.). In reclamation studies of abandoned mine soils containing pyritic spoil, composted paper sludge was roughly equivalent to limed topsoil for successful long-term reclamation (Pichtel et al., 1994). Apparently, a rate of 112 dry MG/ha of primary sludge supplemented with 4 and 1.2 kg of nitrogen and phosphorus, respectively, per dry ton of sludge, gives the best growth response for bermuda grass on mine soil (Feagley et al., 1994b). It has also been suggested that to change soil characteristics significantly at least 45 dry MG/hectare be applied at once, or that repeated applications of lower dose rates be used over time (NCASI, 1984).

Researchers in Ontario have shown that the addition of papermill sludge can be beneficial to container nursery production (Chong and Cline, 1993; Chong et al., 1987) and to soil mixes growing tomatoes, cucumbers, and peppers in the greenhouse and corn and grapes in the field (Chong and Cline, 1991). The growth of the various crops is usually related to the amount of nitrogen available to the plants (Chong and Cline, 1991; Fierro et al., 1997). Paper sludge has also been used as a soil amendment for forest soils (Henry, 1991; Thacker, 1986). Primary sludges were found to increase growth of cottonwood 52–131% when surface applied, apparently having a beneficial mulching effect in comparison to controls, while additional nitrogen fertilizer supplements improved growth 151–223% over control plants (Henry, 1991). Deinking sludges mixed with mineral soil or sand increased growth of several grasses when properly supplemented with nitrogen, and of several legume species when supplemented with phosphorus (Fierro et al., 1997). Secondary sludges further increased growth 112–319% over control plants due to nutritional additives to facilitate microbial activity in the removal of secondary sludges from the wastewaters during papermaking. Recycled newsprint can also be successfully mixed with poultry litter for land applications (Edwards, 1992; Edwards et al., 1993).

Many studies have demonstrated beneficial effects of papermill sludge amendments to plant growth and soil chemistry, especially when supplemented with inorganic fertilizers to improve nitrogen and phosphorus deficits (Chong et al., 1988; Feagley et al., 1994a,b). Concern has already been raised concerning the state of Quebec soils, particularly in monocultures of potato, cereals, and corn. Papermill sludges from the Daishowa Inc. mill in Quebec are currently being used to offset depletions in soil organic matter and soil erosion in potato culture (Trépanier et al., 1996). Decomposition of organic matter by indigenous microorganisms also appears to continue after sludge addition without significant disruption (Elkins et al., 1984; Schuman and Belden, 1991; Madsen, 1991). Erosion control was found to be excellent with the incorporation of papermill sludge while not interfering with plant emergence (Watson and Hoitink, 1985). While fresh sludges have a greyish color, there is little trace of the residue after three years.

Around Eau Claire, Wisconsin, up to 65% of a local papermill deinking sludge is spread on neighboring farmers' fields, and during months of high demand, up to 100% of their sludges are land applied (American Papermaker, 1992). Through this project, the mill has reduced its dependency on landfilling by 85% and is backlogged on orders for sludge from many area farmers. Other mills are also successfully applying fresh paper sludges to agricultural soils (Enzor, 1988) or composting sludges (Smyser, 1982).

VI. EXPERIMENTAL CROPS

In this section we examine reported research on plant responses to paper sludge and other lignocellulosic amendments. Sensitive crops such as tomatoes, peppers, and cucumbers have been grown successfully in the greenhouse in several paper sludge mixtures provided there was no deficiency in available nitrogen (Chong and Cline, 1991).

Considerable research has been done on degraded soil reconditioning for reclamation purposes. According to Smith et al. (1986), plant species grown for revegetation purposes on amended soils (degraded) should possess the following characteristics: sod-forming morphology, ease of establishment from seed, drought and salt tolerance, and adaptability to shallow, poorly drained soil. Many of the plant species tested do not have all of these characteristics but do possess several of them. For example, ponderosa pine, flat sagebrush, many wheat grass species, smooth brome, and saltbush have all been tried with various degrees of success in land reclamation studies (Smith et al., 1985). Black spruce seedlings have been grown successfully on pulpmill wood waste (West and Bishop, 1987), and other tree species grown on poor soils include jack pine, willow (Prégant et al., 1987), and several species of alder (Camiré et al., 1983; Prégant and Camiré, 1985a,b).

Fescues, lespedeza, love grass, and Bermuda grass have been grown successfully on bark, hardwood chips, straw, hay, and hydromulch mixtures (wood fiber product composed largely of cellulosic fibers; Dyer et al., 1984). These authors conclude that mulches high in organic matter tend to encourage the growth of legumes over grasses. This may be due to low available nitrogen favoring the establishment of species capable of nitrogen fixation. Tall fescue (Roberts et al., 1988b) has also been used for site rehabilitation in addition to Engelmann spruce seedlings (Grossnickle and Reid, 1984). Soybeans and corn have also been found to give improved yields in response to the addition of papermill sludges although the addition of supplemental nitrogen was used in the latter case (American Papermaker, 1992).

In greenhouse evaluations at Laval University, we found that with moderate fertilization (20–20–20, weekly), germination and plant establishment were possible on soil composed of more than 50% sludge (Table 2). In other greenhouse tests, desert globemallow, fourwing saltbrush, and rubber rabbitbrush were the most promising species grown in bentonite mine spoils amended with sawdust (Uresk and Yamamoto, 1986). Container experiments have shown that cotoneaster, dogwood, forsythia, spiraea, and weigela have grown well in paper sludge mixtures but that growth was directly related to the amount of nitrogen in the raw sludge. These tests were done as part of the Ontario Nursery Research Program (Chong et al., 1991). Moreover, cotoneaster, dogwood, forsythia, and weigela grew as well or better than control plants in pine bark amended with either 15 or 30% paper sludge (Chong and Cline, 1993). The Ontario Ministry of Agriculture and Food is continuing research into organic waste recycling for container crops.

A. Research at Laval University, Quebec City

As part of an ongoing research program into sludge, Laval University has successfully grown several grass, legume, and small bush and tree species under both greenhouse and field conditions. In preliminary germination tests, many grass and legume species were grown in soils containing up to 50% paper sludge; from these screening trials, grass mixtures were selected for carryover into field experiments (Table 2). Moreover, these preliminary experiments determined several promising paper sludge rates that could be tried in field experiments.

Part of the paper sludge research program at Laval University in Quebec City is examining deinking and primary paper sludge as a principle constituent of several substrate mixtures used as soil amendments in landscape horticulture. Several paper sludge/organic soil/sand mixtures (to a maximum of 50% sludge)

Table 2 Germination Rates for Grass and Legume Species Evaluated in Screening Tests Using either Sand or Soil, Mixed with up to 50% Deinking Paper Sludge

		% Germination					
		Media mixture					
Plant type	Common name	Control	50 ds 50 sand	75 ds 25 sand	50 ds 50 soil	75 ds 25 soil	100 ds
Grasses	Red fescue	61.3	62.7	60.0	57.3	58.7	48.7
	Tall fescue	74.7	78.0	68.7	69.3	67.3	64.7
	Bentgrass	93.3	68.7	68.7	91.3	63.3	80.7
	Timothy	81.3	72.0	72.7	88.0	84.0	82
	Bromegrass	77.3	62.0	70.0	71.3	58.7	56
Legumes	Vetch	74.7	57.7	58.7	62.7	60	61.3
	Birdsfoot treefoil	68.8	44.7	44.7	57.3	64.7	54.7
	Sweet clover	69.3	50.0	50.0	71.3	75.3	75.3
	White clover	84.0	70.0	70.0	85.3	76.7	78
	Alsike clover	86.7	61.3	61.3	90.7	84.0	84.7

Note: $N = 3$ for each species. Abbreviations: ds, deinking sludge.

were compared to an organic soil/sand control. Sludge rates were equivalent to approximately 22.5, 67.5, and 112.5 dry MG/hectare with a 15 cm layer of each mixture incorporated into existing soil to about 30 cm. A strip plot treatment was placed horizontally across each main plot sludge treatment, to evaluate the need for supplemental fertilizer to elevate low substrate C/N ratio. Kentucky bluegrass (seed and sod) and a Kentucky bluegrass/perennial ryegrass (seed) mixture were grown with each treatment. Preliminary results indicated poor ground cover and nitrogen deficiency in plants grown in unfertilized plots (Fig. 1). For sod and seeded grasses, control plots were slightly more healthy than sludge amended plots, which is likely due to a greater concentration of available nitrogen from the organic soil. It was concluded that supplemental fertil-

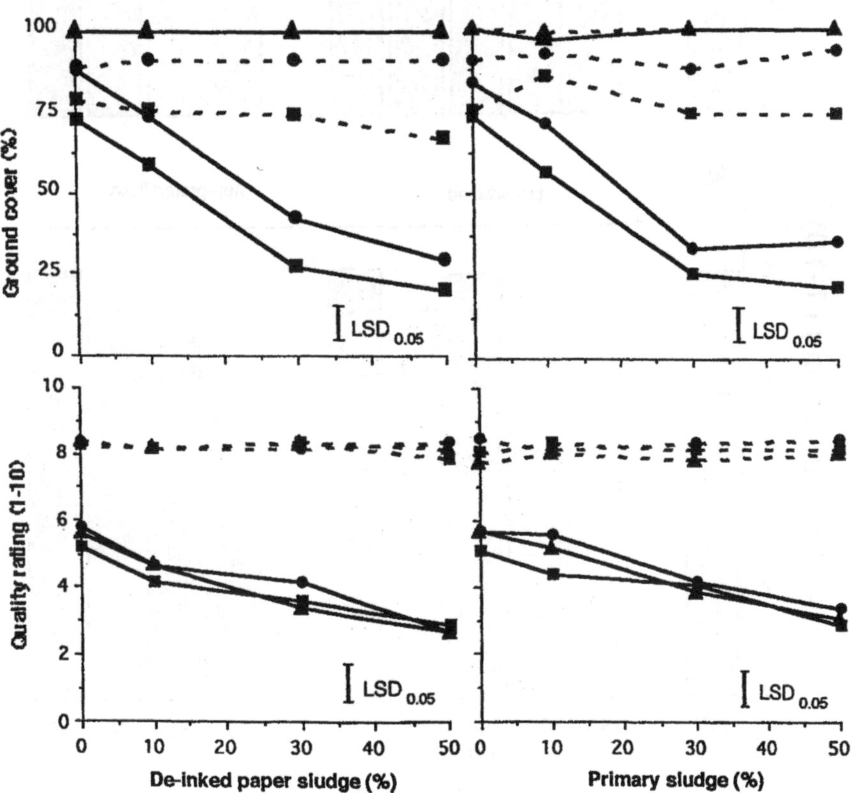

Figure 1 Evaluation of the percentage ground cover and stand quality of grasses cultivated on sludge amended soils without (solid line) or with (dotted line) fertilization for Kentucky bluegrass seed (■), Kentucky bluegrass sod (▲) and Kentucky bluegrass/perennial ryegrass (●) in fall of establishment year.

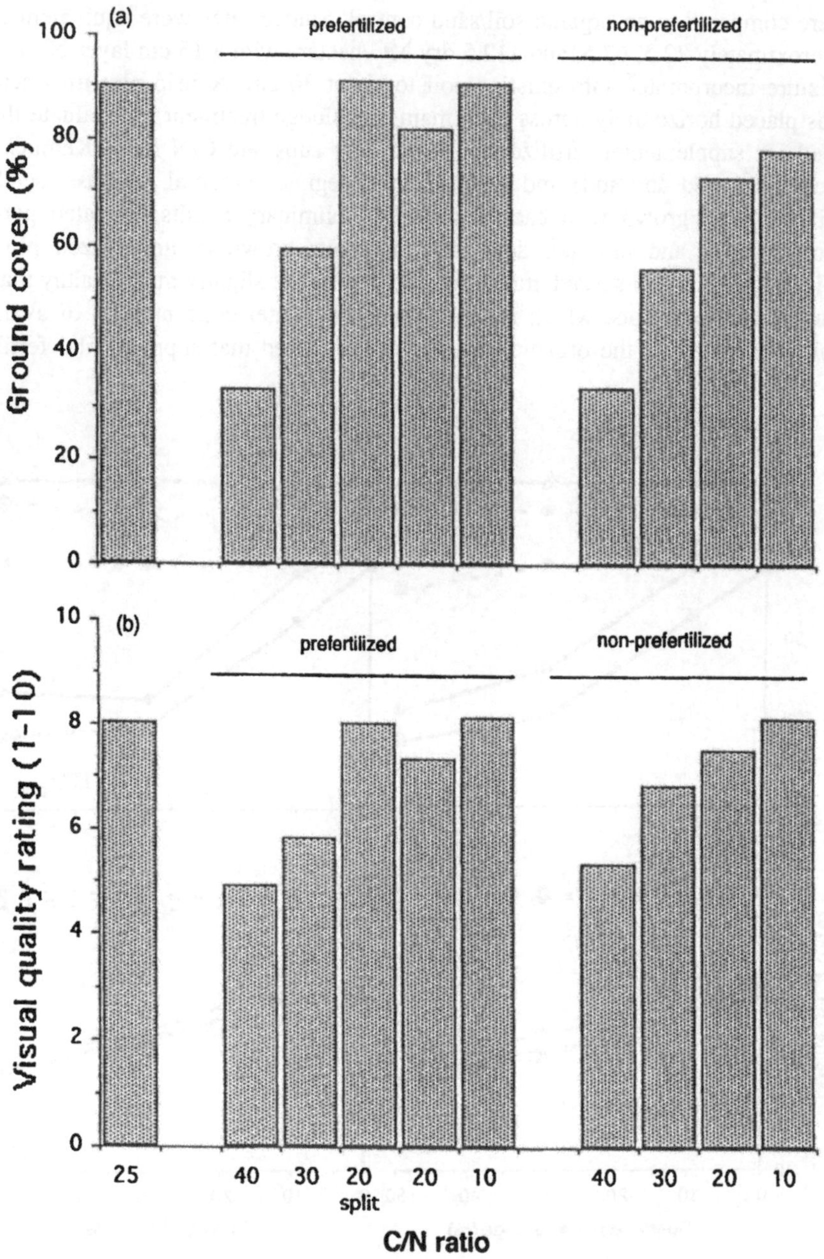

Figure 2 Evaluation of the percentage ground cover and stand quality of Kentucky bluegrass cultivated on sludge amended soils with or without prefertilization to different C/N ratios in fall of establishment year.

ization is needed to decrease the C/N ratio of these substrates to about 20 to 30 for sustained plant growth regardless of sludge amendments.

To determine more accurately the nutritional requirements, experiments were continued to evaluate several C/N ratio levels in one selected sludge/organic soil/sand mixture. Urea was added to mixtures of 30% sludge, 50% sand, and 20% organic soil, to adjust the C/N ratios to between 10 and 40. Further, nitrogen was given either as a prefertilization 3 weeks before each treatment substrate was spread or as a direct fertilization just after substrate application. Kentucky bluegrass (Fig. 2) and a roadside mixture (Fig. 3; recommended by the Provincial Department of Highways) were tried in these amended plots. Unlike the previous experiments, a 20 cm layer of sludge mixture was simply surface applied. Results from our first year indicated that a C/N ratio of about 20 allowed for the most efficient ground cover, plot cover, weed repression, and overall stand quality (Fig. 2 and 3). By adding urea to lower C/N ratios to 10 was not economically feasible from a commercial standpoint, and C/N ratios of 30 and 40 did not give satisfactory results. However, slightly superior results were found by applying nitrogen in several split applications, to arrive at a C/N of 20. This supports the current practice of applying lower fertilizer rates more frequently as opposed to applying less frequent, larger doses. Further experiments should examine the use of slow-release fertilizers and the potential use of nutrient-rich secondary sludges, for C/N ratio adjustment.

A study on the revegetation of an abandoned sand pit has shown a dramatic improvement in the establishment and subsequent growth of the vegetative cover, with a heavy one-time application of deinking sludge. Among the reasons for this success is higher water and nutrient (especially nitrogen) retention by amended soils.

VII. FUTURE RESEARCH AVENUES

Owing to the increasing popularity of paper sludge use, there are several tangent projects that can contribute information on important advantages and disadvantages of medium–and long–term sludge applications. Not only are paper sludge applications of interest to the greenhouse, nursery, and container crop producers, field crop producers also have a direct interest in improving their soils with alternative sources of organic matter. Such practical applications will require research on rate of application and fertilizer requirements, in addition to specific evaluations of crop behavior and nutrition. One crop of specific interest is potato, because of ever decreasing levels of indigenous soil organic matter levels caused by heavy fertilizer management practices. Moreover, direct analysis of the various paper sludges of various origins necessitates study of the composition, heavy metal content, organic constituents, leaching, and effects on local flora and fauna. Microbiological studies of sludge decomposition, water

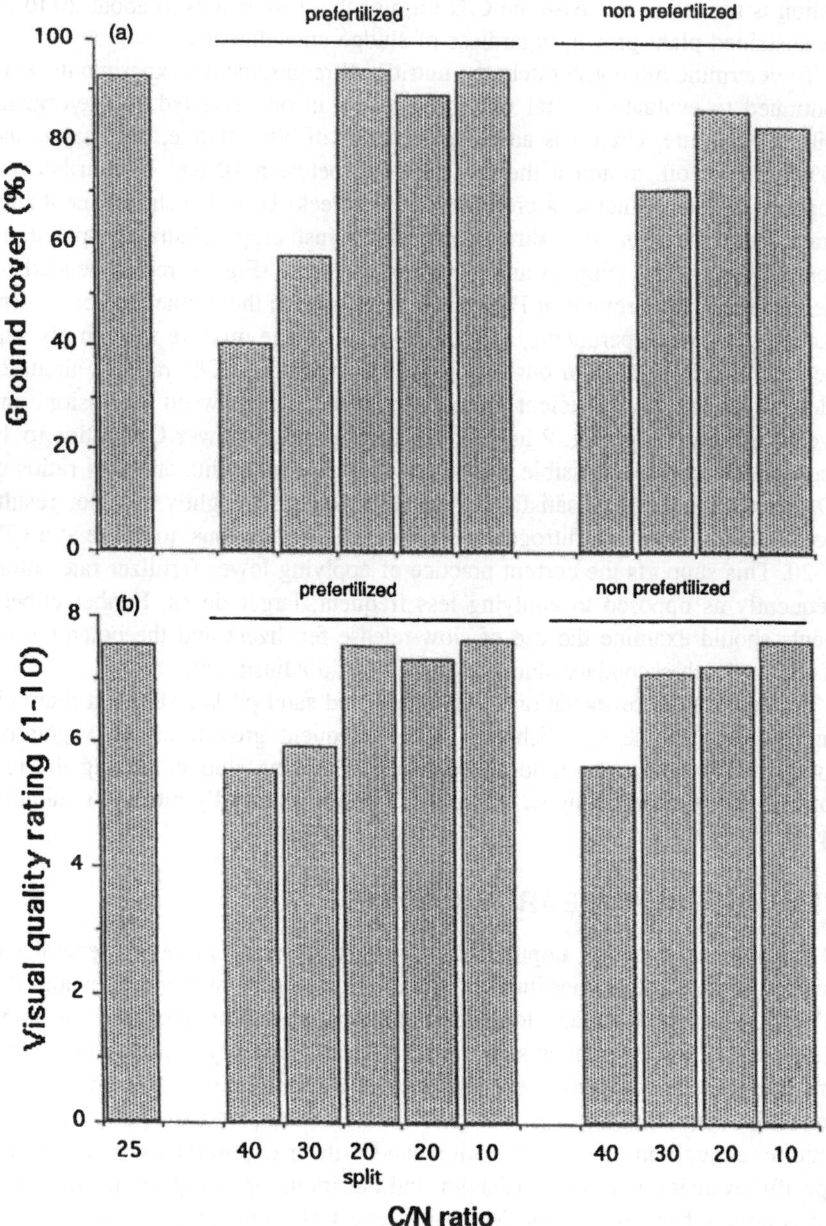

Figure 3 Evaluation of the percentage ground cover and stand quality of Kentucky bluegrass, bentgrass, creeping fescue, and ryegrass mixture cultivated on sludge amended soils with or without prefertilization to different C/N ratios in fall of establishment year.

and nutrient retention, and longer-term effects are also fundamental to a thorough understanding of this potentially wide-reaching organic amendment.

VIII. SUMMARY

As we become more aware of potential benefits from waste products that were once disposed of without consideration for recycling or reuse, paper sludges can become a valued source of organic matter. Owing to this environmentally friendly potential, this chapter has attempted to present an overview of alternative uses for sludges, their composition, effective application rates, potential benefits, and several pertinent research issues concerning uses, applications, and plant behavior in sludge amended soils. Many of the authors, researchers, and organizations cited in this chapter who have promoted sludges for years are only now seeing paper sludge research reach the forefront of soil amendment research. The fruits of these labors are being appreciated within many sectors of the agricultural and horticultural industries. However, continued research is required to identify specific application rates for specific crops, in addition to necessary fertilizer management and other potential benefits, e.g., disease and nutrition, in crops planted on sludge amended soil. Through continued research, the potential of paper sludges in agriculture will be further recognized, and in turn paper sludges will be more widely understood as an effective tool for improved soil conditioning.

REFERENCES

American Papermaker (Staff). Recycling sludge helps agriculture in Pope and Talbot landspreading program. *American Papermaker*, 46–47 (April 1992).

Barclay, H. G. Heavy metal content of deinking and other mill sludges: a literature review. *Pulp and Paper Research Institute of Canada. Miscellaneous Reports MR211.* Montreal, 1991.

Bellamy, K. L., and C. Chong. Paper sludge utilization in agriculture and container nursery culture. *J. Environ. Qual. 24*(6):1074–1083 (1995).

Bockheim, J. G., T. C. Benzel, L.-R. Lu, and D. A. Theil. Groundwater and soil leachate inorganic nitrogen in a Wisconsin Red Pine plantation amended with paper industry sludge. *J. Environ. Qual. 17*(4):729–734 (1988).

Camiré, C., L. Bérard, and A. Villeneuve. Relations station-nutrition-croissance de l'aulne crispé (*Alnus crispa* var. Mollis (ait.) Pursh) en plantation sur les bancs d'emprunt de la région LG-2, Baie James, Québec. *Naturaliste Can. (Rev. Écol. Syst.) 110*:185–196 (1983).

Canarutto, S., G. Petruzzelli, L. Lubrano, and G. Vigna Guida. How composting affects heavy metal content. *Biocycle*, 48–50 June (1990).

Chesire, M. V., and S. J. Chapman. Influence of the N and P status of plant material and of added N and P on the mineralization of 14C–Labelled ryegrass in soil. *Biol. Fert. Soils. 21:*166–170 (1996).

Chong, C. Recycling Ontario papermill sludge in nursery culture. *Proc. Pulpmill Waste Utilization in the Forest.* Clear Lake Ltd., Edmonton, 1993, pp. 59–60.

Chong, C., and R. A. Cline. Recycling organic wastes in ornamental crop culture. *The Grower 41*(1):6 (1991).

Chong, C., and R. A. Cline. Response of four ornamental shrubs to container substrate amended with two sources of raw paper mill sludge. *HortScience 28*(8):807–809 (1993).

Chong, C., R. A. Cline, and D. L. Rinker. Use of papermill sludge in container crop culture. *Landscape Trades. 10*(7):17–18 (1988).

Chong, C., R. A. Cline, and D. L. Rinker. Spent mushroom compost and papermill sludge as soil amendments for containerized nursery crops. *Proc. Int. Plant Prop. Soc. 37*:347–353 (1987).

CQVB (Centre québecois de valorisation de la biomasse) and MICT (ministère de l'industrie du commerce et de la technologie). La technologie des boues activés adopté aux effluents de l'industrie des pâtes et papiers. Fiche technique F-93-03 (1993).

Diehn, K. Recycling a deinking mill's waste through landspreading. Proceedings of *TAPPI Env. Conf.*, 739–46 (1991).

Diehn, K., and B. Zuercher. A waste management program for paper mill sludge high in ash. *TAPPI J.*, 81–86 (April 1990).

Dyer, K. L., W. R. Curtis, and J. T. Crews. Response of vegetation to various mulches used in surface mine reclamation in Alabama and Kentucky: 7-year case history. *Northeastern Forest Experimental Station. General Technical Report NE-93.* U.S. Dept. of Agriculture, Forest Services (1984).

Edwards, J. H. Recycling newsprint in agriculture. *Biocycle*, (1992).

Edwards, J. H., E. C. Burt, R. L. Raper, and D. T. Hill. Recycling newsprint on agricultural land with the aid of poultry litter. *Compost Science and Utilization. 1*(2):79–92 (1993).

Elkins, N. Z., L. W. Parker, E. Aldon, and W. G. Whitford. Responses of soil biota to organic amendments in strip mine spoils in northwestern New Mexico. *J. Environ. Qual. 13*(2):215–219 (1984.)

Enzor, S. B. Virginia fiber disposes of sludge by land application. *American Papermaker 8*:33–35 (1988).

Feagley, S., M. S. Valdez, and W. Hudnall. Bleached, primary papermill sludge effect on bermudagrass grown on a mine soil. *Soil Sci. 157* (1994a).

Feagley, S., M. S. Valdez, and W. Hudnall. Papermill sludge, phosporous, potassium, and lime effect on clover grown on a mine soil. *J. Environ. Qual. 23*:759–765 (1994b).

Ferguson, L. D. Deinking chemistry: part 1. *Tappi J. 76*:75–83 (1992).

Fierro, A., J. Norrie, A. Gosselin, and C. J. Beauchamp. Deinking sludge influences biomass, nitrogen and phosphorus status of several grass and legume species. *Can. J. Soil Sci.* (accepted, 1997).

Grossnickle, S. C., and C. P. P. Reid. Water relations of Engelmann spruce seedlings on a high-elevation mine site: an example of how reclamation can alter microclimate and edaphic conditions. *Reclamation and Revegation Research 3*:199–221 (1984).

Hatch, C. J., and R. G. Pepin. Recycling mill wastes for agricultural use. *Tappi J. 68*:70–73 (1985).

Henry, C. L. Nitrogen dynamics of pulp and paper sludge amendment to forest soils. *Soil Sci. Tech. 24*(3/4):417–425 (1991).

Honeycutt, C. W., L. M. Zibilske, and W.M. Clapham. Heat units for describing carbon mineralization and predicting net nitrogen mineralization. *Soil Sci. Soc. Am. J. 52*: 1346–1350 (1988).

Keenan, R. E., M. M. Sauer, M. S. P. H., and F. H. Lawrence. Examination of potential risks from exposure to dioxine in paper mill sludge used to reclaim abandoned Appalachian coal mines. *Chemosphere 18*(1–6):1131–1138 (1989).

Kraske, C. R., and I. J. Fernandez. Variability factors involved with land application of papermill sludge. *Maine Agricultural Experiment Station 138*:1–15 (1990).

Latua-Somppi, J., H. N. Tran, D. Barham, and M. A. Douglas. Characterization of deinking sludge and its ashed residue. *Pulp and Paper Canada 95*:381–385 (1994).

LRN (Land Resources Network Ltd.). Organic materials as soil amendments in reclamation: a review of literature. *Alberta Conservation and Reclamation Council Report No. RRTAC93-4*, 1993, p. 228.

Madsen, E. Determining in situ biodegradation. *Environ. Sci. Tech. 25*(10):1663–1673 (1991).

Melillo, J. M., J. D. Aber, and J. F. Muratore. Nitrogen and lignin control of hardwood leaf litter decomposition dynamics. *Ecology 63*:621–626 (1982).

NCASI. Pulp and papermill waste disposal by irrigation and land application. Stream Improvement Technical Bulletin No. 124. New York, 1959.

NCASI. Nature and environmental behavior of manufacturing-derived solid wastes of pulp and paper origin. Stream Improvement Technical bulletin No. 319. New York, 1979.

NCASI. The land application and related utilization of pulp and paper mill sludges. Technical Bulletin No 439. New York, 1984.

NCASI. Characterization of wastes and emissions from mills using recycled paper. Tech. Bulletin No. 613. New York, 1991.

NCASI (National Council for Air and Stream Improvement). Solid waste management and disposal practices in the U.S. paper industry. Technical Bulletin No. 641. New York, 1992.

Pichtel, J. R., W. A. Dick, and P. Sutton. Comparison of amendments and management practices for long-term reclamation of abandoned mine lands. *J. Environ. Qual. 23*: 766–772 (1994).

Prégant, G., and C. Camiré. Biomass production by alders on four abandoned agricultural soils in Québec. *Plant and Soil 897*:185–193 (1985a).

Prégant, G., and C. Camiré. Mineral nutrition, dinitrogen fixation, and growth of Alnus crispa and Alnus glutinosa. *Can. J. For. Res. 15*:855–861 (1985a).

Prégant, G., C. Camiré, J. A. Fortin, P. Arsenault, and J. G. Brouillette. Growth and nutritional status of green alder, jack pine, and willow in relation to site parameters of borrow pits in James Bay Territory, Quebec. *Reclamation and Revegation Research 6*:33–48 (1987).

Pridham, N. F., and R. A. Cline. Paper mill sludge disposal: completing the ecological cycle. *Pulp and Paper Canada 89*:2 (1988).

Recycled Paper News (RPN), Recycled paper and sludge: is recycling hurting the environment? *Recycled Paper News 2*(5):1–4 (1992).

Roberts, J. A., W. L. Daniels, J. C. Bell, and D. C. Martens. Tall Fescue production and nutrient status on southwest Virginia mine spoils. *J. Environ. Qual. 17*(1):55–62 (1988).

Schuman, G. E., and S. E. Belden. Decomposition of wood-residue amendments in revegetated bentonite mine spoils. *Soil Sci. Soc. Am. J. 55*:76–80 (1991).

Smith, J. A., G. E. Schuman, E. J. DePuit, and T. A. Sedbrook. Wood residue and fertilizer amendment on bentonite mine spoils I. Spoil and general vegetation responses. *J. Environ. Qual. 14*(4):575–580 (1985).

Smith, J. A., E. J. DePuit, and G. E. Schuman. Wood residue and fertilizer amendment on bentonite mine spoils II. Plant species responses. *J. Environ. Qual. 15*(4):427–435 (1986).

Smyser, S. Compost paying its way for paper producer. *Biocycle 23*(3):25–26 (1982).

Thacker, W. E. Silvicultural land application of wastewater and sludge from the pulp and paper industry. In *The Forest Alternative for Treatment and Utilization of Municipal and Industrial Wastes* (D. W. Cole, C. L. Henry, and W. L. Nutter, eds.). Univ. of Washington Press, Seattle, 1986.

Trépanier, L., J. Caron, S. Yelle, G. Thériault, J. Gallichand, and C. J. Beauchamp. Impact of deinking sludge amendment on agricultural soil quality. *Proc. TAAPI Int. Environ. Conf.*, May 1996.

Uresk, D. W., and T. Yamamoto. Growth of forbs, shrubs, and trees on bentonite mine spoil under greenhouse conditions. *J. Range Management 39*(2):113–117 (1986).

Watson, M. E., and H. A. J. Hoitink. Utilizing papermill sludge: longterm effects in stripmine reclamation. *Biocycle 26*(7–8):51–3 (1985).

West, R. C., and D. L. Bishop. Can pulpmill wood waste be used for land reclamation? Can. Forestry Service Technical Note No. 170 1987.

Zhang, X., A. G. Campbell, and R. L. Mahler. Newsprint pulp and paper sludge as soil additive/amendment for alfalfa and bluegrass: greenhouse study. *Commun. Soil Sci. Plant Anal. 24*:1371–1388 (1993).

Zibilske, L.M. Dynamics of nitrogen and carbon in soil during papermill sludge decomposition. *Soil Sci. 143*:26–33 (1987).

4

Use of Manures for Soil Improvement

Marcello Pagliai and Nadia Vignozzi *Ministry of Agriculture, Florence, Italy*

I. INTRODUCTION

Long-term intensive arable cultivation has negative effects on soil physical properties, particularly on soil structure, with resulting effects on soil erodibility and crop yields. The need to check the degradation of soil structure has caused farmers to consider reduced tillage management as an alternative to conventional tillage and the utilization of waste organic materials. Abandoning traditional farming rotations and adopting intensive monocultures, without applications of farmyard manure or organic materials to the soil, has decreased the soil organic matter content with evident degradation of soil structure. The resulting soil porosity conditions are often unfavorable for crop growth (Pagliai et al., 1983b, 1984, 1989; Shipitalo and Protz, 1987).

The gradual decrease in organic matter content in intensively cultivated soils, which may lead to deterioration of the soil physical status and possible erosion, is particularly worrying, especially in areas where climatic conditions cause the rapid decomposition of organic matter.

The decrease of soil organic matter control can therefore be prevented by additions of organic materials, which can also contribute to improved soil productivity. An abundant source of organic materials is animal manures from intensive livestock production. Application of animal manures to cultivated soils has been increasing steadily in the last decades because of the need for lower disposal costs and for the recycling of nutrient elements in the soil crop system. Some details of the main characteristics of some of these materials are reported in Table 1.

Few studies are available on the effects of long-term application of animal manures on soil physical properties that may influence soil productivity and prevent soil degradation. The agronomic utilization of animal manures may in-

Table 1 Main Characteristics of Some Animal Manures

Properties	Pig slurry	Cattle slurry	Poultry manure	Farmyard manure
pH	6.9	7.0	7.1	6.8
Water (%)	96.0	92.0	14.0	79.1
Organic matter[a]	50.0	52.0	34.8	64.1
Total N[a]	7.7	3.4	5.0	2.0
Total P[a]	2.5	1.0	2.2	0.8
Total K[a]	4.9	3.0	3.3	2.4
Ash[a]	22.2	28.0	37.0	25.0

[a]Expressed as % dry weight.
Source: Modified from Pagliai et al., 1987; Pagliai and Vittori Antisari, 1993.

duce changes in the physical conditions and chemical composition of the soil, and knowledge of these processes is necessary in order to obtain dependable application systems. To evaluate the impact of management practices on the soil environment it is necessary to quantify the modifications of soil structure. According to Greenland (1981), porosity is the best indicator of soil structure conditions. In fact, adequate "storage pores" (0.5–50 μm equivalent pore diameter (e. p. d.), which allows the storage of water for plants and microorganisms) and adequate "transmission pores" (continuous pores ranging from 50 to 500 μm e.p.d., which allow water movement and easy root growth) are necessary for crops. Until now, not only has the necessary proportion of these pores generally been inadequately defined but also there has been a lack of knowledge of how porosity relates to crop yield due to the difficulty of finding an adequate method that allows the complete characterization of soil porosity. Micromorphometric analysis overcomes some of the problems associated with other methods of porosity analysis, because porosity observed by means of an image analyzer in two-dimensional thin soil sections can be related to three-dimensional pore size by utilizing the stereology principle (Ringrose-Voase, 1991).

II. SOIL PORE SYSTEM

Pore space measurements are increasingly being used to quantify soil structure, because it is the size, shape, and continuity of pores that affect many of the important processes in soils (Lawrence, 1977). Possibilities now exist for improving shape models and for creating more accurate models of pore space in soils by using micromorphology coupled with image analysis of photographs of thin soil sections or of impregnated soil blocks (Murphy et al., 1977a,b; Pagliai et al., 1983a,b; Moran et al., 1989; Moran and McBratney, 1992). This technique has the advantage that the measurement and characterization of the pore space can be combined with a visual appreciation of the type and distribution of the pores.

Although it only analyzes 2-dimensional pictures, this method provides useful information on the complexity of pore patterns in soils, not obtainable using other common methods such as mercury intrusion, water retention, and nitrogen sorption. Nowadays, however, with the improvement of software programs, it is possible to apply a mathematical program of stereology to the image analysis in order to characterize soil porosity in three dimensions (Ringrose-Voase and Bullock, 1984; Ringrose-Voase, 1991). This micromorphometric method based on image analysis can be used not only on thin soil sections but also on polished faces of large soil blocks impregnated directly in the field with fairly cheap materials such as paraffin wax (Dexter, 1988), or plaster of Paris (FitzPatrick et al., 1985).

Parameters such as pore size distribution, pore shape, and relative position of aggregates and pores are very important for evaluating induced modifications of soil structure, e.g., by addition of organic materials (Pagliai et al., 1981).

The improvement of soil physical properties and particularly soil porosity may reflect on soil biological and biochemical activities such as, for example, enzyme activity, which is a fundamental property determining the biological fertility of soils. However, very few have considered their reciprocal correlation (Sequi et al., 1985), though it is of utmost importance, since the poral spectrum of soil plays a determining role in the possible habitats of microbes, as it determines hydric conditions, aeration conditions, trophic conditions, and relation between organisms (predation and competition) (Couteaux et al., 1988). Pagliai and De Nobili (1993) have demonstrated that important biological activity such as that of soil enzymes is positively correlated with the amount of pores ranging from 30 to 200 μm. The quantification not only of the changes of soil porosity but also the relation of such changes with chemical and biochemical properties may give useful data for the evaluation of the efficiency of management practices like soil tillage (Pagliai, 1987b), addition of waste organic materials, etc., in order to maintain soil fertility to prevent the soil resource from degradation phenomena.

For a thorough characterization of soil macropores, the main aspects to be considered are not only the pore shape but also the pore size distribution, especially of elongated continuous pores, because many of these pores directly affect plant growth by easing root penetration and storage and transmission of water and gases. For example, according to Russell (1978) and Tippkötter (1983), feeding roots need pores ranging from 100 to 200 μm to grow into. According to Greenland (1977), pores of equivalent pore diameter ranging from 0.5 to 50 μm are the storage pores, which provide the water reservoir for plants and microorganisms, while transmission pores ranging from 50 to 500 μm (elongated and continuous pores) are important both in soil–water–plant relationships and in maintaining good soil structure conditions. Damage to soil structure can be recognized by decreases in the proportion of transmission pores. Pores larger than 500 μm, called fissures, are useful for root penetration and for drainage especially in fine-textured soils. Adequate proportion of all these types of pores are needed for good soil quality and plant growth.

A. Microporosity

In a long-term field experiment established on a silty clay soil (Vertic Cambisol, according to the FAO world map legend) planted with corn, the application of pig slurry and farmyard manure increased the microporosity (Fig. 1). Pig slurries were surface applied by a liquid manure spreader at the end of May, and the addition rates were 100, 200, and 300 metric tons (MG) ha^{-1}; the farmyard manure (50 MG ha^{-1}) was incorporated into the soil at the ploughing time (winter) and the control soil was given chemical fertilizers. Sampling for porosity measurement was carried out at the end of September after ten years of manure application. The microporosity was measured by the combination of scanning electron microscopy and image analysis (Pagliai et al., 1983b; Pagliai and Vittori Antisari, 1993). All type of pores increased, but while the increase of regular and irregular micropores was approximately the same in all treated soils, the increase of elongated micropores was more relevant. In soil treated with pig slurry such an increase seemed to be related to the amount of slurry added to the soil. The increase of elongated micropores was associated with the new formation of microaggregates, and moreover the increase of pores in this size range was important, since these pores are related to the amount of water available to plants. There are few data in the literature about the effect of manure on microporosity, but similar conclusions were also draw by Schjønning et al. (1994) on a sandy loam soil.

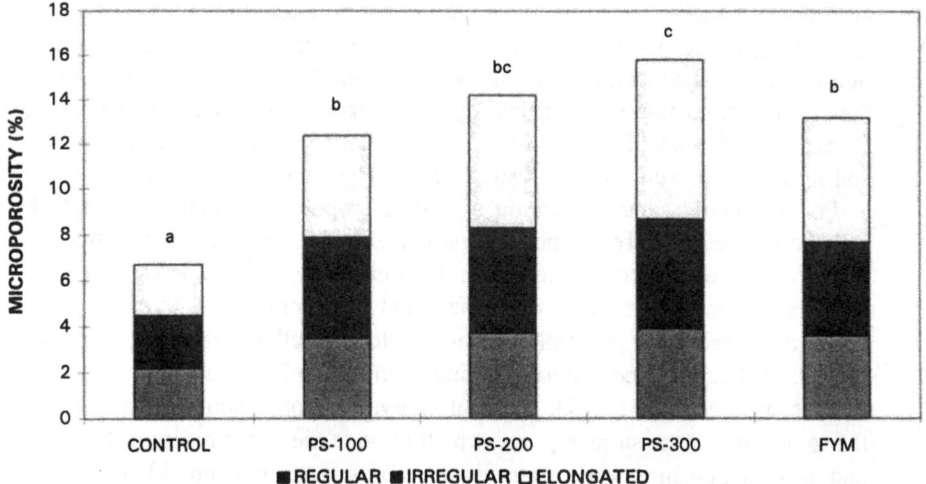

Figure 1 Effect of surface application of pig slurry (PS, 100, 200, and 300 MG ha^{-1}) and incorporation of farmyard manure (FYM) on soil microporosity (0.2–50 μm) on a silty clay soil expressed as a percentage of total area of pores per thin section. Total microporosity values followed by the same letter are not significantly different at the 0.05 level. (Part of these data are modified from Pagliai and Vittori Antisari, 1993.)

B. Macroporosity

In the same long-term field experiment on silty clay soil, mentioned above, total macroporosity (pores larger than 50 μm) increased in the soil treated with pig slurry and farmyard manure with respect to the control soil treated only with chemical fertilizers (Fig. 2). The increase of macroporosity was proportional to the amount of pig slurry added to the soil, while the effect of farmyard manure incorporation was equivalent to the application of 200 MG ha^{-1} of pig slurry. Differences between the control soils and soils treated with the highest amount of slurry (300 MG ha^{-1}) were highly significant (Pagliai and Vittori Antisari, 1993). This could be an important result, because this experimental field was located in an area of high density of pig livestock, and the soil was representative of this area. Therefore the agronomic utilization of such livestock effluents, besides the beneficial effect on the soil, resolves the great problem of the disposal of these wastes.

The effect of liquid manure (livestock effluents) on soil physical properties strongly depends on the time of landspreading. Figure 3 summarizes results obtained in the same field experiment where pig slurries were surface applied in February (instead of May) on the ploughed soil, approximately one month before the corn seeding. In comparison with Fig. 2 it is clear that, at the sampling time in September, i.e., in the period of corn ripening when good soil conditions

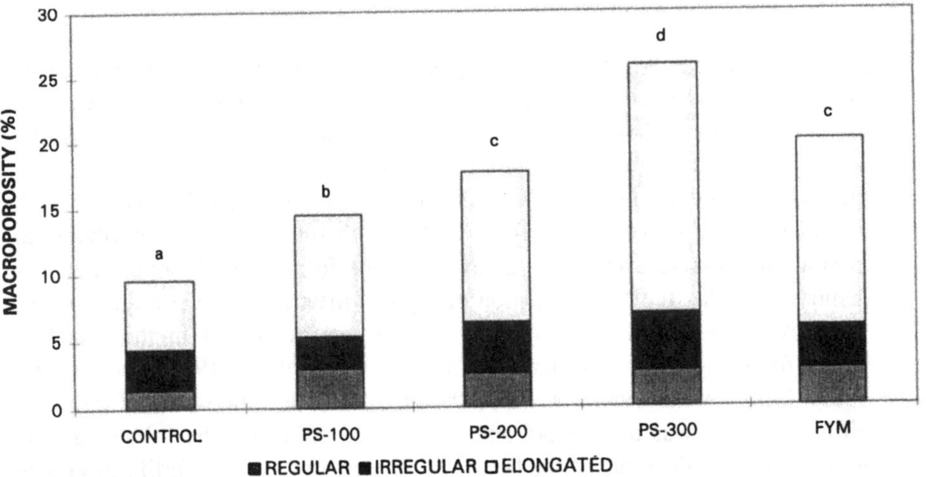

Figure 2 Effect of surface application of pig slurry (PS, 100, 200, and 300 MG ha^{-1}) and incorporation of farmyard manure (FYM) on soil macroporosity (>50 μm) on a silty clay soil expressed as a percentage of total area of pores per thin section. Total macroporosity values followed by the same letter are not significantly different at the 0.05 level. (Part of these data are modified from Pagliai and Vittori Antisari, 1993.)

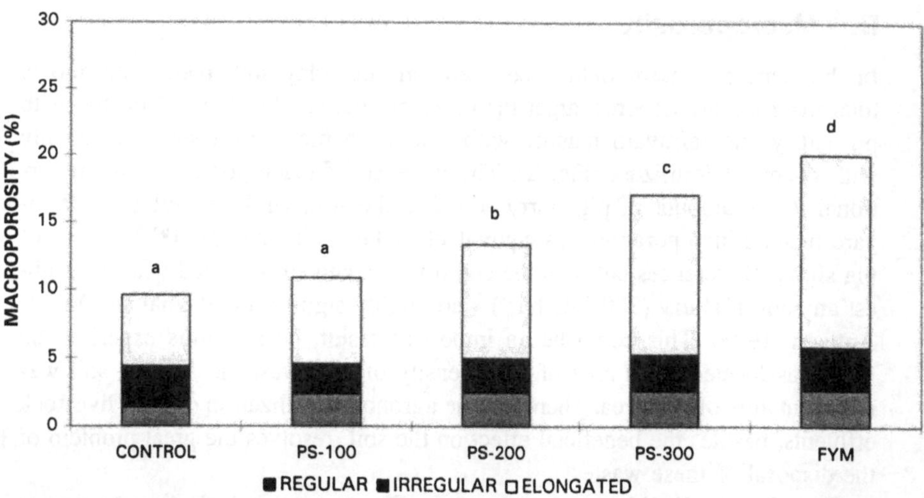

Figure 3 Effect of surface application of pig slurry (PS, 100, 200, and 300 MG ha^{-1}), applied in winter, and incorporation of farmyard manure (FYM) on soil macroporosity (>50 μm) on a silty clay soil expressed as a percentage of total area of pores per thin section. Total macroporosity values followed by the same letter are not significantly different at the 0.05 level. (Part of these data are modified from Pagliai and Vittori Antisari, 1993.)

are critical for crop development, the porosity in soil treated with pig slurry was lower than in soil where slurries were applied at the end of May. The lower addition rates did not show significant difference compared to the control, and the effect of the highest rate was lower than that of the farmyard manure. In May the soil temperature and humidity were optimal for biological activity, which was higher than in February, and the biological activity strongly contributed to improve soil structure through, for example, the formation of biopores. It was demonstrated that the decomposition of organic materials (like sewage sludges, livestock effluents) due to microbial activity was maximum in the first few weeks following their incorporation in soil (Tester et al., 1977; Terry et al., 1979). Pagliai et al. (1985) have also shown that the decomposition of livestock effluents in soil was quite rapid and the residual effect for the following crop was much lower than that of the farmyard manure. For this, the fertilization with animal slurries needs continuous applications of adequate rates of these materials. The amount of the addition rate depends on the type of soil and on its hydrological properties. The limiting factor to the utilization of a large amount of slurry is represented by the nitrogen losses by surface runoff or through the leaching along the soil profile (Brogan, 1981). This may cause a negative environmental impact due to the possible pollution of surface or ground waters.

Other limitations to the agronomic use of waste materials could be represented by the salinity of both soil and products used. Sartori et al. (1985) showed that the application of a compost, supplied by a municipal treatment plant, on a saline clay soil caused a strong decrease of soil porosity and thus leading, as a consequence, to a deterioration of soil physical properties. It is well known that compost and sewage sludges are characterized by relevant saline concentration (this aspect is developed in Chapter 7, 8 and 9 of this book). However, some animal manures may also create problems when applied to saline soil. On the contrary, when these manures are applied to non saline soil, they generally improve the soil physical properties.

Figure 4 summarizes results of laboratory trials to determine the effects of poultry manure applications to a nonsaline clay loam soil on soil porosity following wetting and drying cycles (Pagliai et al., 1987). In control samples the total porosity increased until the 11th wetting and drying cycles, and at the end of the 16th it showed a slight decrease. This means that the studied soil had a good suitability to soil structure regeneration. In poultry-manure treated samples the total porosity was higher than in the control samples. During the wetting and drying cycles it increased until the 11th cycle and then slightly decreased. This slight decrease of porosity after the 11th cycle in both treated

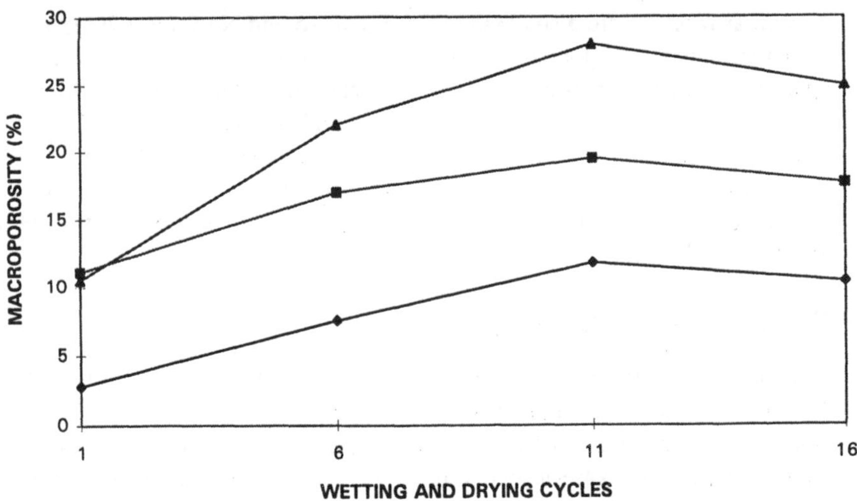

Figure 4 Effects of the addition of poultry manure (PM) on soil porosity, expressed as percentage of total area occupied by pores larger than 50 μm per thin section, following wetting and drying cycles. (Modified from Pagliai et al., 1987.) The poultry manure was applied at two rate levels: low rate 0.75% by weight of soil (both weight of poultry manure and soil were calculated on dry-matter basis) and high rate 7.5%.

and control samples may be ascribed to a decrease of the microbial activity in soil during the continuous repetition of wetting and drying cycles. Utomo and Dexter (1982) demonstrated the importance of microbial activity in the formation of soil aggregates during wetting and drying cycles, and the formation of soil aggregates is related to soil porosity. A comparison of data from the samples treated with the low rate and those treated with the high rate of poultry manure shows that the porosity in these samples was not proportional to the increase of the poultry manure added to soil. Therefore these data indicate that adding poultry manure to soils can improve the physical properties of soil, such as porosity. This parameter is very important for evaluating the effect of the application of waste materials on soil structure, and therefore it is also very important for evaluating the quality of these materials to be applied to soils.

C. Pore Shape and Size Distribution

Figure 5 represents an example of the variations of pore size and shape distribution in soil treated with organic materials compared to untreated soil. Generally, in soils supplied with animal manure the increase of elongated pores mainly involved those in the range of transmission pores (50–500 μm). Such an increase together with those of storage pores (<50 μm), was a clear sign of improved soil physical conditions and therefore of more favorable soil conditions for plant development.

Similar results were also obtained by Kladivko and Nelson (1979a), who reported an increase of porosity in the range of transmission pores following the application of sewage sludges on silty and clay loam soils, thus confirming an improvement of soil pore system. Such results also indicated the similar action of the organic matter contained in sewage sludges and in animal manure. However, several authors reported an increase of soil porosity and a decrease of bulk density after addition of animal manure (Khaleel et al., 1981; Anderson et al., 1990; Rose, 1991).

D. Type of Soil Structure

The variation of soil porosity following manure application directly reflects the soil structure. Indeed, the microscopic observations of thin sections, prepared from undisturbed soil samples collected in field experiments where the applications of manures were tested, generally reveal differences between samples of untreated soil and those treated with several types of manures. The increase of porosity and, overall, the increase of elongated pores in the range of transmission pores (50–500 μm) in soils given manures generally produced a subangular blocky structure homogeneously distributed along the Ap horizon (Fig. 6), which allowed, as already mentioned, more favorable conditions for plant development. The microscopic examinations of thin sections of treated soils also revealed the presence of organic matter such as wall coating of elongated

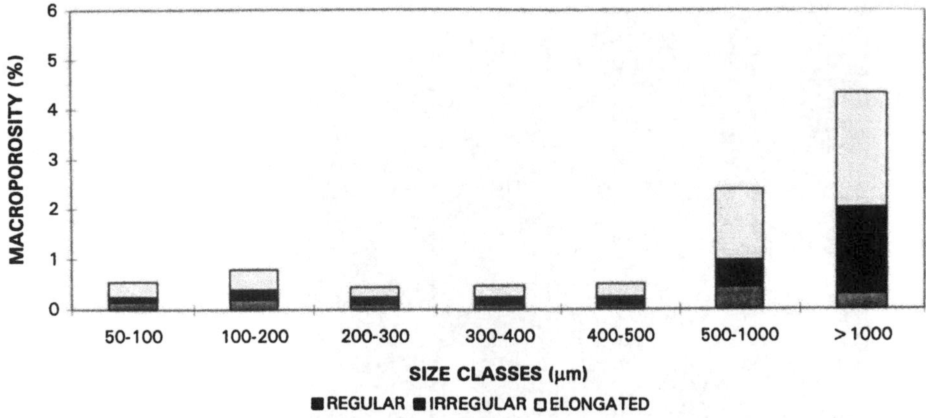

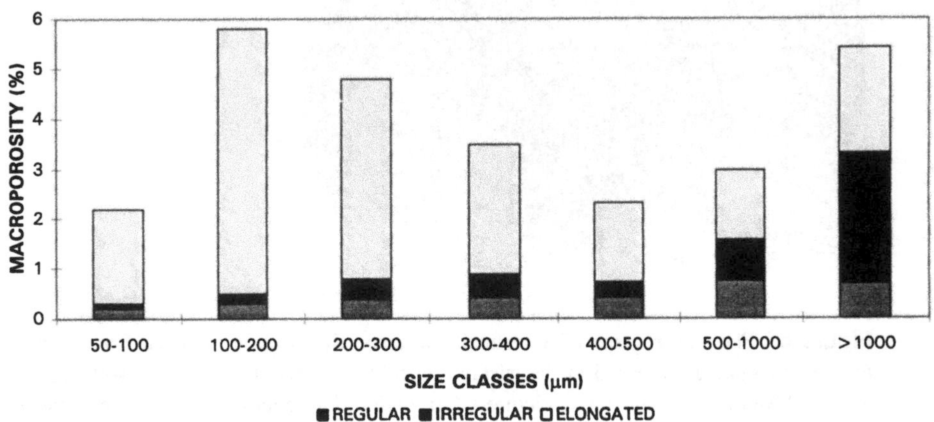

Figure 5 Pore size distribution of pores larger than 50 μm, determined by image analysis on thin section according to the equivalent pore diameter for regular and irregular pores and width for elongated pores, in control plots of the same silty clay soil of Fig. 2 and in plots treated with 300 MG ha^{-1} of pig slurry applied at the end of May. (Modified from Pagliai and Vittori Antisari, 1993.)

pores (Fig. 7). Such organic matter coatings could effectively seal pores from the adjacent soil matrix, stabilizing soil structure against the destructive action of water, and assuring the functionality of the pores. In untreated soils the lower porosity, especially in the size range of transmission pores, produced a more compact soil structure with larger aggregates rather compact inside. In many

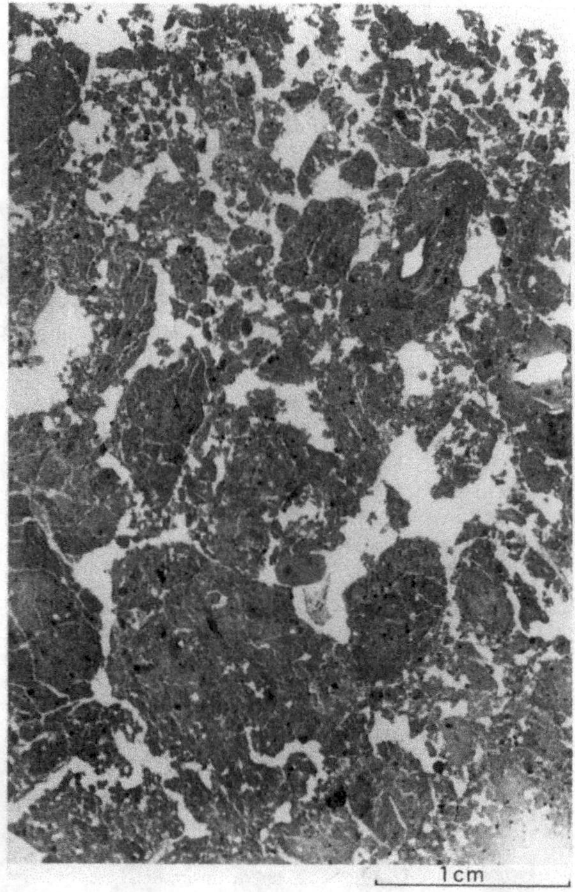

Figure 6 Photomacrograph of a vertically oriented thin section prepared from undisturbed soil samples collected in the surface layer (0–10 cm) of a silty clay soil treated with 300 MG ha^{-1} of pig slurry. Picture taken under plain polarized light (pores appear white). A subangular blocky structure is evident.

cases, such as clay and silty loam soils, the large aggregates were separated by elongated pores parallel to the soil surface originating a platy structure (Fig. 8).

III. AGGREGATE STABILITY

The binding together of individual soil particles forms aggregates, and the strongly bounded aggregates give rise to what is known as well-structured soil that has greater resistance to the destructive forces of water (raindrop impact, etc.) and consequently has more resistance to erosion and degradation. Gener-

Figure 7 Photomicrograph of a vertically oriented thin section prepared from undisturbed samples collected in the surface layer (0–10 cm) of a sandy loam soil treated with farmyard manure. Plain polarized light. The presence of organic matter such as wall coating of elongated pores can be noticed.

ally, the other soil physical properties, such as porosity, water movement, and ease of root penetration will increase with increasing aggregation and aggregate stability.

The intensive crop systems of modern-day agriculture, based on continuous conventional tillage, complete vegetation removal, and copious administration of chemical fertilizers, can decrease the degree of aggregation at the surface and, on more vulnerable soils, can completely destroy surface soil structure. In these situations it is important to increase not only the aggregation but also the

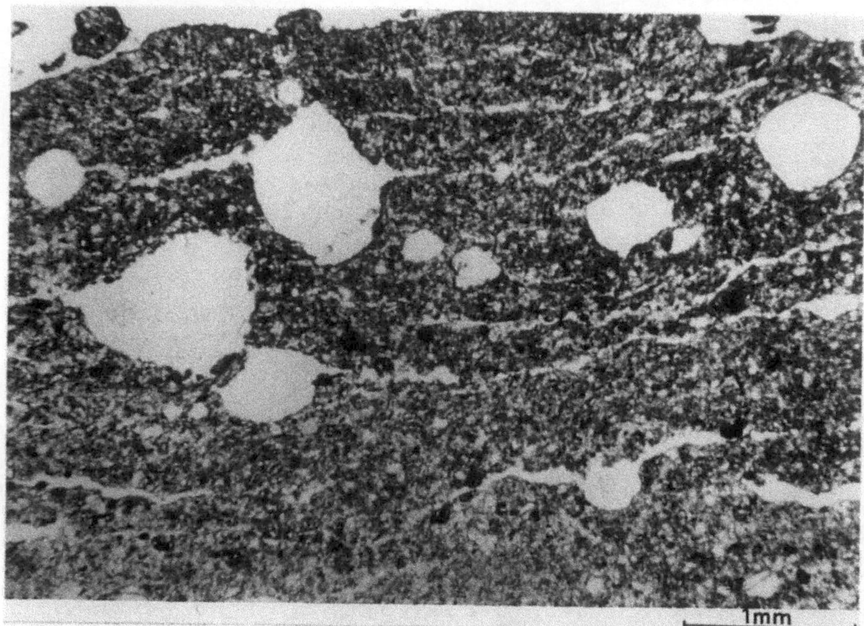

Figure 8 Photomicrograph of a vertically oriented thin section prepared from undisturbed soil samples collected in the surface layer (0–10 cm) of silty clay soil not treated with organic manures (control soil). Plain polarized light. More compact soil structure is visible than in Fig. 6. The thin elongated pores are parallel to the soil surface, thus originating a rather compact platy structure. These pores are not continuous in a vertical sense and therefore not useful for water infiltration. The rounded pores are originated by entrapped air during the drying processes, and the presence of such pores is an indicator of soil structure degradation.

stability of the aggregates against the destructive forces of water. Increasing aggregation is very important on finer-textured soils such as clays, clay loams, silty clays, and silt loams. In these soils, a good water infiltration and drainage is possible if clay and silt particles are bound together into aggregates.

The addition of manures to soil has been found to be an effective management practice not only to increase the aggregation but also to increase the aggregate stability. Figure 9 shows the increase of aggregate stability in a silty clay soil following addition of pig slurry and farmyard manure. As was the case for macroporosity, the highest increase in aggregate stability was found in soils supplied with 300 MG ha^{-1} of pig slurry in May. The winter applications of pig slurry were less efficacious in improving aggregate stability. The same trend was also observed in a sandy loam soil amended with sewage sludges, compost, and farmyard manure (Pagliai et al., 1981). Increases in aggregation and aggregate stability following long-term application of animal

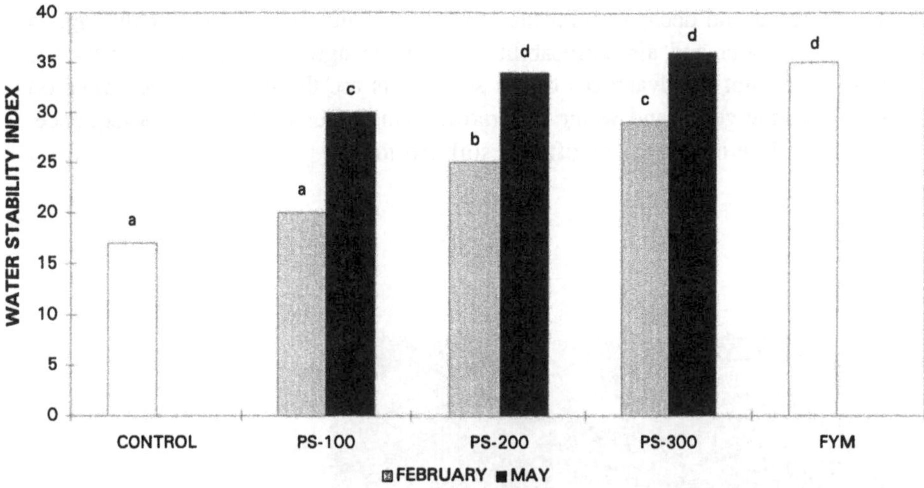

Figure 9 Effect of surface application of pig slurry (PS) in February and in May and incorporation of farmyard manure (FYM) before ploughing on water stability index determined by wet-sieving. Water stability index values followed by the same letter are not significantly different at the 0.05 level. (M. Pagliai, 1993, personal communication.)

manure on sandy loam and silt loam soils were also reported by Darwish et al. (1995). Similar results were summarized by Clapp et al. (1986) following the addition of organic materials such as sewage sludge on different types of soils. The magnitude of the improvement of aggregate stability and of the other physical properties depends not only on the amount of organic matter but also on the type and quality of organic materials present in the manures added to the soil. The most efficacious in producing such an improvement is traditional farmyard manure, but the application of livestock effluents can also improve in the same way the aggregate stability even though their action is less long lasting than that of farmyard manure (Ekwue, 1992; Pagliai and Vittori Antisari, 1993). This is because of the rapid microbial degradation of these materials in soil.

IV. SOIL CRUSTING

Soil crusts are specific physical modifications in the topsoil that can be formed following natural events like raindrop impact, which cause the mechanical destruction of surface soil aggregates and the dispersion of clay materials. In the following drying processes, the formation of hard thin layers occurs. Surface crusts are very widespread especially in cultivated loam and silty loam soils of arid and semiarid regions. Their thickness usually ranges from 0.5 mm to 5 cm. When dry, these features are more compact, hard, and brittle than underlying soil material and decrease both the size and the number of pores, reducing, in this way, water and air permeability. From the agronomic point of view the most important disadvantages of the soil crusts are the influence they have on seedling emergence and water infiltration. This latter leads, as a consequence, to an increase of surface runoff and soil erosion.

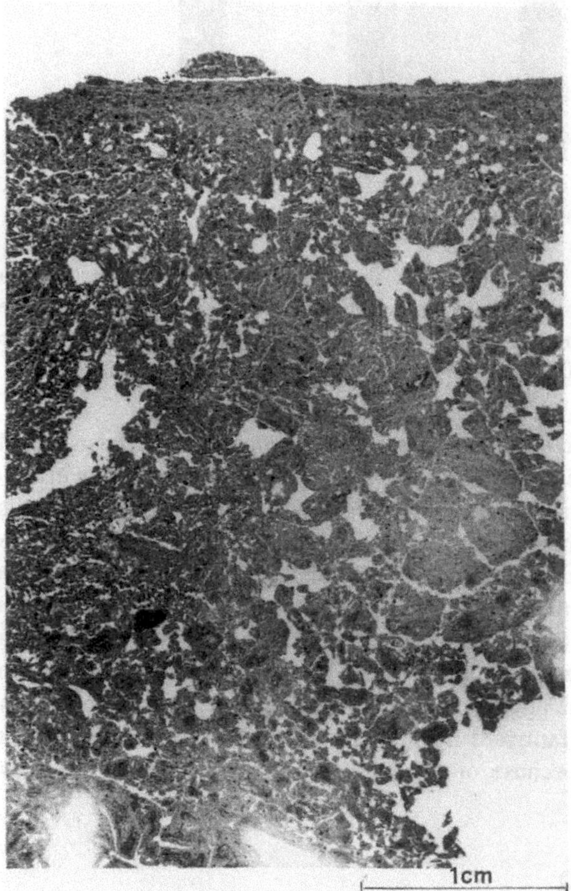

(a)

Figure 10 Photomacrograph of a vertically oriented thin section prepared from undisturbed soil samples collected in the surface layer (0–6 cm) of a silty clay soil. (a) Control soil; (b) soil treated with with 300 MG ha^{-1} of pig slurry. Plain polarized light. A well developed surface crust is evident in the top of the control soil. It is also evident that in treated soil the addition of organic manures can reduce the surface crust formation.

Many results reported in the international literature (Pagliai, 1987a; Pagliai et al., 1985, 1987) show that the formation of soil crusts can be reduced or prevented by the landspreading of livestock effluents (Fig. 10) or the application of farmyard manure. This preventive action can be ascribed both to the organic matter that as said, increases the aggregate stability, and to the organic materials such as straw fragments that remain in the soil surface reducing the direct action of raindrop impact on surface soil aggregates. These organic materials could also break existing surface soil crusts reducing their compactness and improving water intake rates (Fig. 11).

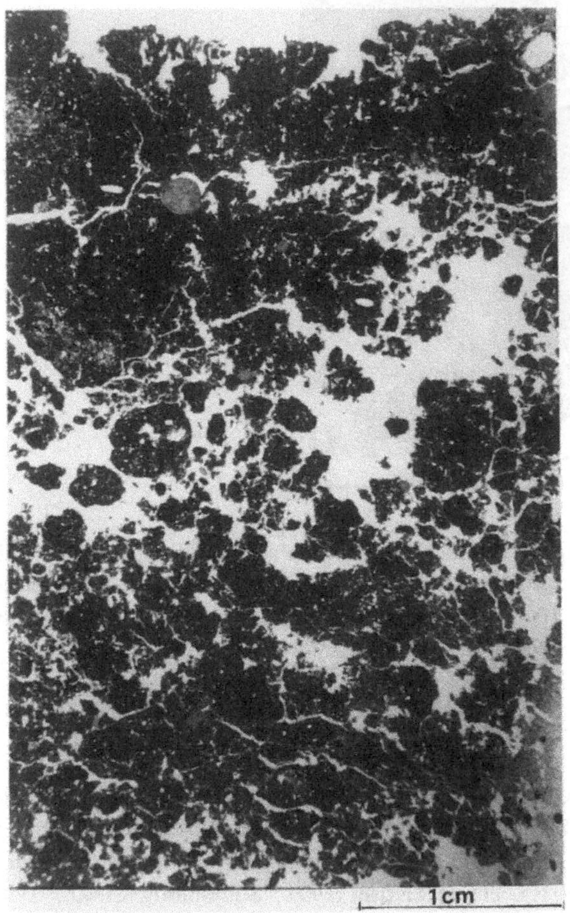

(b)

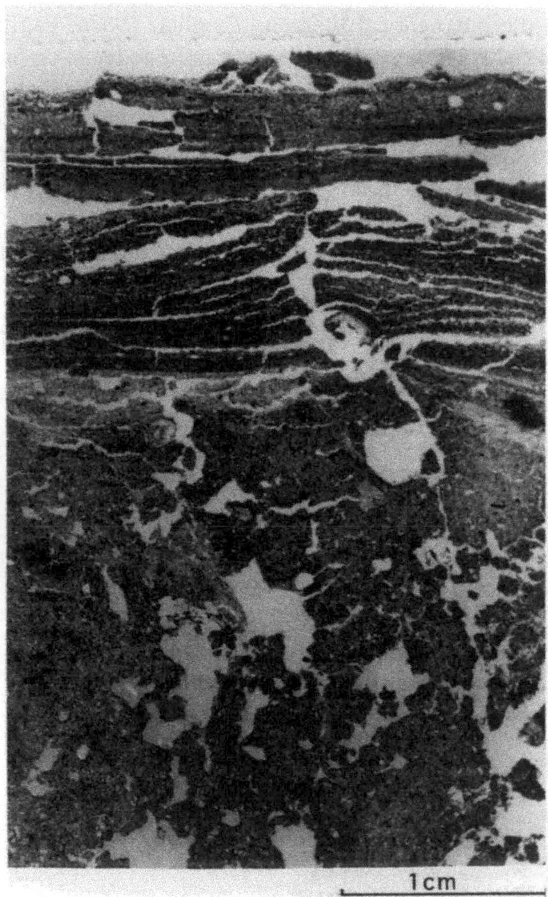

Figure 11 Photomacrograph of a vertically oriented thin section prepared from undisturbed soil samples collected in the surface layer (0–6 cm) of a silty clay soil treated with 100 MG ha^{-1} of pig slurry. Plain polarized light. The rest of the organic materials that break the layers of the surface crust can be noticed.

V. WATER RETENTION

The increase of microporosity in the range of storage pores (0.5–50 μm) in soil given organic material like animal manure leads, as a consequence, to an increase of water retention at both field capacity and wilting point and therefore the availability of water for plants (Rose, 1991; Schjønning et al., 1994; Giusquiani et al., 1995). The increase of water retention can be ascribed, first of all, to the improvement of the pore system in soil, which leads to better soil structure conditions, but also to the water adsorption capacity of organic matter (Metzger and Yaron, 1987).

VI. WATER MOVEMENT

Studies on the effects of soil manure application on water movement are relatively few, owing to the difficulty and time consumption of such determinations. Moreover, results present considerable variability. However many papers report that generally in well structured soils the hydrological properties are good. Therefore the improvement of structural conditions following manure application can lead, as a consequence, to an improvement of water infiltration. Specific studies reported by Metzger and Yaron (1987) clearly show a close correlation between the increase of saturated hydraulic conductivity and the improvement of aggregate stability and porosity in soils amended with organic materials like animal manure and sewage sludge. In some cases it has been shown that the addition of organic materials with hydrorepellent substances can decrease the unsaturated hydraulic conductivity (Metzger and Yaron, 1987).

Water infiltration through the soil surface is strongly dependent on the hydraulic conductivity and increases following the addition of manures. This is an important finding, because a low water infiltration rate in the case of heavy rains leads, as a consequence, to flooding or runoff and erosion, depending on the slope. Kladivko and Nelson (1979b) described a reduction of runoff, and consequently of soil erosion following the administration of organic materials. Similar results were found by Pagliai et al. (1983a), which showed the importance of manure applications to improve aggregate stability, to reduce the formation of surface crusts, and therefore to increase water infiltration.

VII. SOIL STRENGTH, PENETRATION RESISTANCE, SOIL COMPACTION, BEARING CAPACITY

The effects of manure application on the other soil physical properties have been less investigated partly because such properties are strictly connected with the structural condition determined by the principal properties, discussed above, like porosity, aggregate stability, water retention, and water movement.

Generally, the addition of organic manures to soil significantly reduced penetration resistance when compared with control and chemical-fertilizer amended soils (Tester, 1990). The same trend was observed for the modulus of rupture, which decreased with organic matter application, especially in more cohesive soils, i.e., silt loam and clay loam soils (Darwish et al., 1995), thus indicating a reduction in soil cohesion and in the susceptibility of soil to compaction. Decrease in soil strength with organic matter application was obtained in several studies carried out with soils amended with various types of organic matter additions (Ekwue, 1992). However, other studies have reported an increase of soil strength with organic matter addition (Schjønning et al., 1994) including a soil fertilized with farmyard manure which increased its cohesion with decreased water potential.

VIII. CONCLUSIONS

Considering soil physical parameters, such as porosity, soil structure, water retention, and water movement in soil, it can be concluded that generally the administration of manures improves the soil physical properties. Particularly, the addition of such materials causes an increase of "storage" (0.5–50 μm) and "transmission" (50–500 μm) i.e., pores that are necessary for the storage of water for plants and microorganisms and that allow water movement and root growth. Moreover, the application of manures improves the aggregate stability and reduces the formation of surface crusts. The maintenance of good soil structural conditions can be carried out not only with traditional farmyard manure but also with manures derived from livestock effluents. In this case continuous and moderate applications are necessary.

For a correct and efficient organic fertilization with manures it is fundamentally important to take into consideration the pedological environment in which such a fertilization is performed. The above-mentioned positive effects can at times be transformed into the dangerous phenomena of soil degradation. The salinity, for example, is one of the aspects that must not be undervalued in the additions of manures to soils.

REFERENCES

Anderson, S. H., Gantzer, C. J., and Brown, J. R. (1990). Soil physical properties after 100 years of continuous cultivation. *Journal of Soil and Water Conservation 45*: 117–121.

Brogan, J. C., ed. (1981). *Development in Plant and Soil Sciences.* Nitrogen Losses and Surface Runoff from Landspreading of Manures, Vol. 2 (J. C. Brogan, ed.). Martinus Nijhoff/Dr. W. Junk, London, p. 471.

Clapp, C. E., Stark, S. A., Clay, D. E., and Larson, W. E. (1986). Sewage sludge organic matter and soil properties. In *The Role of Organic Matter in Modern Agriculture* (Y. Chen and Y. Avnimelech, eds.). Martinus Nijhoff, Dordrecht, pp. 209–253.

Couteaux, M. M., Faurie, G., Palka, L., and Steimberg C. (1988). La relation prédateur proie (protozoaires–bactéries) dans le sols: rôle dans la régulation des populations et conséquences sur le cycles du carbone et de l'azote. *Rev. Ecol. Biol. Sol. 25*: 1–31.

Darwish, O. H., Persaud, N., and Martens, D. C. (1995). Effect of long-term application of animal manure on physical properties of three soils. *Plant and Soil 176*: 289–295.

Dexter, A. R. (1988). Advances in characterization of soil structure. *Soil Tillage Res. 11*: 199–238.

Ekwue, E. I. (1992). Effect of organic and fertilizer treatments on soil physical properties and erodibility. *Soil Tillage Res 22*: 199–209.

FitzPatrick, E. A., Makie, L. A., and Mullins, C. E. (1985). The use of plaster of Paris in the study of soil structure. *Soil Use Manag. 1*: 70–72.

Guisquiani, P. L., Pagliai, M., Gigliotti, G., Businelli, D., and Benetti, A. (1995). Urban waste compost: effects on physical, chemical and biochemical soil properties. *J. Environ. Qual. 24*: 175–182.

Greenland, D. J. (1977). Soil damage by intensive arable cultivation: temporary or permanent? *Philos. Trans. R. Soc. London 281*: 193–208.

Greenland, D. J. (1981). Soil management and soil degradation. *J. Soil Sci. 32*: 301–322.

Khaleel, R., Reddy, K. R., and Overcash, M. R. (1981). Changes in soil physical properties due to organic waste applications: a review. *J. Environ. Qual. 10*: 133–141.

Kladivko, E. J., and Nelson, D. W. (1979a). Changes in soil properties from application of anaerobic sludge. *J. Water Pollution Control Federation 51*: 325–332.

Kladivko, E. J., and Nelson, D. W. (1979b). Surface runoff from sludge-amended soils. *Jo. Water Pollution Control Federation 51*: 100–110.

Lawrence, G. P. (1977). Measurement of pore size in fine textured soil: a review of existing techniques. *J. Soil Sci. 28*: 527–540.

Metzger, L., and Yaron, B. (1987). Influence of sludge organic matter on soil physical properties. *Advances in Soil Science 7*: 141–163.

Moran, C. J., McBratney, A. B., and Koppi, A. J. (1989). A rapid method for analysis of soil macropore structure. I. Specimen preparation and digital binary image production. *Soil Sci. Soc. Am. J. 53*: 921–928.

Moran, C. J., and McBratney, A. B. (1992). Acquisition and analysis of three-component digital images of soil pore structure. II. Application to seed beds in a fallow management trial. *J. Soil Sci. 43*: 551–566.

Murphy, C. P., Bullock, P., and Turner, R. H. (1977a). The measurement and characterization of voids in soil thin sections by image analysis: Part I. Principles and techniques. *J. Soil Sci. 28*: 498–508.

Murphy, C. P., Bullock, P., and Biswell, K. J., (1977b). The measurement and characterization of voids in soil thin sections by image analysis: Part II. Applications. *J. Soil Sci. 28*: 509–518.

Pagliai, M. (1987a). Effects of different management practices on soil structure and surface crusting. *Soil Micromorphology* (N. Fedoroff, L. M. Bresson, and M. A. Courty, eds.). AFES, Paris, pp. 415–421.

Pagliai, M. (1987b). Micromorphometric and micromorphological investigations on the effect of compaction by pressures and deformations resulting from tillage and wheel traffic. *Soil Compaction and Regeneration* (G. Monnier and M. J. Goss, eds.). A. A. Balkema, Rotterdam, pp. 31–38.

Pagliai, M., and De Nobili, M. (1993). Relationships between soil porosity, root development, and soil enzyme activity in cultivated soils. *Geoderma 56*: 243–256.

Pagliai, M., and Vittori Antisari, L. (1993). Influence of waste organic matter on soil micro and macrostructure. *Bioresource Technology 43*: 205–213.

Pagliai, M., Guidi, G., La Marca, M., Giachetti, M. and Lucamante, G. (1981). Effect of sewage sludges and composts on soil porosity and aggregation. *J. Environ. Qual. 10*: 556–561.

Pagliai, M., Bisdom, E. B. A., and Ledin, S. (1983a). Changes in surface structure (crusting) after application of sewage sludges and pig slurry to cultivated agricultural soils in northern Italy. *Geoderma 30*: 35–53.

Pagliai, M., La Marca, M., and Lucamante, G. (1983b). Micromorphometric and micromorphological investigations of a clay loam soil in viticulture under zero and conventional tillage. *J. Soil Sci. 34*: 391–403.

Pagliai, M., La Marca, M., Lucamante, G., and Genovese, L. (1984). Effects of zero and conventional tillage on the length and irregularity of elongated pores in a clay loam soil under viticulture. *Soil Tillage Res 4*: 433–444.

Pagliai, M., La Marca, M., and Lucamante, G. (1985). Relationship between soil structure and time of landspreading of pig slurry. *Long-term Effects of Sewage Sludge and Farm Slurries Applications* (J. H. Williams, G. Guidi, and P. L'Hermite, eds.). Elsevier, London, pp. 45–56.

Pagliai, M., La Marca, M., Lucamante, G. (1987). Changes in soil porosity in remoulded soils treated with poultry manure. *Soil Science 144*: 128–140.

Pagliai, M., Pezzarossa, B., Mazzoncini, M., and Bonari, E. (1989). Effects of tillage on porosity and microstructure of a loam soil. *Soil Technology* 2: 345–358.

Ringrose-Voase, A. J. (1991). Micromorphology of soil structure: description, quantification, application. *Aust. J. Soil Res. 29*: 777–813.

Ringrose-Voase, A. J., and Bullock, P. (1984). The automatic recognition and measurement of soil pore types by image analysis and computer programs. *J. Soil Sci. 35*: 673–684.

Rose, D. A. (1991). The effect of long-continued organic manuring on some physical properties of soils. *Advances in Soil Organic Matter Research*, Special Publication No. 90 (W. S. Wilson, ed.). Royal Society of Chemistry, Cambridge, pp. 197–205.

Russel, E. W. (1978). Arable agriculture and soil deterioration. Transactions of the 11th International Congress of Soil Science, 19–27 June 1978 at University of Alberta, Edmonton, Vol. 3. Canadian Society of Soil Science, Alberta, pp. 216–227.

Sartori, G., Ferrari, G. A., and Pagliai, M. (1985). Changes in soil porosity and surface shrinkage in a remoulded saline clay soil treated with compost. *Soil Science 139*: 523–530.

Schjønning, P., Christensen, B. T., and Carstensen, B. (1994). Physical and chemical properties of a sandy loam receiving animal manure, mineral fertilizer, or no fertilizer for 90 years. *European J. Soil Sci. 45*: 257–268.

Sequi, P., Cercignani, G., De Nobili, M., and Pagliai, M. (1985). A positive trend among two soil enzyme activities and a range of soil porosity under zero and conventional tillage. *Soil Biol. Biochem. 17*: 255–256.

Shipitalo, M. J., and Protz, R. (1987). Comparison of morphology and porosity of a soil under conventional and zero tillage. *Can. J. Soil Sci. 67*: 445–456.

Terry, R. E., Nelson, D. W., and Sommers, L. F. (1979). Decomposition of anaerobically digested sewage sludge as affected by soil environmental conditions. *J. Environ. Qual. 8*: 342–346.

Tester, C. F., Sikora, C. J., Taylor, J. H., and Parr, J. F. (1977). Decomposition of sewage sludge compost in soil: 1. Carbon and nitrogen transformation. *J. Environ. Qual. 6*: 459–463.

Tester, C. F. (1990). Organic amendment effects on physical and chemical properties of a sandy soil. *Soil Sci. Soc. Am. J. 54*: 827–831.

Tippkötter, R. (1983). Morphology, spatial arrangement and origin of macropores in some hapludalfs, West Germany. *Geoderma 29*: 355–371.

Utomo, W. M., and Dexter, A. R. (1982). Changes in soil aggregate water stability induced by wetting and drying cycles in non-saturated soil. *J. Soil Sci. 33*: 623–637.

5
Biosolids and Their Effects on Soil Properties

Alan Olness *Agricultural Research Service, U. S. Department of Agriculture, Morris, Minnesota*

C. E. Clapp *Agricultural Research Service, U. S. Department of Agriculture, St. Paul, Minnesota*

Ruilong Liu *University of Minnesota, St. Paul, Minnesota*

Antonio J. Palazzo *Geochemical Sciences Division, U.S. Army, Hanover, New Hampshire*

I. INTRODUCTION

The global population, now estimated at about 5.8 billion, is expected, barring some catastrophe, to reach 10 billion within the next 20 or so years. An increasing population in and of itself would pose growing pains if it were uniformly distributed over the land areas. However, at the same time that the global population has been increasing exponentially, much of that population has been concentrating in large urban centers. This continuing concentration creates serious waste problems.

Food needs alone lead to export from production areas and import into rather dense urban centers. Wastes produced as a result of the natural consumption process, historically having little value, have been discharged as sewerage into waterways and the oceans. Initially, the volumes were small and caused little environmental perturbations. However, as urban centers and populations continue to expand, these waste volumes have increased enormously.

Global industrialization also contributes to the waste disposal problem through the use of large amounts of water in a variety of industrial processes and its eventual elimination. The water supply has been used as a means of eliminating industrial waste in the form of soluble metals, organic and inorganic solvents, and other residues.

Contribution from the USDA Agricultural Research Service, North Central Soil Conservation Research Laboratory, Morris, Minnesota; USDA-ARS, Soil and Water Management Research Unit and Department of Soil, Water and Climate, St. Paul, Minnesota; and U.S. Army Cold Regions Research and Engineering Laboratory, Hanover, New Hampshire.

Recognition of the hazards of such practices has led to significant reduction but not elimination of most of the more toxic materials from the waste stream. Still, disposal of municipal sewage wastes remains an important dilemma for many large urban areas. One option for sewage biosolids utilization is land application, an ecologically appealing option in that this is the natural method of disposal in the absence of urban concentration. Land application is potentially very beneficial in that it returns fertility in the form of mineral nutrients to the soil. However, land application has some important limitations, and formulation of general guidelines for this purpose is straightforward.

II. PRACTICAL LIMITS OF SEWAGE WASTES AND OTHER BIOSOLIDS

With regard to the application of biosolids (municipal sewage sludges and composts) to agricultural lands as a means of utilizing waste, three approaches seemed prominent as research with these materials has progressed. The first of these is the yield approach. Most experiments with land application of sewage sludges have shown favorable crop responses over a brief period of time. In early tests, application rates that resulted in yield gains were regarded as acceptable and those that seemed to have no adverse effects on crop yields were regarded as permissible (Keeney et al., 1975). This attitude was expressed in rates of sewage sludge applied in various field tests; these rates often exceeded 100 Mg ha^{-1}. Flaws in this approach were quickly recognized when application rates that resulted in yield gains led to unacceptable accumulations of metals such as cadmium (Cd), zinc (Zn), nickel (Ni), and copper (Cu). Only a few long-term experiments are still being monitored (Clapp et al., 1994; Peterson et al., 1994; Linden et al., 1995). Emphasis should be placed on availability of nitrogen (N) and phosphorus (P) and uptake of metals after sludge application has stopped. Additional advice regarding long-term applications can be extrapolated from soil studies and application of livestock wastes.

Later, application rates were deemed acceptable if the extractable levels of potentially toxic metals did not increase beyond threshold levels. By maintaining the extractable levels of metals below threshold levels it was expected that accumulations in edible portions of vegetables greater than World Health Organization standards (FAO, 1972) would be avoided. Flaws in the second attitude were exposed with observations of differential varietal accumulations of metals (for example, Harrison, 1986; Yuran and Harrison, 1986; Xue and Harrison, 1991; Chaney, 1994) and the recognition that interactions of various metals and nonmetals could effect serious and near permanent damage to the environment. Soil and plant species interactions markedly affected plant uptake and relative toxicities of Zn, Cu, Ni, and Cd (Bingham et al., 1975; Kirkham, 1975; Mitchell et al., 1978; Mullins and Sommers, 1986; Smith, 1994). For example, Chang and Donald (1952) noted little relationship between extractable or total soil Ni content, but guava (*Psidium guajava* L.) leaf Ni concentration increased as the pH of the soil became less acidic (Fig. 1).

The third attitude might be regarded as an economic approach; it consists of matching exports of metals with their imports in the form of biosolids. The amount of metal applied is equal to that removed by the crop. The economic approach can be summarized as 1.

$$Y^*C_{ex} = R^*(C_{im} - k) \tag{1}$$

where

Y = Yield of crop removed from the land
C_{ex} = Concentration of the element of interest in the exported crop yield
R = Acceptable application rate of the biosolid
C_{im} = Concentration of the element of interest imported in the biosolid
k = other loss mechanisms such as erosion, volatilization, etc.

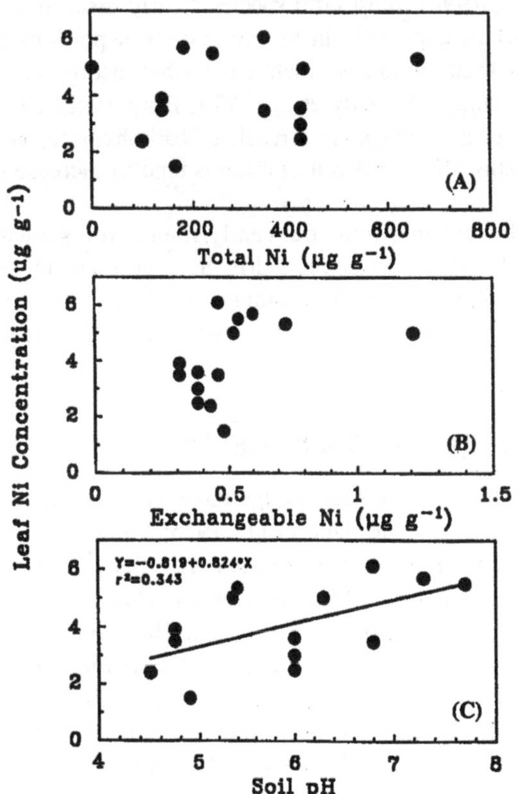

Figure 1 Nickel concentrations of guava leaves as functions of (A) total soil Ni content; (B) exchangeable Ni content; and (C) soil pH. (Data from Chang and Donald, 1952.)

For most elements, k is negligible and can be ignored. This equation can be rearranged to give the acceptable application rate as

$$R = Y^{*}C_{ex}{}^{*}C_{im}{}^{-1} \tag{2}$$

This approach avoids problems of accumulation of metals, but as Table 1 shows, it severely limits the application of biosolids to minimal levels. It points out that maximal rates of application based on this approach seldom exceed 1 Mg ha^{-1} yr^{-1} even when nutrient limiting elements such as N are considered. In those instances in which the trace metal content of the soil is in inadequate supply, additions of biosolids serve as a very effective source of the metal for crop production. Although it seems unlikely, eventually metal concentrations may be minimized in biosolids and elements such as P, potassium (K), sodium (Na), and N might ultimately limit application rates to the land. Metals tend to be concentrated in waste streams.

Although hyperaccumulator plants are known (Baker and Proctor, 1988; Baker, 1989; Martens and Boyd, 1994), most crop species clearly discriminate against the accumulation of metals, especially in the grain or fruit portions of the plants (Keefer et al., 1986). Accumulation of metals is further discriminated against when the grains are consumed (Dowdy et al., 1978; Bray et al., 1985; McKenna et al., 1992; Dowdy et al., 1994). As a result of carbohydrate, protein, and fat extraction from grains, the metal concentrations tend to increase in the waste streams.

Interestingly, even N, the element that most frequently limits crop growth, also limits the application of biosolids to about 1 Mg ha^{-1} yr^{-1} on an export/import basis. Nitrogen contained in biosolids is mineralized and converted to nitrate-N, the form used in greatest amounts by most plants grown on arable soils.

III. EFFECTS OF BIOSOLIDS ON CROP YIELDS

Yield responses of crops to additions of sewage sludge have varied with crop and soil. Yields of corn (*Zea mays* L., Giordano et al., 1975; Pierzynski and Jacobs, 1986; Cripps et al., 1992), soybean (*Glycine max* (L.) Merr.; Pierzynski and Jacobs, 1986; Reddy and Dunn, 1986), snap beans (*Phaseolus vulgaris* L.; Dowdy et al., 1978), carrots (*Daucus carota* L.; Harrison, 1986), bluegrass (*Poa pratensis* L.; Zhang et al., 1993) and barley (*Hordeum vulgare* L.; Miller et al., 1995) generally increased as composted garbage or sewage sludge application rate increased. Much of the yield increase for many crops has undoubtedly stemmed from the large additions of N, P, and K contained in these biosolids. However, Giordano et al. (1975) and Cripps et al. (1992) showed yield increases apparently due to factors other than N and K in their studies.

Table 1 Calculated Permissible Application Rates of Sewage Sludges to Agricultural Soils Based on Export and Import of Elements

Element	Crop	Concentration Sludge	Concentration Crop	Crop yield	Calculated sludge application rate	Data source
		mg kg^{-1}		Mg ha^{-1} yr^{-1}		
Cd	Bean	6	0.08	19	0.25	Dowdy et al., 1978
Zn	Bean	1080	40	19	0.73	Dowdy et al., 1978
N	Corn	84000	14000	5.9	1.0	Soon et al., 1978
Zn	Barley	4070	51	6.3	0.079	Miller et al., 1995
Cu	Barley	1367	5.6	6.3	0.026	Miller et al., 1995
Ni	Barley	591	0.9	6.3	0.0095	Miller et al., 1995
Zn	Ryegrass[a]	1740[b]	300	14	2.41	Smith, 1994
Zn	Ryegrass[a]	1740[b]	400	10	2.30	Smith, 1994
Zn	Ryegrass[a]	1740[b]	200	11	1.26	Smith, 1994
Zn	Ryegrass[a]	1740[b]	500	6	1.72	Smith, 1994
Cd	Corn	71	0.09	3.6	0.005	Kelling et al., 1977 a,b,c
Zn	Corn	3000	23.8	3.6	0.028	Kelling et al., 1977 a,b,c
Cu	Corn	1430	0.8	3.6	0.002	Kelling et al., 1977 a,b,c
Ni	Corn	700	1.2	3.6	0.06	Kelling et al., 1977 a,b,c
N	Corn	91400	16600	3.6	0.65	Kelling et al., 1977 a,b,c
P	Corn	24100	3000	3.6	0.45	Kelling et al., 1977 a,b,c
Cd	Rye[a]	71	0.09	3.5	0.004	Kelling et al., 1977 a,b,c
Zn	Rye[a]	3000	30.6	3.5	0.036	Kelling et al., 1977 a,b,c
Cu	Rye[a]	1430	15	3.5	0.037	Kelling et al., 1977 a,b,c
Ni	Rye[a]	700	1.2	3.5	0.006	Kelling et al., 1977 a,b,c
Cr	Rye[a]	1830	0.1	3.5	0.0002	Kelling et al., 1977 a,b,c
N	Wheat	28000	28900	0.7	0.72	Sabey and Hart., 1975
Cd	Wheat	3	0.066	0.7	0.015	Sabey and Hart., 1975
Zn	Wheat	4000	52.5	0.7	0.092	Sabey and Hart., 1975
Cu	Wheat	1300	4.46	0.7	0.0024	Sabey and Hart., 1975
Ni	Wheat	240	1.1[c]	0.7	0.03	Sabey and Hart., 1975
Cd	Carrot	16[b]	0.631	8.5	0.335	Harrison, 1986
Zn	Carrot	1740[b]	25.3	8.5	0.124	Harrison, 1986
Cu	Carrot	850[b]	7.4	8.5	0.074	Harrison, 1986
P	Carrot	23000[b]	3540	8.5	1.31	Harrison, 1986
K	Carrot	3000[b]	27000	8.5	76.5	Harrison, 1986

[a]Ryegrass (*Lolium perenne* L.); Rye (*Secale cereale* L.).
[b]Mean value for sewage sludges from Sommers (1977).
[c]Based on averages of other grasses in the table.

Some yield enhancements may be more a consequence of the waste production process than from the content of organic matter and nutrients. For example, additions of limed sludge increased the pH of a Metea loamy sand from 4.8 to 7.1 (Pierzynski and Jacobs, 1986), and yield increased as pH increased. Increasing soil pH from a very acidic condition to near neutrality often effects favorable growth response. Unfortunately, no control studies of lime ($CaCO_3/MgCO_3$) and N additions to non-sludge-amended soil were reported.

Yield decreases have been noted for wheat (*Triticum aestivum* L.; Sabey and Hart, 1975), corn (Cunningham et al., 1975; Mench et al., 1994), and bush beans (*Phaseolus vulgaris* L.; Giordano et al., 1975) for sewage sludge additions and especially at high loading rates. Yield decreases appear to be toxicity reactions incurred as a result of elevated salt and metal concentrations. Yield decreases can be quite large; for example, vegetative yields decreased by 27% with sewage sludge loadings of ≥ 300 Mg ha^{-1} (Mench et al., 1994) on a very sandy soil with mildly acidic pH. These authors attributed yield reductions to induced manganese (Mn) deficiency and elevated concentrations of Ni and Cd in plant tissues.

IV. EFFECTS OF BIOSOLIDS ON SOIL PHYSICAL CHARACTERISTICS

The ability of plants to thrive in a given soil environment depends on soil physical, chemical, and biological (especially microbiological) characteristics. Use of biosolids as soil amendments has some predictable effects on soil properties and processes (Larson and Clapp, 1984). Biosolids are characterized by their organic nature, and Sommers (1977) has provided a summarization of chemical characterization of a broad range of sewage sludges and composts. Organic carbon (C) contents of sewage sludges range from 6 to 48% with a median content of about 32%; by comparison, dried plant matter often has an organic C content of 39 to 44%. Digested sludges and composts tend to have less organic C (median 28 and 32% for anaerobic and aerobically digested sludges, respectively, versus 33% for other forms) as a result of microbial decay of the organic matter, but the differences are relatively small.

A. Effect of Water-Holding Capacity

One of the most important soil characteristics for crop production is the water-holding capacity of the soil. Plants use between 200 to 400 g of water to produce a gram of dry matter, so water storage is a key to crop yield in most soils. Two components generally determine water-holding capacity of most soils: clay content and organic matter content. Field capacity (θ_{fc}) is defined as the amount of water that a soil will hold against the pull of gravity. The limit of water that can be withdrawn by plants is defined as the wilting point (θ_{wp}). The difference between field capacity and wilting point is defined as the available water (θ_a):

$$\theta_a = \theta_{fc} - \theta_{wp} \quad [3]$$

The relationship between soil clay content and soil water-holding capacity is complex (Fig. 2) and depends somewhat on the mineralogy of the clay and the fineness of the soil fractions. On average, organic C residues resulting from plant decay will retain about 2.5 times their weight in water against the pull of gravity. Because a portion of the organic matter is adsorbed by the soil mineral surfaces, soil organic matter is less efficient in serving as a reservoir for available water than free organic forms. Still, on a weight basis, organic matter is more efficient at retaining available water than soil mineral particles alone. Many soils have a less than optimal organic matter content and this has led to casual claims that increases in organic matter are always desirable. Little thought has been given to optimal contents of soil organic matter.

B. Effect of Bulk Density

Optimizing soil organic matter implies an objective, and in soils, enhanced microbial production of nitrate-N (NO_3^--N) is one objective. An optimal aeration of soils for maximal aerobic soil biological activity has been estimated by Skopp et al. (1990) as that bulk density and water content in which about 66% of the pore space is filled with water. We can use this observation to predict the

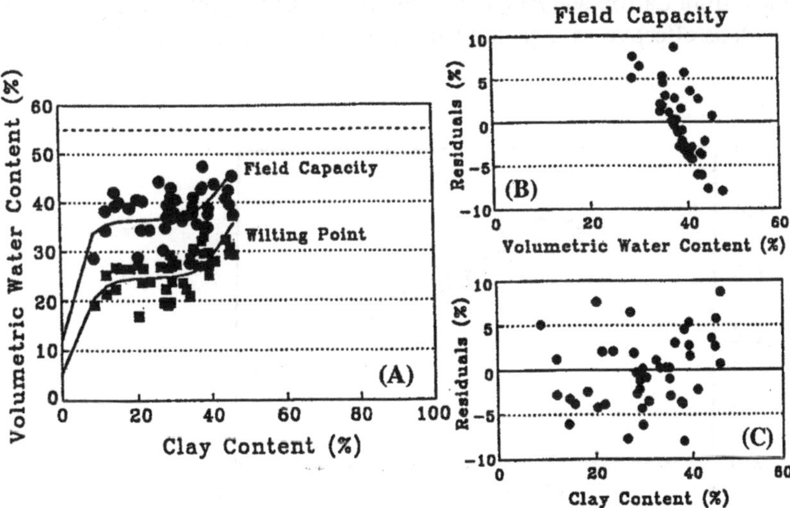

Figure 2 The relationship between (A) clay content and soil volumetric water content at wilting point and field capacity; (B) model residuals for field capacity with respect to volumetric water content; and (C) model residuals for field capacity with respect to soil clay content. (Data from Olson, 1970, and models by Olness, 1977.)

optimal application rate and accumulation of organic matter in soils; but first we need to know the impact of added organic matter on an important soil property, bulk density (ρ_{db}).

As a general estimate, an increase of 1% naturally produced organic C in the surface soil horizon will decrease the ρ_{db} by about 0.0215 Mg m^{-3} (Larson and Allmaras, 1971; Fig. 3). Only part of this decrease is due directly to the increased organic C; the rest is due to increased biological burrowing and mixing by mesofauna and rooting by plants as they forage through what is often a microbially enriched medium for nutrients and energy sources. Much of the organic matter in sewage sludge has undergone significant microbial decomposition, and that which remains is less readily mineralized to CO_2 than freshly incorporated plant material. As a result, sewage sludge additions of organic C produce changes in soil ρ_{db} similar to naturally produced organic matter.

Data compiled by Clapp et al. (1986) suggest that for every Mg ha^{-1} of organic C added in the form of sewage sludge, the ρ_{db} will decrease by about 0.0037 Mg m^{-3} and this effect will last for two or more years (Fig. 4). The variance in this estimated effect is large, probably due to the types of tillage, elapsed time between application and measurement, climate, etc. The estimate itself is developed from a very broad range of sludge types and application rates having a mean organic C content of 30.4%. While the estimate of the effect is crude, the general effect, namely that the ρ_{db} decreases as application rate of sewage sludge increases, is quite clear. By comparing the data from Clapp et al. (Fig. 4) with that from Larson and Allmaras (Fig. 3), we see that sewage sludge is about one-sixth as effective as organic matter from long-term grass production.

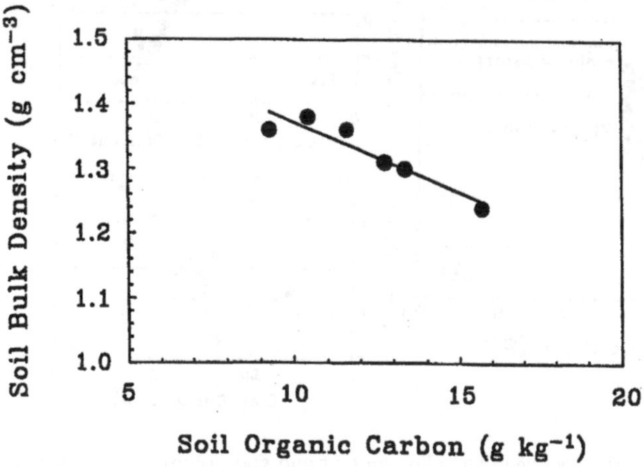

Figure 3 An example of the change in soil bulk density as a function of soil organic carbon content. (Data from Larson and Allmaras, 1971.)

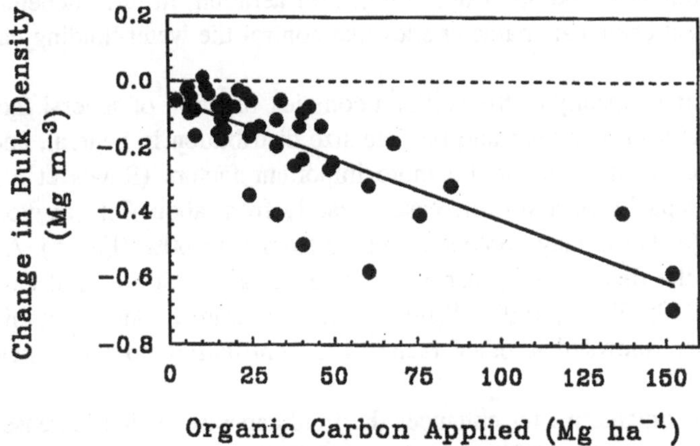

Figure 4 The general effect of adding organic carbon in the form of biosolids on the change in soil bulk density. (Data from a compilation by Clapp et al., 1986.)

From these relationships, we now have a means of developing optimal physical conditions by amending soils with biosolids. Given a soil texture and ρ_{db} we can estimate the percentage of water-filled pore space at field capacity; these two soil characteristics are readily measured.

V. EFFECTS OF BIOSOLIDS ON SOIL CHEMICAL CHARACTERISTICS

The organic materials in these wastes have rather large amounts of organic N (range <0.1 to 5.0%, median 3.9%) as well as total P (range <0.1 to 14.3%, median 3.9%), total sulfur (S) (range 0.6 to 1.5%, median 1.1%), and a host of other elements including nutrients and metals in amounts that vary widely. Thus biosolids should make ideal fertilizers for crop production, and in many cases they effect large crop yield increases. The relatively rich supply of N in biosolids was noted very early and, because N is the nutrient element most often limiting plant growth and crop yield, this contributed much to the encouragement for application of biosolids to agricultural lands.

A. Effect of Soil Aeration

The addition of such a nutritious material to soil has a complex effect on the soil biological activity, and this effect varies with the soil physical character. Mineralization from organic sources depends on soil aeration. Optimal aeration for mineralization occurs when about 66% of the soil pore space is filled with water (Doran et al., 1990; Skopp et al., 1990). Soil aeration, in turn, depends on complex physical-chemical characteristics that control the water-holding capacity of soil.

The water-holding capacity of the soil is a complex function of several factors, but clay content in particular and particle size distribution in general, and organic matter content are two of the more important factors (Rawls et al., 1991). The field capacity of a soil increases steadily from about 7% (by volume) to about 53% as the clay content increases from 0 to 60% (Fig. 5). An approximation of the field capacity can be obtained as a function of soil clay content (Olson, 1970; Olness, 1997). While the model residuals are randomly scattered, their size shows that other factors are contributing to the water-holding capacity.

The presence of organic matter enhances the field capacity, which increases much more rapidly from about 12% to a plateau at 33% as the clay content increases from 0 to 10% (Fig. 5). With organic matter, the field capacity changes very little as the clay content increases from 10 to 35%, but the field capacity increases steadily as the clay content increases at clay contents > 35%, until a second plateau is achieved at about 53% volumetric water content at 60% clay content. Also, with organic matter present, the wilting point curve, that limit at which plants are able to extract water from the soil, tends to parallel the field capacity curve.

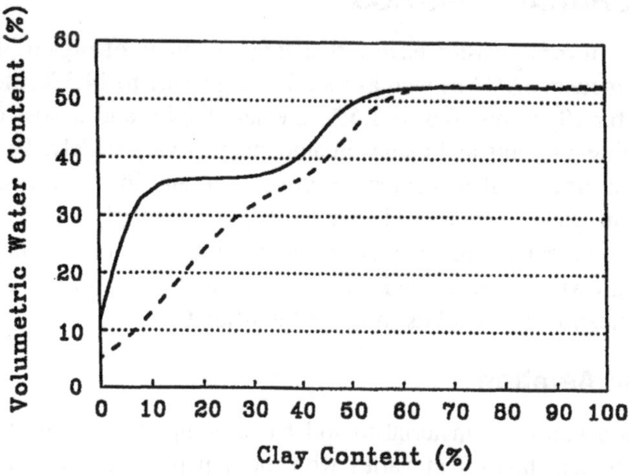

Figure 5 Field capacity as a function of clay content with (solid line) and without (dashed line) organic matter. (Data from Olson, 1970, and models by Olness, 1997.)

B. Effect of Bulk Density

A final determinant of aeration is bulk density (Fig. 6). At a bulk density of 1.0 Mg m^{-3}, a value often approached after moldboard plow tillage, about 66% of the pore space is filled with water with a clay content of about 40% with organic matter, or about 46% without organic matter. At clay contents <40% and a bulk density of 1.0 Mg m^{-3}, the soil remains essentially too dry to maximize microbial activity and thus N mineralization. At clay contents >47%, the soil usually remains too wet to achieve maximal N mineralization when the bulk density is 1.0 Mg m^{-3}.

As the bulk density increases, the optimal clay content for achieving 66% water-filled pores decreases. For example, at a bulk density of 1.2 Mg m^{-3}, optimal aeration is achieved with about 12% clay content in the presence of organic matter and about 38% clay content in the absence of organic matter. A bulk density of 1.2 Mg m^{-3} is about the value obtained after secondary tillage at planting time. At this bulk density, the soil tends to be too dry for maximal N mineralization at clay contents <12% and too wet at clay contents of >38%.

Both primary tillage of the soil and additions to or increases in the organic matter content of the soil tend to decrease soil bulk density. Also, increases in soil organic matter increase the field capacity of a soil at a given clay content.

The complex interaction of bulk density and clay content on relative soil aeration has been modeled in a N decision aid developed by Cordes et al. (1995),

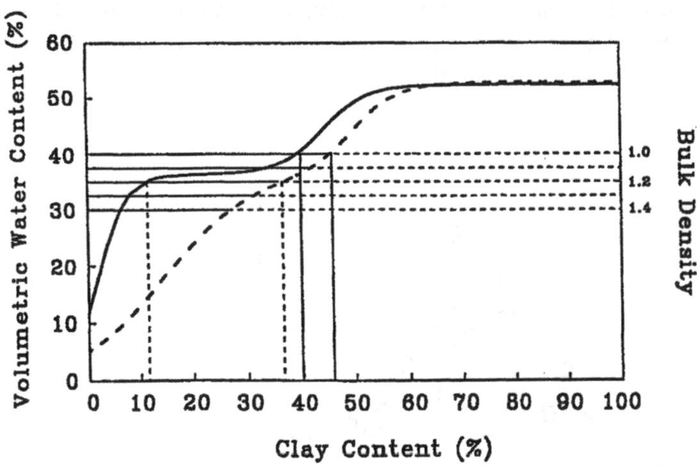

Figure 6 The interaction between organic carbon content, volumetric water content, clay content, and bulk density on soil aeration. Those points at which the bulk density lines (horizontal) intersect the field capacity lines are the points at which two-thirds of the soil pore space will be filled with water. Field capacity data obtained from Olson, 1970, with models by Olness, 1997.

who included two additional factors that strongly affect N mineralization in soil; these are temperature and pH. This model was extended by Sweeney et al. (1996) to include effects of leaching.

C. Effect of pH

The complex pH effect (Fig. 7) on the relative rate of NO_3^--N formation, nitrification, has been modeled by Olness (1997). The rate of nitrification appears minimal in soils at pH values of <5.0 and >8.0. An apparent optimal pH of 6.7 is needed to attain a maximal rate of nitrification. Thus, additions of biosolids having pH values different from the soil will affect the rate of nitrification. A probable example of such an effect was the application of garbage compost to an acidic soil by Giordano et al. (1975). In their study, the soil pH was raised from about 4.9 to 6.3 by application of compost, and corn yields were increased with application rates of 112 or 224 Mg ha^{-1}.

Nitrogen in sludges and composts is rather rapidly mineralized and nitrified. Voos and Sabey (1987) found 850 kg N ha^{-1} was mineralized from an application of 4890 kg N ha^{-1} within 16 weeks on loam mine spoil soils with pH values of 6.0 to 7.3. Their result illustrates one of the potential adverse effects of adding excessive amounts of N in the form of sludge or compost to agricultural lands, the excessive production and accumulation of NO_3^--N. Kelling et al. (1977c) found accumulations of >100 ppm of nitrate in the soil solution to depths of 120 to 150 cm after application of 30 Mg sewage sludge ha^{-1}; this accumulation occurred within 10 weeks of application on a silt loam in Wisconsin. They were unable to recover more than 19% of the applied N in a corn crop over a 4-yr period (Kelling et al., 1977b).

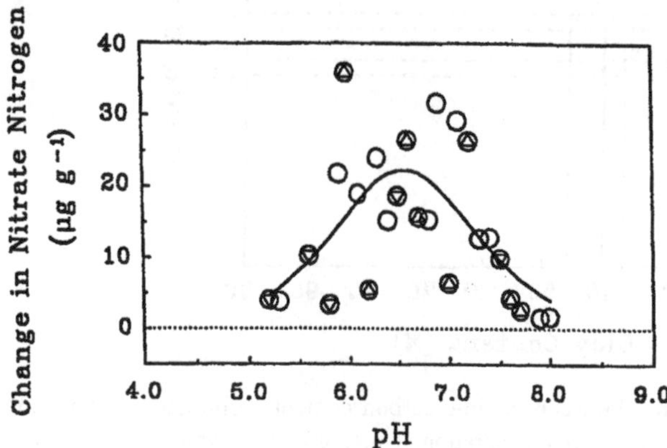

Figure 7 The general effect of pH on change in nitrate-nitrogen concentration in the upper 30 cm of soil. (Data from CSRS/NC-201 Report, 1997; model by Olness, 1997.)

Kelling et al. (1977b) obtained a near steady-state mineralization of 180 and 250 mg NO_3^--N kg^{-1} yr^{-1} at 30 and 60 Mg ha^{-1} sludge application rates, respectively, for a period of up to 25 months after an initial 21-day period during which as much as 50% of the organic N was mineralized. This translates to a potential NO_3^--N production of 360 to 500 kg N ha^{-1} yr^{-1}, an amount that exceeds most crop removal capacities. At the 60 Mg ha^{-1} sludge application rates, about 5480 kg N ha^{-1} were applied; this amount of N is far in excess of any crop's ability to recover N in grain and forage within 20 years if the system were conservative. The concentration of N in corn grain at the greatest sludge application rates ranged from about 1.58 to 1.66%; this is at the limit of N concentrations in corn grain. Yields increased from about 4.78 to 7.13 Mg ha^{-1} on silt loam and sandy loam soils and equals an N removal in the grain of <120 kg N ha^{-1}.

D. Phosphorus and Potassium Availability

Little appears to have been done to separate the effects of P and K in biosolids from the effects of other metals and nutrients. Potassium availability is generally unaffected by its source, and thus K in biosolids should be equally effective at remediating deficiencies as other K sources. McCoy et al. (1986) noted that composts were much less effective than triple superphosphate when the application rate provided >100 kg P ha^{-1}; the soils used (both ultisols) ranged from sandy loam to clay loam textures with pH values ranging from 4.9 to 5.4. These researchers did add N (NH_4NO_3) and K to untreated plots to minimize effects of contributions of these elements from the biosolids.

VI. EFFECTS OF BIOSOLIDS ON SOIL BIOLOGICAL CHARACTERISTICS

Qualitative analysis of soil microbial populations in the rhizosphere after sewage sludge treatments has been mostly focused on human pathogens such as fecal coliform, *Streptococcus* or *Salmonella* (Ibiebele et al., 1985; Ottolenghi & Hamparian, 1987; Goldstein et al., 1988; Straub et al., 1992; Tsai et al., 1993). Basically, these pathogens can survive from days (bacteria) to months (viruses) depending on environmental conditions (Straub et al., 1992). Sewage sludge is a complex mixture of organic and inorganic nutrient compounds with high populations of living organisms both microflora and macrofauna; however, the study of general population dynamics and population shifts in the rhizosphere has been less focused. Since sewage sludge is being recycled on land as a soil amendment, soil responses in indigenous microbial population shifts should have important concerns. These may be directly related to the level of soil fertility and the rate of soil organic matter turnover.

A. Soil Microorganisms

Burton et al. (1990) found that significant net nitrification occurred in soil cores receiving surface applications of anaerobically digested sewage sludge. Their results indicate NO_3^--N leaching following sludge application to acid forest soils in Michigan. This finding agreed with a previous study by Higgins (1984) who found annual applications of sludge at 44.8 Mg of dry solids ha^{-1} resulted in gross contamination of the ground water by NO_3^--N. Soil nitrifiers are responsible for the net nitrification in these systems. The use of soil microbial acitivity as an indicator for soil fertility has been studied by Brendecke et al. (1993). In a 4-yr study with annual sludge applications to a cotton crop on a clay loam soil, they found an increase in soil PO_4-P but no significant adverse effects on soil microbial populations or activity for soil bacteria, actinomycetes, and fungi. Soil, as a medium, has very strong buffering capacity to balance the microbial population, no matter how much of an initial addition of microbes from sewage sludge treatments. This result is seen in Fig. 8, where a field survey on microbial population was averaged over a 20-yr sludge-treated soil experiment. Stabilized soils have bacterial, actinomycete, and fungal populations of about 10^7, 10^6 and 10^4 g^{-1} of soil, respectively. These results are very close to the soils without sludge amendments (data not shown). Soil provides buffering environments for microbes to grow, to multiply and to metabolize. When the nutrient supply in the environment becomes limited, the population gradually returns to the normal level.

Application of sewage sludge to soil does not cause much change in the size of the general population over the long term, but it may change some specific groups of microbes in soil. Coppola et al. (1988) indicated that the ammonifiers in a volcanic soil were particularly affected by sludge containing Cd. Another depressive effect of Cd was on free-living aerobic N fixers. Hamlett (1986), in a study of changes in a salt marsh bacterial community, found that the number of colony forming units (CFU), the percentage of mercury (Hg)- and Cd-resistant bacteria, and the percentage of antibiotic-resistant bacteria all increased in sludge-fertilized plots. Kinkle et al. (1987) suggested that the application of heavy metal–containing sludge did not have a long-term detrimental effect on soil rhizobial numbers, nor did it result in a shift in nodule serogroup distribution. The addition of sewage sludge to soil would increase the buffering capacity of soil by the concurrent addition of organic matter. Chander and Brookes (1991) suggested that the most probable order of increasing metal toxicity to microbial biomass is Cu > Zn > Ni or Cd. No single metal (Cu, Zn, Ni, or Cd) at or below current European Communities–permitted total soil metal limits decreased the soil microbial biomass.

Altered relationships after sludge application between general populations and specific groups of microbes, such as soil nitrifiers or some antibiotic-resistant groups, have been given less attention. Douglas (1980) has noted the

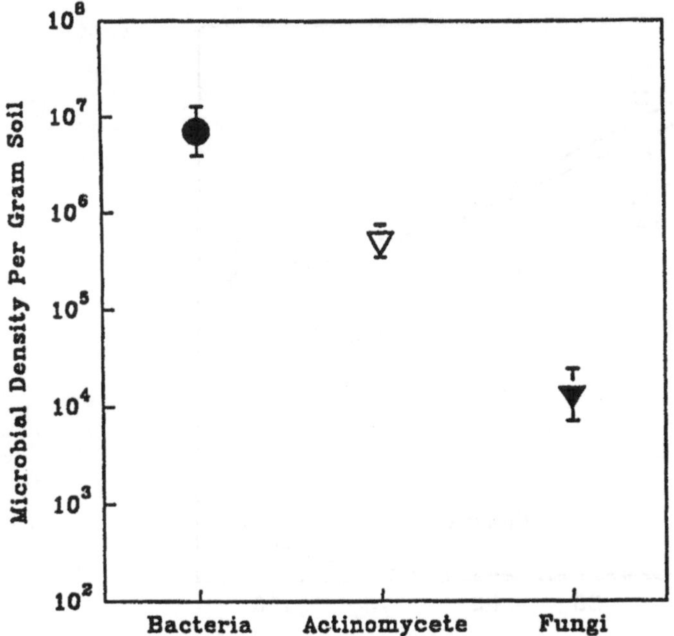

Figure 8 Field survey of microbial populations for 4 loam soils with 20-yr sludge treatment at Rosemount, MN (Linden et al., 1995), and 4 sandy soils from Fort Drum, NY. The last application of sludge was made two years before the samples were taken.

evidence of simultaneous or alternating nitrification–denitrification processes observed in his sludge incubation experiments. Such relations may pose long-term impacts on environmental quality and soil fertility.

Application of sewage sludge to soils may temporarily increase the general population of microflora over a short period of time, compared to soils not treated with sludge. This is because fresh sludge contains large amounts of living microorganisms. However, the bacterial population will drop quickly after sludge is applied to the soil. This result can be seen from a minirhizotron experiment where the bacterial population dropped about 1 to 2 orders of magnitude in 75 days (Liu et al., 1994). Figure 9 shows a parallel type of population drop under three different sludge mixing depths. This decrease may indicate a nutrient and environmental stress on the bacteria after sludge application. Control samples remained at a much lower level of CFU at the beginning of the experiment but increased with turfgrass growth. It can be predicted that all bacteria CFU will eventually reach a balanced level, as shown in Fig. 8, after a long period of time. A significant gain in actinomycete populations was also found during the first 75 days after sludge application (data not shown). This

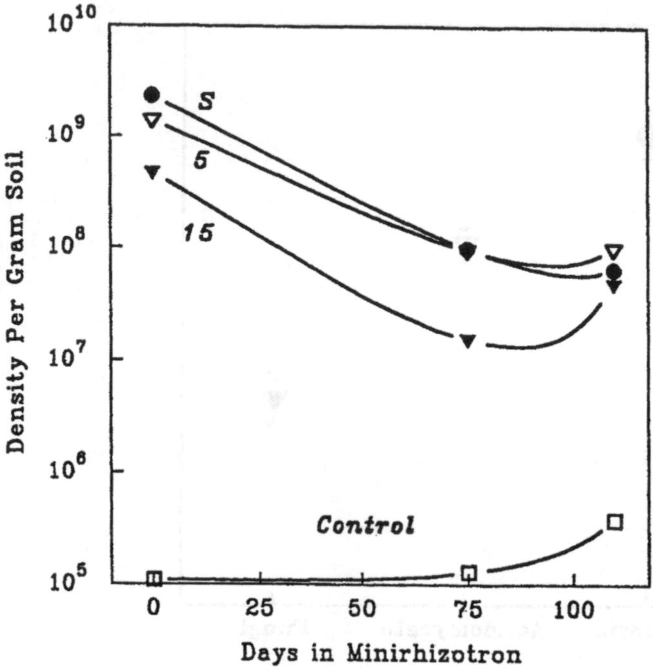

Figure 9 Bacteria population changes after turfgrass growth in minirhizotron (Liu et al., 1994). Sludge rate was 18 Mg ha^{-1} for surface application (S), 5-cm depth of mixing (5) and 15-cm depth of mixing (15).

fact may indicate active organic decomposition of the sewage sludge by soil actinomycetes.

B. Earthworms

Studies on the effects of biosolids on soil macrofauna are mainly focused on earthworms, although the use of sewage sludge as an effective method for nematode control has been studied (Akhtar and Alam, 1993). Earthworm populations are important in determining soil conditions by mixing and aerating the soil. Feeding sewage sludge to soil earthworms can greatly enhance indigenous microbial activity, as shown by Hornor and Mitchell (1981). They found that the earthworm *E. foetide* feeding stimulated aerobic decomposition, particularly the C and S constituents in sewage sludge. Sludge treatment to a mixed hardwood forest soil and old field plots increased earthworm *L. terrestris* both in numbers and in individual biomass (Hamilton and Dindal, 1989).

Since earthworms are important food sources for wildlife such as reptiles, birds, and mammals, accumulation of heavy metals from sewage sludge through earthworms poses a risk of biological concentration throughout the food chain. There are several reports (Table 2) on the relationship of trace metal accumulation in earthworms and sludge treatments (Beyer et al., 1982; Kruse and Barrett, 1985; Levine et al., 1989; Brewer and Barrett, 1995). Beyer et al. (1982) suggested that Cd, Cu, and lead (Pb) were concentrated by earthworms but not Ni or Zn. Kruse and Barrett (1985) suggested that possible biological regulation to Cu and Pb by earthworms may reduce accumulation of these metals. Brewer and Barrett (1995) later showed that the concentrations of Cd and Zn were significantly greater in earthworms collected from sludge treatments in a 5-yr study. However, a decrease of organic matter content in soil may result in greater concentrations of metals in free ionic form in solution available for both plants and earthworms, therefore reducing the share of accumulation in earthworms. Table 2 shows that Cd was highly concentrated in earthworms in all cases, while Cu was less concentrated, and Ni, Pb, and Zn varied with different applications and types of sludge. Clearly, earthworms are very sensitive indicators of soil quality and useful in assessment of trace metal influences after sewage sludge and applications.

Table 2 Trace Metal Concentrations in Earthworms After Sludge Application

Element	Rate	Years	Concentration		Data source
			Control	Sludge	
	Mg ha^{-1}yr^{-1}		μg g^{-1} dry weight		
Ni	22.4	8	10	15	Beyer et al., 1982
Cd	22.4	8	17	46	Beyer et al., 1982
Cu	22.4	8	11	29	Beyer et al., 1982
Pb	22.4	8	11	23	Beyer et al., 1982
Zn	22.4	8	442	475	Beyer et al., 1982
Cd	9.0	4	15	136	Kruse and Barrett, 1985
Cu	9.0	4	11	21	Kruse and Barrett, 1985
Pb	9.0	4	2.5	8.8	Kruse and Barrett, 1985
Zn	9.0	4	924	1087	Kruse and Barrett, 1985
Cd	9.0	5	1.0	81	Brewer and Barrett, 1995
Cu	9.0	5	10	42	Brewer and Barrett, 1995
Pb	9.0	5	25	22	Brewer and Barrett, 1995
Zn	9.0	5	70	900	Brewer and Barrett, 1995

VII. METAL ACCUMULATIONS AND INTERACTIONS

The relative importance of metal concentrations in biosolids is seen in comparing them with concentrations in soils and food crops. Most soils have metal concentrations much less than comparable concentrations in biosolids (Holmgren et al., 1993; Table 3). In some instances, metal concentrations in soil may be less than those required to promote optimal plant growth. Under these circumstances, additions of metal enriched biosolids has a beneficial effect of mitigating a deficiency. This is particularly true for elements such as Zn, Cu, and Ni.

Dahiya et al. (1993) obtained increased dry matter yields of wheat in pot experiments with increased soil Ni concentrations of ≤10 mg kg^{-1}; the soil was impoverished in both organic C and N and had a pH of 8.0. They also obtained an interaction with added N. Unfortunately, in many studies, the soil resource has been much less well characterized than the applied biosolids, and the exact source of some benefits cannot be assigned to a single element.

A. Interactions of pH

Soil pH is a significant factor in sorption of Cd, Zn, Ni, and cobalt (Co) and in the fixation of Ni (King, 1988b). Generally, as pH becomes more alkaline, exchangeable Cd in soils decreases by rather large amounts. However, as was seen with Ni, the relationship between extractable amounts and plant uptake as affected by pH is not always straightforward.

B. Interactions with Other Elements

Soil iron (Fe) is a significant factor in Cd and Zn sorption and fixation, Cu and Pb sorption, and Co and Ni fixation and minimization of antimony (Sb) fixation (King, 1988a). These and other cations often interact with both Fe- and Mn-oxides such that prediction of sorption and subsequent availability to plants becomes quite complex. As the soil pH becomes alkaline, carbonates begin to play an increasingly important role in relative solubilities of some elements. Interactions between Cd and Zn, P and vanadium (V), molybdenum (Mo) and tungsten (W), are known. Both V and W depress cellular metabolism and usually depress yields when present in sufficient amounts. Because Cd and Zn apparently compete for similar sorption sites in soils, the availability of Cd increases as the amount of Zn (Sikora and Wolt, 1986), Cu (Kuo and Baker, 1980) or calcium (Ca) (McBride et al., 1981) in the biosolid increased. Thus, interactions with cations in general and divalent cations in particular can have large effects on the availability and uptake of Cd by plants.

Relative aeration and redox potential are also important factors in sorption, fixation, and release of not only the redox sensitive elements such as Fe, Mn, Cu, etc. but also of the elements that interact with the oxides of these elements. Under anoxic conditions, Fe and Mn oxides, which often combine with other

Table 3 Some General Characteristics and Concentrations of Metals in Agricultural Soils of the United States, Municipal Sludges, and Common Foods

	Metal (ppm)							
	Cd	Pb	Zn	Cu	Ni	pH	CEC mmol (+) kg^{-1}	Organic carbon (%)
Soils								
Min.	<0.005	0.5	1.5	0.3	0.7	3.9	0.6	0.09
Max.	2.0	135	264	495	269	8.9	204	63
5th perctle.	0.036	4.0	8.0	3.8	4.1	4.7	2.4	0.36
50th perctle.	0.20	11	53	18.5	18.2	6.1	14	1.05
95th perctle.	0.78	23	126	95	57	8.1	135	33.3
Sludges[b]								
Min.	3	13	101	84	2	—	—	6.5
Max.	3410	19700	27800	10400	3520	—	—	48.0
Median	16	500	1,740	850	82	—	—	30.4
Foods/ Grains[c]								
Min.	0.05	—	14.6	1.7	0.7	—	—	—
Max.	1.5	—	28.5	7.7	0.7	—	—	—
Mean	0.09	—	18.6	4.6	1.8	—	—	—
Fruits, seeds and berries[d]								
Min.	<0.001	<0.005	0.11	0.21	0.006	—	—	—
Max.	0.98	0.200	62	20	6.0	—	—	—
Mean	0.08	0.018	17.5	4.2	0.8	—	—	—

[a]Adapted from Holmgren et al. (1993).
[b]Adapted from Sommers (1977).
[c]Adapted from Chen et al. (1994); data from China.
[d]Adapted from Jorhem and Sundstrom (1993); data from Sweden.

cations, become soluble and as a consequence affect the solubilities of many other elements, including Ca and magnesium (Mg), which are otherwise unaffected by changes in redox potential.

VIII. SUMMARY

Biosolids obtained from municipal and industrial sources have properties that make them desirable amendments for soils. However, because these materials affect physical, chemical, and biological soil properties, practical limits of application (PLA) rates must be observed. Trace metals remain a significant source of concern because they are incorporated into edible fractions of harvested plants and they concentrate in the food web.

Even without a concern for trace metals, the concentration of N, the element most often deficient in crop production systems, places a PLA in the order of 1 to 2 Mg ha^{-1}, depending on soil texture, in most land areas. Application rates based on total metal content of soil or total available metal content of soil as determined by current measures are unreliable indicators of the tolerance for metal additions contained in sludges. Changes in availability with changes in soil pH or redox potential are also unreliable guides for prediction of metal availability or accumulation by plants, because even though extractability in the laboratory may change, accumulation by the plant may be independent of the laboratory method used to determine availability.

Because organic matter comprises large fractions of most sludges and because it affects the soil bulk density and the field capacity, these characteristics further affect the PLA to agricultural land. Many soils will benefit from the application of organic matter contained in sewage sludges or biosolids.

Soils have very strong buffering capacities to balance microbial populations. Recycling sewage sludge to soil does not change very much the general populations over the long-term, although short-term increases can occur, and the populations of soil bacteria, actinomycetes, and fungi can be stabilized. Effects on specific populations showed changes for nitrifiers and some antibiotic-resistant groups of microorganisms in soil. Much less attention has been paid to the relations between specific groups and general populations of microbes after sludge treatment. Soil earthworms are very sensitive indicators for soil quality and trace metal influences. The trace metal Cd was highly concentrated in earthworms, while Cu was somewhat less concentrated. The metals Zn, Pb, and Ni varied with different case studies.

Many of the gains in crop productivity with application of large amounts of biosolids, >5 Mg ha^{-1} annually, are undoubtedly related to the supply of N, P, and K to soils in which the availability of these elements was marginal or inadequate. Complex interactions between elements within the plant are also known that affect the plant's ability to acquire essential elements or enhance undesirable accumulations of metals. Thus, a need exists for the development of appropriate tests and guidelines for the use and application of biosolids to lands. Of the current guidelines, the economic approach, which measures the amounts of metals, etc. removed from agricultural lands and the gains from application of biosolids, provides an initial estimate of the practical limits of application.

REFERENCES

Akhtar, M., and Alam, M. M. (1993). Utilization of waste materials in nematode control: a review. *Bioresource Technol. 45*:1–7.

Baker, A. J. M. (1989). Terrestrial higher plants which accumulate metallic elements—a review of their distribution, ecology and phytochemistry. *BioRecovery 1*:81–126.

Baker, A. J. M., and Proctor, J. (1988). Ecological studies on forest over ultrabasic rocks in the Philippines. *Bull. Br. Ecol. Soc. 19*:29–34.

Beyer, W. N., Chaney, R. L., and Mulhern, B. M. (1982). Heavy metal concentrations in earthworms from soil amended with sewage sludge. *J. Environ. Qual. 11*:381–385.

Bingham, F. T., Page, A. L., Mahler, R. J., and Ganje, T. J. (1975). Growth and cadmium accumulation of plants grown on a soil treated with a cadmium-enriched sewage sludge. *J. Environ. Qual. 4*:207–211.

Bray, B. J., Dowdy, R. H., Goodrich, R. D., and Pamp, D. E. (1985). Trace metal accumulations in tissues of goats fed silage produced on sewage sludge-amended soil. *J. Environ. Qual. 14*:114–118.

Brendecke, J. W., Axelson, R. D., and Pepper, I. L. (1993). Soil microbial activity as an indicator of soil fertility: long-term effects of municipal sewage sludge on an arid soil. *Soil Biol. Biochem. 25*:751–758.

Brewer, S. R., and Barrett, G. W. (1995). Heavy metal concentrations in earthworms following long-term nutrient enrichment. *Bull. Environ. Contam. Toxicol. 54*:120–127.

Burton, J. A., Hart, J. B., Jr., and Urie, D. H. (1990). Nitrification in sludge-amended Michigan forest soils. *J. Environ. Qual. 19*:609–616.

Chander, K., and Brookes, P. C. (1991). Effects of heavy metals from past applications of sewage sludge on microbial biomass and organic matter accumulation in a sandy loam and silty loam U.K. soil. *Soil Biol. Biochem. 23*:927–932.

Chaney, R. L. (1994). Trace metal movement: soil–plant systems and bioavailability of biosolids applied metals. *Sewage Sludge: Land Utilization and the Environment.* (C.E. Clapp et al., eds.) SSSA Misc. Publ., ASA-CSSA-SSSA, Madison, WI, pp. 27–31.

Chang, A.-T., and Donald, S. G. (1952). The nickel content of some Hawaiian soils and plants and the relation of nickel to plant growth. *Univ. Hawaii Agric. Expt. Sta. Tech. Bull. No. 9.*

Chen, F., Cole, P., Wen, L., Mi, Z., and Trapido, E. J. (1994). Estimates of trace element intakes in Chinese farmers. *J. Nutrition 124*:196–201.

Clapp, C. E., Dowdy, R. H., Linden, D. R., Larson, W. E. Hormann, C. M., Smith, K. E., Halbach, T. R., Cheng, H. H., and Polta, R. C. (1994). Crop yields, nutrient uptake, soil and water quality during 20 years on the Rosemount sewage sludge watershed. *Sewage Sludge: Land Utilization and the Environment.* (C. E. Clapp et al., eds.) SSSA Misc. Publ., ASA-CSSA-SSSA, Madison, WI, pp. 137–148.

Clapp, C. E., Stark, S. A., Clay, D. E., and Larson, W. E. (1986). Sewage sludge organic matter and soil properties. *The Role of Organic Matter in Modern Agriculture.* (Y. Chen and Y. Avnimelech, eds). Martinus Nijhoff, Dordrecht, The Netherlands, pp. 209–253.

Cooperative States Research Service. (1997). *NC-201 Committee Report* (in press).

Coppola, S., Dumontet, S., Pontonio, M., Basile, G., and Marino, P. (1988). Effect of cadmium-bearing sewage sludge on crop plants and microorganisms in two different soils. *Agric. Ecosyst. Environ. 20*:181–194.

Cordes, J., Olness, A. E., Lopez, D., and Voorhees, W. B. (1995). Modeling changes in nitrate-nitrogen in the natural environment. *Abstr. Minn. Acad. Sci. Ann. Meetings. Morris. MN.* April 28–30, 1995, p. 17.

Cripps, R. W., Winfree, S. K., and Reagan, J. L. (1992). Effects of sewage sludge application method on corn production. *Commun. Soil Sci. Plant Anal. 23*:1705–1715.

Cunningham, J. D., Keeney, D. R., and Ryan, J. A. (1975). Yield and metal composition of corn and rye grown on sewage sludge-amended soil. *J. Environ. Qual. 4*:448–454.

Dahiya, D. J., Singh, J. P., and Kumar, V. (1993). Nitrogen uptake in wheat as influenced by the presence of nickel. *Arid Soil Res. Rehab. 8*:51–58.

Doran, J. W., Mielke, L. N., and Power, J. F. (1990). Microbial activity as regulated by soil water filled pore space. *Proc. 14th Intern. Congr. Soil Sci., Kyoto, Japan. 3*:94–99.

Douglas, L. A. (1980). Chemical and microbiological impact of sewage sludge on soil systems. *New Jersey Water Resources Research Institute. Final Tech. Comp. Rpt.*, OWRT Project No. A-038-N. J.

Dowdy, R. H., Clapp, C. E., Linden, D. R., Larson, W. E., Halbach, T. R., and Polta, R. C. (1994). Twenty years of trace metal partitioning on the Rosemount sewage sludge watershed. *Sewage Sludge: Land Utilization and the Environment.* (C. E. Clapp et al., eds.) SSSA Misc. Publ., ASA-CSSA-SSSA, Madison, WI, pp. 149–155.

Dowdy, R. H., Larson, W. E., Titrud, J. M., and Latterell, J. J. (1978). Growth and metal uptake of snap beans grown on sewage sludge-amended soil: a four-year field study. *J. Environ. Qual. 7*:252–257.

Food and Agriculture Organization—World Health Organization. (1972). Evaluation of certain food additives and the contaminants mercury, lead and cadmium. *16th Rpt. Joint FAO/WHO Expert Committee on Food Additives.* FAO Nutrition Mtg. Rpt. 51; WHO Tech Rpt. 505. FAO, Rome.

Giordano, P. M., Mortvedt, J. J., and Mays, D. A. (1975). Effect of municipal wastes on crop yields and uptake of heavy metals. *J. Environ. Qual. 4*:394–399.

Goldstein, N., Yanko, W. A., Walker, J. M., and Jakubowski, W. (1988). Determining pathogen levels in sludge products. *Biocycle May/June 1988*:44–67.

Hamilton, W. E., and Dindal, D. L. (1989). Impact of landspread sewage sludge and earthworm introduction on established earthworms and soil structure. *Biol. Fertil. Soils 8*:160–165.

Hamlett, N. (1986). Alteration of a salt marsh bacterial community by fertilization with sewage sludge. *Appl. Environ. Microb. 52*:915–923.

Harrison, H. C. (1986). Carrot response to sludge application and bed type. *J. Am. Soc. Hort. Sci. 111*:211–215.

Higgins, J. A. (1984). Land application of sewage sludge with regard to cropping systems and pollution potential. *J. Environ. Qual. 13*:441–448.

Holmgren, G. G. S., Meyer, M. W., Chaney, R. L., and Daniels, R. B. (1993). Cadmium, lead, zinc, copper, and nickel in agricultural soils of the United States of America. *J. Environ. Qual. 22*:335–348.

Hornor, S. G., and Mitchell, M. J. (1981). Effect of the earthworm, *Eisenia foetide (oligchaeta)*, on fluxes of volatile carbon and sulfur compounds from sewage sludge. *Soil Biol. Biochem. 13*:367–372.

Ibiebele, D. D., Inyang, A. D., Lawrence, C. H., Coleman, R. L., and Pees, N. (1985). Some characteristics of the behavior of indicator bacteria in sewage-amended soil. *Environ. Pollution (Series A) 39*:175–182.

Jorhem, L., and Sundstorm, B. (1993). Levels of lead, cadmium, zinc, copper, nickel, chromium, manganese, and cobalt in foods on the Swedish market. *J. Food Compos. Anal. 6*:223–241.

Keefer, R. F., Singh, R. N., and Horvath, D. J. (1986). Chemical composition of vegetables grown on an agricultural soil amended with sewage sludges. *J. Environ. Qual. 15*:146–152.

Keeney, D. R., Lee, K. W., and Walsh, L. M. (1975). Guidelines for the application of wastewater sludge to agricultural land in Wisconsin. *Wisconsin Dept. Nat. Resour. Tech. Bull. No. 88*. Madison, WI.

Kelling, K. A., Keeney, D. R., Walsh, L. M., and Ryan, J. A. (1977a). A field study of the agricultural use of sewage sludge: III. Effect on uptake and extractability of sludge-borne metals. *J. Environ. Qual. 6*:352–358.

Kelling, K. A., Peterson, A. E., Walsh, L. M., Ryan, J. A., and Keeney, D. R. (1977b). A field study of the agricultural use of sewage sludge. I. Effect on crop yield and uptake of N and P. *J. Environ. Qual. 6*:339–345.

Kelling, K. A., Walsh, L. M., Keeney, D. R., Ryan, J. A., and Peterson, A. E. (1977c). A field study of the agricultural use of sewage sludge. II. Effect on soil N and P. *J. Environ. Qual. 6*:345–352.

King, L. D. (1988a). Retention of metals by several soils in the southeastern United States. *J. Environ. Qual. 17*:239–246.

King, L. D. (1988b). Retention of cadmium by several soils of the southeastern United States. *J. Environ. Qual. 17*:246–250.

Kinkle, B. K., Angle, J. S., and Keyser, H. H. (1987). Long-term effects of metal-rich sewage sludge application on soil populations of *Bradyrhizobium japonicum*. Appl. Environ. Microb. 35:315–319.

Kirkham, M. B. (1975). Uptake of cadmium and zinc from sludge by barley grown under four different sludge irrigation regimes. *J. Environ. Qual. 4*:423–426.

Kruse, E. A., and Barrett, G. W. (1985). Effects of municipal sludge and fertilizer on heavy metal accumulation in earthworms. *Environ. Pollution (Series A) 38*:235–244.

Kuo, S., and Baker, A. S. (1980). Sorption of copper, zinc, and cadmium by some acid soils. *Soil Sci. Soc. Am. J. 44*:969–974.

Larson, W. E., and Allmaras, R. R. (1971). Management factors and natural forces as related to compaction. *Compaction of Agricultural Soils*. Am. Soc. Agric. Eng., St. Joseph, MI, pp. 367–427.

Larson, W. E., and Clapp, C. E. (1984). Effects of organic matter on soil physical properties. *Organic Matter and Rice*. Int. Rice Res. Inst., Los Baños, Laguna, Philippines, pp. 363–385.

Levine, M. B., Hall, A. T., Barrett, G. W., and Taylor, D. H. (1989). Heavy metal concentrations during ten years of sludge treatment to an old-field. *J. Environ. Qual. 18*:411–418.

Linden, D. R., Larson, W. E., Dowdy, R. H., and Clapp, C. E. (1995). Agricultural utilization of sewage sludge. *Minnesota Agricultural Experiment Station Bulletin 606–1995*.

Liu, R., Palazzo, A. J., Clapp, C. E., Cheng, H. H., and Iskandar, I. K. (1994). Impact of rhizosphere microbial dynamics on sludge-treated soils. *Agron. Abstr.*

Martens, S. N., and Boyd, R. S. (1994). The ecological significance of nickel hyperaccumulation: a plant chemical defense. *Oecologia 98*:379–384.

McBride, M. B., Tyler, L. D., and Hovde, D. A. (1981). Cadmium adsorption by soils and uptake by plants as affected by soil chemical properties. *Soil Sci. Soc. Am. J. 45*:739–744.

McCoy, J. L., Sikora, L. J., and Weil, R. R. (1986). Plant availability of phosphorus in sewage sludge compost. *J. Environ. Qual. 15*:403–409.

McKenna, I. M., Chaney, R. L., Tao, S. H., Leach, R. M., Jr., and Williams, F. M. (1992). Interactions of plant zinc and plant species on the bioavailability of plant cadmium to Japanese quail fed lettuce and spinach. *Environ. Res. 57*:73–87.

Mench, M. J., Martin, E., and Solda, P. (1994). After effects of metals derived from a highly metal-polluted sludge on maize (*Zea mays* L.). *Water Air Soil Pollution 75*:277–291.

Miller, R. W., Azzari, A. S., and Gardiner, D. T. (1995). Heavy metals in crops as affected by soil types and sewage sludge rates. *Commun. Soil Sci. Plant Anal. 26*:703–711.

Mitchell, G. A., Bingham, F. T., and Page, A. L. (1978). Yield and metal composition of lettuce and wheat grown on soils amended with sewage sludge enriched with cadmium, copper, nickel, and zinc. *J. Environ. Qual. 7*:165–171.

Mullins, G. L., and Sommers, L. E. (1986). Characterization of cadmium and zinc in four soils treated with sewage sludge. *J. Environ. Qual. 15*:382–387.

Olness, A. E. (1997). Unpublished model data (in press).

Olson, T. C. (1970). Water storage characteristics of 21 soils in eastern South Dakota. *USDA-Agricultural Research Service Bulletin # ARS-41-166*.

Ottolenghi, A. C., and Hamparian, V. V. (1987). Multiyear study of sludge application to farmland: prevalence of bacterial enteric pathogens and antibody status of farm families. *Appl. Environ. Microb. 53*:1118–1124.

Peterson, A. E., Speth, P. E., Corey, R. B., Wright, T. H., and Schlecht, P. L. (1994). Effect of twelve years of liquid digested sludge application on the soil phosphorus level. *Sewage Sludge: Land Utilization and the Environment* (C. E. Clapp et al., eds.). SSSA Misc. Publ. ASA-CSSA-SSSA, Madison, WI, pp. 237–247.

Pierzynski, G. M., and Jacobs, L. W. (1986). Molybdenum accumulation by corn and soybeans from a molybdenum-rich sewage sludge. *J. Environ. Qual. 15*:394–398.

Rawls, W. J., Gish, T. J., and Brakensiek, D. L. (1991). Estimating soil water retention from soil physical properties and characteristics. *Adv. Soil Sci. 16*:213–234.

Reddy, M. R., and Dunn, S. J. (1986). Heavy-metal absorption by soybean on sewage sludge treated soil. *J. Agric. Food Chem. 34*:750–753.

Sabey, B. R., and Hart, W. E. (1975). Land application of sewage sludge: I. Effect on growth and chemical composition of plants. *J. Environ. Qual. 4*:252–256.

Sikora, F. J., and Wolt, J. (1986). Effect of cadmium- and zinc-treated sludge on yield and cadmium–zinc uptake of corn. *J. Environ. Qual. 15*:341–345.

Skopp, J., Jawson, M. D., and Doran, J. W. (1990). Steady-state aerobic microbial activity as a function of soil water content. *Soil Sci. Soc. Am. J. 54*:1619–1625.

Smith, S. R. (1994). Effect of soil pH on availability to crops of metals in sewage sludge-treated soils. I. Nickel, copper and zinc uptake and toxicity to ryegrass. *Environ. Pollution 85*:321–327.

Sommers, L. E. (1977). Chemical composition of sewage sludges and analysis of their potential use as fertilizers. *J. Environ. Qual. 6*:225–232.

Soon, Y. K., Bates, T. E., Beauchamp, E. G., and Moyer, J. R. (1978). Land application of chemically treated sewage sludge: I. Effects on crop yield and nitrogen availability. *J. Environ. Qual. 7*:264–269.

Straub, T. M., Pepper, I. L., and Gerba, C. P. (1992). Hazards from pathogenic microorganisms in land-disposed sewage sludge. *Rev. Environ. Contam. Toxicol. 132*:55–91.

Sweeney, C., Olness, A. E., Lopez, D., Cordes, J., and Voorhees, W. B. (1996). A model for predicting leaching of nitrate-nitrogen. *Abstr. Minn. Acad. Sci. Annual Meetings. St. Paul. MN.* April 26–27, 1996, p. 14.

Tsai, Y., Palmer, C. J., and Sangermano, L. R. (1993). Detection of *Escherichia coli* in sewage and sludge by polymerase chain reaction. *Appl. Environ. Microb. 59*:353–357.

Voos, G., and Sabey, B. R. (1987). Nitrogen mineralization in sewage-sludge-amended coal mine spoil and topsoils. *J. Environ. Qual. 16*:231–237.

Xue, Q. and Harrison, H. C. (1991). Effect of soil zinc, pH and cultivar on cadmium uptake in leaf lettuce (*Lactuca sativa* L., var. *crispa*). *Comm. Soil Sci. Plant Anal. 22*:975–991.

Yuran, G. T., and Harrison, H. C. (1986). Effects of genotype and sewage sludge on cadmium concentration in lettuce leaf tissue. *J. Am. Soc. Hort. Sci. 3*:491–494.

Zhang, X., Campbell, A. G., and Mahler, R. L. (1993). Newsprint pulp and paper sludge as a soil additive/amendment for alfalfa and bluegrass: greenhouse study. *Comm. Soil Sci. Plant Anal. 24*:1371–1388.

6

Cheese Whey as a Soil Conditioner

Charles W. Robbins and Gary A. Lehrsch *Agricultural Research Service, U.S. Department of Agriculture, Kimberly, Idaho*

I. WHEY PRODUCTION, COMPOSITION, AND CHARACTERISTICS

Whey is the liquid by-product of cheese and cottage cheese manufacture from milk. Each kg of cheese produced results in the production of about 9 kg of whey. In 1993, the U.S. cheese and cottage cheese industry produced approximately 23×10^6 m^3 (6×10^9 gal) of whey (National Agricultural Statistics Service, 1994). Most of this is used directly as livestock feed or concentrated or dehydrated and used in human food and animal feed manufacture. Depending on the locality and economic factors, 20 to 100% of the whey produced is applied for beneficial effects on soils, or is land applied as a disposal procedure.

Fresh whey composition varies depending on the cheese manufacture process used (Table 1). Most cheeses are made by biological culture processes that coagulate the milk proteins, and the resulting whey is often called "sweet" whey. Some cottage and creamed cheeses are made by coagulating the milk proteins using an equivalent of 3 g H_3PO_4 per kg of milk, and the resulting whey is often called "acid" whey. Sodium chloride is also added in some cheese-making processes. Even without salt or acid additions, whey is very salty due to the salts that come from the milk.

Ryder (1980) discusses eleven useful whey byproduct separation methods that separate usable carbohydrates or proteins from the liquid phase. Most of these processes still produce large volumes of liquid waste with essentially the same mineral composition as fresh whey. As a consequence of differences in methods, whey and these other wastes vary in composition somewhat from one

Table 1 Typical Fresh Whey Composition

	Sweet Whey	Acid Whey
Water	92%	92%
Total solids	8%	8%
COD[a]	5–7.5%	5%
pH	3.8–4.6	3.3–3.8
Electrical conductivity	7–12 dS m^{-1}	7–8 dS m^{-1}
	mg kg^{-1}	
Total nitrogen	900–2200	900–2200
Total phosphorus	300–600	1100
Calcium	430–1100	840
Magnesium	90–120	100
Sodium[b]	360–1900	600
Potassium	1000–1400	1000–1400
SAR[b]	4–16	3–4

[a]COD is the chemical oxygen demand.
[b]The sodium concentration and the sodium adsorption ratio (SAR) vary with the amount of salt used in the various cheese manufacturing processes and the fraction that ends up in the whey.

type of cheese to the next as well as the subsequent wastes generated by processing whey to remove butterfat, lactose, or casein for other uses.

Whey is mostly water with only about 8% solids. It is a mild acid with high soluble salt, COD (chemical oxygen demand), and fertilizer nutrient contents compared to most waste waters. Because of these traits, cheese whey is a potential soil amendment or conditioner for many soils if the distance from the production plant to the use site is minimal. If applied in excess, whey can decrease soil productivity and cause environmental degradation.

II. WHEY AS A PLANT NUTRIENT SOURCE

A 10 mm deep (100 m^3 ha^{-1}) whey application applies 90 to 220 kg N ha^{-1}, 30 to 60 kg P ha^{-1} from sweet whey, about 110 kg P ha^{-1} from acid whey and 100 to 140 kg K ha^{-1}, using the concentrations from Table 1.

The main disadvantage of using whey as a fertilizer source is the cost of transporting a material that is 92 to 93 percent water and contains less than 2.5 kg N Mg^{-1} (5.6 lb N ton^{-1}), 0.3 to 1.1 kg P Mg^{-1} (0.7 to 2.5 lb P ton^{-1}) and 1.0 to 1.4 kg K Mg^{-1} (2.2 to 3.1 lb K ton^{-1}) of whey. Unless the cheese manufacturer is willing to accept most of the transportation costs as a whey disposal cost, whey as a fertilizer, or any other amendment for that matter, is not economical. A second disadvantage of using whey as a fertilizer is that whey is produced on a year-round basis. Many crop uses of whey are limited to seasonal application conditions, especially where very wet or frozen soil conditions exist for part of the year. On the other hand, successful year-round application systems have been developed where application rates have been limited to crop fertilizer needs and more than one crop type is treated throughout the year from a particular cheese plant.

The fertilizer value and use potential of whey has been recognized for some time and has been demonstrated on acid soils in high to moderate rainfall areas in Scotland (Berry, 1922), New Zealand (Radford et al., 1986), Nova Scotia (Ghaly and Singh, 1985), Michigan (Peterson et al., 1979), and Wisconsin (Sharratt et al., 1962; Watson et al., 1977). The plant nutrient benefits of land applied whey have more recently been demonstrated on calcareous soils in the 7.6 to 8.8 pH range under irrigation in an arid climate (Robbins et al., 1996; Robbins and Lehrsch, 1992).

Nitrogen in fresh whey is present primarily as proteins, however, nitrate measurement in field soils and laboratory column soils receiving 50, 100, 200, and 300 mm deep applications of whey to a Miami silt loam in Wisconsin showed that under aerobic conditions the organic nitrogen was readily converted to nitrates by soil microflora (Sharratt et al., 1962). The initiation of conversion to nitrate was measured within two weeks of application and continued throughout the first corn (*Zea mays* L.) growing season. Nitrates continued to be produced at reduced rates during the second corn growing season. The nitrification rate appeared to be controlled by the carbon:nitrogen ratio of the whey and treated soil.

Fresh cottage cheese acid whey applied in sodic soil reclamation studies (Jones et al., 1993b) contained 79% ortho-P and 21% organic P. Fresh sweet whey from a plant making swiss and mozzarella cheese used in a whey land disposal study (Robbins et al., 1996) contained 58% ortho-P and 42% organic P. Fresh whey samples collected in 1994 from a cheddar type hard cheese plant, a processed cheese plant, and a plant that produces creamed and mozzarella cheese all contained about 63% ortho-P and 37% organic P.

Acid whey was applied to two sodic soils by Jones et al. (1993b). The first soil was in field plots and the second soil was in greenhouse lysimeters. The acid whey contained an equivalent of 0.3% phosphoric acid on a wet basis and contained 1.05 g P kg^{-1} whey. One-time 0, 25, 50, and 100 mm deep (0, 250, 500, and 1,000 m^3 ha^{-1}) whey applications added 263, 525, and 1,050 kg P ha^{-1}. After the whey infiltrated into the soil, 100, 75, 50, and 0 mm of water was applied to the respective treatments to bring all treatments to the same water content. Seven days later the soil surfaces were tilled to mix the whey into the upper 0.10 m of soil. They were then planted to barley (*Hordeum vulgare* L. cv. Ludd) and irrigated, as needed, until the barley matured.

Table 2 Bicarbonate Extractable ortho-P Concentrations in a Cottage Cheese (Acid) Whey Treated Calcareous Freedom Silt Loam Soil in Greenhouse Lysimeters

Whey (mm)	Total P added (kg ha^{-1})	P extracted (mg P kg^{-1})			
		0–0.15 m	0.15–0.30 m	0.30–0.60 m	0.60–0.90 m
0	0	4.9 a	4.8 a	5.0 a	6.2 a
25	263	14.1 b	6.8 a	4.1 a	5.2 a
50	525	28.9 c	10.2 ab	5.5 a	6.0 a
100	1050	29.6 c	11.5 b	6.3 a	6.8 a

Numbers in a column followed by the same letter are not different at the $P \leq 0.05$ level.

The first part of the study consisted of applying these treatments to a slightly sodic (sodium adsorption ratio (SAR) of 13.3, pH of 8.2, and saturation paste extract electrical conductivity $(EC)_{se}$) of 1.1 dvSm^{-1}) Freedom silt loam (fine-silty, mixed, Xerollic Calciorthids) soil in greenhouse weighing lysimeters (1.0 m deep by 0.30 m diameter, Robbins and Willardson, 1980). The four treatments were randomly replicated three times. The initial bicarbonate extractable ortho-P concentrations (5 mg P kg^{-1} soil) were very low (Table 2). The lysimeter soils were irrigated at a 0.25 leaching fraction until the barley had matured and 0.5 pore volumes of water had drained from the bottom of each lysimeter. At 104 days after planting, the soils were sampled at 0–0.15, 0.15–0.30, 0.30–0.60, and 0.60–0.90 m depth increments. The 0.5 M $NaHCO_3$ extractable PO_4-P concentration in each depth increment was determined using an ascorbic acid method (Watanabe and Olsen, 1965) (Table 2). In the second study (same treatments as the first), treatments were applied to a saline-sodic (SAR of 21, pH of 8.8, and EC_{se} of 27 dS m^{-1}) Declo loam (coarse-loamy, mixed, mesic, Xerollic Calciorthids) soil in 2.0 by 2.0 m field basins. The bicarbonate extractable ortho-P was initially very low (2 to 4 mg P kg^{-1} soil) throughout the sampled profile (Table 3). The four unreplicated treatments were randomly located in a previously nonirrigated grazed range site. Four 150 mm flood irrigations were applied to the basins during the barley growing season.

After the barley matured (59 days after planting), four samples were taken from each basin at 0–0.01, 0.01–0.05, 0.05–0.15, 0.15–0.25, 0.25–0.50, 0.50–0.75, and 0.75–1.00 m depth increments and air dried. The bicarbonate extractable ortho-P concentration in each of the four samples at each depth increment was determined as described above. The study methods are described in greater detail in Jones et al. (1993b).

In the greenhouse study, the bicarbonate extractable ortho-P concentrations were increased in the surface 0.15 m by all whey applications and by the 525 and 1050 kg P ha^{-1} treatments in the 0.15–0.30 m depth increment (Table 2). Below 0.30 m, the bicarbonate extractable ortho-P concentrations were not sig-

Table 3 Bicarbonate Extractable ortho-P Concentrations at Different Depths in a Cottage Cheese (Acid) Whey Treated Saline Sodic Declo Loam Field Soil

Whey (mm)	Total P added (kg ha^{-1})	P extracted (mg P kg^{-1})						
		0.0–0.01 m	0.01–0.05 m	0.05–0.15 m	0.15–0.25 m	0.25–0.50 m	0.50–0.75 m	0.75–1.00 m
0	0	3.8 ± 0.6	4.0 ± 0.2	4.0 ± 0.6	3.6 ± 0.6	2.2 ± 0.7	2.1 ± 0.6	2.9 ± 0.6
25	263	12.2 ± 1.4	13.8 ± 0.8	11.2 ± 2.2	5.9 ± 0.6	2.4 ± 0.6	1.7 ± 0.2	3.0 ± 0.4
50	525	28.3 ± 4.5	32.0 ± 1.6	13.4 ± 3.5	8.3 ± 1.9	3.5 ± 2.1	2.5 ± 1.3	3.3 ± 0.9
100	1050	30.4 ± 1.2	28.1 ± 3.5	15.5 ± 5.8	10.2 ± 1.1	5.9 ± 0.7	4.9 ± 0.1	4.3 ± 1.5
	Initial pH	8.5	8.5	8.5	8.8	9.6	9.7	9.9

Each value shown is the mean of four samples taken at each depth from a single nonreplicated plot.

nificantly changed. In the field plots, the bicarbonate extractable ortho-P concentrations were increased in the surface 0.25 m by all whey application rates (Table 3). There also appears to be a slight increase for the 1050 kg P ha^{-1} treatment down to at least 0.75 m.

The 25 mm whey treatment increased the bicarbonate extractable P to adequate levels of these two very low P soils down to a depth of 0.15 m. It appears that the 50 and 100 mm treatments, upon mixing in the 0.15 to 0.30 m depths, would also bring the surface 0.30 m of soil up to adequate P fertility levels. It does not appear that sufficient P is moving below 0.5 m to be of environmental concern, even though 21% of the original P was in organic forms.

Measurement of saturation extract K movement and exchangeable K changes in acid and calcareous soils suggest that whey K is either mostly inorganic or that it is rapidly released from organic compounds upon whey application to soils and the K becomes readily involved in cation exchange and adsorption reactions (Peterson et al., 1979; Robbins et al., 1996). Trace element concentrations of Al, Fe, B, Cu, Zn, Mn, and Cr are essentially that of whole milk and are too dilute to be of plant nutrient value at reasonable whey application rates (Peterson et al., 1979).

When whey was applied to a Wisconsin Miami silt loam at 0, 50, 100, 200, and 300 mm depth increments on field corn plots, the maximum stover and grain yields were achieved with the 50-mm application the first year after whey additions (Table 4) (Sharratt et al., 1962). The 200-mm whey application produced the greatest stover production the second year, while the 300-mm whey application produced the highest grain yield. Both stover and grain yields de-

Table 4 Effects of Applying Whey to a Miami Silt Loam in the Spring of 1959 on Corn Stover and Grain Yield and Soil Salinity at Planting in 1959 and 1960

Whey added[a] (mm depth)	Corn stover[b] (kg ha^{-1})		Corn grain[b] (kg ha^{-1})		Saturation extract EC (dS m^{-1})	
	1959	1960	1959	1960	1959	1960
0	4030	3070	4870	3930	1.1	1.1
50	6600	5020	7260	5730	3.0	1.8
100	6520	5730	7050	6340	4.4	2.1
200	5870	6920	6630	6620	5.1	3.1
300	5470	6518	6400	7060	6.5	3.5

[a]Each 100 mm of whey added 740 kg N ha^{-1}, 250 kg P ha^{-1}, and 900 kg K ha^{-1}.
[b]Average of duplicate plots.
Source: Adapted from Sharratt et al., 1962.

creased the first year when more than 50-mm of whey was applied and the stover decreased on the 300-mm treatment the second year. Prior to the first planting after the whey application, the saturation extract EC values were drastically increased by the whey additions and were still elevated at the time of the next planting date. Corn forage yield reduction due to soil salinity starts at an EC of about 3 dS m^{-1}, and corn grain yield starts to decrease at an EC of about 2.5 dS m^{-1} (Bresler et al., 1982). Additionally, seedlings are usually more salt sensitive than plants at later growth stages. The yield decreases at the higher whey application rates appear to be salinity induced (see Section VIII). The corn grain N, P, and K concentrations continued to increase with increased whey application rates, even though yields decreased at the higher whey rates (Table 5).

Phosphorus and K leaf concentrations continued to increase with increased whey application on Plano silt loam in Arlington, Wisconsin, when 0, 100, 200, 400, and 800 mm of whey was applied prior to the first crop in a five-year whey treatment study (Peterson et al., 1979). The maximum corn yields were produced with the 200-mm whey rate for the first three years, and the 800-mm whey rate produced the greatest yields the fourth and fifth years. The 100-mm treatment increased the corn yields 2.5, 2.2, 2.2, 1.7, and 2.0 times that of the untreated plots for the first through fifth years of the study. The 200-mm whey plot yields were only slightly greater than the 100-mm plot yields.

These two studies were intended as whey disposal method evaluations and show that approximately 50 to 100-mm (500 to 1000-m^3 ha^{-1}) whey applications provide the needed plant nutrients for maximum crop yields on soils in rainfed crop areas. Neither study gave any soil pH data or indicated whether lime had been applied to the soils.

Table 5 Effects of Whey Nutrients Applied to a Miami Silt Loam in the spring of 1959 on the N, P, and K Contents of Shelled Corn Grown in 1959 and 1960

Whey (mm)	Added nutrients (kg ha^{-1})			Nitrogen (g kg^{-1})		Phosphorus (g kg^{-1})		Potassium (g kg^{-1})	
	N	P	K	1959	1960	1959	1960	1959	1960
0	0	0	0	14.2	13.0	2.4	2.2	3.3	2.9
50	740	250	900	15.9	15.5	3.2	3.2	3.4	3.4
100	1480	490	1790	16.6	15.9	3.3	3.3	3.4	3.3
200	2960	990	3590	17.9	17.2	3.5	3.5	3.6	3.5
300	4430	1480	5370	18.4	16.7	3.5	3.5	3.7	3.5

Source: Adapted from Sharratt et al., 1962.

III. WHEY AS AN AMENDMENT FOR SODIC AND SALINE-SODIC SOILS

Salt affected soils are categorized as normal, saline, sodic, and saline-sodic (Bresler et al., 1982; Robbins and Gavalak, 1989) (see Chapters 7, 8, 9). Normal soils, in this context, are those soils that do not contain sufficient soluble salts or a sufficiently high exchangeable sodium percentage (ESP) or sodium adsorption ratio (SAR) to limit plant growth of salt or high pH sensitive plants. Saline soils contain sufficient soluble salts in the upper root zone to reduce yields of most cultivated or ornamental plants. Total soluble salts are estimated by measuring the electrical conductivity (EC) of saturated soil extracts (Robbins and Wiegand, 1990). A soil with an EC greater than 2 to 4 dS m^{-1} (depending on soil type and plants grown) is considered to be saline due to salt effect on growing plants. Sodic soils have saturation extract ECs less than 2 to 4 dS m^{-1} but have sufficiently high ESPs to destroy soil structure, which in turn reduces aeration and water infiltration rates. When the SAR or ESP values exceed 10 to 15 (depending on soil texture, clay type, and irrigation method), soil physical properties deteriorate, and the soils are said to be sodic. Saline-sodic soils have ECs greater than 2 to 4 dS m^{-1} and SAR or ESP values greater than 15 or 13, respectively. Saline-sodic soils limit plant growth due to high soluble salts; however, if they are leached with low salt water, they will convert to sodic soils with the associated poor physical characteristics.

Because of the high soluble salt concentration in whey, whey should not be applied to saline soils or to normal soils with shallow water tables. Whey should be applied sparingly (25 mm $year^{-1}$) where salt sensitive crops are to be grown (see Bresler et al., 1982, or Robbins and Gavlak, 1989, for salt sensitivity data).

Acid wheys and sweet wheys that contain less than 1000 mg Na kg^{-1} (SAR less than 10) are ideal for reclaiming sodic and saline-sodic soils. These wheys are mild acids that will lower the soil pH by neutralizing soil solution carbonates and bicarbonates and consequently increase the solubility of calcium carbonates (lime) which, in turn, decreases the soil SAR and ESP (Robbins 1985). Whey contains about 5% readily decomposable organic matter (measured as chemical oxygen demand of COD), and its decomposition contributes to the lowering of soil pH by generating additional organic acids and mineralization of nitrogen to nitrate. All wheys are rich in Ca, Mg, and especially K, relative to the Na concentrations, and will replace the exchangeable Na, thus decreasing the SAR and ESP (Robbins, 1984). The high ionic concentration in whey also acts as a flocculating agent in sodic soils, increasing infiltration rates and allowing the Na to be more readily leached from the root zone.

Acid whey applied to a sodic Freedom silt loam (fine-silty, mixed mesic, Xerollic Calciorthids) soil in leaching columns was shown to decrease pH and SAR while increasing aggregate stability (Table 6). The whey was applied at

Table 6 Whey Effects on pH, EC, SAR, ESP, and Aggregate Stability of a Sodic Freedom Silt Loam

Whey depth (mm)	pH	EC (dS m^{-1})	SAR	ESP	Aggregate stability (%)
		0–150 mm soil depth			
0	8.5 a	0.9 a	10.7 a	11.3 a	11 a
20	7.2 b	1.9 b	3.4 b	5.5 b	12 a
40	7.2 b	2.4 c	2.7 c	4.5 c	18 ab
80	6.7 c	3.8 d	1.9 d	2.6 d	22 b
		150–300 mm soil depth			
0	8.5 a	1.4 a	14.9 a	13.3 a	8 a
20	7.4 b	2.9 b	10.5 b	9.2 b	9 a
40	8.3 b	3.7 c	8.9 c	9.1 b	7 a
80	6.8 c	4.2 d	6.2 d	6.2 c	8 a
Original soil	8.3	3.8	16.3	14.9	

Numbers in the same column for the same depth increment followed by the same letter are not significantly different at the p = 0.05 level.
Source: Adapted from Robbins and Lehrsch, 1992.

0, 20, 40, and 80-mm depth (0, 200, 400, and 800 m^3 ha^{-1}) treatments. The 20 mm treatment reclaimed the surface 150 mm of soil and additional leaching with low EC, low SAR water, the 150–300 mm soil depth would be reclaimed by the 40 mm whey application, if not by the 20 mm application (Robbins and Lehrsch, 1992).

Jones et al. (1993b) treated saline-sodic Declo loam (coarse-loamy, mixed, mesic, Xerollic Calciorthids) field plots at 0, 25, 50, and 100-mm whey depths (0, 250, 500, 1,000 m^3 ha^{-1}) (Table 7). The whey was applied and then tilled into the surface followed by planting barley (*Hordeum vulgare* L. *Ludd) and four irrigations with high-quality water. Leaching this soil with high-quality water, without any whey treatment, decreased the pH, EC, SAR, and ESP, but the process caused the soil surface to disperse and seal, reducing air and water entry. Addition of the whey prior to the first irrigation, plus the four irrigations, further reduced pH, EC, SAR, and ESP and increased the infiltration rate. The 50-mm whey treatment reclaimed the surface 50 mm of soil, while the 100-mm whey treatment reclaimed the soil down to at least 150 mm. The irrigation water used on this soil has an EC of 0.2 dS m^{-1} and an SAR of less than 0.5. Consequently, a one-time application of 100 mm of acid whey will permanently reclaim this surface soil. There is not a shallow water table associated with the salinity problem in this soil.

Table 7 Whey Effects on pH, EC, SAR, and ESP at Two Depth Increments and Time to Infiltrate 120 mm of Low EC (0.2 dS m^{-1}) Irrigation Water for a Saline Sodic Declo Silt Loam

Whey depth (mm)	pH	EC	SAR	ESP	Time (h)
		(dS m^{-1})			
		10–50 mm			
0	8.5	1.3 b	9.5 c	11.0 e	54 b
25	8.1	0.9 a	3.7 b	7.9 b	18 a
50	8	1.8 c	3.6 b	5.9 a	17 a
100	7.7	1.2 ab	3.0 a	5.4 a	14 a
		50–150 mm			
0	8.2	1.4 b	9.3 c	9.6 b	
25	8	1.0 a	3.5 a	6.6 a	
50	8	2.1 b	6.0 b	8.6 b	
100	7.8	1.2 a	3.0 a	5.5 a	
Original soil	8.8	27	21	20	

Numbers in the same column for the same depth increment followed by the same letter are not significantly different at the P = 0.05 level.
Source: Adapted from Jones et al., 1993.

IV. WHEY EFFECTS ON AGGREGATE STABILITY

Adding whey decreases soil pH and increases Ca solubility. This, along with the soluble salts in the whey, increases the ionic strength of the soil solution, reducing the diffuse double-layer thicknesses next to the clay, and causes clay flocculation (Lehrsch et al., 1993). This improves aggregation and increases the soil's pore size distribution, allowing increased air and water movement within the soil profile (Hillel, 1982). Aerobic microorganisms that decompose lactose and whey proteins produce polysaccharides that help to stabilize these newly formed aggregates (Allison, 1968). Kelling and Peterson (1981) noted that most whey solids are milk sugars and proteins and are quite susceptible to microbial decomposition. The resultant products of such decomposition substantially improve soil aggregation and tilth.

Improvements in soil structure make soils easier to manage and less susceptible to erosion. As aggregation increases, more large pores (macropores) are formed throughout soil profiles. When these macropores occur at or near the soil surface, infiltration rates increase and runoff rates decrease. Watson et al. (1977) measured increased infiltration rates into a fallow soil about 3 months after sweet whey was surface applied. They attributed the infiltration increases to improved soil structure. Stable aggregates at the soil surface resist fracturing due to raindrop or sprinkler drop impact and slaking as water accumulates on the soil surface. Since fewer surface pores become obstructed with aggregate fragments and primary particles, infiltration rates decrease more slowly and runoff rates are kept relatively low. Low runoff rates minimize offsite sediment movement. As aggregate stability increases, erosion commonly decreases (Luk, 1979). Robbins and Lehrsch (1992) found that the aggregate stability of the uppermost 150 mm of a sodic, Freedom silt loam in laboratory columns doubled from 11 to 22%, when 80 mm of an acid whey was surface applied and incorporated (Table 6). In a subsequent study, Lehrsch et al. (1994) found aggregate stability to increase from 25% to 80% when 80 mm of acid whey was surface applied, and incorporated into sodium-affected soils.

From a soil management standpoint, larger soil aggregates are preferred over smaller ones. Sharratt et al. (1962) applied whey in the spring of 1959 and measured the aggregate size distribution in the 0.18 m plow layer in the fall of 1959 and 1960. The percent of aggregates with diameters >0.25 mm increased with whey application rate. In a field study, Kelling and Peterson (1981) found that the proportion of water-stable aggregates increased as whey application rates increased from 50 to 300 mm. In a greenhouse soil aggregate size distribution study, Kelling and Peterson (1981) found soil aggregation to improve as much from an application of 25 mm of whey (250 m^3 ha^{-1}) as from an application of 22.4 Mg ha^{-1} corn residue or 11.2 Mg ha^{-1} cow manure.

V. CONTROLLING FURROW EROSION IN IRRIGATED AGRICULTURE WITH WHEY

Furrow irrigation-induced erosion is a major problem threatening agricultural productivity in the western U.S. (Carter, 1993). A variety of techniques have been developed and are available to control this erosion (Carter, 1990; Lentz et al., 1992).

When Brown et al. (1996) applied whey to irrigation furrows in the spring prior to the first irrigation, furrow erosion was effectively controlled until the soil was disturbed by cultivation. They measured sediment losses from 91-m furrow lengths with 2.3% slopes that were untreated or treated with straw, whey or straw plus whey (Fig. 1). The whey was applied at about 200 l/min until about 300 l had flowed about 75% of the row length. The whey continued to flow to near the end of the furrow but did not leave the field. The plot area had been planted to sweet corn (*Zea mays* L.). Eight irrigations were applied during the growing season, and the soil was not cultivated after the treatments had been applied. Sediment loss during the eight irrigations for the whey alone was 14% that of the control, 16% for the straw alone, and less than 2% for the straw plus whey. If cultivation of weeds is necessary, the treatment effect is lost and the whey and/or straw must be reapplied to be effective. On shorter, steeper (4.4%)

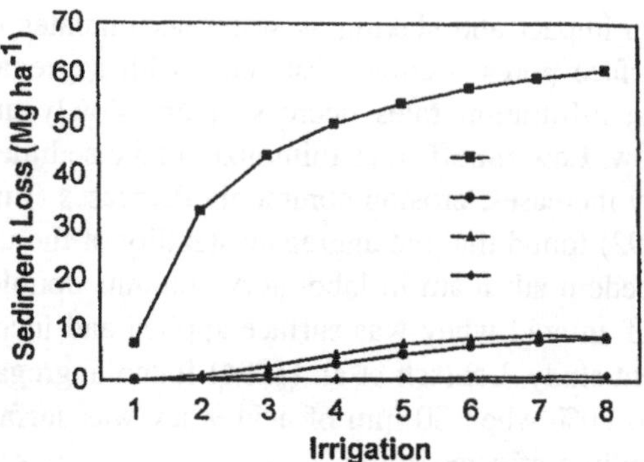

Figure 1 Sediment loss reductions from Portneuf silt loam as a result of treating furrows with whey (●), straw (▲), straw plus whey (◆), and check (■). (Adapted from Brown et al., 1996.)

slopes, the whey alone only reduced the sediment loss by one-third to one-half that of the control, while, straw plus whey still reduced the sediment loss to 3% that of the untreated furrows.

Whey's marked ability to decrease furrow erosion may be a consequence of its ability to increase aggregate stability (Lehrsch et al., 1994) and stabilize soil along wetted perimeters (Brown et al., 1996). When applied to the straw, the sticky nature of the whey also appears to cause the straw particles to stick together and to stick to the soil surface. Lehrsch et al. (1997) found that a single spring whey application increased a Portneuf silt loam's aggregate stability in the uppermost 15 mm of soil in furrow bottoms from 64% in control furrows to more than 83% in treated furrows by early July. These stability increases were correlated with measured decreases in furrow erosion. These research findings confirm what Kelling and Peterson (1981) had observed on acid soils, i.e., that whey-induced increases in aggregate stability (and infiltration) could significantly reduce both runoff and erosion.

VI. INFILTRATION AND HYDRAULIC CONDUCTIVITY CHANGES WITH WHEY APPLICATION

Depending on application rates, soil conditions, and time since last application, whey can increase or decrease infiltration and hydraulic conductivity rates.

Excessive whey applications may decrease infiltration rates and/or hydraulic conductivities in the short term owing to organic overloading. Barnett and Upchurch (1992) noted that high whey applications and the resulting organic matter loading on fine-textured soils caused rapid slime-producing bacteria growth which reduced infiltration rates. They recommended a 2 to 3 week rest period between whey applications for New Zealand soils. On these soils, 10–12 days were required for bacteria to decompose the whey and return the site infiltration rate to its previous rate prior to the next whey application. McAuliffe et al. (1982) measured saturated hydraulic conductivity decreases of 46% within 2 days after they applied 350 m^3 ha^{-1} of sweet whey (also in New Zealand). The hydraulic conductivities increased back to previous rates in 1 to 3 weeks after the whey was applied. Extreme whey applications may adversely affect hydraulic properties for long periods. Plots treated with 2000 m^3 sweet whey ha^{-1} in August 1972 and again in June 1973 by Watson et al. (1977) still had reduced infiltration rates into a dry Wisconsin prairie soil in the fall of 1974. Infiltration into wet soil was not reduced over the control by this date. Applications of 250, 500, and 1000 m^3 ha^{-1} of acid whey to a southern Idaho sodic Freedom silt loam in 0.3 m diameter by 1.0 m deep greenhouse lysimeters decreased infiltration rates measured 11, 27, and 53 days after whey application (Jones et al., 1993b). The lower infiltration rates appeared to be caused by increased microbial activity stimulated by the added organic matter and relatively warm soil conditions (Jones et al., 1993a).

Whey also affects infiltration rates measured under negative heads, that is, under tension. Infiltration measured at slightly negative water potentials (from -30 to -150 mm of water) excludes water flow through the largest soil pores, thus yielding a quantitative estimate of infiltration through the bulk of the soil matrix. Lehrsch and Robbins (1996) measured negative-head infiltration rates into a Portneuf silt loam following the harvest of a winter wheat crop that had been treated with whey during the growing season. They found that, as whey applications increased from 0 to 80 mm, infiltration rates at potentials of —-60 and −150 mm decreased linearly, but slowly (Figs. 2 and 3). At these potentials, flow occurred only through pores with diameters of 0.5 mm or less. The decreases, most pronounced at the highest whey rate, were thought to be caused by organic clogging and microbiological activity. To maintain negative head infiltration at levels comparable to untreated conditions, they recommended that whey applications, if not incorporated, be limited to 40 mm during the growing season. Siegrist and Boyle (1987) suggest that the accumulation of organic material (particularly carbonaceous and nitrogenous compounds) reduce infiltration rates by clogging soil pores and that reducing the amount of organic material applied was necessary to avoid soil clogging.

Whey additions to soils have also been shown to increase infiltration rates when properly applied and managed. Watson et al. (1977) measured up to fourfold increases in infiltration rates (measured using a sprinkling infiltrometer) into a fallow Plano silt loam about 3 months after they surface applied up to 204 mm of whey. They attributed the marked infiltration increases to improved soil structure strength where whey was applied.

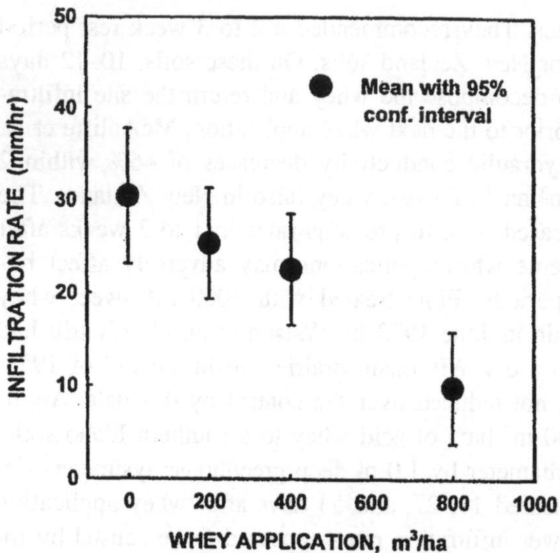

Figure 2 Infiltration rate at a water potential of –60 mm as a function of whey application. (Adapted from Lehrsch and Robbins, 1996.)

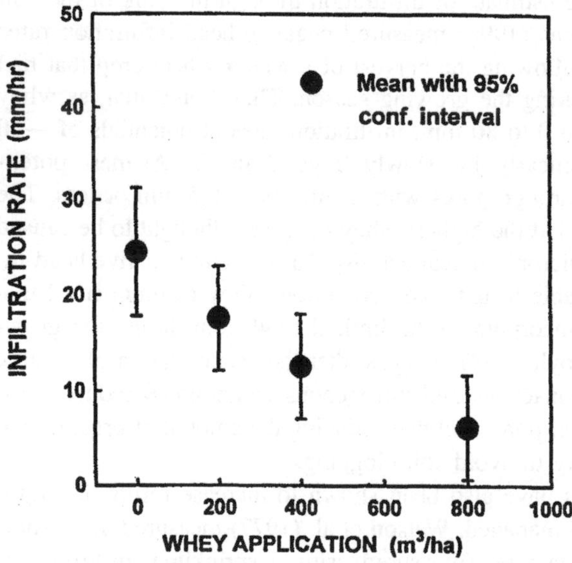

Figure 3 Infiltration rate at water potential of –150 mm as a function of whey application. (Adapted from Lehrsch and Robbins, 1996.)

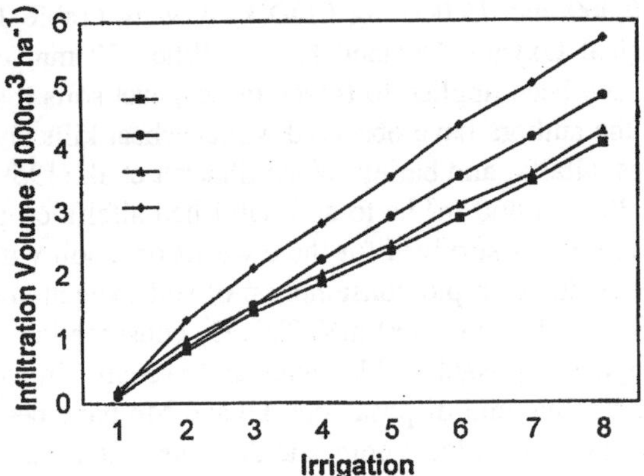

Figure 4 Cumulative infiltration after treating Portneuf silt loam soil furrows with whey (●), straw (▲), straw plus whey (◆), and check (■). (Adapted from Brown et al., 1996.)

Fifty-three days after saline-sodic, Declo loam field plots had been treated with whey, Jones et al. (1993b) measured faster infiltration of high-quality, low-EC (0.20 dS m^{-1}) water into the treated than into the untreated plots (Table 7). The higher infiltration rates were attributed to increased aggregate stability (Lehrsch et al., 1994; Robbins and Lehrsch, 1992) due to lower soil SAR and ESP as a result of the acid whey applications. They concluded that incorporating the whey, followed by an adequate resting period (about four weeks since the whey application), would increase infiltration rates.

Infiltration has also been measured into whey-treated irrigation furrows. Brown et al. (1996) increased furrow infiltration rates by treating furrows with whey, straw, and straw plus whey (Fig.4). Along the wetted perimeter of furrows not treated, a surface seal formed that limited infiltration. Upon drying, cracks appeared in the soil of the whey-treated furrows and remained for the rest of the season. Those cracks provided additional pathways through which irrigation water moved laterally and downward. The straw and straw plus whey placed in the furrows also increased seasonal infiltration.

VII. ENVIRONMENTAL CONCERNS ASSOCIATED WITH EXCESSIVE WHEY APPLICATIONS TO SOILS

As shown, cheese whey has the potential to improve chemical, physical, and possibly microbiological soil conditions. Whey, if applied in excess, also has the potential of degrading soils.

Whey contains between 50,000 and 75,000 mg COD kg^{-1} whey (Table 1) (26,000 to 40,000 mg Biological Oxygen Demand Kg $^{-1}$). When 50 mm or greater applications of whey have been applied to frozen or very wet soils that remain wet for 24 h or more, the authors have observed winter wheat kills and severe crop damage to potatoes, alfalfa, and barley. When Sharratt et al. (1959) weekly applied 0, 140, 290, 430, and 860 m^3 ha to an established alfalfa crop, only the 0 and 140 m^3 ha treated plants survived for three weeks on a soil with a pH of 6.7. The crop damage is due to rapid consumption of soil oxygen and rapid drops in redox potential to as low as −350 mV. This O_2 consumption is due to the oxidation of readily decomposable milk sugars and proteins. In excessive whey applications at a commercial disposal site, Fe and Mn have been solubilized to the extent of contaminating local domestic drinking-water wells. A 10-fold increase in corn leaf Mn and an 8-fold increase in corn leaf Zn concentration measured under 800-mm whey applications by Peterson et al. (1979) suggests that reduced soil redox potential solubilized considerable concentrations of these two metals and made them available for plant uptake and leaching.

Each mm (10 m^3 ha^{-1} of whey applied to the soil adds 400 to 600 kg total salt ha^{-1}. Another way of looking at the salt in whey is that whey would have to be diluted 1:20 with rainwater or distilled water to be considered of acceptable irrigation water quality. The effects of whey application rate on soil EC values are shown in Table 8 from the work of Sharratt et al. (1962). Both that paper and the paper of Peterson et al. (1979) show crop yield leveling off and then decreasing at and beyond 100-mm whey application rates due to increased soil salinity in high rainfall areas. In irrigated areas where soils or irrigation

Table 8 Effects of Whey on Soil EC of a (Miami) Silt Loam Soil Extract During the Two Growing Seasons Following Application

	Saturation extract EC (dS m^{-1}) at 25 C				
Date sampled[a]	Whey applied (mm)				
	0	50	100	200	300
4/20/59	1.20	1.15	1.20	1.20	1.20
5/26/59	1.15	3.00	4.40	5.10	6.45
9/20/59	1.05	1.95	2.10	3.45	3.95
5/27/60	1.10	1.80	2.05	3.05	3.50
9/29/60	1.00	1.65	1.95	2.30	2.50

[a]4/29/59, before whey applied; 5/26/59, 16 days after whey application, just before corn planting; 9/20/59, at corn harvest; 5/27/60, before second corn planting; 9/29/60, at second corn harvest.
Source: Adapted from Sharratt et al., 1962.

water may contain marginal to excessive soluble salt concentrations, whey addition should proceed with caution when salinity is a concern, and soil salinity status should be monitored. Selection of salt tolerant crops (Bresler et al., 1982) should also be part of the management plan for all sites that receive more than 50 mm whey per year.

Acid whey application to a sodic soil effectively lowered the pH and increased its production (Jones et al., 1993b). The acid nature of whey can also adversely lower the pH of acid soils to the point of being injurious to crops. When sweet whey with a pH of 4.0 was added to an acid Spencer silt loam and a near neutral Miami silt loam, the pH of the acid soil was lowered sufficiently for a short period to be injurious to crops (Sharratt et al., 1959). The pH was also lowered in the neutral soil but not sufficiently to cause damage in most crops (Table 9).

When applying whey to irrigation furrows to minimize erosion, precautions should be taken. Because of whey's high COD (Jones et al., 1993a), it should never be released directly into surface waters without treatment. Watson et al. (1977) recommended that whey application rates be kept low enough to prevent runoff from entering surface water bodies. Kelling and Peterson (1981) were even more conservative when they recommended that whey be applied only to rough and/or residue-protected soil surfaces to minimize runoff and control erosion.

Table 9 Soil pH as Affected by Whey Application Rate and Time on Spencer and Miami Silt Loam Soils

Hours after initial application	Spencer silt loam[a]				Miami silt loam[b]			
	Whey applied (m^3 ha)							
	0	90	180	360	0	90	180	360
2	5.2	4.8	4.7	4.6	6.7	6.4	6.2	5.9
4	5.2	4.8	4.7	4.6	6.7	6.3	6.3	6.0
8	5.1	4.8	4.7	4.6	6.8	6.2	6.3	6.6
12	5.2	4.9	4.8	4.6	6.8	5.9	6.2	5.6
24	5.3	5.0	4.9	4.6	6.8	6.0	6.3	5.5
48	5.4	5.2	5.1	4.9	7.1	7.0	6.7	5.7
72	5.3	5.3	5.2	5.0	6.8	7.1	7.1	5.6
96	5.4	5.3	5.2	5.0	6.8	7.4	7.4	6.1
192	5.5	5.4	5.3	5.2	6.8	7.0	7.0	7.1

[a]Initial pH = 5.2, derived from granitic glacial till.
[b]Initial pH = 6.8, derived from limestone glacial till.
Source: Adapted from Sharratt et al., 1959.

VIII. ADDITIONAL RESEARCH NEEDS

Additional research is needed to better characterize the physical and hydraulic properties of soils treated with whey. Physical property changes occurring at and below the soil surface, though not well characterized to date, must be known to apply whey safely to soils for long periods of time. Further research should also examine the effects of whey, incorporated by tillage, on the physical and hydraulic properties of soil surfaces. A likely increase in aggregate stability after tillage (Lehrsch et al., 1994) may offset the infiltration reductions sometimes measured after whey additions.

Ghaly and Singh (1985) also cautioned that continuous applications of whey at high rates could contaminate groundwater with nitrate-N. In the laboratory, they added 32 mm of whey to soil columns. Thereafter, every eight days they applied 100 mm of simulated rainfall, representative of the May through September rainfall received at Halifax, Nova Scotia, Canada. The nitrate-N concentration in the leachate from columns 0.6 to 1.8 m deep ranged from 3.8 to 7.5 mg l^{-1} both 4 and 8 days after the whey was surface applied. Depending upon the climatic regime, irrigation needs, and hydraulic conductivity of the soil on the application site, the potential exists for nitrate-N to be leached from the soil profile to underlying groundwater.

Many crop uses of whey are limited to seasonal application conditions, especially where very wet or frozen soil conditions exist for part of the year. On the other hand, successful year-round application systems have been developed where application rates have been limited to crop fertilizer needs and different crops are treated throughout the year. As an example, hay or pasture sites are well suited to summer and fall whey application, row crops land is suited to winter and spring application, and winter grains are good crops for preplanting fall whey applications.

IX. CONCLUSIONS

Both sweet and acid wheys have been beneficially used to improve physical and chemical soil properties. Whey, especially acid whey, is an ideal amendment for sodic and saline-sodic soils if sodium chloride has not been added during manufacture of the cheese. Incorporated whey increases soil structure and aggregate stability. Whey applied to irrigation furrows, with or without straw mulch, prior to irrigation greatly reduces furrow erosion. Whey is rapidly decomposed when added to soils at moderate (up to 500 m^3 ha^{-1}) rates, and the N, P, and K from whey becomes available to crops within a few days to a few weeks of application. The disadvantages of using whey as a soil amendment or fertilizer include the high water content (92 to 95%), the high COD, and year-round whey production. The high water content limits its value in relation to transportation costs. The high COD limits the application rates to cold or wet soils. Some plant operators are reluctant to allow alternative whey use for short periods, such as soil application, if it interrupts continuous whey flow to livestock feeders or concentrating plants, even though the whey may be more economically disposed of as a soil amendment or fertilizer.

REFERENCES

Allison, F. E., (1968). Soil aggregation—some facts and fallacies as seen by a microbiologist. *Soil Sci. 106*: 136–143.

Barnett, J. W., and Upchurch, G. C. (1992). Irrigation of wastewater from the manufacturing dairy industry onto pasture. The use of wastes and byproducts as fertilizers and soil amendments for pastures and crops. Proceedings, Workshop Fertilizer and Lime Res. (P. E. H. Gregg and L. D. Currie, eds.) Centre, Palmerston North, New Zealand. 19–20 Feb 1992. Occasional Rep. No. 6, Fert. and Lime Res. Centre, Massey Univ., Palmerston North, New Zealand, pp. 195–207.

Berry, R. A. (1992). The production, composition and utilization of whey. *J. Agric. Sci. 13*: 192–239.

Bresler, E., McNeal, B. L., and Carter, D. L. (1982). *Saline and Sodic Soils, Principles–Dynamics–Modeling*. Springer-Verlag, New York.

Brown, M. J., Robbins, C. W., and Freeborn, L. L. (1997). Combining cottage cheese whey and straw reduces erosion while increasing infiltration in furrow irrigation. *J. Soil Water Cons.* (in press).

Carter, D. L. (1990). Soil erosion on irrigated lands. Irrigation of agricultural crops. *Agron. Monogr. 30* (B. A. Stewart, and D. R. Nielson, eds.). ASA, CSSA, and SSSA, Madison, WI., pp. 1143–1171.

Carter, D. L. (1993). Furrow irrigation erosion lowers soil productivity, *J. Irrig. Drainage Eng. 199*:964–974.

Ghaly, A. E., and Singh, R. K. (1985). Land application of cheese whey. Agric. Waste Utilization Management. Proceedings of the Fifth Intl. Symp. on Agric. Wastes. ASAE 16, 17 Dec, 1985, Chicago Illinois. ASAE, St. Joseph, MI., pp. 546–553.

Hillel, D. (1982). *Introduction to Soil Physics*. Academic Press, New York.

Jones, S. B., Hansen, C. L., and Robbins, C. W. (1993a). Chemical oxygen demand fate from cottage cheese (acid) whey applied to a sodic soil. *Arid Soil Res. Rehab. 7*:71–78.

Jones, S. B., Robbins, C. W., and Hansen, C. L. (1993b). Sodic soil reclamation using cottage cheese (acid) whey. *Arid Soil Res. and Rehab. 7*:51–61.

Kelling, K. A., and Peterson, A. E. (1981). Using whey on agricultural land—a disposal alternative. *Cooperative Extension Program Publication Serial No. A3098*, University of Wisconsin, Madison.

Lehrsch, G. A., and Robbins, C. W. (1996). Cheese whey effects on surface soil hydraulic properties, *Soil Use Mgmt. 12*:205–208.

Lehrsch, G. A., Brown, M. J., and Robbins, C. W. (1997). Whey effects on aggregate stability and erosion under furrow irrigation. *J. Soil Water Cons.* (in press).

Lehrsch, G. A., Robbins, C. W., and Hansen, C. L. (1994). Cottage cheese (acid) whey effects on sodic soil aggregate stability. *Arid Soil Res. Rehab. 8*:19–31.

Lehrsch, G. A., Sojka, R. E., and Jolley, P. M. (1993). Freezing effects on aggregate stability of soils amended with lime and gypsum. *Soil Surface Sealing and Crusting. Catena Suppl. 24* (J. W. A. Poesen, and M. A. Nearing, eds.). Catena Verlag, Cremlingen, pp. 115–127.

Lentz, R. D., Shainberg, I., Sojka, R. E., and Carter, D. L. (1992). Preventing irrigation furrow erosion with small applications of polymers. *Soil Sci. Soc. Am. J. 56*:1926–1932.

Luk, S. H. (1979). Effect of soil properties on erosion by wash and splash. *Earth Surface Processes 4*:241–255.

McAuliffe, K. W., Scotter, D. R., MacGregor, A. N., and Earl, K. D. (1982). Casein whey effects on soil permeability. *J. Environ. Qual. 11*:31–34.

National Agricultural Statistics Service (1994). *Agricultural Statistics 1994*. U.S. Government Printing Office, Washington, D.C.

Peterson, A. E., Williams, W. G., and Watson, K. S. (1979). Effect of whey application on chemical properties of soils and crops. *J. Agric Food Chem. 27*:654–658.

Radford, J. B., Galpin D. B., and Parkin, M. F. (1986). Utilization of whey as a fertilizer replacement for dairy pasture. *New Zealand J. Dairy Tech. 21*:65–72.

Robbins, C. W. (1984). Sodium adsorption ratio-exchangeable sodium percentage relationships in a high potassium saline-sodic soil. *Irr. Sci. 5*:173–179.

Robbins, C. W. (1985). The $CaCO_3$–CO_2–H_2O system in soils. *J. Agron. Education 14*:3–7.

Robbins, C. W., and Gavlak, R. G. (1989). Salt and sodium affected soils. *Cooperative Extension Service Bulletin No. 703*, College of Agriculture, University of Idaho, Moscow.

Robbins, C. W., and Lehrsch, G. A. (1992). Effects of acidic cottage cheese whey on chemical and physical properties of a sodic soil. *Arid Soil Res. Rehab. 6*: 127–134.

Robbins, C. W., and Wiegand, C. L. (1990). Field and laboratory measurements. *Agricultural Salinity Assessment and Management* (K. K. Tanji, ed.). ASAE, New York, pp. 201–219.

Robbins, C. W., and Willardson, L. S. (1980). An instrumented lysimeter system for monitoring salt and water movement. *Trans. ASAE 23*: 109–111.

Robbins, C. W., Hansen, C. L., Roginske, M. F., and Sorensen, D. L. (1996). Bicarbonate extractable K and soluble Ca, Mg, Na and K status of two calcareous soils treated with whey. *J. Environ. Qual. 25*: 791–795.

Ryder, D. N. (1980). Economic considerations of whey processing. *J. Soc. Dairy Tech. 33*: 73–77.

Sharratt, W. J., Peterson, A. E., and Calbert, H. E. (1959). Whey as a source of plant nutrients and its effect on the soil. *J. Dairy Sci. 42*: 1126–1131.

Sharratt, W. J., Peterson, A. E., and Calbert, H. E. (1962). Effects of whey on soil and plant growth. *Agron. J. 54*:359–361.

Siegrist, R. L., and Boyle, W. C. (1987). Wastewater-induced soil clogging development. *J. Environ. Eng. 113*:550–566.

Watanabe, F. S., and Olsen, S. R. (1965). Test of an ascorbic acid method for determining phosphorus in water and $NaHCO_3$ extracts from soils. *Soil Science Society of America Proceedings. 29*:677–678.

Watson, K. S., Peterson, A. E., and Powell, R. D. (1977). Benefits of spreading whey on agricultural land. *J. Water Poll. Cont. Fed. 49*: 24–34.

7

Mined and By-Product Gypsum as Soil Amendments and Conditioners

Guy J. Levy *Institute of Soils and Water, Agricultural Research Organization, The Volcani Center, Bet Dagan, Israel*

Malcolm E. Sumner *University of Georgia, Athens, Georgia*

I. INTRODUCTION

A. Use of Gypsum on Cultivated Lands

Gypsum occurs naturally in mineral form and is a by-product of various industrial processes. It has been used for many years as a soil conditioner for sodic and heavy clay soils, as a nutrient source of Ca and S for plant growth, and more recently as an ameliorant for the subsoil acidity syndrome (Shainberg et al., 1989; Sumner, 1993, 1995). These uses of gypsum will be explored in the discussion to follow.

B. Sources of Gypsum and Their Aqueous Chemistry

For centuries, naturally occurring mineral gypsum has been mined for use in agriculture. These deposits have usually resulted from precipitation during the evaporation of inland seas. More recently, nongeological sources of gypsum have become available from industries producing $CaSO_4$ by-products. The most common and important of these is the phosphate fertilizer industry, which produces phosphoric acid from rock phosphate (apatite) by wet-process acidulation:

$$Ca_{10}(PO_4)_6F_2(s) + 10H_2SO_4 + 20H_2O \rightarrow 10CaSO_4.2H_2O(s) + 6H_3PO_4 + 2HF \quad (1)$$

The phosphoric acid is used to manufacture high-analysis P fertilizers, while by-product gypsum accumulates as a waste. This fine-grained, high-purity material, termed phosphogypsum (PG), is produced in large quantities and presents a serious disposal problem.

Many chemical processing and plating industries that produce waste sulfuric acid also commonly generate by-product gypsum by neutralization with hydrated lime:

$$Ca(OH)_2 + H_2SO_4 \rightarrow CaSO_4.2H_2O \quad (2)$$

These materials can be of high purity as long as near-stoichiometric amounts of base are used to prevent contamination of the waste gypsum with other metals in the acid solution. In addition, as a result of environmental legislation to reduce atmospheric acidity, a growing source of waste gypsum arises from the capture of SO_2 from stack gases produced by fossil-fuel-fired electric power plants. This is accomplished by passing the stack gases through a lime slurry, where the solvated SO_2 gas is oxidized to SO_4^{2-}:

$$2H_2O + CaCO_3(s) + SO_2 + 1/2O_2 \rightarrow CaSO_4.2H_2O(s) + CO_2(g) \quad (3)$$

The flue gas desulfurization gypsum (FDG) obtained by this method is of high purity because fly ash and other gas contaminants have been previously removed (Behrens and Hargrove, 1980).

Gypsum is slightly soluble in aqueous solutions (approximately 2.5 g L^{-1}, or 15mM) with the by-product materials dissolving faster than mined gypsum (Fig. 1) due to the larger surface area of the former (Keren and Shainberg, 1981). This level of solubility makes a substantial contribution to the ionic strength of most soil solutions, yet it is low enough to allow continued release of salt into solution over a considerable time period. Other Ca salts are either less ($CaCO_3$) or more soluble ($CaCl_2$, $Ca(NO_3)_2$). Overall dissolution of gypsum in soils is promoted by exchange of Ca for other exchangeable ions, which may have a limited effect on raising equilibrium Ca levels by releasing diverse ions into the soil solution.

II. GYPSUM AS AN AMELIORANT IN ACID SOILS

The use of lime has proved to be highly effective in overcoming acidity in surface soils, but the high costs of incorporation and the fact that it does not move readily down the profile have made it unsuitable for subsoil acidity amelioration. Following the pioneering work of Sumner (1970) and Reeve and Sumner (1972), who first demonstrated the feasibility of using gypsum to counter the effects of subsoil acidity, many cases in which substantial yield responses (10–100%) in a variety of crops (corn, sugarcane, coffee, wheat, alfalfa, soybean, barley) were obtained have been reported (Sumner, 1993, 1994). Examples are presented in Table 1. These types of response have only been obtained on oxic soils in which there are substantial Fe and Al oxide surfaces available to react with the gypsum, as illustrated in Eq. (4). In most cases, the responses, which only appear after the surface incorporated gypsum has been leached into the subsoil, arise as

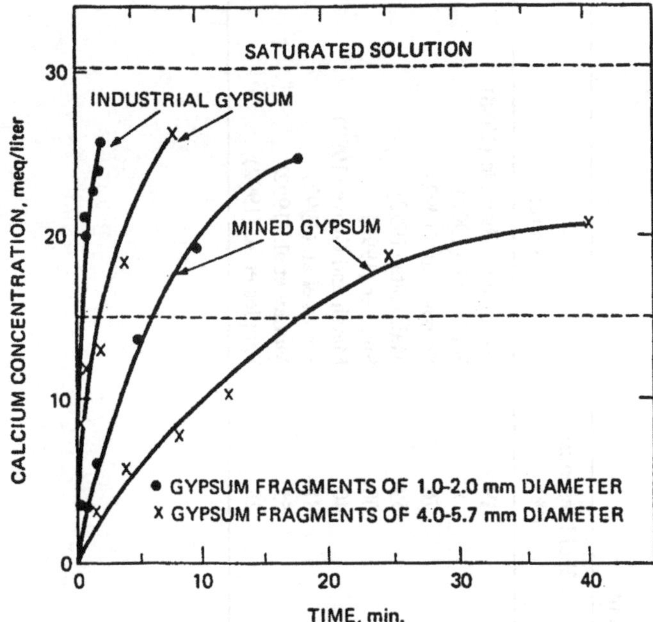

Figure 1 Effect of time and fragment size on the dissolution of various gypsum materials as measured by the Ca concentration in solution. (Keren and Shainberg, 1981.)

a result of increased pH and Ca availability and reduced Al toxicity (Fig. 2). This improved environment allows roots to penetrate the hostile subsoil as illustrated in Table 1. The surface applied gypsum has had a profound effect on root distribution in all crops. In all cases, exchangeable Ca increased and exchangeable Al decreased in the subsoil following gypsum application. As a result of this root proliferation into the acid subsoils, the crops could extract water previously out of the reach of the roots, which translates into improved yields when water is limiting (Table 2). The gypsum effect is more marked under the more severe drought conditions. The effect of a single application of gypsum has been shown to be long-lasting, with the improvement still visible after 13 years (Sumner, 1994). This makes the practice highly economic.

The most likely mechanism responsible for this effect is the so-called "self-liming effect" originally proposed by Reeve and Sumner (1972) as illustrated in the following reactions:

$$2[Fe,Al]\overset{OH}{\underset{OH}{\diamond}}[Fe,Al] + Ca^{2} + SO_4^{2-} \rightarrow 2[Fe,Al]\overset{SO_4^-}{\underset{OH}{\diamond}}[Fe,Al] + Ca(OH)_2 \quad (4)$$

$$2Al^{3+} + 3Ca(OH)_2 \rightarrow 2Al(OH)3 + 3Ca^{2+} \quad (5)$$

Table 1 Yield Responses of Various Crops to Topsoil Incorporated Gypsum

Crop	Location	Soil type	Gypsum rate Mg ha^{-1}	Yield response (%)	Source
Corn	South Africa	Plinthic Paleudult	10	19	Farina and Channon (1988)
Sugarcane	Brazil	Typic Hapludox	6	8	Vitti et al. (1992)
Corn	Brazil	Xanthic Hapludox	6	76	Souza et al. (1992)
Coffee	Brazil	Oxisol	2.6	59	Malavolta (1992)
Alfalfa	Georgia	Typic Halpudult	10	100	Sumner (1990)
Soybean	Kentucky	Typic Hapludult	3.5	40	Marsh and Grove (1992)
Leucena	Brazil	Xanthic Hapludox	6	81	Souza et al. (1992)
Wheat	Australia	Yellow sandplain	9	55	McLay et al. (1992)
Pasture	Australia		2.5	28	Peoples et al. (1992)

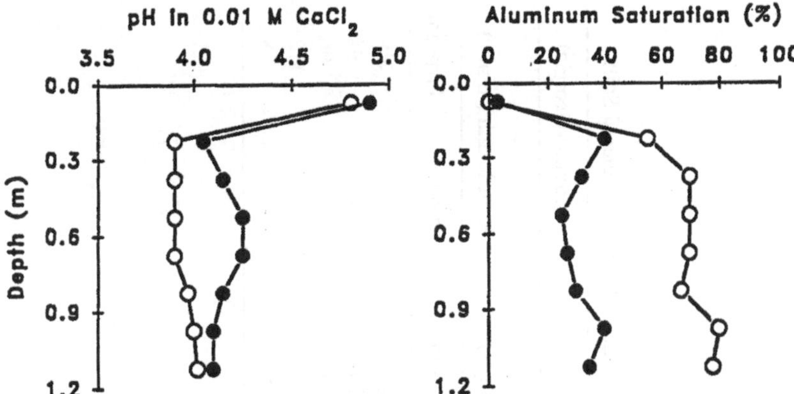

Figure 2 Effect of gypsum (6 Mg ha^{-1}) on soil pH and Al saturation in a Red Yellow Latosol (Xanthic Halpudox) profile from Brazil, (Souza and Ritchey, 1986.)

This basically requires the presence of Al and Fe oxyhydroxides from which the SO_4^{2-} can liberate OH^- by ligand exchange which in turn neutralizes exchangeable Al^{3+}. Other possibilities include the precipitation of aluminum hydroxysulfates (Adams and Rawajfih, 1977), ion pair ($AlSO_4^+$) formation (Pavan et al., 1982, 1984) or cosorption of SO_4^{2-} and Al^{3+} (Sumner et al., 1986). A simple method that involves measuring pH and electrical conductivity of the soil in two dilute salt solutions (0.005 *M* $CaSO_4$ and 0.005 *M* $CaCl_2$) has been developed to assess the potential of a soil to respond to gypsum (Sumner, 1993).

In addition to the beneficial effects on subsoil acidity, gypsum has been shown to increase the recovery of nitrate from subsoil horizons (van Raij et al., 1988) and promote the uptake of nutrients (N, P, K, Ca, S, Cu, and Mn) in an Oxisol (Souza et al., 1992).

Because gypsum elevates the level of Ca in the soil, care should be taken in applying this technology to sandy soils where deficiencies of Mg and K may be induced (Syed-Omar and Sumner, 1991; Alva and Gascho, 1991). The downward movement of Mg and K has been reported as a result of gypsum application (Farina and Channon, 1988; van der Watt and Valentin, 1992; Vitti et al., 1992). To overcome this negative effect, Mg and K applications should be made after the gypsum front has moved into the subsoil.

The use of gypsum to overcome the subsoil acidity syndrome is being commercially applied to sugarcane in South Africa, Brazil, and Australia, to alfalfa in Georgia, USA, and to coffee in Guatemala, with good results in all cases.

Table 2 Effect of Gypsum Application on the Distribution of Roots of Various Crops Down the Profile of Highly Weathered Soils

	Corn (South Africa)		Corn (Brazil)		Apples (Brazil)		Alfalfa (Georgia)	
	Root density (m L^{-1})		Relative root distribution (%)		Root density (cm g^{-1})		Root length (m m^{-3})	
Depth (cm)	Control	Gypsum	Control	Gypsum	Control	Gypsum	Control	Gypsum
0–15	3.10	2.95	53	34	50	119	375	439
15–30	2.85	1.60	27	25	60	104	40	94
30–45	1.80	2.00	10	12	18	89	11	96
45–60	0.45	3.95	8	19	18	89	52	112
60–75	0.08	2.05	2	10	18	89	4	28

Source: Sumner, 1993.

III. GYPSUM AS AN AMENDMENT FOR SOIL PHYSICAL PROPERTIES

A. Introduction

Soil permeability to water depends to a great extent on both the exchangeable sodium percentage (ESP) of the soil and the electrolyte concentration (EC) of the percolating solution, tending to decrease with increasing ESP and decreasing EC (Quirk and Schofield, 1955; McNeal and Coleman, 1966; McNeal et al., 1968). Soil permeability can be maintained even at high ESP levels, provided that the EC of the percolating water is above a critical (threshold) level (Quirk and Schofield, 1955). Conversely, when very good quality water is used, even an ESP of 5 caused a two-order of magnitude decrease in the permeability of soils (McIntyre, 1979).

Soil permeability to water can be characterized by two different types of measurements, hydraulic conductivity (HC) and infiltration rate (IR). Hydraulic conductivity measurements are made under conditions in which the soil surface is not disturbed; conversely, considerable surface disturbance (beating action of raindrops, stirring effect of flowing water) occurs under IR measurements. Disturbance of the soil surface during water infiltration renders the surface more sensitive to low sodicity (ESP < 5), as mechanically disturbed soil particles disperse at lower sodicity than undisturbed soil (Emerson, 1967; Rengasamy et al., 1984). In addition, the EC of the soil solution at the soil surface is determined predominantly by the applied water, whereas that in the soil profile depends also on the rate of dissolution of soil constituents. For instance, in a calcareous soil, the EC in the soil solution in the profile will seldom be less than 2 $mmol_c$ L^{-1} even when leached with distilled water. On the other hand, unless a rapidly soluble source of gypsum (by-product) is available at the soil surface, insufficient electrolyte will be present to prevent dispersion and reduce IR.

Gypsum (both from geological and nongeological sources) is the most commonly used amendment for sodic soil reclamation, primarily because of its low cost. Gypsum added to a sodic soil can increase permeability by means of both EC and cation-exchange effects (Loveday, 1976). The relative importance of the two effects is of interest for several reasons. If the electrolyte effect is sufficiently great to prevent clay dispersion and swelling, surface application of gypsum may be worthwhile. In this case, the amount of gypsum required depends on the amount of high-quality water applied and the rate of gypsum dissolution. It is somewhat independent of the amount of exchangeable Na in the soil profile. Conversely, in soils where the EC effect is of lesser importance than cation exchange, the amount of gypsum required depends on the amount of exchangeable Na in a selected depth of soil. The cation-exchange process has formed the basis of several "gypsum requirement" tests (Dutt et al., 1972; Oster and Frenkel, 1980; U.S. Salinity Laboratory Staff, 1954), whereas EC effects have, in general, received less attention.

The following sections discuss the potential of gypsum products in preventing soil degradation and maintaining adequate soil hydraulic and physical properties in cultivated soils.

B. Effect of Gypsum on Soil Hydraulic Conductivity (HC)

1. Hydraulic Conductivity: Definition and Controlling Mechanisms

Water flow in soils takes place in accordance with Darcy's equation:

$$\frac{Q}{At} = \frac{K\Delta H}{\Delta L} \tag{4}$$

where Q is the volume of water passing through the soil in time t, A is the cross-sectional area of the soil column, ΔH is the hydraulic head difference, ΔL is the soil interval ($\Delta H/\Delta L$ is the hydraulic gradient), and K is the hydraulic conductivity (HC) of the soil. Soil HC is usually measured empirically and includes soil (tortuosity, pore size distribution, etc.) and percolating fluid (viscosity) properties.

Because swelling and/or dispersion of soil colloids alter the geometry of soil pores, they are the two main mechanisms affecting soil HC (Quirk and Schofield, 1955). Swelling of clay occurs when the electrical repulsion forces between two negatively charged particles increase. The electric double layer repulsive force, also called "swelling pressure," can be calculated by means of diffuse double layer theory (van Olphen, 1977). Clay dispersion, which is the formation of a stable suspension of particles in water, is caused by the mutual repulsion between particles to a distance where the attractive forces are no longer strong enough to negate the repulsive forces, and the particles can move in response to an external force. An increase in swelling pressure and/or clay dispersion results in the narrowing or blocking of the water-conducting pores in the soil.

The differences between swelling and dispersion of soil colloids are important. Swelling is a continuous process and decreases gradually with increase in solution EC. It is hardly affected by low ESP levels (below 10–15) but increases significantly as the ESP increases above 15. Swelling is essentially a reversible process (Shainberg and Letey, 1984). On the other hand, dispersion is sensitive to the presence (even at low levels) of highly hydrated monovalent cations (Na^+), increased pH, and low EC in the soil solution. In smectitic soils, clay dispersion depends on the type of exchangeable cation and on the composition and concentration of the soil solution (Quirk and Schofield, 1955). In kaolinitic soils, clay dispersion depends mainly on the pH of the soil solution (Schofield and Samson, 1954) and the presence of dispersive anions (Shanmuganathan and Oades, 1983). Clay dispersion and particle movement are irreversible processes.

2. Effect of Electrolyte Concentration on the HC of Soils

Because the permeability of a soil to water depends on soil ESP and on the percolating solution EC, Quirk and Schofield (1955) developed the concept of "threshold concentration." This is the EC of the percolating solution that causes a 10–15% decrease in soil permeability at a given ESP value. A unique threshold concentration exists for each soil (McNeal and Coleman, 1966).

Felhendler et al. (1974) observed for two smectitic soils (Netanya sandy loam and Nahal Oz calcareous silt loam) that both were only slightly affected by SAR (up to 20) as long as the EC exceeded 10 $mmol_c\ L^{-1}$ (Fig. 3). However, when the percolating solution was replaced by distilled water, simulating rainfall, the response of the two soils differed drastically. At ESP values of 10 and 20, the HC of the calcareous silt loam and sandy loam soils dropped to 42 and 18% and 5 and 0% of the initial values, respectively. Felhendler et al. (1974) also noted that the clay in the sandy loam soil was mobile and appeared in the leachate, whereas no clay dispersion occurred in the silt loam soil.

Differences in the sensitivity of the HC of these soils to changes in EC can be explained on the basis of differential salt release by mineral dissolution during leaching, which determines the EC (Shainberg et al., 1981a,b). Young soils may release 3–5 $mmol_c\ L^{-1}$ of Ca and Mg into the solution as a result of the

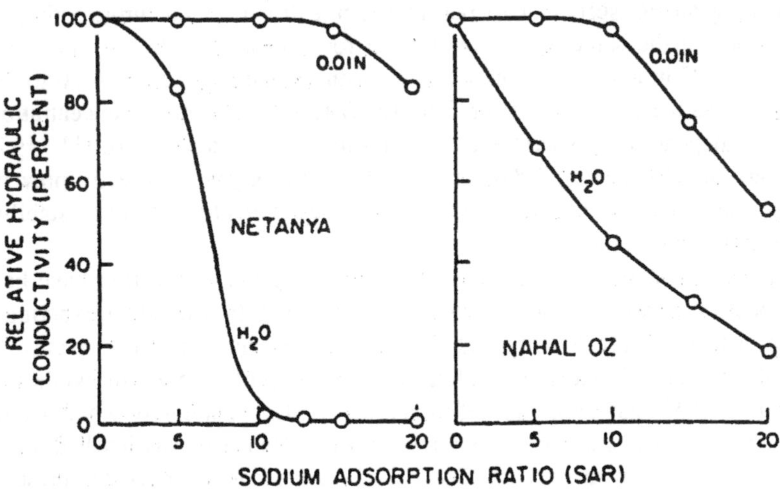

Figure 3 Relative hydraulic conductivities of sandy (Netanya) and silt loam (Nahal Oz) soils from Israel as functions of sodium adsorption ratio (SAR) and concentration of the leaching solutions. (Felhendler et al., 1974.)

dissolution of plagioclase, feldspars, hornblende, and other minerals (Rhoades et al., 1968), while calcareous soils can subtend 2–3 $mmol_c\ L^{-1}$ (Shainberg et al., 1981b). Soils containing minerals that readily release soluble electrolytes will not easily disperse when leached with distilled water (simulating rainwater) at moderate ESP values, because they maintain sufficiently high EC in the soil solution (> 3 $mmol_c\ L^{-1}$) to prevent clay dispersion, behavior exhibited by the Nahal Oz calcareous silt loam (Fig. 3). Conversely, soil solution EC in soils that do not contain readily weatherable minerals is likely to be below the threshold concentration to maintain flocculation, thus rendering the soils more susceptible to clay dispersion and HC decreases, behavior exhibited by the Netanya sandy loam. Similar results for California soils have been reported by Shainberg et al. (1981b).

3. Soil HC Reclamation with Gypsum

Gypsum (mined and by-product) is by far the most commonly used amendment for maintaining adequate hydraulic properties in sodic soils because of its low cost, availability, and ease of handling. Its value has been known for a long time: Hilgard (1906), for example, discussed its role in preventing deflocculation, thus allowing leaching to proceed, and Kelley and Arany (1928) reported successful soil reclamation with gypsum. A large volume of literature has developed over the years leading to a good general understanding of its role in preventing HC deterioration. Gypsum promotes hydraulic properties through its effect on both EC and cation exchange as discussed below.

Effect of Electrolyte Concentration. Gypsum readily dissolves in the soil, maintaining an EC sufficient to permit water entry into and through the soil profile, and at the same time provides Ca for exchange with Na. Potentially achievable EC in a soil solution saturated with gypsum ranges from 30 to 266 $mmol_c\ L^{-1}$ as the ESP varies from 0 to 40 (Oster, 1982). Actual concentrations achieved depend on experimental conditions (Quirk and Schofield, 1955; Chaudhry and Warkentin, 1968; Dutt et al., 1972). Gypsum directly prevents swelling and dispersion and indirectly increases porosity, structural stability, and hydraulic properties.

Water penetration and storage benefits from gypsum application have been described for a variety of soils throughout the world. In a ponding experiment on a brown sodic clay, McIntyre et al. (1982) found that, without gypsum, 292 mm infiltrated in 379 days of ponding, wetting the profile to 2.1 m, while with gypsum at 10 Mg ha^{-1}, 605 mm infiltrated in 145 days, enough of which passed deeper than 2 m to raise the groundwater level. Chloride was leached 1.0 m without and 2.8 m with gypsum. These data clearly demonstrated the presence of a region of low HC in the upper profile that was ameliorated by the addition of gypsum, which in turn may lead to improved crop and pasture yields.

In order to take advantage of the fact that worthwhile soil and crop responses may be obtained from quite modest gypsum applications through the electrolyte

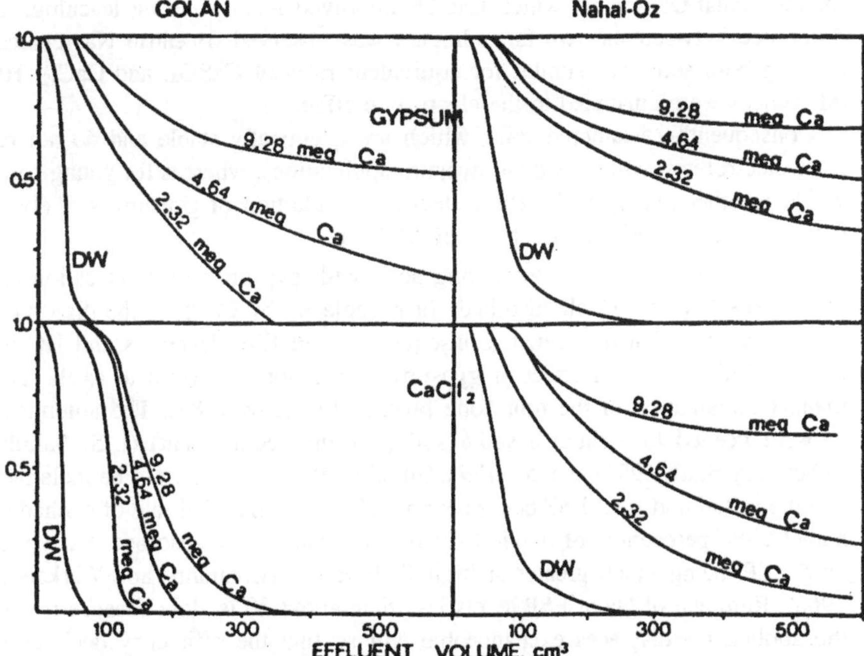

Figure 4 Effect of varying applications of $CaSO_4$ and $CaCl_2$ on the relative hydraulic conductivities of the Golan and Nahal Oz soils as a function of cumulative effluent volume. (Shainberg et al., 1982.)

effect, techniques for the recognition of gypsum-responsive soils are a prerequisite. Shainberg et al. (1982) studied the relative importance of the electrolyte and reclamation (cation exchange) effects on the HC of sodic soils (Fig. 4). The contribution of electrolyte was estimated by comparing the HC of the sodic soils when reclaimed by equivalent amounts of $CaCl_2$ and $CaSO_4$. Although the exchange reclamation was similar with both amendments, the highly soluble $CaCl_2$ was leached out of the soil profile rather quickly. The slightly soluble gypsum, on the other hand, was slowly released into the leaching water. Depending on the rate of water application and gypsum dissolution, sufficient EC could be maintained to prevent dispersion and accompanying HC reductions.

The relative importance of the electrolyte provided by gypsum depends on soil properties (Shainberg et al., 1982), the most relevant of which is the soil's ability to release electrolytes when leached with distilled water. In Fig. 4, the Golan soil, which is $CaCO_3$ free and does not release appreciable electrolyte, was very sensitive to the type of Ca amendment added, with complete sealing taking place in the $CaCl_2$ but not in the gypsum treatments. On the other hand, for the Nahal-Oz soil, in which $CaCO_3$ dissolved readily during leaching, no difference between the two amendments was observed. Because Na replacement in both soils was similar for equivalent rates of $CaSO_4$, and $CaCl_2$, HC differences were attributed to the electrolyte effect.

Consequently, weathered soils, which are chemically stable and do not release electrolytes, will respond to gypsum applications, whereas for young soils, which weather readily and release electrolyte, addition of gypsum is likely to have a limited beneficial effect on soil HC.

Cation Exchange Effects. Amending soils with gypsum results in cation exchange reactions in which adsorbed Na is replaced by Ca from the dissolving gypsum and the Na is leached. Consequently, soil ESP decreases and HC increases. This beneficial effect of gypsum is commonly referred to as the permanent reclamation of the root zone profile (Loveday, 1984). The amount of gypsum needed to reclaim a sodic soil (gypsum requirement) (U.S. Salinity Laboratory Staff, 1954), is *inter alia* a function of the soil depth to be reclaimed and the initial and final ESP being sought. The efficiency and rate of exchange, namely, the percentage of applied Ca that exchanges for adsorbed Na, varies with ESP, being much greater at high ESP values (Chaudhry and Warkentin, 1968). Removal of Na at ESP levels less than about 10 is slow because part of the applied Ca displaces exchangeable Mg so that the efficiency declines to about 30% (Loveday, 1976). Efficiency may also be low (20–40%) in fine-textured soils because of the slow exchange of Na inside the structural elements (Manin et al., 1982).

Tanji and Deveral (1984) described representative modeling approaches (chromatographic and miscible displacement) for the reclamation of sodic soils, appraised their strengths and weaknesses, and suggested future research directions. The models include many of the known chemical reactions occurring in natural soils as a result of $CaCO_3$ and $CaSO_4$ dissolution, salt leaching, and exchange reactions. Chromatographic models (Dutt et al., 1972; Tanji et al., 1972) assume that local equilibrium occurs between the plates for chemical processes within the solution phase, between the solution and exchanger phases, and between the solution and soil mineral phases. They compute monovalent-divalent exchange using the Gapon and/or Davis equations, taking solubilities of gypsum and $CaCO_3$ into consideration. Both models have been successfully field tested.

However, when using models, assumptions must be made about the desired level of reclamation, i.e., to what level must ESP be reduced, and to what soil depth is the ESP reduction required? There is no unique level of ESP that can be regarded as critical. The soil surface, being very susceptible to sodicity (see Section C), may tolerate only very low ESP (<3), whereas at depths below the cultivated layer, ESP >15 may be tolerable. Thus the objective in reclaiming sodic soils might be only to spread gypsum on the soil surface where it will release enough electrolyte to prevent clay dispersion and HC decline both at the surface and within the soil profile. Thus frequent applications (once a year, before the rainy season) of small amounts of gypsum, preferably by-product because it dissolves readily, may be the optimal practice in gypsum reclamation.

C. Effect of Gypsum on Seal Formation and Infiltration Rate (IR)

1. Seal Formation and Infiltration Rate

Seal formation in soils exposed to the beating action of rain or irrigation drops ("structural" seal) is caused by two mechanisms: (a) mechanical aggregate breakdown caused by drop impact, which reduces the average pore size in the surface layer and increases compaction of the uppermost soil layer (McIntyre, 1958; Chen et al., 1980), and (b) physicochemical clay dispersion causing clay migration into the soil with the infiltrating water, clogging of the conducting pores immediately beneath the surface, and formation of a layer of low permeability ("washed-in" zone) (McIntyre, 1958). These mechanisms act simultaneously as the first enhances the second. The seal formed is characterized by higher density and shear strength, finer pores, and lower saturated HC than the underlying soil (McIntyre, 1958; Bradford et al., 1987).

The degree of seal formation is commonly quantified by the soil infiltration rate (IR) (Shainberg and Letey, 1984), which is defined as the volume flux of water entering the soil per unit surface area, and has the dimensions of velocity. In general, IR is high at first, particularly when the soil is initially dry, and decreases monotonically to approach an asymptotic constant rate—the final IR (FIR). In soils with stable surface structures, decreases in IR arise from the inevitable decrease in matric suction gradient as infiltration proceeds (Baver et al., 1972). Decreases in IR from an initially large value can also result from gradual deterioration of surface structure due to the formation of a surface seal. When a seal with a very low HC forms, the IR determines the permeability of the soil (Baver et al., 1972; Morin and Benyamini, 1977).

Changes in IR, and hence susceptibility to seal formation, are significantly affected by the EC of the soil solution at the surface (i.e., that of the applied water) and the ESP of the soil (Agassi et al., 1981; Kazman et al., 1983).

2. Controlling Seal Formation with Gypsum

Chemical dispersion of surface soil during rain can be prevented by maintaining soil solution EC at the soil surface above a critical level, termed the critical flocculation concentration (CFC), which ensures that clay remains flocculated. Application of by-product gypsum (or any readily available electrolyte source) at the soil surface is an effective measure to prevent clay dis-

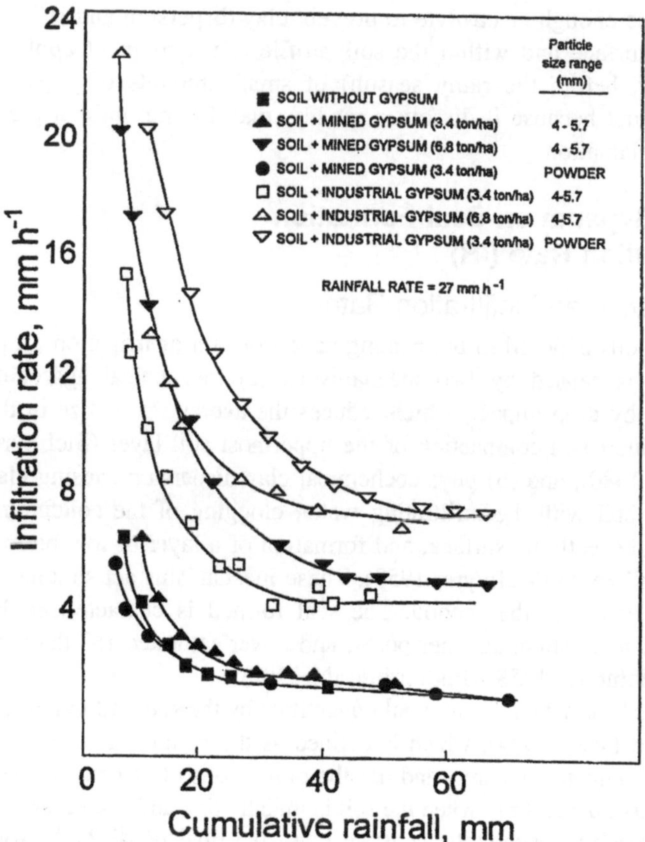

Figure 5 Effects of varying rates and particle sizes of mined and by-product gypsum materials on the infiltration rate of a loess soil as a function of cumulative rainfall. (Keren and Shainberg, 1981.)

persion. Mined gypsum is usually not suitable for this purpose because of its slower dissolution.

The effect of gypsum source, amount, and particle size on the IR of a loess with ESP 30 is illustrated in Fig. 5 (Keren and Shainberg, 1981). Without gypsum, the IR decreased sharply to a constant value of 2 mm h^{-1} (Fig. 5). When 3.4 Mg ha^{-1} of powdered PG and mined gypsum were spread on the soil surface, FIRs were 7.5 and 5.5 mm h^{-1}, respectively. When coarser particles were used (4–5.7 mm), mined gypsum had no effect on the IR irrespective of the amount applied, while PG was effective in maintaining high IR, which increased with PG rate.

Phosphogypsum rates and application methods for controlling IR were studied by Agassi et al. (1982) on three Israeli smectitic soils. Surface applied PG rates of 3, 5, or 10 Mg ha^{-1} had very similar effects on IR. Spreading PG over the soil surface was more effective than mixing it to a depth of 5 mm, because when it was mixed, only 20% was available in the upper 1 mm of soil where the seal forms. Moreover, the effect of surface applied PG was still evident after six consecutive storms of 35 mm at 3–5 days intervals. Thus in soils susceptible to seal formation, incorporating by-product gypsum into the soil should be discouraged. Aggregate size at the soil surface (0–3 mm and 0–10 mm) affected PG efficiency in controlling seal formation, which was higher with small than with large aggregates. This result, which was unexpected, being the reverse of the behavior in the control treatment, was explained by the PG's erosion from the large aggregates to the intervening depressions, which promoted seal formation on aggregate surfaces, and enhanced IR in the depressions (Agassi et al., 1982). This effect was verified by Agassi et al. (1985b) who found PG requirements of 10 and 5 Mg ha^{-1} for coarse and smooth reliefs, respectively.

Phospgypsum application is highly effective in reducing seal formation and maintaining IR over a wide range of ESP values in both calcareous and noncalcareous soils (Kazman et al., 1983) (Fig. 6). As soil ESP increases in the absence of PG, the FIR and depth of rain required to reach it decrease due to dispersion caused by sodicity. Surface application of 5 Mg ha^{-1} PG drastically improved IR and FIR even at low ESP levels (2.2), indicating that some chemical dispersion can take place even in soils almost devoid of Na.

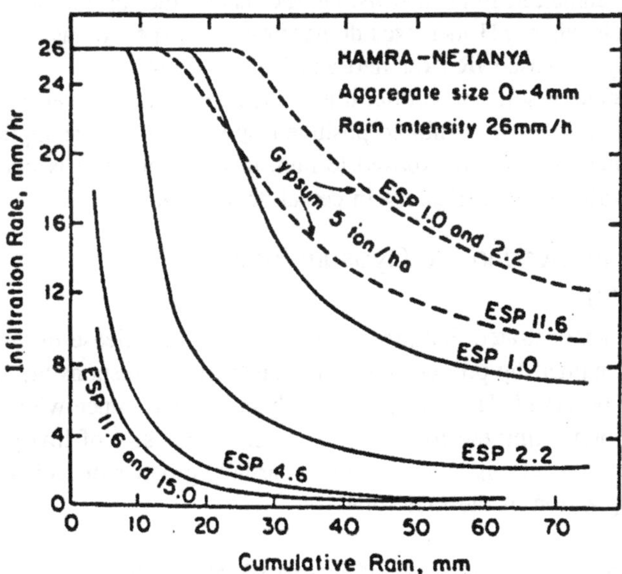

Figure 6 Effect of exchangeable sodium percentage (ESP) on the infiltration rate of a sandy loam (Netanya) soil. (Kazman et al., 1983.)

The beneficial effect of surface application of PG relative to incorporation (5 and 10 Mg ha^{-1}) in maintaining high IR and reducing runoff has been confirmed in the field with natural rainfall in Israel on two bare calcareous Haploxeralfs (ESP 4.6 and 19.3) (Keren et al., 1983). However, in a separate experiment, increased plot length reduced runoff due to water storage in surface depressions, and PG only increased wheat yields from 0.48 to 0.69 Mg ha^{-1}, which is not economic (Agassi et al., 1985b). Consequently, under field conditions, the PG effect is likely to be smaller than that measured in the laboratory.

By-product gypsum (5 Mg ha^{-1}) has also proved to be effective in controlling seal formation and maintaining high IR on kaolinitic soils from both humid (southeastern United States) and semiarid (South Africa) regions (Miller, 1987; Miller and Scifres, 1988; Stern et al., 1991a, b; Zhang and Miller, 1996). The presence of smectitic impurities in the clay fraction promoted dispersion, and the efficiency of PG in reducing runoff was inversely related to rain intensity. The effect of PG was relatively short-lived and disappeared after about 700 mm of rainfall (Stern et al., 1991a).

Norton et al. (1993) demonstrated that the efficacy of various FDG materials in promoting IR and reducing runoff on three humid soils of the eastern United States depended on their relative solubilities. However, gypsum materials from fluidized bed processes (FBC) that are alkaline increase pH and enhance dispersion, particularly in variable-charge soils. Consequently, the EC subtended by the FBC gypsum may be insufficient to flocculate the clay with seal formation being enhanced in such soils. On the other hand, in dominantly permanent-charge soils where pH increases do not promote dispersion as much, FBC gypsum actually increased IR (Reichert and Norton, 1994).

Because a wide range of soils (smectitic, illitic, and kaolinitic) in diverse climatic zones (humid to arid) have exhibited positive responses to surface applied by-product gypsum materials when exposed to rain, this phenomenon is likely to be encountered worldwide, particularly in cereal growing areas.

3. Mechanisms Involved in the Gypsum Effect on Soil Sealing

Agassi et al. (1986) demonstrated that processes other than the release of electrolyte from surface applied by-product gypsum were partially responsible for the beneficial effects observed. These included mechanical interference with the structure of the seal preventing the formation of a continuous seal of low permeability, and the gypsum serving as a mulch that partially protects the soil surface from the beating action of the raindrops.

D. Effect of Gypsum on the IR of Depositional Seals

Depositional seals form by translocation of soil particles a certain distance from their original location and their deposition as the water infiltrates (Arshad and

Mermut, 1988; Chen, et al., 1980). Depositional seals typically form under furrow or basin irrigation systems where sediment originates from erosion of the furrow sides by hydraulic shear forces during irrigation. Deposition of suspended material and formation of depositional seals on furrow or basin bottoms proceeds progressively during irrigation and accounts for decreased infiltration, which leads to major problems in crop production (Oster and Singer, 1984).

Only recently have the mode of formation of depositional seals and their effect on water infiltration been studied. The HCs of laboratory depositional crusts were 2 to 3 orders of magnitude lower than that of the bulk soil when the EC of the applied suspension was less than 3 $mmol_c L^{-1}$ (Shainberg and Singer, 1985). At EC values >5 $mmol_c L^{-1}$, crust HC was only one order of magnitude lower than that of the bulk soil. Similarly, when 5 Mg ha^{-1} PG was spread on the soil surface before low-electrolyte irrigation water was used, high rates of water penetration were maintained. The beneficial effect of the PG was obtained for three 130-mm suspension applications indicating a lasting effect (Shainberg and Singer, 1985).

Because EC plays similar critical roles in both structural (Agassi et al., 1981) and depositional seal formation, Shainberg and Singer (1985) postulated that the dispersion–flocculation status of the suspended sediments determined the structure and hence the hydraulic properties of the depositional seal formed, which was confirmed by micromorphological studies of depositional seals formed from suspensions with various EC values (Southard et al., 1988). Seals formed with low EC water exhibited high bulk density (1.95 Mg m^{-3}), which was reduced (1.41 Mg m^{-3}) when PG was surface applied. Suspensions in distilled water penetrated at least 5 mm into the bulk soil and produced crusts with highly birefringent layers of clay oriented parallel to the soil surface. At the higher EC levels produced by PG, clays existed as floccules which formed more porous seals because of the random orientation of particles (not parallel to the soil surface).

E. Effect of Gypsum on Soil Erosion

1. Introduction

Erosion by water begins when raindrops strike a bare soil, detaching soil particles, which are subsequently transported downslope by splash or overland flow (runoff) once rainfall exceeds IR. When runoff is limited, water flows in thin sheets, and sheet (or interrill) erosion results, but when water velocity exceeds 0.30 m s^{-1} (Ellison, 1947), flow becomes turbulent and causes rills to form. The hydraulic shear forces developed by the concentrated and confined water flow lead to efficient detachment and transport of soil particles, thus causing rill erosion (Lane et al., 1987).

The susceptibility of soils to water erosion depends on a number of factors. Soils with aggregates stabilized by organic and inorganic colloids are less prone to interrill erosion than others because seals do not form (limited runoff) and the aggregates are too large to be transported by thin overland flow (Wischmeier and Smith, 1978). Conversely, soils high in silt (Wischmeier and Mannering, 1969), or Na (Singer et al., 1982) or lacking structural cements (Miller, 1987) are susceptible to erosion. Because strong positive correlations exist between soil erodibility and clay dispersion (Miller and Baharuddin, 1986; Wischmeier and Mannering, 1969), dispersive soils are more likely to be erodible than nondispersive soils.

2. Gypsum and Soil Loss from Gentle Slopes (<10%)

The effects of by-product gypsum on erosion from rain simulation on gentle slopes for kaolinitic and smectitic soils are summarized in Table 3. Rain energy, intensity, and depth varied for the various studies, so a quantitative comparison is impossible. Nevertheless, all soils developed seals with low FIR (<10 mm h^{-1}) and runoff accounted for 40–80% of the rain applied in the untreated soils. By-product gypsum application (5 Mg ha^{-1}) resulted in decreased runoff (0.3–2.5-fold) and soil loss (≈ 50%) compared with the control (Table 3). Under field conditions, PG controlled soil erosion even more effectively (3–5-fold less than control) (Agassi et al., 1985b).

Because interrill erosion (Lane et al. 1987) predominates on gentle slopes (Walker et al., 1977), benefits of gypsum are relatively small over controls, as in both cases soil detachment is restricted, on the one hand, by high seal shear strengths in untreated soils (Bradford et al., 1987), and on the other by the stabilized surface in gypsum amended soils (Warrington et al., 1989).

Because the detachment capacity of runoff is small and depends on runoff volume and velocity, which in turn is a function of slope and surface roughness, runoff causes less interrill erosion than raindrop impact (Young and Wiersma, 1973). Gypsum application reduces runoff volume to approximately half that of untreated soil, and by increasing surface roughness and flowpath tortuosity it decreases runoff velocity.

Because gypsum increases EC in both runoff and percolating water, clay dispersion is prevented (Agassi et al., 1981), which results in larger aggregates at the soil surface that are less susceptible to erosion in runoff. The higher EC of runoff water from gypsum-amended soil reduces sediment concentration in the runoff water and promotes deposition under gravity during an erosion event (Rose, 1985). Miller (1987) found no clay-size particles in runoff water from PG-treated soils, whereas in untreated soils, 15–30% of the sediment was clay. Thus gypsum treatment increases the size of the suspended material, leading to enhanced sediment deposition and less soil loss.

Table 3 Effect of Phosphogypsum (PG) on Infiltration Rate (IR), Percent Runoff, and Soil Loss from Soils with Gentle Slopes

Soil classification	Dominant clay minerals[a]	Final IR ($mm\ h^{-1}$)		Runoff (%)		Soil loss ($Mg\ ha^{-1}$)	
		Control	PG	Control	PG	Control	PG
Cecil[b]	K	10.0	25.0	42	12.7	0.266	0.096
Wedowee[b]	K	5.0	17.1	72	31.7	1.135	0.442
Worsham[b]	K	4.0	12.0	74	55.7	1.315	0.732
Rhdoxeralf[c]	St, K	2.9	8.3	66	32.3	9.350	4.140
Maimi[d]	Expandable	4.3	8.4	71	66	0.180	0.150
Opequon[d]	Mica	2.1	7.5	74	65	0.240	0.080

[a]K = kaolinite, St = smectite, Expandable = smectite, vermiculite, etc.
[b]Georgia, U.S. soils (Miller, 1987).
[c]Israeli soil (Warrington et al., 1989).
[d]Indiana and Maryland, U. S. soils (Norton et al., 1993)

3. Gypsum and Soil Loss from Steep Slopes

The fact that both empirical-based (Universal Soil Loss Equation [USLE]) and process-based (Water Erosion Prediction Project [WEPP]) models for erosion prediction include slope as a factor indicates its significance. For soils susceptible to sealing, slope has a complicated effect on IR, runoff, and erosion (Poesen, 1987). As slope increases, runoff velocity and shear forces increase, which leads to seal erosion resulting in subsequent increases in IR and decreases in runoff. The effects of increasing slope were attributed to (1) more splash and sheet erosion, which removed the seal continuously; (2) fewer raindrop impacts per unit surface area; (3) decreases in the normal component of drop impact force; and (4) the presence of a thin water film, which reduces compaction due to raindrop impact (Palmer, 1964).

Warrington et al. (1989) investigated the effect of PG on erosion from a smectitic sandy loam soil at slopes ranging between 5% and 30%. The FIR increased for both treated and untreated soil with increasing slope (Fig. 7a), confirming the findings of Poesen (1987) and supporting the explanation that the

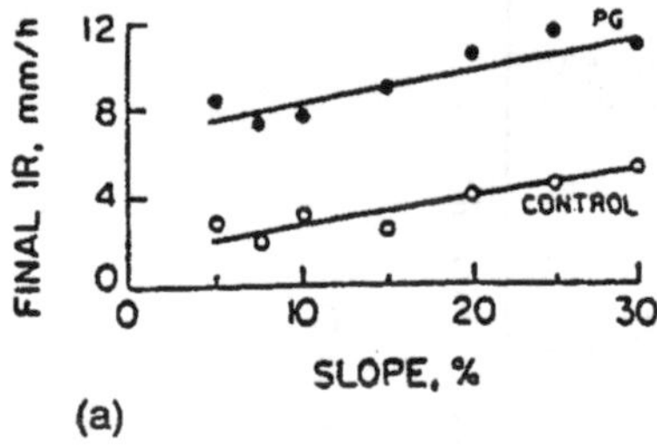

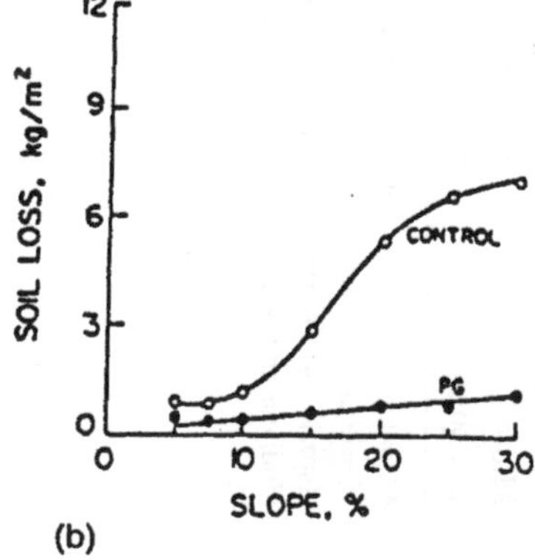

Figure 7 Effects of slope and phosphogypsum (PG) on (a) final infiltration rate, and (b) soil loss. (Warrington et al., 1989.)

IR increases owing to removal of the seal by erosion. Because erosion is not so pronounced on gypsum-treated soils, a different explanation for the increase in IR with increased slope is required. In the presence of gypsum, clay dispersion is prevented, and the seal is formed mainly by drop impact. With increase in slope, both the number of drops impacting a unit surface area and the normal component of their impact decrease. Consequently, a less well developed seal is formed and the IR increases (Agassi et al., 1985a).

The striking effect of PG on soil loss at various slopes is illustrated in Fig. 7b. Increasing slope from 5% to 25% increased soil loss by 668% and 190% for the untreated and PG-treated soils, respectively (Warrington et al, 1989). Large increases in soil loss in the control soil occurred only at slopes >10% (Fig. 7b). At gentle slopes (<10%), runoff velocities are low, and the main cause of soil loss is soil detachment by raindrop impact. As raindrop detachment is insensitive to slope, erosion changes only slightly with slope <10%, but once this is exceeded, runoff velocity is high enough to produce hydraulic shear forces capable of detaching soil particles from the soil surface, and thus enhance erosion. At high slopes (>10%), large (7.5 × 20 mm) rills formed in the untreated soil, accelerating soil loss. In the PG-treated soil, on the other hand, no rills formed, and the soil surface maintained a certain degree of roughness at all slope angles, which restricted runoff velocity and prevented high soil loss on steep slopes (Warrington et al., 1989).

Under field conditions on a Chromoxerert in Israel, PG reduced soil loss to 5–10% and 1–3% of that in the control on gentle and steep slopes, respectively (Agassi el al., 1990). Addition of PG reduced runoff from 50% in the control to 15% in the treated soil and was only slightly influenced by slope. The beneficial effect of PG in reducing soil loss was due not only to the decrease in runoff but also to decreased sediment concentration, which supports results from laboratory studies using rainfall simulators (Warrington el al., 1989). Further field experiments on a Typic Rhodoxeralf (Agassi and Ben Hur, 1991) showed on a 48% slope that, although PG reduced runoff by 23% irrespective of runoff plot length, erosion was 5.4 times greater in long (10 m) than short (1.5 m) plots with PG efficiency in reducing erosion being smaller in long than in short plots. In short plots, erosion was determined by interrill erosion processes, whereas in long plots, rill erosion is the predominant mechanism. Thus once the slope factor (slope angle × length) is large enough, rills form and accelerate erosion, partially negating the beneficial effects of PG. On three kaolinitic soils from the eastern U.S., Norton et al. (1993) observed that by-product gypsum products were not as effective in reducing soil loss on 30% slopes as on 5% slopes, thus questioning the use of by-product gypsum on high slopes.

The differential efficacy of by-product gypsum (5 Mg ha^{-1}) in controlling soil loss on steep slopes in smectitic and kaolinitic soils may be explained by the results of Stern (1990), who separated the soils into stable (no swelling minerals) and unstable (some swelling minerals) soils. The latter were more erodible, because they dispersed and formed a less permeable seal that generated higher levels of runoff. Phosphogypsum addition significantly reduced soil loss from both stable and unstable soils. However, the effect of PG was more pronounced in the more erodible soils, especially on the steeper slopes. This phenomenon is attributed to the greater susceptibility of unstable soils to clay dispersion, which is offset by gypsum treatment. Under less erosive conditions (gentle slopes or stable soils), gypsum efficacy in controlling erosion is less dramatic, though still important (Stern, 1990).

3. Mechanism of Gypsum Effects on Soil Erosion

Surface spread by-product gypsum releases electrolytes at the soil surface to the percolating and runoff water and thus decreases soil erosion by the following mechanisms:

1. The fraction of rain that penetrates the soil increases, and hence runoff depth decreases.
2. The stability of surface aggregates increases, and fewer particles are detached by either raindrop impact or overland flow.
3. Velocity of runoff flow decreases by maintaining roughness at the surface.
4. Flocculation and deposition of suspended clay-size particles is enhanced.

The efficiency of by-product gypsum materials in reducing soil erosion from various slopes extends to soils of varying mineralogy. Thus these products are universally useful in stabilizing cultivated soils on gentle slopes and natural or artificial steep slopes.

F. Effect of Gypsum on Mechanical Strength

1. Surface Impedance

The formation of surface soil seals that restrict water penetration may also have an adverse effect on surface mechanical impedance. Upon drying, sealed surfaces become much harder than nonsealed surfaces, and thus emergence and early development of seedlings, especially vegetable crops, are frequently impaired, resulting in poor stands that often require replanting. The occurrence of crusts (the term crust is introduced to distinguish between wet phase seal formation and dry phase crust formation) of high mechanical strength that inhibit emergence has been widely documented (Cary and Evans, 1974). Crusts with high shear strength are promoted by clay dispersion at the soil surface; the formation of washed-in layers in dispersive soils enhances crust strength and thereby often severely limits emergence. Nondispersive soils may also form emergence-inhibiting crusts because of aggregate breakdown and compaction under rainfall. These often occur in humid areas where high intensity rains are common, as crust impedance depends strongly on rain intensity (Morrison et al., 1985).

Farmers often irrigate to reduce crust strength and promote emergence, but this is expensive and consumes water. Surface applied by-product gypsum has been effective in reducing crust strength and increasing emergence on dispersive soils in India and Georgia (Chaudhri and Das, 1977; Bennett et al., 1964). In Australia, Loveday (1974) showed a general relationship between crust strength, emergence, and HC for a range of dispersive soils and demonstrated increases of 100% in emergence with gypsum treatment. Further work by Shanmuganathan and Oades (1983b) has confirmed the positive effect of gypsum on emergence and showed the importance of reducing clay dispersion in establishing good stands. These conclusions have been implemented in a recommendation for gypsum application on dispersive, hardsetting wheat soils in Australia to improve both emergence and water acceptance (Howell, 1987).

Most of the reported emergence responses to applied gypsum have been in controlled laboratory studies, with only limited field testing. Observations in the southeastern United States (W. P. Miller, unpublished data) have shown that rainfall conditions after planting are a crucial factor determining the shear strength of the seal formed upon drying and the probability of a response to by-product gypsum additions. Crusts of low impedance are promoted by high-intensity rainfall immediately after planting, followed by intense drying. Improved emergence from gypsum application is only likely in years when such conditions are prevalent.

2. Subsoil Impedance

Poor subsoil root growth has often been considered as one of the main causes for limited crop yields, especially in weathered soils from humid regions. The lack of root proliferation in subsoil horizons has been attributed, in part, to unfavorable chemical conditions (Al toxicity and/or Ca deficiency) (Adams and Moore, 1983) (see Section II), and high subsoil impedance. Root growth is severely limited when soil mechanical impedance rises above 2.0 MPa (Taylor et al., 1966), a value frequently exceeded in many soils (Cassel et al., 1978; Chancy and Kamprath, 1982).

The ability of surface applied gypsum to improve these unfavorable subsoil chemical conditions is well documented (Reeve and Sumner, 1972; Pavan et al., 1984; Oates and Caldwell, 1985) (see Section II). The ability of surface applied gypsum to promote root growth in subsoils with high mechanical impedance has been demonstrated on a Typic Hapludult with deep rooted crops (alfalfa and peaches) (Radcliffe et al., 1986). The gypsum on entering the subsoil promotes aggregation in conjunction with the deep rooted crop, thereby reducing mechanical impedance.

IV. CONCLUSIONS

Both mined and by-product gypsum have a role to play in ameliorating the chemical and physical properties of soils. Because gypsum promotes clay flocculation, it reduces the tendency of soils to form seals and thereby prevents serious infiltration rate declines, improves internal soil drainage, and reduces the mechanical impedance of surface crusts and subsoil hardpans. In addition, gypsum ameliorates the subsoil acidity syndrome by supplying soluble Ca and reducing toxic levels of Al, thereby permitting the proliferation of roots into previously hostile subsoils. In applications where electrolyte is required at the soil surface, such as in preventing seal formation, the use of by-product gypsum is essential owing to its more rapid dissolution rate. For all other applications, such as subsoil amelioration (both chemical and physical), either form is suitable.

REFERENCES

Adams, F., and Z. Rawajfih. Basaluminite and alunite: a possible cause of sulfate retention by acid soils. *Soil Sci. Soc. Am. J. 41*:686 (1977).

Adams, F., and B. L. Moore. Chemical factors affecting root growth in subsoil horizons of Coastal Plain soils, *Soil Sci. Soc. Am. J. 47*:99 (1983).

Agassi, M., and M. Ben-Hur. Effect of slope length, aspect and phosphogypsum on runoff and erosion from steep slopes. *Aust. J. Soil Res. 29*:197 (1991).

Agassi, M., I. Shainberg, and J. Morin. Effect of electrolyte concentration and soil sodicity on infiltration rate and crust formation. *Soil Sci. Soc. Am. J. 45*:848 (1981).

Agassi, M., J. Morin, and I. Shainberg. Infiltration and runoff control in the semi-arid region of Israel. *Geoderma 28*:345 (1982).

Agassi, M., J. Morin, and I. Shainberg. Effect of raindrop impact energy and water salinity on infiltration rates of sodic soils. *Soil Sci. Soc. Am. J. 49*:186 (1985a).

Agassi, M., J. Morin, and I. Shainberg. Infiltration and runoff in wheat fields in the semi-arid region of Israel. *Geoderma 36*:263 (1985b).

Agassi, M., I. Shainberg, and J. Morin. Slope, aspect, and phosphogypsum effects on runoff and erosion. *Soil Sci. Soc. Am. J. 54*:1102 (1990).

Alva, A., and G. Gascho. Differential leaching of cations and sulfate in gypsum amended soils. *Commun. Soil Sci. Plant Anal. 22*:1195–1206 (1991).

Arshad, M. A., and A. R. Mermut. Micromorphological and physico-chemical characteristics of soil crust types in northwestern Alberta, Canada. *Soil Sci. Soc. Am. J. 52*:724 (1988).

Baver, L. D., W. H. Gardner, and W. R. Gardner. *Soil Physics*. Wiley, New York, 1972.

Behrens, G. P., and O. W. Hargrove. *Evaluation of Chiyoda Thoroughbred 121 FGD Process and Gypsum Stacking*. EPRI Project 536-4. Electric Power Research Institute, 1980.

Benites, J. R. Transfer of acid tropical soils management technology. *Management of Acid Tropical Soils for Sustainable Agriculture* (P. A. Sanchez. E. R. Stoner, E. Pushparajah, and C. L. Garver, eds.). IBSRAM, 1985, p. 245.

Bennett, O. L., D. A. Ashley, and B. D. Doss. Methods of reducing soil crusting to increase cotton seedling emergence. *Agron. J. 56*:162 (1964).

Bradford, J. M., J. E. Ferris, and P. A. Remley. Interrill soil erosion processes: I. Effect of surface sealing on infiltration, runoff, and soil splash detachment. *Soil Sci. Soc. Am. J. 51*:1566 (1987).

Cameron, K. C., N. P. Smith, C. D. A. McLay, P. M. Fraser, R. J. McPherson, D. F. Harrison, and P. Harbottle. Lysimeters without edge flow: an improved design and sampling procedure. *Soil Sci. Soc. Am. J. 56*:1625 (1992).

Cary, J., and D. D. Evans. *Soil Crusts*. Tech. Bull. No. 214. University of Arizona, 1974.

Cassel, D. K., H. D. Bowen, and L. A. Nelson. An evaluation of mechanical impedance for three tillage treatments on Norfolk sandy loam. *Soil Sci. Soc. Am. J. 42*:116 (1978).

Chancy, H. F., and E. J. Kamprath. Effects of deep tillage on nitrogen response by corn on a sandy Coastal Plain soil. *Agron. J. 74*:657 (1982).

Chaudri, K. G., and D. K. Das, Physical characteristics of soil crusts. *J. Soc. Exp. Agric.* 2:40 (1977).

Chaudhry, G. H., and B. P. Warkentin, Studies on exchange of sodium from soils by leaching with calcium sulphate. *Soil Sci. 105*:190 (1968).

Chen, J., J. Tarchitzky, J. Morin, and A. Banin. Scanning electron microscope observations on soil crusts and their formation. *Soil Sci. 130*:135 (1980).

Deshpande, T. L., D. J. Greenland, and J. P. Quirk. Role of iron oxides in the bonding of soil particles. *Nature 201*:107 (1964).

Dutt, G. R., R. W. Terkeltoub, and R. S. Rauschkolb. Prediction of gypsum and leaching requirements for sodium-affected soils. *Soil Sci. 14*:93 (1972).

Ellison, W. D. Soil erosion studies. IV. Soil detachment by surface flow. *Agric. Eng. 28*:442 (1947).

Emerson, W. W. A classification of soil aggregates based on their coherence in water. *Aust. J. Soil Res. 5*:47 (1967).

Farina, M. P. W., and P. Channon. Acid-subsoil amelioration: II. Gypsum effects on growth and subsoil chemical properties. *Soil Sci. Soc. Am. J. 52*:175 (1988).

Felhendler, R., I. Shainberg, and H. Frenkel. Dispersion and hydraulic conductivity of soils in mixed solution. *Trans. 10th Interna. Congress Soil Science 1*:103 (1974).

Hilgard, E. W. *Soils, Their Formation, Properties, Composition and Relation to Climate and Plant Growth in the Humid and Arid Regions*. Macmillan, New York, 1906.

Howell, M. Gypsum use in the wheatbelt. *J. Agric. W. Aust. 28*:40 (1987).

Kazman, S., I. Shainberg, and M. Gal. Effect of low levels of exchangeable Na and applied phosphogypsum on the infiltration rate of various soils. *Soil Sci. 135*:184 (1983).

Kelley, W. P., and A. Arany. The chemical effect of gypsum, sulphur, iron sulphate and alum on alkali soil. *Hilgardia 3*:393 (1928).

Keren, R., and I. Shainberg. Effect of dissolution rate on the efficiency of industrial and mined gypsum in improving infiltration of a sodic soil. *Soil Sci. Soc. Am. J. 45*:103 (1981).

Keren, R., I. Shainberg, H. Frenkel, and Y. Kalo. The effect of exchangeable sodium and gypsum on surface runoff from loess soil. *Soil Sci. Soc. Am. J. 47*:1001 (1983).

Lane, L. J., G. R. Foster, and A. D. Nicks. Use of fundamental erosion mechanics in erosion prediction. ASAE Paper No. 87-2540, 1987.

Loveday, J. Recognition of gypsum-responsive soils. *Aust. J. Soil Res. 25*:87 (1974).
Loveday, J. Relative significance of electrolyte and cation exchange effects when gypsum is applied to a sodic clay soil. *Aust. J. Soil. Res. 14*:361 (1976).
Loveday, J. Amendments for reclaiming sodic soils. *Soil Salinity Under Irrigation* (I. Shainberg, and J. Shalhevet, eds.). Springer-Verlag, 1984, p. 220.
Malavolta, E. O gesso agricola no ambiente e na nutricao da planta: perguntas e respostas. *II Sem. Uso Gesso Agricultura*, 1992, p. 41.
Manin, M. A. Pissarra, and J. W. Van Hoorn. Drainage and desalinization of heavy clay soil in Portugal. *Ag. Water Manag. 5*:227 (1982).
Marsh, G. H., and J. H. Grove. Surface and subsurface acidity: soybean root response to sulfate-bearing spent lime. *Soil Sci. Soc. Am. J. 56*:1837 (1992).
McIntyre, D. S. Permeability measurements on soil crusts formed by raindrop impact. *Soil Sci. 85*:185 (1958).
McIntyre, D. S. Exchangeable sodium, subplasticity and hydraulic conductivity of some Australian soils. *Aust. J. Soil Res. 17*:115 (1979).
McIntyre, D. S., J. Loveday, and C. L. Watson. Field studies of water and salt movement in an irrigated swelling clay soil. I. Infiltration during ponding. II. Profile hydrology during ponding. III. Salt movement during ponding. *Aust. J. Soil Res. 20*:81 (1982).
McNeal, B. L., and N. T. Coleman. Effect of solution composition on soil hydraulic conductivity. *Soil Sci. Soc. Am. J. 30*:308 (1966).
McNeal, B. L., D. A. Layfield, W. A. Norvell, and J. D. Rhoades. Factors influencing hydraulic conductivity of soils in the presence of mixed-salt solutions. *Soil Sci. Soc. Am. J. 32*:187 (1968).
Miller, W. P. Infiltration and soil loss of three gypsum-amended Ultisols under simulated rainfall. *Soil Sci. Soc. Am. J. 51*:1314 (1987).
Miller, W. P., and M. K. Baharuddin. Relationship of soil dispersibility to infiltration and erosion of southeastern soils. *Soil Sci. 142*:235 (1986).
Miller, W. P., and J. Scifres. Effect of sodium nitrate and gypsum on infiltration and erosion of a highly weathered soil. *Soil Sci. 145*:304 (1988).
Morin, J., and Y. Benjamini. Rainfall infiltration into bare soils. *Water Resour. Res. 13*:813 (1977).
Morrison, M. W., L. Prunty, and J. F. Giles. Characterizing strength of soil crusts formed by simulated rainfall. *Soil Sci. Soc. Am. J. 49*:427 (1985).
Norton, L. D., I. Shainberg, and K. W. King. Utilization of gypsiferous amendments to reduce surface sealing in some humid soils of the eastern USA. *Soil Surface Sealing and Crusting* (W. A. Poesen, and M. A. Nearing, eds.). Catena Verlag, 1993, p. 77.
Oates, K. M., and A. G. Caldwell. Use of by-product gypsum to alleviate soil acidity. *Soil Sci. Soc. Am. J. 49*:915 (1985).
Oster, J. D. Gypsum use in irrigated agriculture: a review. *Fert. Res.* 3:73 (1982).
Oster, J. D., and I. Shainberg. Flocculation value and gel structure of Na/Ca montmorillonite and illite suspensions. *Soil Sci. Soc. Am. J. 44*:955 (1980).
Oster, J. D., and M. J. Singer. *Water Penetration Problems in California Soils*. University of California, 1984.
Palmer, R. S. The influence of a thin water layer on water drop impact forces. *Int. Assoc. Hydro. Sci. 65*:141 (1964).

Pavan, M. A., F. T. Bingham, and P. F. Pratt. Toxicity of aluminum to coffee (*Coffea arabica* L.) in Ultisols and Oxisols amended with CaCo3, $MgCO_3$, and $CaSO_4{\cdot}2H_2O$. *Soil Sci Soc. Am. J. 46:*1201–1207 (1982).

Pavan, M. A., F. T. Bingham, and P. F. Pratt. Redistribution of exchangeable calcium, magnesium, and aluminum following lime or gypsum applications to a Brazilian Oxisol. *Soil Sci. Soc. Am. J. 48*:33 (1984).

Peoples, M. B., V. F. Burnett, A. M. Ridley. Evaluation of strategies to ameliorate subsurface acidity in pasture soils. *Aust. Soil Acid. Res. Mewlet 9*:36 (1992).

Poesen, J. A. W. The role of slope angle in surface seal formation. *International Geography* (V. Gardner, ed.). Wiley, New York, 1987.

Quirk, P. P., and R. J. Schofield. The effect of electrolyte concentration on soil permeability. *J. Soil Sci. 6*:163 (1955).

Radcliffe, D. E., R. L. Clark, and M. E. Sumner. Effect of gypsum and deep-rooting perennials on subsoil mechanical impedance. *Soil Sci. Soc. Am. J. 50*:1566 (1986).

Reeve, N. G., and M. E. Sumner. Amelioration of subsoil acidity in Natal Oxisols by leaching of surface-applied amendments. *Agrochemophysica 4*:1 (1972).

Reichert, J. M., and L. D. Norton. Aggregate stability and rain-impacted sheet erosion of air-dried and prewetted clayey surface soils under intense rain. *Soil Sci. 158*:159 (1994).

Rengasamy, P., R. S. B. Greene, G. W. Ford, and A. H. Mehanni. Identification of dispersive behaviour and the management of red-brown earths. *Aust. J. Soil Res. 22*:413 (1984).

Rhoades, J. D., D. B. Kruger, and M. J. Reed. The effect of soil mineral weathering on the sodium hazard of irrigation waters. *Soil Sci. Soc. Am. Proc. 32*:643 (1968).

Ritchey, K. D., D. M. G. Souza, E. Lobato, and O. Correa. Calcium leaching to increase rooting depth in a Brazilian Savanna oxisol. *Agron. J. 72*:40 (1980).

Rose, C. W. Development in soil erosion and deposition models. *Adv. Soil Sci. 2*:1 (1985).

Schofield, R. K., and H. R. Samson. Flocculation of kaolinite due to attraction of oppositely charged crystal faces. *Dis. Far. Soc. 18*:135 (1954).

Shainberg, I., and J. Letey. Response of soils to sodic and saline conditions. *Hilgardia 52*:1 (1984).

Shainberg, I., J. D. Rhoades, and R. J. Prather. Effect of low electrolyte concentration on clay dispersion and hydraulic conductivity of a sodic soil. *Soil Sci. Am. J. 45*:273 (1981a).

Shainberg, I., J. D. Rhoades, D. L. Suarez, and R. J. Prather. Effect of mineral weathering on clay dispersion and hydraulic conductivity of sodic soils. *Soil Sci. Soc. Am. J. 45*:287 (1981b).

Shainberg, I., R. Keren, and H. Frenkel. Response of sodic soils to gypsum and calcium chloride application. *Soil Sci. Soc. Am. J. 46*:113 (1982).

Shainberg, I., M. E. Sumner, W. P. Miller, M. P. W. Farina, M. A. Pavan, and M. V. Fey. Use of gypsum on soils: a review. *Adv. Soil Sci. 9*:1 (1989).

Shanmuganathan, R. T., and J. M. Oades. Influence of anions on dispersion and physical properties of the A horizon of a red-brown earth. *Geoderma 29*:257 (1983a).

Shanmuganathan, R. T., and J. M. Oades. Modification of soil physical properties by addition of calcium compounds. *Aust. J. Soil Res. 21*:285 (1983b).

Singer, M. J., P. Janitzky, and J. Blackard. The influence of exchangeable sodium percentage on soil erodibility. *Soil Sci. Am. J. 46*:117 (1982).

Southard, R. J., I. Shainberg, and M. J. Singer. Influence of electrolyte concentration on the micromorphology of artificial depositional crust. *Soil Sci. 145*:278 (1988).

Souza, D. M. G. and K. D. Ritchey. Uso do gesso no solo de Cerrado. Anais do I Seminario sobre o Uso do Fosfogesso ne Agricultura. pp. 119–144. EMBRAPA, Brasilia DF, Brazil (1986).

Souza, D. M. G., E. Lobato, K. D. Ritchey, and T. A. Rein. Resposta de culturas anuais e leucena a gesso no Cerrado. *II Sem. Uso Gesso Agricultura*, 1992, p. 277.

Stern, R. Effects of soil properties and chemical ameliorants on seal formation, runoff and erosion. D. Sc. thesis, University of Pretoria, Pretoria, South Africa, 1990.

Stern, R., M. Ben-Hur, and I. Shainberg. Clay mineralogy effect on rain infiltration, seal formation and soil losses. *Soil Sci. 152*:455 (1991a).

Stern, R., B. E. Eisenberg, and M. C. Laker. Correlation between micro-aggregate stability and soil surface susceptibility to runoff and erosion. *S. Afr. J. Plant Soil 8*:136 (1991b).

Sumner, M. E. *Gypsum as an Ameliorant for the Subsoil Acidity Syndrome*. Florida Institute of Phosphate Research, Bartow, F L., 1990, p. 1.

Sumner, M. E. Gypsum and acid soils: the world scene. *Adv. Agron. 51*:1 (1993).

Sumner, M. E. Amelioration of subsoil acidity with minimum disturbance. *Subsoil Management Techniques* (N. S. Jayawardane, and B. A. Stewart, eds.). Lewis Publishers, 1994, p. 147.

Sumner, M. E., W. P. Miller, D. E. Radcliffe, and J. M. McCray. Use of phosphogypsum as an amendment for highly weathered soil. *Proc. 3rd Sem. Phosphogypsum*. Florida Inst. Phosphate Res., Bartow. FL., 1986.

Syed-Omar, S. R., and M. E. Sumner. Effect of gypsum on soil portassium and magnesium status and growth of alfalfa. *Comm. Soil Sci. Plant Anal. 22*:2017 (1991).

Tanji, K. K., and S. J. Deverel. Simulation modeling for reclamation of sodic soils. *Soil Salinity Under Irrigation* (I. Shainberg, and J. Shalhevet, eds.). Springer-Verlag, 1984, p. 238.

Tanji, K. K., L. D. Doneen, G. V. Ferry, and R. S. Ayers. Computer simulation analysis on reclamation of salt-affected soil in San Joaquin Valley, California. *Soil Sci. Soc. Am. J. 36*:127 (1972).

Taylor, H. M., G. M. Roberson, and J. J. Parker. Soil strength-root penetration relations for medium to coarse textured soil materials. *Soil Sci. 101*:18 (1966).

USSL staff. *Diagnosis and Improvement of Saline and Alkali Soils*. Washington D. C., 1954.

van der Watt, H. v. H., and C. Valentin. Soil crusting: the African view. *Soil Crusting: Chemical and Physical Processes* (M. E. Summer, and B. A. Stewart, eds.). Lewis Publishers, 1992, p. 301.

van Olphen, H. *An Introduction to Clay Colloid Chemistry*. John Wiley, New York, 1977, p. 1.

van Raij, B., H. Cantarella, and P. R. Furlani. Efeito na reacao do solo da absorcao de amonio e nitrato pelo sorgo na presenca e na ausencia de gesso. *R. Bras. Ci. Solo 12*:131 (1988).

Vitti, G. C., J. A. Mazza, H. S. Pereira, and J. L. I. Dematte. Resultados experimentais do uso de gesso na agricultura. *II Sem. Uso Gesso Agricultura*, 1992, p. 191.

Walker, P. H., J. Hutka, A. J. Moss, and P. I. A. Kinnell. Use of a versatile experimental system for soil erosion studies. *Soil Sci. Am. J. 41*:610 (1977).

Warrington, D., I. Shainberg, M. Agassi, and J. Morin, Slope and phosphogypsum's effects on runoff and erosion. *Soil Sci. Soc. Am. J. 53*:1201 (1989).
Wischmeier, W. H., and J. M. Mannering. Relation of soil properties to its erodibility. *Soil Sci. Soc. Am. Proc. 33*:131 (1969).
Wischemeier, W. H., and D. D. Smith. *Predicting Rainfall Erosion Losses*. Agric Handbk. No. 537, U. S. Dept. of Agriculture, 1978.
Young, R. A., and J. L. Wiersma. The role of rainfall impact in soil detachment and transport. *Water Resour. Res. 9*:1629 (1973).
Zhang, X. C., and W. P. Miller. Physical and chemical processes affecting runoff and erosion in furrows. *Soil Sci. Soc. Am. J. 60*:860 (1996).

8
Use of Acids and Acidulants on Alkali Soils and Water

S. Miyamoto *Texas A&M University, El Paso, Texas*

I. INTRODUCTION

The use of acidulants on irrigated soils of the western USA dates back to the 1930s. Prior to that time, gypsum was considered the amendment for sodium-affected soils. Hilgard (1889), for example, noted that "in the case of black alkali, the impenetrability of the surface soil itself renders underdrains ineffective, unless sodium carbonate, with its compacting effects on the soil, is first destroyed by the use of gypsum." By the mid 1930s, the use of sulfur was proposed along with the understanding of microbial oxidation to H_2SO_4, and subsequent dissolution of $CaCO_3$ and removal of exchangeable Na by Ca (Kelley and Brown, 1934; McGeorge and Green, 1935). When the USDA *Handbook 60* was published in 1954, a number of inorganic chemicals, including acids and acidulants, were cited as being potentially effective amendments (USSL Staff, 1954). However, the experimental data to substantiate these possibilities were limited. The first extensive field evaluation of sulfuric acid, along with sulfur and gypsum as an amendment for sodic alkali soils, was conducted in 1951 in the Central Valley of California with some impressive results (Overstreet et al., 1951). Starting in the late 1960s, pollution abatement activities at various industries intensified, and sulfuric acid became readily available from SO_2 scrubbing from smelter gases in the southwestern states (McKee, 1969). Sulfur recovery from natural gas and oil refining has also increased, and so has research to utilize these products on sodic or alkali soils and water (e.g., Miyamoto et al., 1975a).

Today, sulfuric acid and elemental S are used singly or in combination with various nitrogen compounds for improving sodic or alkali soils and irrigation water. These uses may be categorized into two types: conditioning of alkali or sodic irrigation water, mainly to control Ca precipitation and/or to maintain soil structure and water infiltration; and treatment of sodic calcareous or sodic alkali soils, mainly to lower soil sodicity, pH, and/or to improve water infiltration, and at times, availability of certain micronutrients. The areas where these types of treatments are potentially beneficial are large worldwide (e.g., Dudal and Purnell, 1986; Bower and Fireman, 1957), and the applications range from intensive irrigated crop production to revegetation of sodium-affected lands. However, treatment effects are highly soil and site specific. The purpose of this chapter is to outline the conceptual basis for using acids and acidulants on sodic or alkali soils and water, and to provide some guidelines for their usage.

II. SOURCES AND PROPERTIES OF ACIDS AND ACIDULANTS

A. Acids

1. Sulfuric Acid

The sources of sulfuric acid include the acid produced from the oxidation of elemental S, by-product acids from smelting and wet scrubbing of SO_2, spent sulfuric acid from the gasoline alkylation process (Weiss et al., 1953), and waste sulfuric acid from various industrial processes (Tisdale, 1968; Beaton and Fox, 1971). The concentration of sulfuric acid varies with the source, but sulfur-burn acid and smelter acid are usually marketed at ingredient concentrations of 93 to 95%. Sulfuric acid with a concentration exceeding 98% readily freezes at low temperatures and generates hazardous SO_2 fumes at room or elevated temperatures. All types of sulfuric acid are highly corrosive, and its handling must be carried out in accordance with applicable safety, transportation, storage, and handling regulations, which are becoming stringent. Some of the properties of sulfuric acid are shown in Table 1.

2. Phosphoric Acid

The main source of phosphoric acid is from sulfuric acid digestion of phosphorus containing rocks, such as apatites or fluorapatites (Phillips and Webb, 1971). A by-product, gypsum ($CaSO_4 \cdot 2H_2O$) is separated out from phosphoric acid, while another by-product, HF, is usually recovered as sodium fluorosilicate. The concentrations of phosphoric acid from the wet digestion process are usually low but are elevated to 55 to 75% (17 to 24% P) prior to shipment. Superphosphoric acid produced through a special process contains higher concentrations, but the resulting high viscosity can present handling difficulties. Phosphoric acid is also corrosive and must be handled in accordance with applicable safety and handling regulations.

Table 1 Properties of Acids and Acidulants Used for Conditioning or Treating Alkali Soils and Irrigation Water

Materials	Formula	Molecular wt.	Ingredient % wt	Density kg L^{-1}	Ca release mol/mol
Acids					
Sulfuric acid	H_2SO_4	98	93–95	1.82–1.83	1.0
Phosphoric acid	H_3PO_4	98	55–75	1.45–1.6	1.5
Urea-sulfuric acid	$CO(NH_2)_2\ H_2SO_4$	60 + 98	94–95	1.50–1.55	1.0[a]
Ammonia-sulfuric acid	$(NH_4)_x{\cdot}H_2SO_4$	$(18)_{1.0} + 98$	—	—	1.0[a]
Acidulants					
Elemental sulfur	S	32	99	2.0	1.0
Ammonium polysulfide	$NH_4S_x^a$⌋	$18 + (32)_{1.0}$	94	1.17	2.0
Iron sulfate	$FeSO_4{\cdot}7H_20$	278	98	1.89	1.0
Aluminum sulfate	$Al_2(SO_4)_3{\cdot}18H_20$	666	98	1.69	3.0
Acidic gases					
Sulfur dioxide	SO_2	64	99	1.43	1.0
Carbon dioxide	CO_2	44	99	1.10	1.0

[a]Assumed to be 1:1 molar mixture, and the contribution of N compounds on Ca release is included.

3. Acid Blends

Urea, $(NH_2)_2CO$, a polar compound, is soluble in sulfuric acid and is easily blended to form urea-sulfuric acid solutions. The reaction is endothermic, and it lowers the corrosiveness of sulfuric acid somewhat. The blending ratio varies, but a typical blend consists of one mole of urea against one mole of acid, which provides a product 17-0-0-19(S). This product was originally used for alleviating drip emitter clogging but is now used as a water conditioning chemical and a liquid acid-N fertilizer in sodic or alkali soils.

Urea is also added to phosphoric acid to form a solid granule compound, 17.5-44.4-0. This is a form of N-P fertilizer and has the advantage of easier handling than phosphoric acid, yet it retains the characteristics of acidic compounds when applied to either water or soils (Achorn, 1984).

Another form of acid application is a simultaneous injection of sulfuric acid and ammonia gas (NH_3) into irrigation water flow. The simultaneous injection idea was originally proposed to reduce NH_3 volatilization and Ca precipitation induced by water application of NH_3 or NH_4OH (Miyamoto and Ryan, 1976). Premixing of NH_3 and H_2SO_4 results in violent reactions and precipitation of $(NH_4)_2SO_4$. Thus the two compounds are applied to irrigation water separately. The application ratio is usually one mole of acid against two moles of NH_3 plus additional acid to remove approximately 90% of the bicarbonate concentration of irrigation water. This provides for complete neutralization of NH_3 (or NH_4OH) and the necessary conversion of HCO_3^- and CO_3^{-2} to H_2CO_3.

4. Other Acids

Theoretically, both nitric and hydrochloric acids can also be used for alkali soil and water treatments. However, these acids are rarely used, for various reasons including high costs, volatility, and handling difficulties. High dosages of HCl can also pose potential Cl toxicity to some crops, and excessive applications of HNO_3 can cause ground water contamination with NO_3^-.

B. Acidulants

1. Elemental Sulfur

Elemental sulfur (S) is among the oldest soil acidifying compounds used commercially. Sources include natural deposits of brimstone, sulfur recovered from natural gas containing H_2S, and petroleum desulfication (Beaton and Fox, 1971). Some sulfur is produced through the reduction of SO_2 waste gases and sulfate salts. Elemental sulfur is marketed in a wide array of forms, ranging from flakes, aggregates, and powders to suspended slurry (Tisdale, 1968). Sulfur dust is an irritant to eyes and subject to fire hazards. The purity of elemental sulfur was rather variable, but today has improved to 99% plus. Solid forms of elemental sulfur are stable in storage and provide high acidification potential per unit weight. Iron pyrite (FeS) is another source of S and is discussed in a separate chapter (Chapter 9).

2. Polysulfides of Ammonium or Calcium

Ammonium polysulfide, NH_4S_x, is manufactured by reacting sour gas (H_2S) with aqueous ammonia. The ratios of NH_3 and S in commercial products vary somewhat but are usually around an equal molar ratio of NH_4 and S, which provides a product characterized as 20-0-0-40 (Table 1). It is corrosive, strongly alkaline (pH > 10) and generates fumes of both H_2S and NH_3. The material must be kept in a closed system made of stainless steel or black iron, but not in copper-containing alloys such as brass or bronze. Ammonium polysulfide is a strong reducing agent.

Another compound allied with ammonium polysulfide is ammonia-sulfur, which is produced by dissolving or suspending powder or colloidal sulfur into liquid anhydrous ammonia (Phillips and Scott, 1996). A commercial product may contain a typical composition of 74-0-0-10. This product has not been used extensively due to certain application difficulties associated with thiosulfate precipitation.

Calcium polysulfide, CaS_x, is manufactured by reacting H_2S with calcium hydroxide at a Ca to S molar ratio of about 1 to 3. A commercial product may contain 24% sulfur and 9% Ca. This product is also strongly alkaline but acts as an acidulant upon sulfide oxidation in soils.

3. Sulfates of Iron and Aluminum

These granule products are among the oldest soil acidifiers and have the advantage of ease of handling compared to strong acids. However, the quantities required to achieve sufficient dissolution of $CaCO_3$ are substantially greater than most other products (Table 1), so that these products suffer from high costs of transportation. Another old product, $(NH_4)_2SO_4$, is still used widely, but mostly as a N-fertilizer rather than as a soil amendment with soil conditioning value.

C. Acidic Gases

1. Sulfur Dioxide

Sulfur dioxide (SO_2) used for conditioning irrigation water is usually produced by burning elemental sulfur on site with specialized equipment. The SO_2 is then injected into irrigation water. The use of SO_2 is confined mostly to some areas of California (Leclercq et al., 1972).

2. Carbon Dioxide

Carbon dioxide (CO_2) generated through the combustion of fuel at irrigation pump stations can, in theory, be scrubbed with water to prevent Ca precipitation. However, the solubility of CO_2 in water is low, and little is documented about its effectiveness as a conditioner of alkaline water. Carbon dioxide, however, plays a major role in $CaCO_3$ dissolution in soils.

III. REACTIONS OF ACIDS AND ACIDULANTS IN ALKALI SOILS AND WATER

A. Reactions in Water

1. Dissolution

Strong acids (H_2SO_4, and H_3PO_4), and acid blends (urea/sulfuric acid and urea/phosphoric acid) dissociate almost instantaneously in irrigation water, releasing H^+, urea, plus the anions. The dissolution of these compounds is exothermic, creating considerable heat at the point of contact with water. The dissolution of polysulfides into irrigation water generates sulfur plumes of milky appearance. Sulfides precipitate as colloidal S, as pH decreases from the original level of about 10, whereas ammonium and Ca ions dissolve readily into water.

Acidic gases (SO_2 and CO_2) dissolve into water and form sulfurous acid or carbonic acid.

$$SO_2 + H_2O \Leftrightarrow H_2SO_3 \Leftrightarrow 2H^+ + SO_3^{-2} \quad (1)$$

$$CO_2 + H_2O \Leftrightarrow H_2CO_3 \Leftrightarrow H^+ + HCO_3^- \quad (2)$$

The dissociation constant of H_2SO_3 is much higher than that of H_2CO_3.

2. Neutralization Reactions

Hydrogen ions dissociated from acids or dissolved acidic gases react readily with OH^-, either initially present in the water or added in the form of ammonium hydroxide or ammonia gas.

$$NH_4OH + H^+ \Leftrightarrow NH_4^+ + H_2O \quad (3)$$

3. Reactions with Carbonates

Hydrogen ions also react with carbonate species, primarily HCO_3^- present in alkali irrigation water, and convert them to carbonic acid.

$$H^+ + HCO_3^- \Leftrightarrow H_2CO_3 \quad (4)$$

Under a system open to the atmosphere, carbonic acid decomposes to CO_2 and H_2O as shown by the reverse reaction of Eq. (2). The pH of the water remains above neutral, until about 90% of the HCO_3^- ions present are converted to H_2CO_3, and eventually to CO_2 and H_2O (Miyamoto et al., 1975b). The concentration of H_2CO_3 in an open system is largely fixed by the partial pressure of CO_2. Under a closed system, which may exist in water flowing through water-filled pipes or tubes, carbonic acid accumulates in the system. This causes the pH of the water to drop below neutral, even when the acid application is as low as 10% of the concentration of HCO_3^- (Miyamoto et al., 1975b).

When Ca ions are present in the water, the reduction in HCO_3^- concentration reduces Ca precipitation.

$$Ca^{++} + 2HCO_3^- \Leftrightarrow CaCO_3\ (\downarrow) + H_2CO_3 \tag{5}$$

or

$$Ca^{++} + NH_3 + HCO_3^- \Leftrightarrow CaCO_3\ (\downarrow) + NH_4^+ \tag{6}$$

The reaction given by Eq. (5) is strongly dependent of the partial pressure of CO_2 in gas phase. The precipitation of Ca in ammoniated water is extensive and nearly instantaneous (Miyamoto and Ryan, 1976).

B. Reactions in Soils

1. Hydroxylation

Iron or aluminum sulfates, which are usually applied directly to soils, undergo hydroxylation in contact with water and release H^+.

$$FeSO_4 + 2H_2O \Leftrightarrow Fe\ (OH)_2 + 2H^+ + SO_4^{-2} \tag{7}$$

$$Al_2(SO_4)_3 + 6H_2O \Leftrightarrow 2Al(OH)_3 + 6H^+ + 3SO_4^{-2} \tag{8}$$

These reactions allow H^+ to react with soil carbonates, principally $CaCO_3$ and Na_2CO_3 if present. Both Fe and Al hydroxides precipitate in calcareous or alkali soils.

2. Oxidation Reactions

Elemental S applied to the soils is subject to microbial oxidation through the activity of Thiobacillus.

$$S + 3/2\ (O_2) + H_2O \Leftrightarrow 2H^+ + SO_4^{-2} \tag{9}$$

The rate of oxidation is affected by soil moisture, soil temperature, and the surface area (or particle sizes) of elemental S. The oxidation of S requires elevated soil temperatures, typically above 25°C, and preferably as high as 40°C (Paulina, 1951), and wet to moist soils with a soil water suction range of 0.01 to 0.1 bar (Moser and Olson, 1953). The rate of oxidation increases with the fineness of S particles (Burns, 1967), and upon wetting and drying (Haynes, 1928). A powder form of S can be oxidized in several months, whereas flake sulfur may require a year or more for oxidation, especially in sandy dry soils with little organic matter. The oxidation of large S particles (>0.4 mm) may be enhanced by inoculation with *Thiobacillus thiooxidans* (McCaskill and Blair, 1987).

The oxidation of sulfides, applied either as NH_4S_x or CaS_x, is believed to be considerably faster than that of elemental S because of its colloidal size, provided that soil conditions are favorable for Thiobacillus activities. In order to satisfy the comparatively high soil moisture requirement for oxidation, colloidal sulfur, which forms in irrigation water through the application of sulfides, must penetrate into the soil to some depth. This requirement is usually met when these products are applied to plowed and disked fields.

Urea or ammonium applied with sulfuric or phosphoric acid is also known to undergo microbial oxidation to NO_3^-, which generates H^+ (e.g., Morrill and Dawson, 1967).

$$NH_4^+ + 2O_2 \Leftrightarrow NO^-_3 + 2H^+ + H_2O \quad (10)$$

Prior to nitrification, urea must be first decomposed to NH_4^+ through urease activities, and this reaction usually begins in a matter of several days in the presence of organic matter. The microsoil environments adjacent to urea temporarily become alkaline (Fenn and Miyamoto, 1981).

3. Dissolution of Soil Carbonates and Oxides

Hydrogen ions (H^+) present in excess of neutralizing NH_4OH or HCO_3^- react with soil carbonates.

$$CaCO_3 + 2H^+ \Leftrightarrow Ca^{+2} + H_2CO_3 \quad (11)$$

The extent of dissolution depends on the system in question. Under an open system, two moles of H^+ are usually required to dissolve one mole of $CaCO_3$, and H_2CO_3 is decomposed to CO_2 and H_2O. In a closed system, H_2CO_3 can release additional Ca (Miyamoto et al., 1975b).

Once soil carbonates are removed, hydrogen ions react with oxides (or hydroxides) of Fe and Al. Al and Fe oxides help develop stable oil aggregates under low pH, but not necessarily under alkaline conditions (Vodyanitskiy and Dokuchayev, 1985). Some DTPA extractable Mn and Fe may appear at acid application rates below the neutralization point, but water soluble Mn, Zn and Fe appear mostly after the complete removal of soil carbonates, usually in the order of Mn, Zn, then Fe (Ryan et al., 1974).

IV. ACID AND ACIDULANT APPLICATION FOR CONDITIONING IRRIGATION WATER

The concern over the use of sodic water on irrigated lands is probably best characterized by the statement of Scofield and Headly (1921): "Hard water makes soft land, and soft water makes hard land". Soft water rich in sodium carbonate tends to disperse soil particles and lowers aggregate stability and soil permeability. Carbonates in water have been recognized as a factor contributing to soil structural degradation as they precipitate Ca (Eaton, 1950). In spite of these well-established facts, the use of acids for water conditioning had been discouraged by some, partly because of the concern that acid applications might corrode irrigation systems and partly because of handling difficulties. With the introduction of drip irrigation systems, this perception has changed, and acid and acidifying N fertilizer applications became a common practice to reduce emitter clogging (Gilbert et al., 1979). Likewise, the use of acid or acid-based N fertilizers for altering sodicity of irrigation water became a common practice in surface irrigated areas (Stroehlein et al., 1978). The type of irrigation water treated today is no longer confined to "soft" water, but includes some "hard" ones.

A. Treatments for Preventing Ca Precipitation

1. Amendment Requirements

The quantity of acids, acid blends, or acidic gases required to prevent Ca precipitation is highly dependent upon the system in question. In a system open to the atmosphere, it equals the acid required to convert about 90% of HCO_3^- to CO_2 and H_2O, and it is practically independent of Ca concentrations. At the rate of acid application, the pH of the water remains above neutral. However, once the acid concentration exceeds the equivalent concentration of HCO_3^-, the pH of the water sharply drops below neutral (Miyamoto et al., 1975b). Table 2 shows examples of acids or acidic gas (SO_2) required to prevent Ca precipitation.

The quantities of acids, acid blends, and acid gases required to prevent Ca precipitation under a closed system are low, usually no more than 10% of the equivalent concentrations of HCO_3^-. The requirement is further reduced with decreasing Ca concentrations in the water. Acid applications exceeding about 10% of HCO_3^- concentration in a closed system bring the pH of water below 7, but not below 6, until the acid application reaches about 90% of the HCO_3^- concentrations (Miyamoto et al., 1975b).

When NH_3 or NH_4OH is to be applied simultaneously, the quantity of acid required will increase by the equivalent concentration of NH_3 or NH_4OH (Miyamoto and Ryan, 1976). Acids or acidic gases applied to ammoniated irrigation water not only prevent Ca precipitation but also reduce ammonia volatile losses from irrigation water and soil surfaces (Miyamoto et al., 1975d). Table 2 includes the concentration of H^+ and quantities of acid, acid blends, or SO_2 required to prevent Ca precipitation in ammoniated irrigation water.

2. Application Systems

Open Ditch Flow. Acid applications to open irrigation ditches are usually made from a storage tank through flexible tubing. Since the density of acid is considerably greater than that of water (Table 1), acids tend to settle to the bot-

Table 2 Quantities of Acids and Acidic Gases Required to Prevent Ca Precipitation in Irrigation Water

Conditions	HCO^-_3 or NH_3 mg L^{-1}		Acid Required[a] H_2SO_4[b]	H_3PO_4[b],	Urea-acid[c]	SO_2
Open system			mg L^{-1}			
	50	0	38	34	62	24
	100	0	76	68	124	47
	150	0	114	103	184	71
	200	0	152	137	248	94
	250	0	190	172	308	118
Closed system (approximate)						
	50	0	3	3	6	2
	100	0	7	6	12	4
	150	0	11	10	18	7
	200	0	15	13	24	9
Open and closed systems						
	0	25	76	69	123	47
	0	50	152	137	246	94
	0	75	228	206	370	141
	0	100	304	275	493	188

[a]Expressed in mg L^{-1}, which equals kg per 10 cm water × ha.
[b]The concentration of sulfuric acid was assumed to be 95% and that of phosphoric acid 70%.
[c]Assumed to contain 1 mole of urea and 1 mole of H_2SO_4.

tom of the ditch, especially at a low flow rate. This problem is usually alleviated through acid applications to a turbulent flow portion or by providing a simple device to enhance mixing. Acid storage tanks should be placed in a manner whereby accidental spills can be contained. The most common incident of acid spills is caused by valve or joint failures. Care must also be taken to avoid water entry, including water vapor condensation in an empty storage tank, which can cause a violent reaction in contact with acids.

Pipe Flow. Acid applications to pressurized pipe flow require a corrosion resistant injection device commonly used for liquid fertilizer injection. In the case of drip irrigation systems made of plastic materials, acids can be applied with little concern for corrosion. The precipitation of Ca generally occurs at emitters where the water is exposed to the atmosphere. In sprinkler systems made of Al or copper containing alloys, acid application is not recommended, unless reliable precision application systems for NH_3 and H_2SO_4 are available or these chemicals are premixed in a water reservoir.

B. Treatments for Reducing Sodicity and Soil Structural Degradation

1. Reducing Sodicity

The use of acids and acidulants for reducing sodicity can be classified into two types. The first type is to reduce sodicity of irrigation water (and consequently that of the irrigated soils), and the second type is to reduce sodicity of surface soils. The first type is usually used when sodicity of irrigation water is high enough (or salinity of the water is too low) to cause soil structural degradation, unless amended constantly, or NH_3 (or NH_4OH) is applied through water. The second type is used to keep sodicity of surface soils in check when irrigation water has salinity sufficient to avoid serious structural degradation of the soils without constant application.

The quantities of acid, acidic gases, or acid blends required to prevent an increase in sodicity (associated with Ca precipitation) are similar to those used for preventing Ca precipitation discussed earlier (Table 2). At that rate of amendment application, both the sodium adsorption ratio (SAR) and the salinity of water remain the same as those of the irrigation water prior to Ca precipitation. Hydrogen ions are consumed for decomposing HCO_3^-, and SO_4^{-2} or PO_4^{-3} ions take the place of HCO_3^-. Acids or acidic gas applications to irrigation water are thus to prevent an increase in sodicity of irrigation water or a reduction in salinity associated with the precipitation of Ca^{+2} and HCO_3^- (Table 3). When NH_3 (or NH_4OH) is applied with acids or acidic gases, however, sodicity of the surface soil is actually reduced following NH_4^+ oxidation to NO_3^-. The use of

Table 3 Changes in Sodium Adsorption Ratio (SAR) of Irrigation Water Following Aeration, Ammonia Application, and Ammonia Plus Sulfuric Acid Application

Water sources Location	Well Gila Bend	Well Safford	Salt R Phoenix	Colorado R Yuma
Initial quality				
Salinity (dS m^{-1})	2.39	0.74	1.25	1.1
Ca^{+2} (meq L^{-1})	3.2	2.5	1.0	4.4
HCO_3^- (meq L^{-1})	3.7	2.8	2.7	2.7
Sodicity (SAR) changes				
Initial	13.8	7.2	6.3	2.3
Aeration[a]	17.	9.5	7.6	2.4
with NH_3 (50 mg L^{-1})	26	12	10	3.6
with $NH_3 + H_2SO_4$ (100 mg L^{-1})	20	9.5	8.1	3.0

[a]Increase in SAR associated with aeration is commonly referred to as HCO_3 effects.
Source: Miyamoto and Ryan, 1976.

urea-acid blends follows the same principle, but their ability to reduce sodicity of soil surfaces is lower than that of ammonium-acid mixtures, because uncharged urea moves into the soil profile following the wetting front, and the acid release potential of urea is less than NH_4^+. Nonetheless, acids used as a solvent of urea help prevent Ca precipitation in irrigation water and an increase in sodicity of the soil (Ali, 1989).

Ammonium or calcium polysulfide applied through water does not prevent Ca precipitation in irrigation water, but it helps reduce Ca precipitation in soils or, if applied in excess of the concentrations of HCO_3^- in irrigation water, it can solubilize soil $CaCO_3$ when sulfides are oxidized to SO_4^{-2} or NH_4^+ is oxidized to NO_3^-. The oxidation of S and NH_4^+ or the addition of Ca through Ca polysulfide applications can help lower sodicity of the surface soils.

An alternative to water application is the direct soil application of acids and acidulants. These are discussed later in the section dealing with reclamation of sodium-affected soils. Water application is usually more efficient if the water is sodic, or low in salinity, and improvements in water infiltration is the primary goal. However, if the goal is to lower soil sodicity, soil hardness, or crusting, soil application of sulfuric acid, phosphoric acid, urea-phosphoric acid, iron sulfate, or sulfur provides options.

2. Improving Soil Permeability

Hydraulic Conductivity. The primary goal of reducing sodicity of irrigation water and/or surface soils is usually to obtain improved water transmission through the soils. There are several studies that evaluated the effectiveness of these compounds in improving the saturated hydraulic conductivity, but the results are rather mixed, especially in water having salinity greater than about 1 dS m^{-1}. A laboratory percolation test performed with three soils from the San Joaquin Valley, California, using water having low salt (EC = 0.05 and 0.7 dS m^{-1}) and low sodium (SAR = 0.4 and 1.7), has shown significant increases in percolation rate when SO_2 or gypsum was applied to the water (Mohammed, 1972). However, the response to Ca polysulfide applications was inconsistent, and at times hydraulic conductivity decreased with increasing application rates, presumably because of pore plugging caused by precipitated colloidal sulfur. In addition, the extent of percolation improvements depended on soil types. A similar laboratory study conducted with five soil types from southern Arizona using four types of simulated irrigation water having a salinity range of 0.7 to 3.5 dS m^{-1} has shown that sulfuric acid applied at a rate equal to approximately 90% of the HCO_3^- concentrations in chemical equivalents provided significant improvements in percolation rates in some soils only when the SAR of the irrigation water was higher than 7.9 (Gumaa et al., 1976). In a simulated water having salinity of 0.7 dS m^{-1} and a SAR of 9.7, the percolation rate increases ranged from 12 to 66% depending on soils. In saline sodic water having salinity of 3.5 dS m^{-1} and sodicity of 10.5, the increase ranged from 0 to 14%. In moderately saline water having salinity of 1.1 dS m^{-1} and sodicity of 7.9, the increase over nontreated ranged from none to 48%, depending on soil type. There seemed to be no definite correlation between hydraulic conductivity improvements and clay minerals. The difference in hydraulic conductivity among the five soils varied by a factor of 5, while the difference among the different water sources was within a factor of less than 2.

Judging from these results, the saturated hydraulic conductivity is primarily a function of soil types and salinity of irrigation water. When salinity of irrigation water is low, e.g., less than 1 dS m^{-1} and the soil is dispersible, acids or polysulfide treatments of irrigation water are likely to yield a significant increase in saturated hydraulic conductivity, especially when the SAR exceeds about 7. This general guideline is consistent with other observations of water quality effects on saturated hydraulic conductivity.

Water Infiltration. Water infiltration into irrigated soils is usually affected by water quality more sensitively than saturated hydraulic conductivity (Oster and Schroer, 1979). The reasons include soil aggregate destruction and soil particle dispersion induced by slaking of dry soils and physical force of water (Abu-Sharar et al., 1987; Shainberg and Letey, 1984). Another reason relates to precipitation of Ca at the soil surface when salts are brought to the soil surface through upward capillary flow. This process can increase sodicity of the soil surface many times greater than that of irrigation water or the soil solutions below the surface, especially in sandy or loamy soils with low cation exchange capacities (Miyamoto and Cruz, 1987). All these contribute to the formation of a thin layer of soil having reduced permeability at the soil surface, which then becomes a flow-limiting layer during water infiltration.

Application of acids, acidic gases, or acidulants is a way to minimize soil structural degradation and associated reductions in water infiltration. In a field infiltration test under ponding, acid applications at a rate of 2.2 meq L^{-1} to low-salt and high-sodium water (EC = 0.62 dS m^{-1} and SAR = 30) produced a 75% increase in water infiltration rate into irrigated silt loam soil (Miyamoto and Stroehlein, 1975). In a similar field test with low-salt and low-sodium water (EC = 0.42 dS m^{-1} and SAR = 2.5), infiltration improvement was not significant (Stroehlein et al., 1978). Application of acid to ammoniated water increased the infiltration rate of low salt water (EC = 0.6 dS m^{-1} SAR = 8.9) by 28 to 83% depending on soils (Miyamoto and Ryan, 1976). The effect of the acid treatment of ammoniated water persisted in subsequent irrigations without any chemical additives, probably because of NH_4 oxidation which releases Ca in calcareous soils.

Water infiltration responses to chemical treatments of water vary with the compound used. In a field infiltration experiment performed in the Imperial Valley of California, water applications of ammonium polysulfide and sulfuric acid at a sulfur rate of 70 kg ha^{-1} provided infiltration increases of 28 and 13%, respectively (Robinson et al., 1968). The quality of the irrigation water used was not given, but the Colorado River water at the location usually has an EC of 1.0 to 1.4 dS m^{-1} and a SAR value of 2 to 6. In the same test, calcium polysulfide applied at a sulfur rate of 70 kg ha^{-1} did not produce any significant increase, which is consistent with the finding from laboratory hydraulic conductivity measurements mentioned earlier. Pore plugging caused by sulfur precipitation is believed to have masked out the effect of calcium polysulfide.

In a subsequent field infiltration test at the same location, ammonium polysulfide applied through water at a S rate of 70 kg ha^{-1} provided a twofold increase in infiltration rate, while ammonium nitrate applied at an equivalent rate yielded only a 10% increase. However, neither treatment gave any residual effect when untreated water was subsequently applied. The small increase in infiltration rate observed from sulfuric acid or ammonium nitrate treatments in these tests is within the range commonly observed for this type of water. The superior performance of ammonium polysulfide treatment is beyond the range that is ordinarily expected under nonsodic conditions, and it led the authors to conclude that the increase was evidently caused by a unique property of this compound, which is not yet understood.

In our recent laboratory infiltration experiments, water application of ammonium polysulfide, when compared to $CaCl_2 \cdot 2H_2O$, urea/acid, or ammonia/acid applied at a chemically equivalent rate, did not produce any superior results (Table 4). Instead, the effect of NH_4S_x was similar to that of ammonia/sulfuric acid mix. Improvements in infiltration rates from any of the chemical amendments were fairly small, as can be expected from the types of water tested. The Rio Grande water is low in sodicity, whereas the test water from Pecos had high salinity to counter the elevated level of sodicity. This experiment, however, does not rule out the possibility of improved water infiltration into dry soils with an application of ammonium polysulfides, which are widely speculated to have a property of improving soil wettability. This experiment seems to indicate that NH_4/sulfuric acid and ammonium polysulfides can improve water infiltration as much as, if not more than, the equivalent concentration of $CaCl_2 \cdot 2H_2O$. It seems that the choice of water conditioning chemicals must include nitrogen values.

From a basic research point of view, the effect of pH on soil permeability warrants additional study. A laboratory permeability study involving three noncalcareous soils in California indicates that Na-induced hydraulic conductivity reductions in low salt water (e.g., EC < 1 dS m^{-1}) at a SAR of 20 can be prevented by lowering the pH to 6.0 (Suarez et al., 1984). One possible reason is thought to be the changes in soil particle bonding orientation when the charge characteristics of oxides of Fe and Al and kaolinite change from negative to positive with lowering pH. The positively charged oxides and kaolinite can form a

Table 4 The Depth of Water Intake of Chemically Treated Water into Saneli and Hoban Silty Clay Loam when the Intake of Unconditioned Water Reached 10 cm

Water sources Location	Chemicals added	Rio Grande El Paso		Well Water Pecos	
EC (dS m^{-1})		1.1		3.9	
SAR		3.5		14.2	
pH		8.3		7.9	
Soil types		Saneli S.C.L.		Hoban S.C.L.	
Soil conditions		Moist	Wet	Moist	Wet
Chemicals	meq L^{-1}	Intake Depth (cm)			
None	0	10.0	10.0	10.0	10.0
$CaCl_2\cdot$	3.8	10.8	11.6[a]	10.6	10.9
Urea/H_2SO_4	3.8	—	—	10.5	10.6
NH_4/H_2SO_4	3.8	10.5	11.6[a]	11.6[a]	11.6[a]
NH_4S_x	3.8	11.5[a]	11.2	11.7[a]	11.6[a]
PAM (ppm)	5.0	10.7	11.3	11.8[a]	11.9[a]
$CaCl_2\cdot 2H_20$	7.8	—	—	11.6[a]	11.0
$CaCl_2$ Soln	38.0	11.5[a]	11.7[a]	11.7[a]	11.7[a]

[a]Significant at a 5% level.
Source: Unpublished data, author's laboratory.

bond with the negatively charged clay particles. Another possible reason relates to a reduction in the thickness of the diffused double layer with lowering pH. In calcareous soils, the pH reduction caused by acidification is buffered. However, the pH of water in subsurface drip systems and soil solutions can be reduced below neutral when carbonic acid accumulates. The pH of irrigation water high in sodium carbonate or injected with ammonia ranges from 9 to 10, which is readily reduced with acid to near neutral. Under these circumstances, reported increases in soil permeability or water infiltration with acid application may be partly accounted for by lowering pH, besides preventing Ca precipitation.

A recent field infiltration study conducted on noncalcareous sandy loam (pH = 6.3) in California, however, indicated that water infiltration improvements obtained through water application of Ca acetate, Ca nitrate, and Ca chloride seemed to have increased with increasing pH of the soils from 6.3 to a range of 6.7 to 7.3 (Wildman et al., 1988). These Ca compounds were applied to low salt water (0.1 dS m^{-1}) at a rate of 3 meq L^{-1}. The reason for this behavior is not clear but seems to be related to the increase in Ca retention in surface soils presumably due to the increase in the cation exchange capacity caused by the modest increase in pH.

From a practical point of view, several aspects warrant some attention. One relates to the assessment of the need for water conditioning. This may include the effect of water conditioning on infiltration of rain water, which has not yet been studied adequately. In assessing the need for water conditioning, salinity control is also a part of the picture in saline areas. Even a small improvement in infiltration rates can affect the salt balance in the root zone. Water testing for salinity, sodicity, and pH certainly helps, but water infiltration or penetration should be field-tested to verify the effectiveness of the intended conditioning. Compacted soils, for example, usually respond very poorly to chemical treatments (Baumhardt et al., 1992) until the compaction is removed by physical means. Sodding or cropping can also alter infiltration response to chemical treatments. Another aspect relates to selection of chemical amendments. From the view of containing production costs, it seems that greater attention should be given to the use of NH_4 containing acid fertilizers as a source for water conditioning. The frequency of chemical application is another issue. The current indication is that low salt water requires constant treatment unless the compound provides sufficient and steady release of H^+ at the soil surface. This requirement, however, can change with the addition of polyacrylamide, which is known to provide some persisting effects as discussed in a separate chapter (15). There is, of course, the bottom line question of how to project crop response to water conditioning. A rule of thumb is that a positive response is likely if there is clear evidence to indicate that water and/or salt stress related to poor water infiltration and penetration are limiting crop performance. Much-publicized effects of lowering pH of water and resulting increases in crop performance are, however, questionable at best in most calcareous soils, unless the pH of water is well above the equilibrium pH of $CaCO_3$ systems.

V. ACID AND ACIDULANT APPLICATION FOR RECLAIMING SODIC SOILS

The soils containing high levels of exchangeable Na are found widely both in irrigated and nonirrigated lands or arid and semiarid regions of the world. Most of these soils are calcareous or alkaline, although some sodic soils are acidic at times. In either case, poor soil permeability induced by high levels of exchangeable Na presents a major obstacle for reclamation that involves not only the replacement of Na but also the leaching of replaced Na and soluble salts. Chemical amendments, especially gypsum, were recognized to help reduce exchangeable Na and improve water infiltration necessary for reclamation. Acids and acidulants were thus originally considered simply as an alternative or a substitute for gypsum, as they convert $CaCO_3$ to Ca^{+2} and SO_4^{-2} in sodium-affected calcareous soils.

Subsequent studies, however, have shown that acid treatments can result in a greater crop response than gypsum treatments under certain conditions. The primary reason seems to be related to improved availability of certain plant nutrients, especially micronutrients and phosphorus (Stroehlein et al., 1978). Recent studies have also shown that in highly sodic alkaline soils, soil application of sulfuric acid can provide significantly better water infiltration and crop response than gypsum application (Miyamoto and Stroehlein, 1986). We will review and examine these unique characteristics of acids and acidulants as amendments for reclaiming sodic soils.

A. Reclamation of Moderately Sodic Calcareous Soils

Moderately sodium-affected calcareous soils with exchangeable Na percentages (ESP) of 15 to 35% are readily reclaimed for crop production, provided that soil texture is not impervious clay and that drainage outlets are available. Indeed, the current reclamation activities in the western USA deal primarily with this type of soil. The target of the reclamation is to lower exchangeable Na and to improve water infiltration necessary for leaching. In some cases, acids or acidulants may be applied for a dual purpose of lowering ESP and improving nutrient availability in Na-affected calcareous soils.

1. Acid and Acidulant Requirements

Reducing Exchangeable Na. For the removal of exchangeable Na, $CaCO_3$ present in the soils must first be solubilized with acids or acidulants as discussed earlier. Once Ca is solubilized, it undergoes the well-known exchange reaction.

$$Ca^{2+} + 2\,NaX \Leftrightarrow CaX_2 + 2Na^+ \tag{12}$$

where X denotes the cation exchange sites. This reaction is largely governed by the law of mass reaction, and one mole of Ca added may or may not replace two moles of exchangeable Na, depending on the dissolved cation concentrations and the exchange constants. For simplicity, however, the quantity of amendments required to replace the exchangeable Na is computed with an assumption of quantitative replacement.

$$AR = (ESP_0 - ESP)\ CEC\ (\alpha\rho\ D) \tag{13}$$

where AR is the quantity of acids or Ca required to replace exchangeable Na to a soil depth of D, ESP_0 the initial exchangeable Na percentage, ESP the target exchangeable Na percent, CEC the cation exchange capacity, ρ the soil bulk density, and α the unit conversion factor, which is 100 if AR, ESP, CEC, ρ, and D are expressed in eq ha^{-1}, %, eq kg^{-1}, kg m^{-3}, and m, respectively. The amendment required in terms of material weights per ha can then be estimated by multiplying applicable equivalent weights shown in Table 1.

The target ESP is traditionally considered to be 15%. However, it usually requires some adjustments depending on soil type, crops to be grown, water quality, and the irrigation system used. The depth of the soils to be reclaimed is somewhat arbitrary but is usually referenced to the depth of the major root zone. The estimated acid requirements to lower ESP_0 to 10 and 15% from the assumed depths of the soils with a CEC of 0.25 eq kg^{-1} are shown in Table 5. In clayey soils with a greater CEC, the requirement increases in proportion to the CEC. According to the estimates, the amendment required (expressed in H_2SO_4 equivalent) is in the range of 2.5 to 13 Mg ha^{-1} in moderately sodic soils with an ESP of 30%. In highly sodic soils or in clayey soils with a greater CEC, the estimated quantities may become impractically large. In such cases, alternative methods of reclamation may be required as discussed later.

The above procedures, originally developed for Ca compounds, are adapted here for acids and acidulants with an assumption that one mole of acids or acidulants solubilizes one mole of Ca from $CaCO_3$. This assumption is realistic in an open system where the acid and carbonate reactions take place at the soil surface following ground applications. In the case of sulfur incorporated into moderately sodic calcareous soils, a greater quantity of Ca can be solubi-

Table 5 The Quantities of Acids, Acidulants, and Gypsum (All Assumed to be 100% Ingredient) Required to Lower the Exchangeable Na Percent (ESP) to Desired Levels from Various Soil Depths at a Cation Exchange Capacity of 0.25 eq kg^{-1}

Initial ESP	Target ESP	Amendment types	Quantities required for soil depths		
			15	30	60 cm
%	%			Mg ha^{-1}	
30	15	H_2SO_4	2.5	5.0	10.1
		NH_4S_x[a]	1.3	2.6	5.1
		S	0.8	1.6	3.3
		$CaSO_4{\cdot}2H_2O$	4.4	8.8	17.6
		$FeSO_4{\cdot}7H_2O$	7.1	14.2	28.4
30	10	H_2SO_4	3.3	6.6	13.2
		NH_4S_x[a]	1.7	3.4	6.8
		S	1.1	2.2	4.3
		$CaSO_4{\cdot}2H_2O$	5.6	11.6	23.2
		$FeSO_4{\cdot}7H_2O$	9.4	18.7	37.4
50	15	H_2SO_4	8.5	17.6	35.3
		S	2.8	5.6	11.1
		$CaSO_4{\cdot}2H_2O$	14.9	29.8	59.7
70	15	H_2SO_4	15.1	30.2	60.4
		S	4.9	9.8	19.7
		$CaSO_4{\cdot}2H_2O$	26.5	53.0	106.0

[a]The contribution of NH_4 oxidation is included, and the product assumed is NH_4S.

lized, at least in theory, because of soil entrapment of CO_2. This possibility, however, has not yet been fully verified, partly owing to highly variable oxidation rates of elemental S.

Improving Water Infiltration. The quantities of acids or acidulants required to improve water infiltration are usually superseded by the quantities required for Na replacement. However, the idea of computing the amendment requirements based on permeability improvements, instead of Na replacement, has merits in reclamation of highly sodic soils where the amendment requirements for Na replacement become impractically large. This topic is discussed later. Even in moderately sodic soils, this concept may deserve some attention when nonsodic saline water is to be used for irrigation. Under such circumstances, the exchangeable Na can be replaced eventually, as long as amendments are applied at a quantity needed to maintain water infiltration. The quantity of amendments required to maintain water infiltration is soil dependent, and usually 2 to 5 Mg ha^{-1} in moderately sodic soils.

Improving Nutrient Availability. The quantities of acids or acidulants required to improve availability of certain micronutrients and phosphorus are usually determined empirically. As demonstrated by Ryan et al. (1974), it requires almost a complete removal of $CaCO_3$ to release most micronutrients, although this rule does not seem to apply to phosphorus. Acid applications usually solubilize Ca-bound phosphorus, but not Fe or Al-bound phosphorus (Khorsandi, 1994). In most calcareous soils, large quantities of acids or acidulants are required if one attempts to dissolve $CaCO_3$ from the entire root zone. (Note that it will require 10 Mg ha^{-1} of sulfuric acid to remove $CaCO_3$ from 1 cm of soil layer containing 7.5% $CaCO_3$ by weight.) However, experience has shown that the removal of $CaCO_3$ from a small portion of the root zone is usually sufficient to provide necessary micronutrients (e.g., Wallace and Mueller, 1978; Fenn et al., 1990). The quantities of acids and acidulants necessary for sufficient release through soil injection of sulfuric acid typically range from 0.5 to 2 Mg ha^{-1}, which are usually lower than those required for Na replacement in calcareous sodic soils.

2. Amendment Application

The quantities of acids or acidulants required to reclaim sodic soils are usually too large to be applied through irrigation water, not only because the pH of water is reduced to the levels that pose corrosion hazard but also because of CO_2 entrapment in soil profiles, which slows water infiltration (Miyamoto et al., 1973). In addition, amendments applied through water usually result in a greater amendment application in permeable portions, and a lesser application in less permeable portions of the fields, which is not compatible with the objective of uniform reclamation. Thus amendments are usually broadcast directly to the soils or selectively to the portion of the field that requires most reclamation. Prior to amendment applications, soil compaction or petrogenetic pans, which limit water infiltration and penetration, must be removed through appropriate mechanical or physical measures. Chemical amendments are not a substitute for chiseling or subsoiling.

Surface Broadcast Application. If sulfuric acid or phosphoric acid is to be applied, the ground surface has to be dry. Otherwise vigorous reactions with moist soils generate noxious acid fumes. During the ground application of strong acids, care must be taken to reduce acid splatters, especially in windy areas. Some custom applicators have used flexible tubing, instead of drip nozzles, to apply acid on the ground, and others have used a sheet of rubber around the spray boom to reduce splatters. As soon as acids are broadcast, the field should be irrigated without delay, or otherwise be disked. Exposure to ground-applied acid is the prime source of injuries. Bedding of acid-treated ground enhances solubilization of certain micronutrients and phosphorus in row crops but is usually not recommended for achieving uniform Na replacement and leaching until the acid-treated ground is irrigated. If the field is to be irrigated with mobile sprinkler systems, such as side-rolls or center pivots, acid-treated fields should be disked to avoid acid splatters onto the sprinkler systems. Needless to say, aluminum pipes should not be laid on acid-treated fields. If surface irrigation methods are to be used, double borders should be placed at low ends so as to contain acidic water in the case of accidental break-off of the primary borders. If surface irrigation systems are designed for tail water reuse, the acid-treated ground should be disked to avoid corrosion of the reuse system. If the tail water becomes acidic, injection of ammonia into the collection reservoir may be necessary prior to pumping for reuse. The acidic tail water may not be discharged to drain ditches.

Ground Injection. Soil injection of strong acids is preferred to surface broadcast applications in terms of lowering hazards from acid application and improving availability of micronutrients efficiently in calcareous soils. Soil injection also allows the use of SO_2 in a manner used for NH_3 injection into crop beds (Roberts and Koehler, 1965). However, acid or acidic gas injection to crop beds (used for improving availability of certain nutrient elements) lowers the uniformity of Na replacement and salt leaching. An injection device designed to apply acid at a shallow depth with close spacings on a flat ground is a good compromise. Alternatively, ground injection can be combined with water application of polysulfides to improve the coverage of amendment applications.

3. Performance of Amendments

Reducing Exchangeable Na. It was noted earlier that the efficiency of Na replacement is subject to the law of mass reactions. Experimental data obtained from surface-applied amendments indicate that the removal of exchangeable Na by dissolved Ca is curtailed in saline sodic soils, owing to the presence of high concentrations of dissolved Na initially present plus those replaced from the exchange sites. With the leaching of the dissolved Na, the exchange reaction then proceeds in a more efficient manner, provided that soluble Ca compounds are left at the soil surface. Soil column experiments have shown that the quantities of exchangeable Na replaced after leaching with 1.5 pore volume of water amounted to 79 to 100% of the Ca (or H_2SO_4) added to the soils as amendments plus those present in irrigation water (Table 6). These numbers, however, include the exchangeable Na removed by leaching alone without chemical additions. If the fraction is subtracted, the exchangeable Na removed amounted to 42 to 84% of the Ca added as amendment. An additional quantity of exchangeable Na may be removed through leaching by the process known as the dilution effect.

The replacement of Na is greater near the soil surface, and it becomes negative below certain depths due to Na translocation to deeper depths (Miyamoto et al., 1975c). Consequently, a typical pattern of exchangeable Na distribution after chemical applications and leaching is lowest near the soil surface. The ESP of the top 10 cm of soils can be lowered to as low as 5% when acids are broadcast with an intent to reduce the ESP to a range of 10 to 15% from the top 30 cm (Miyamoto et al., 1975c; Miyamoto and Enriquez, 1990). The type of amendments used, except for elemental S, does not seem to affect this pattern to any significant degree. However, the small differences in ESP at the soil surface coupled with the difference in soluble salts left at the soil surface can make a significant difference in the rate of water infiltration and leaching during reclamation in structurally weak soils (Miyamoto and Enriquez, 1990).

Improving Water Infiltration. The rate of water infiltration directly affects the rate of reclamation and is an important factor in reclamation of poorly permeable soils. Water infiltration in moderately sodic soils (ESP = 11 to 25%) following broadcast applications of concentrated sulfuric acid at 3 to 5 Mg ha^{-1} has increased roughly twofold (Yahia et al., 1975). When application rates exceeded 5 Mg ha^{-1}, however, no added benefit was observed, and at an application rate of 10 or 15 Mg ha^{-1}, the rate of water infiltration was actually reduced in moderately Na-affected soils. This apparent reduction in water infiltration is often related to the formation of a thin soil layer developed by reprecipitation of Fe and Al when they are transported from the surface to the alkaline zone. The depth of acidified layers at an acid application rate of 10 Mg ha^{-1} extends 1 to 5 cm in soils having the titratable basicity of 100 to 20 kg m^{-3} or $CaCO_3$ contents of 7.5 to 1.5% by weight, respectively. Thus the flow restricting layer usually develops near the soil surface, and physical disturbance of this thin layer restores water infiltration. As mentioned earlier, CO_2 evolved as a result of acid reaction with $CaCO_3$ also interferes with water infiltration at high rates of acid application (Miyamoto et al., 1973).

According to a recent soil column study (Miyamoto and Enriquez, 1990), sulfuric acid broadcast and incorporated into 5 cm of moderately Na-affected soils provided about the same rate of water infiltration as powder gypsum in-

Table 6 Removal of Exchangeable Na with Application of Sulfuric Acid, Gypsum, and Calcium Chloride Followed by Leaching at 1.5 Pore Volume in Two Moderately Na-affected Calcareous Soils

Soil type	Amendment eq kg^{-1}	Amendment Mg ha^{-1}	Exch. Na removed eq kg^{-1}	Final ESP %	Removal ratio[a]	Leaching duration days
Saneli silty clay loam (Initial ESP = 17.5%, exch. Na = 0.035 eq kg^{-1}) (CEC = 0.20 eq kg^{-1})						
H_2SO_4	10.7	4.3	12	11.5	1.0	13
$CaSO_4 \cdot 2H_2O$	10.7	7.5	10	12.5	0.85	13
$CaCl_2 \cdot 2H_2O$	10.7	6.4	10	12.4	0.84	24
Control	0.0	0	3	16.2	—	55
Glendale silty clay (Initial ESP = 13.7%, exch. Na = 0.049 eq kg^{-1}) (CEC = 0.36 eq kg^{-1})						
H_2SO_4	24.0	8.5	23	7.2	0.91	31
$CaSO_4 \cdot 2H_2O$	24.0	15.0	23	7.4	0.91	32
$CaCl_2 \cdot 2H_2O$	24.0	12.8	20	8.2	0.79	42
Control	0	0.0	13	10.0	—	>100

[a]Defined as the exchangeable Na removed by Ca added in chemical equivalent.

Source: Miyamoto and Euriquez, 1990.

corporated to the same depth at an equivalent rate (Table 6). The test soil was moderately Na-affected saline calcareous soil (CEC = 0.20 eq kg^{-1} ESP = 17.5% and EC_e = 6.0 dS m^{-1}), and the water used for leaching had a salt composition similar to that of the Rio Grande (EC = 1.1 dS $^{-1}$ and SAR of 3.5). The data also show that the soil application of $CaCl_2 \cdot 2H_2O$ at a chemically equivalent rate was not nearly as effective as acid or gypsum treatment, presumably due to rapid leaching of $CaCl_2 \cdot 2H_2O$ from the soil surface layer. A similar study involving another moderately Na-affected Glendale silty clay (CEC = 0.36 eq kg^{-1}, ESP = 14% EC_e = 6.0 dS m^{-1}) has shown essentially the same results (Table 6).

These soil column studies also included iron and aluminum sulfate treatments, incorporated to a depth of 5 cm at the same equivalent rate as sulfuric acid or gypsum. Iron and aluminum sulfate treatments provided infiltration rates similar to or slightly less than those from the gypsum or sulfuric acid treatment. Some other studies (e.g., Tobia and Pollard, 1958), however, indicate that aluminum sulfate applied to alkali or calcareous soils provided much higher infiltration rates than sulfuric acid applied through water at an equivalent rate. This is probably not a fair comparison, however. As indicated earlier, acid applied through water in excess of HCO_3 concentration generates CO_2, which interferes with water percolation. The flow of acidic water can completely cease when CO_2 is entrapped in a soil profile.

A field infiltration study conducted at Safford Arizona (Alawi, 1977) using a moderately Na-affected Pima clay loam (ESP of 20 to 25%) has shown twofold increases in infiltration rate when sulfuric acid was broadcast at a rate of 2.2 Mg ha^{-1} and leached with water having salinity of 0.51 dS m^{-1}. The effect of gypsum (a commercial grade) applied at an equivalent rate (3.8 Mg ha^{-1}) was not nearly as prounounced as the sulfuric acid treatment, probably due to a slower dissolution rate of commercial granule products as compared to powder gypsum (Keren and O'Connor, 1982). When saline water having an EC of 3.2 dS m^{-1} was used for leaching, the results were essentially the same, except that the infiltration rates were overall higher than in the case of low salinity water. In a separate field infiltration test performed in Arizona (Miyamoto and Stroehlein, 1975) using saline sodic Gothard clay loam (ESP = 15 to 18%, and EC_e = 10 dS m^{-1}), both sulfuric acid and powder gypsum applied at 5 Mg ha^{-1} of sulfuric acid equivalent rate doubled the infiltration rate (Table 7). The leaching water used had low salinity (5 meq L^{-1}) and a moderate level of sodicity (SAR = 5.7).

Judging from some of these reports, it appears that water infiltration during the initial leaching of moderately Na-affected soils can be improved by a factor of about 2 at acid application rates of 2 to 5 Mg ha^{-1}. These rates are roughly equal to or smaller than the acid requirements estimated for the Na replacement from the top 15 to 30 cm at the target ESP of 10 to 15%. In many of the soils tested, the actual ESP at and near the surface was reduced to much

Table 7 Effect of Sulfuric Acid and Gypsum Applications on Water Infiltration and Yields of Grain Sorghum in Gothard Clay Loam (ESP = 17%, CEC = 0.42 eq kg^{-1}, and EC_e = 10 dS m^{-1}) Under Field Conditions

Amendments	Rate Mg ha^{-1}	Infiltration cm d^{-1} 1st	2d	Sorghum head wt. kg ha^{-1}
No treatment	0	—	—	2000
Leaching only	0	9	8	3100
Sulfuric acid	3	19	17	8400
Sulfuric acid	5	21	20	7300
Gypsum (powder)	9	20	21	8000

Source: Stroehlein et al., 1978.

lower values, typically around 5%. The flow limiting layer for infiltration into plowed and disked sodic soils usually develops within the top 10 cm where aggregate breakdown occurs first (Miyamoto and Enriquez, 1990). Thus lowering ESP and maintaining elevated levels of salinity in the top layer is critical for maintaining water infiltration. Such requirements can be met with the surface application of sulfuric acid or sulfates of Ca, Fe, and Al, but not always with calcium chloride when applied at a rate commonly used.

Sustainability of adequate levels of infiltration after acid treatment is a separate issue and would depend to a large extent upon quality of irrigation water used, besides soil texture and compaction. If needed, water treatments discussed earlier will provide an option for infiltration maintenance. Soil application of elemental S is another option for maintenance, but the performance has been rather inconsistent (e.g., Overstreet et al., 1951; McGeorge et al., 1956). What is obvious is that the exchangeable Na should be lowered during reclamation at least to a depth of the plow layer. Otherwise, plowing will bring the soil with higher sodicity to the soil surface, thus reintroducing infiltration problems.

There must be a word of caution against indiscriminate uses of acids and acidulants. Ground application of sulfuric acid or gypsum, for example, has been practiced on orchard floor for the purpose of improving water infiltration. Orchard floors are usually highly compacted, and our study has shown that chemical amendments are of little value, unless the compacted soil is first dealt with using physical measures (Helmers and Miyamoto, 1990). Likewise, chemical amendments are usually of limited value in improving poorly permeable clays with low ESP.

Crop Response. Crop responses to acid or acidulant applications may result from a combined effect of reduced ESP, improved water infiltration, reduced salinity and alkalinity, and improved nutrient availability. In a greenhouse pot experiment using two moderately sodic saline soils from Arizona (Pima clay loam, ESP = 23%, EC_e = 15 dS m^{-1}) and Gothard clay loam (ESP = 17% EC_e = 12 dS m^{-1}), sulfuric acid was broadcast at 10 and 3 Mg ha^{-1}, respectively, then were leached with two pore volumes of water before planting kidney beans (*Phaseolus vulgaris*). These treatments provided vegetative growth nearly equal to those obtained from nonsaline and nonsodic soil used as a control (Stroehlein et al., 1978). This simple experiment may indicate that these saline sodic soils can be reclaimed with acid applications and leaching to a level that salt and sodium sensitive crops, such as beans, can be grown satisfactorily. In another greenhouse experiment using the same soil, the effect of sulfuric acid broadcast at rates of 3 and 6 Mg ha^{-1} was compared against 7 and 10 Mg ha^{-1} rates of gypsum along with an Fe chelate application at 1 kg ha^{-1} using grain sorghum (*Sorghum bicolor*) as a test crop (Stroehlein et al., 1978). In Pima clay loam, sulfuric acid and Fe-chelate treatments provided greater vegetative growth than the gypsum treatment, indicating that the superior crop response from the acid treatments might have been a result of improved availability of Fe. The effectiveness of acids to improve Fe availability has been demonstrated in both greenhouse and field trials (e.g., Mathers, 1970; Ryan et al., 1975b). In Gothard silt loam with an adequate level of Fe, acid and gypsum treatments provided an equal crop response. In a subsequent field test on Gothard silt loam, both acid and gypsum treatments provided dramatic increases in sorghum yields (Table 7), presumably because of improved water infiltration and salt and Na leaching. The test site was in an area commonly known as "salt spots" by practical farmers. Yield response from the high rate of acid application (5 Mg ha^{-1}) was somewhat lower than the low rate of acid application, presumably because of inadequate leaching of salts generated by the acid treatment. Note that this soil was highly saline (EC_e = 12 dS m^{-1}). A field study conducted in Israel (Oron and DeMalach, 1989) has shown that sulfuric acid applied between furrows at a rate of 4 Mg ha^{-1} has increased yields of cotton in slightly Na-affected soils (ESP of 7% in the top 60 cm) irrigated with sewage water having an EC of 1.8 to 1.9 dS m^{-1}.

There are many studies where improvements in nutrient availability were tested in nonsodic calcareous soils. Such subjects are beyond the scope of this chapter, and one may refer to a separate review article (e.g., Stromberg and Tisdale, 1979). These studies have shown that acid applications were effective in alleviating Fe, P, and Zn deficiency in various crops. The rate of acid or sulfur application to achieve the desired nutritional responses varies from 500 kg ha^{-1} to several Mg per ha, depending on the $CaCO_3$ contents of the soil, the crops and the methods of application. The longevity of such treatments has not yet been determined adequately. If the acid band remains undisturbed, a study with a tree crop, pecans (*Carya illinoensis*), shows that the effect of subsurface acid bands can last at least for 9 years after application in supplying Zn to the trees

(Fenn et al., 1990). A similar treatment with sulfur was also reported to be effective in correcting chlorosis in citrus (McGeorge, 1939). The longevity factor has a significant impact on cost-effectiveness of acid or sulfur application as compared to conventional means of fertilization or foliar spray in providing micronutrients.

B. Reclamation of Highly Sodic Alkaline Soils

Alkaline soils containing high levels of exchangeable Na (e.g., ESP of 35 to 100%) are distributed throughout the arid and semiarid regions of the world. Some of these soils also occur in coal mining areas affected by disposal of oilfield brine. These soils present not only the common problem of low permeability but also high pH and, at times, high salinity and are often void of vegetation, structurally unstable, and highly erodible. In reclaiming these soils, the primary targets are to lower the exchangeable Na, to improve water infiltration and soil structural stability, and to lower soil pH. The quantities of chemical amendments required are usually larger, in theory, than those used for moderately sodic soils, and in many cases high costs of amendments make reclamation unattractive. However, if the soil is sandy, it can be readily reclaimed.

1. Reclamation of Highly Sodic Sandy Alkaline Soils

Sandy soils, even when saturated with Na, have been reclaimed with the use of amendments with minor adjustments in crop selection. These soils have inherently large water conducting pores and low cation exchange capacities (CEC), the feature favorable for Na replacement and leaching. Highly sodic Fresno silt loam reclaimed in the Central Valley of California was saturated with Na, but the CEC ranged from 0.06 to 0.12 eq kg^{-1} (Overstreet et al., 1951). Likewise, the alkaline soils reclaimed in northwestern states are mostly silt loam in texture (Robbins, 1986). Prior to amendment application, the land to be reclaimed for crop production may require subsoiling, leveling, or grading so as to avoid ponding of water in low lying areas. The first crop during or after reclamation usually consists of forages and grains, most of which are salt-tolerant, easy to grow, and produce substantial biomass.

Amendment Requirements. The quantities of acid and acidulants required for Na replacement in sodic sandy soils are lowered by the low CEC. If the CEC of the soil is as low as 0.10 eq kg^{-1}, as is the case with Fresno silt loam, the quantity of sulfuric acid required to lower ESP from 100 to 15% in the top 30 cm is, according to Eq. (13), 12 Mg ha^{-1}, or 6 Mg ha^{-1} for 15 cm of soil thickness. These rates are not greatly different from the quantities required for reclamation of moderately sodic soils with high CEC (Table 5).

The quantities of acids or acidulants required to lower the pH of alkaline soils to 8.3 (an equilibrium pH of calcareous soils) are usually lower than those needed for Na replacement, unless the soils contain a large amount of alkali generating minerals, such as soda ($Na_2CO_3 \cdot 10H_2O$) and trona ($Na_2CO_3 \cdot NaHCO_3 \cdot 2H_2O$).

Improving Water Infiltration. There has been concern that the quantities of amendment required to replace Na from the surface soil may not be adequate to provide water infiltration necessary for leaching in Na-saturated soils. This concern is certainly warranted with elemental or even colloidal sulfur, as the oxidation reaction is too slow to be effective for initial leaching. In the case of H_2SO_4, soil column studies have shown that acid broadcast applications at a rate of 10 to 15 Mg ha^{-1} are usually sufficient to attain water infiltration necessary for initial leaching, even in clay loam soils which have a CEC as high as 0.40 eq kg^{-1} (Table 8). In the sandy textured Fresno soil series, satisfactory water infiltration was attained from sulfuric acid applied at 3 Mg ha^{-1} (Overstreet et al., 1951). When acid application rates approach 10 to 20 Mg ha^{-1}, infiltration rates may actually slow for the reasons stated earlier. Gypsum applications were not nearly as effective as sulfuric acid applications in improving water infiltration in highly sodic soils (Table 8), presumably because of limited solubility. Gypsum powder applications in excess of about 30 Mg ha^{-1} actually reduced water infiltration (Table 8) as a result of soil pore plugging

Table 8 Effects of Sulfuric Acid and Gypsum Application on Water Penetration into Dry Sodic Calcareous or Alkaline Soils

Soil types and properties						
Acid (Mg ha^{-1})	0	1.0	5.0	10.0	15.0	20.0
Gypsum (Mg ha^{-1})	0	1.75	8.75	17.5	26.2	35.0
Pima clay loam (CEC = 0.3 eq kg^{-1}, ESP = 25%, ECe = 17 dS m^{-1})						
Penetration (cm/5 hrs)						
Acid	13	16	19	21	17	—
Gypsum	13	17	18	18	17	—
Fresno silt loam (CEC = 0.09–0.11 eq kg^{-1}, ESP = 80–100%, EC_e = 8–13 dS m^{-1})						
Penetration (cm/5 weeks)						
Acid	31	40	52	—	—	—
Gypsum	31	—	4	—	—	—
Stewart silt loam (CEC = 0.26 eq kg^{-1}, ESP = 100%, EC_e = 56 dS m^{-1})						
Penetration (cm/5 hrs)						
Acid	10	12	33	57	80	62
Gypsum	10	11	19	24	29	26
Playa clay loam (CEC = 0.51 eq kg^{-1}, ESP = 100%, EC_e = 40 dS m^{-1})						
Penetration (cm/5 hrs)						
Acid	3	7	16	25	26	26
Gypsum	3	6	7	11	12	13

Source: Yahia et al., 1975; Overstreet et al., 1951.

with fine undissolved gypsum particles (Keren et al., 1980). In a field trial conducted on sodic alkaline sandy clay loam (ESP = 93–95% and pH of 10) in India, however, sulfuric acid applied at 7.5 and 13 Mg ha^{-1} did not yield significantly higher rates of water infiltration than gypsum or aluminum sulfate applied at chemically equivalent rates (Chand et al., 1977). In this experiment, sulfuric acid was apparently spread on a thin layer of standing water. If sulfuric acid was broadcast on dry soils as were other amendments, the results could have been different.

It should be kept in mind that these data were obtained with air-dry soils (except for the acid treatment in the field test in India), which are well-aggregated owing to the presence of high levels of salts. Once these soils, especially clay loam, are wet with low salt water, water infiltration is usually severely reduced because of aggregate slaking and destruction. Data related to the effect of chemical amendments on wet sodic soils under high water tables are currently scarce.

Soil and Crop Response. In the field trial on Na-saturated silt loam in California (Overstreet et al., 1951), acid and gypsum broadcast treatments lowered ESP from an initial value of 100 to 22 and 42%, and soluble Na concentrations from about 300 to 130 meq L^{-1} in the soil extract (Table 9). In addition, the pH measured in 1:5 extract was reduced from 9 to 8.6 and 7.7 with the gypsum and the acid treatments, respectively. The treatment with an equivalent rate of an unspecified form of elemental S decreased neither ESP nor soluble Na concentrations. These soil samples were collected from a soil depth of 0 to 30 cm at one monitoring location per plot about 8 months after the treatments. Because of the limited number of soil samples analyzed, these data should be considered merely an indication. Nonetheless, the observed reduction in ESP seems to have fallen well below the target. Application of sulfuric acid at 12.5 Mg ha^{-1} or gypsum at 22 Mg ha^{-1} should, according to Eq. (13), have lowered ESP to 15% instead of the observed range of 22 to 42%. In the field test in India (Chand et al., 1977), the application of sulfuric acid, gypsum, and aluminum sulfate reduced the ESP of the top 15 cm soil below the target level of 15%. The pH of the soil was reduced to below 8.6 at the high rate of application and below 9.0 at the low rate of application, both in acid- and in gypsum-treated plots. The high rates of application were 13 Mg ha^{-1} of sulfuric acid and 22 Mg ha^{-1} of gypsum, and the low rate were equal to half of the high rates.

In the field experiment with Fresno silt loam (Overstreet et al., 1951), yields of mixed pasture consisting of seven species of moderately to high salt tolerant grass species and clover (*Trifolium hybridum*) have shown consistently superior performance of the acid treatment over the gypsum treatment in all cuttings (Table 9). A good stand of clovers was reportedly obtained only in acid-treated plots. The sulfur treatment did not perform well, and this was thought to be a

Table 9 Effects of Three Amendments Applied to Na-Saturated Fresno Silt Loam on Properties of Top 30 cm Soil and Forage Crop Yields

Amendments	Control	Elem S	$CaSO_4 \cdot 2H_2O$	H_2SO_4
Rate (Mg ha^{-1})	0	4.1	22	12.5
Soil properties				
CEC, initial (meq kg^{-1})	118	111	93	86
EC_e, initial (dS m^{-1})	14	8.0	15.0	7.0
ESP, initial (%)	89	77	97	101
After 8 months	65	76	22	42
Na, initial (meq L^{-1})	254	168	304	154
After 8 months	265	173	133	132
pH (1.5) initial	9.5	9.6	9.0	9.1
After 8 months	9.6	9.4	8.6	7.7
Forage production		Mg ha^{-1}		
1st cut 9 months	3.23	3.76	5.28	7.63
2nd cut 1 year	2.33	2.57	3.38	4.33
3rd cut 1.5 years	1.83	1.69	2.60	4.86
Total 1.5 years	7.39	8.03	11.40	16.83

Source: Overstreet et al., 1951.

result of poor oxidation under the high-pH and high-salt environment. The superior performance of sulfuric acid treatment in this test remains unexplained, especially in light of comparable performance of gypsum in lowering ESP as well as soluble Na. Among the soil properties measured, the soil pH seems to be the only parameter that has changed differently between the two treatments. In a separate greenhouse experiment conducted on Na-saturated Playa silt loam with a pH of 10.3, acid treatments at a rate of 10 Mg ha^{-1} lowered the pH to 8.3 and produced respectable alfalfa (*Medicago sativa*) yields, while gypsum applications at the equivalent rate (17.5 Mg ha^{-1}) resulted in little production (Stroehlein et al., 1978). Soil analyses for P and Fe revealed no deficiency. It is probable that the superior performance of acid treatments in both of these experiments may be related to alleviating hydroxide toxicity (Thorup, 1969). In the field experiment conducted in India (Chand et al., 1977), sulfuric acid, gypsum, and aluminum sulfate treatments have provided about equal soil pH and yields of barley (*Hordeum vulgare*). In rice (*Oriza sativa*), however, the sulfuric acid treatment provided significantly greater yields and Zn concentrations in tissue. Rice yields as well as Zn concentrations from aluminum sulfate treatment were, however, significantly lower than those of the gypsum treatment, presumably owing to Al induced Zn deficiency.

2. Reclamation of Highly Sodic Alkaline Soils Involving Reclamation Crops

When soil texture changes from silt loam to clay loam, the cation exchange capacity (CEC) usually doubles or triples. Accordingly, the quantities of amendments required for the same level of ESP will, in theory, double or triple, leading to acid requirements in a range of 30 Mg ha^{-1} to as high as 100 Mg ha^{-1}. These rates are beyond the current investment strategy of most farmers. A classic strategy to deal with this situation is the use of reclamation crops.

Amendment Requirements. There is currently no accepted procedure to determine the amendment requirements for reclamation involving reclamation crops. However, the quantities can be determined appropriately based on the need to provide water infiltration sufficient to obtain initial leaching. The experimental data reviewed earlier indicate that the application rates of 10 to 15 Mg ha^{-1} in H_2SO_4 equivalent should be adequate to provide initial water intake, even in clay loam soils having a CEC as high as 0.4 eq kg^{-1}, provided that the soil is not wet. Application rates can be reduced in sandy soils or when saline water is used for irrigation. Such a rate of amendment application, however, may not lower ESP beyond the top 10 to 30 cm, depending primarily on the CEC. Removal of the exchangeable Na in deeper depths must then rely on cropping effects. These application rates, if sulfuric acid is to be used, should also be sufficient to lower soil pH to a satisfactory level in most alkaline soils.

Reclamation Crop Selection. Selection of reclamation crops has been primarily based on practicality and adaptability; it extends from rice and forage crops to trees (e.g., Gupta and Abrol, 1990). In essence, the underlying idea has been to grow whatever may grow in highly saline and/or alkaline soils with hope of soil improvements and some economic return. The presence of organic matter in reclamation of sodic alkaline soils has been viewed as having positive effects on soil structure, although some studies (e.g., Gupta et al., 1984) do not necessarily support this idea in sodic soils. A recent lysimeter study with Freedom silt loam (ESP = 68%, ECe = 2.4 dS m^{-1}) is among the few that have provided a rational basis for selecting crops for reclamation of Na-affected calcareous soils. The study examined the role of reclamation crops in improving water infiltration and removing exchangeable Na. Results indicate that the crops that produce a large root mass and CO_2, such as sorghum and sudan (*Sorghum sudanese*), provided the largest removal of exchangeable Na and improvements in water infiltration as compared to crops of lower root mass such as cotton (*Gosypium hirsutum*). The lysimeter study has also shown that the exchangeable Na removed by live plants was greater in extent and has reached deeper into the soil profile than the results attained by the incorporation of applicable crop residues. The field study conducted in India (Chand et al., 1977) also indicates that surface incorporation of organic matter, such as manure, had a limited value in reducing ESP, even though the decomposition of organic matter generates CO_2. A study conducted on nonsodic soils shows appreciable effects of crops on water infiltration (e.g., Ali and Swartzendruber, 1994). These studies seem to indicate that both CO_2 generation rates and root mass would be important criteria for selecting reclamation crops.

One can certainly wonder how any crops can be grown adequately without chemical amendments in the soil with ESP as high as 68%, because of the well-known difficulties with water infiltration. In the lysimeter experiment, the soil used (Freedom silt loam) was apparently sandy enough to permit adequate water infiltration. If not, an application of acids should provide the necessary initial water infiltration. An alternative may be to use saline water for leaching while maintaining soil permeability. In such cases, crops of higher salt tolerance than ordinary crops may be required. There are many plant species that have higher salt tolerance and are capable of producing large biomass, as reviewed elsewhere (Miyamoto et al., 1994). Improvements in water infiltration per se, however, may not lower soil pH to a desirable level for crop growth in sodic alkaline soils, unless acids or acidulants are used as amendments. Selection of reclamation crops based on their adaptability to high pH conditions then becomes a consideration (Kumar Ashok and Abrol, 1986).

Soil and Crop Response. There are many reports that claim that highly sodic alkaline soils were reclaimed by the use of reclamation crops. In many instances, such claims seem to refer to the success of establishing reclamation crops, because of their adaptability to saline sodic or alkaline conditions, but not necessarily improving the soil for subsequent cropping with a higher economic value. The lysimeter experiment with Freedom silt loam (Robbins, 1986) is among the few that actually demonstrated the major effects of reclamation crops on reducing ESP. The ESP of the soil was, for example, reduced from 68 to 30% to a depth of 0.7 m through cropping with sorghum or sudan in a matter of one season, which was more than the ESP reduction caused by the gypsum treatment without cropping. The effect of the gypsum treatment in removing the exchangeable Na was confined to the top 0.3 m. There are also other reports documenting favorable changes in soil properties following vegetation establishment in alkaline soils (e.g., Singh et al., 1990), although these results are not as spectacular as the results of the lysimeter experiment with Freedom silt loam.

It is, however, necessary to point out that the lysimeter study with Freedom silt loam is in apparent contradiction with the field study conducted on Fresno silt loam, where the reduction in ESP was minimal with cropping alone (Overstreet et al., 1951). Poor growth of crops in the field experiment and general slowness of reclamation with cropping may partly account for the apparent discrepancy. It is also possible that the difference in the extent of drainage between the two experiments may account for the apparent discrepancy. The lysimeter experiment used a large quantity of water and deliberate leaching, while the field experiment with Fresno silt loam did not. It seems that the effective removal of exchangeable Na cannot be achieved without adequate leaching, regardless of how the Ca was solubilized, either through CO_2 respiration or $CaCO_3$ dissolution with acids.

The complete picture of reclamation involving reclamation crops with or without chemical amendments cannot be ascertained without a comprehensive reclamation model. At the moment, it seems that the principal role of acids and acidulants (or other chemical amendments) in reclamation of highly sodic alkaline soils is to facilitate crop establishment by improving surface soil conditions (including soil pH) and to provide water intake necessary for leaching, until cropping itself can carry out such functions.

3. Use of Acids on Revegetation of Highly Sodic Soils

Highly sodic soils are found more frequently under dryland conditions than in irrigated lands in most parts of arid and semiarid regions. Sodic lands are also being formed through saline seeps, strip mining activities (Sandoval and Gould, 1978), and as a result of disposal of oil field brine or drilling mud (McFarland, et al., 1987). Reclamation of these sodic lands is considered important, not only for soil and habitat conservation, but also for minimizing water quality degradation caused by suspended soil particles and salts. These concerns prompted the passage of various restoration regulations. Reclamation of sodic mine spoils is an example of regulatory measures placed on strip mining areas in the western states. A similar measure, though less stringent, is beginning to be enforced in the restoration of rangelands affected by disposal of oifield brine.

In the case of strip mining, a standard measure for reclamation is to top-dress salt or sodium affected spoils with native topsoil. This measure is costly but effective, provided that the topsoil is available in sufficient quantity. Otherwise, it is necessary to use measures to improve the properties of sodium-affected spoils or bring in exotic species that can adapt to the saline sodic soil conditions. In the case of revegetation of lands contaminated by oil field brine, top-dressing is usually not mandatory. Chemical amendments (especially the use of gypsum) have been thought to be an option in these cases (Doering and Willis, 1975), yet its uses are essentially at an experimental stage. In a preliminary pot experiment (Scholl and Miyamoto, 1984) involving coal mine spoils derived from Na-affected sandstone (initial ESP of 21% with EC_e of 6.7 dS m^{-1}), broadcast applications of sulfuric acid at a rate of 2.5 and 5.0 Mg ha^{-1} reduced ESP and improved water infiltration, availability of P, and top dry weights of two salt tolerant range plants (Table 10). In Na-affected spoils derived from shale (ESP = 38%, EC_e = 5.2 dS m^{-1}), however, acid applications at a rate of 5.0 Mg ha^{-1} caused a significant reduction in seedling emergence, presumably owing to excessive acidification to lower the pH of the surface soil

Table 10 Some Chemical and Physical Properties of Coal Mine Spoils and Performance of Two Range Grass Species as Affected by Sulfuric Acid and Gypsum Treatments of Potted Soils

Amend type	Control	Sulfuric acid		Gypsum		Sand mulch
Rate (Mg ha^{-1})	0	2.5	5.0	5.0	5.0 + P[a]	1.25 cm
Sandstone spoil (initial ESP = 21%, EC_e = 6.9 dS m^{-1}, pH = 8.0)						
ESP (%)	19	7	4	6	6	—
Water penetrat. (cm/h)	3	8	11	5	5	—
EC_e(dS m^{-1})	6.7	7.2	8.1	7.4	7.4	—
pH (0–7.0 cm)	7.9	7.7	7.4	7.9	7.9	—
(0–2.5 cm)	—	—	6.9	—	—	—
$NaHCO_3$ ext P (mg kg^{-1})	7	14	16	7	62	—
Emergence (%)						
Sacaton	71	74	76	71	57	—
Saltbush	13	18	21	14	12	—
Shale spoil (initial ESP = 38%, EC_e = 5.5 dS m^{-1}, pH = 8.3)						
ESP (%)	35	14	7	23	23	15
Water penetrat. (cm/hr)	1	4	6	2	2	—
EC_e (dS m^{-1})	5.2	6.5	7.7	5.7	5.7	4.2
pH (0–7.0 cm)	8.2	7.3	6.9	8.0	8.0	8.2
(0–2.5 cm)	—	—	4.8	—	—	—
$NaHCO_3$ ext. P	2	4	5	2	72	1
Emergence (%)						
Sacaton	56	71	7	57	57	75
Saltbrush	9	15	4	22	22	19
Top dry wt. (g/pot)						
Sacaton	1.3	1.6	0.2	1.0	3.5	1.4
Saltbush	1.2	2.5	1.9	1.7	3.3	2.0

[a]Applied as CaH_2PO_4 at 200 kg ha^{-1}.
Source: Scholl and Miyamoto, 1984.

to 4.8 (Table 10). In both cases, a gypsum application at 5 Mg ha^{-1} resulted in insignificant increases in water infiltration and crop response. However, the gypsum application combined with phosphorus application increased plant growth. This experiment was conducted in a 7 cm deep pot with water application insufficient to cause any drainage. In a separate greenhouse study, many of the range plant species common to the Southwestern US were found to respond to P and Fe applications in calcareous soils (Ryan et al., 1975c).

These exploratory studies seem to indicate that the effects of acid application to mine spoils consisting of Na-affected sandstone are similar to those observed in irrigated calcareous soils. In spoils developed from shale, however, these studies have shown a possibility of excessive acidification and poor release and/or subsequent precipitation of P. Low pH can affect seed germination (Ryan et al., 1975a). An additional concern is poor leaching of dissolved salts in dryland situations because of limited precipitation. The dissolved salt concentrations usually increase with acid application, whereas gypsum applications usually result in only modest increases in salinity owing to its limited solubility. The limited solubility of gypsum, however, usually means slow effects, especially under limited rainfall. In order to dissolve 5 Mg ha^{-1} of gypsum, at least 20 cm of water is needed, which is higher than the annual precipitation in most arid areas. The dissolution of gypsum is further curtailed by the presence of high levels of SO_4^{-2} in coal mine spoils.

Band application of sulfuric acid (or gypsum) may be an option to enhance rainfall utilization for leaching. In a soil box study (Yahia et al., 1975), acid bands 10 cm wide were placed every 50 cm at an acid application rate of 2 Mg ha^{-1} on Na-saturated Stewart silt loam and Playa clay loam, followed by leaching using 6.5 cm of water. (The acid application rate on the basis of the treated area amounted to 10 Mg ha^{-1}.) Increased water infiltration in the acid band caused the ponded water to infiltrate predominantly through the band. Consequently, soluble salts were leached in a radial direction from the band, resulting in lower salinity below the acid bands than under the broadcast application (Table 11).

Table 11 Effects of Banded and Broadcast Applications of Sulfuric Acid at 2 Mg ha^{-1} on Initial Salt Leaching and Subsequent Dry Matter Production of Two Range Plant Species in Na-Saturated Stewart Silt Loam with the Initial EC_e of 42 dS m^{-1}.

Application methods	Banded		Broadcast	
Initial soil salinity after 6.5 cm water application				
Soil depth cm	EC_e dS m^{-1}		EC_e dS m^{-1}	
1	8		30	
10	6		26	
20	32		52	
Dry matter production after 4 months				
	Banded kg ha^{-1}		Broadcast kg ha^{-1}	
Irrigation cm	Bermuda	Sacaton	Bermuda	Sacaton
5.5	9.1	8.3	1.7	0.0
11.0	24.0	16.0	5.2	4.8

Source: Unpublished data, author's laboratory.

The acid treated soils were left for one month, and seeded with alkali sacaton (*Sporobolus airoides*) and common bermuda (*Cynodon dactylon*), using 5 and 10 cm of distilled water. Under the broadcast treatment, 30% of the seed germinated at 5 cm of water application and 70% at 10 cm of water application (unpublished report). Under the banded treatment, germination in the acid band exceeded 70%, but it declined to zero 11 cm away from the edge of the acid band. The top dry weights measured after 4 months indicate superior performance of the banded treatment over the broadcast treatment (Table 11).

No published report is currently available concerning the effect of sulfuric acid on revegetation of sodic soils on a field scale. However, judging from the exploratory work, acid treatment may help establish vegetation when combined with various water-catchment measures employed in range management. Obviously, additional research is needed to establish effective uses of acids and acidulants in such applications.

VI. SUMMARY

The use of acidulants on alkali soils began in the 1930s, mostly as a substitute for gypsum. Subsequent studies have shown that acids and acidulants have the unique ability to improve structural stability of sodic soils, to remove carbonates from irrigation water and calcareous soils, to improve availability of micronutrients and phosphorus, and to lower high pH of irrigation water and alkaline soils. In the category of water conditioning, sulfuric acid and its blends with nitrogen fertilizers are used primarily to reduce Ca precipitation and to maintain infiltration of low salt or ammoniated water. Polysulfides have been used to lower sodicity of surface soils and for improving water penetration. The cost-effectiveness of acid uses for water conditioning can be improved through blending with nitrogen fertilizers and improved guidelines for uses on diverse soil and water conditions. In the category of soil treatments, sulfuric acid and sulfur compounds are used primarily to improve water infiltration, availability of certain nutrient elements, and soil reactions in sodic calcareous or sodic alkaline soils. Sulfuric acid seems to hold a definite advantage over gypsum in reclamation of highly sodic alkaline soils, or calcareous sodic soils with micronutrient or phosphorus deficiency, whereas elemental S applications seem to suffer from slow and often inconsistent oxidation. In spite of these potential advantages of acids, handling difficulties present a challenge. The use of acids and acidulants for soil treatments can be made effective through improved understanding of pH effects on soil structural stability, nutrient availability, hydroxide toxicity, and the role of reclamation crops in reclaiming highly sodic soils. With increasing costs of crop production, and increasing regulations and costs of handling acids and acidulants, there is a continuing need to improve application strategies and methods. Acids and acidulants should not be regarded simply as substitutes for gypsum.

ACKNOWLEDGMENTS

Many of the studies cited here were conducted by Dr. J. L. Stroehlein, who has retired from the University of Arizona, and Dr. John Ryan, who has left for an assignment in the Middle East. Their contributions to the advancement of the use of sulfuric acid as a soil amendment and conditioner is acknowledged. Dr. Rami Keren at the Institute of Soils and Water, The Volcani Center, and Dr. Lloyd Fenn, my colleague, have provided constructive reviews of this paper. The funding for the preparation of this paper came from the Texas Agricultural Experiment Station with a matching grant from the Binational Agricultural Research and Development (BARD) fund.

REFERENCES

Abu-Sharar, T. M., Bingham, F. T., and Rhoades, J. D. (1987). Reduction in hydraulic conductivity in relation to clay dispersion and disaggregation. *Soil Sci. Soc. Am. J. 57*:342–346.

Achorn, F. P. (1984). Acid fertilizers. *Solution 3*:33–39.

Alawi, B. J. (1977). Effect of irrigation water quality, sulfuric acid and gypsum on plant growth and on some physical and chemical properties of Pima soil. Ph.D. diss., Univ. of Arizona.

Ali, A. M. (1989). Reactions of urea phosphate in calcareous and alkali soils:ammonia volatilization and effects on soil sodium and salinity. Ph.D diss. Univ. of Arizona.

Ali, A. S. I., and Swartzendruber, D. (1994). An infiltration equation to assess cropping effects on soil water infiltration. *Soil Sci. Soc. Am. J. 58*:1218–1223.

Baumhardt, R. L., Wendt, C. W., and Moore, J. (1992). Infiltration in response to water quality, tillage and gypsum. *Soil Sci. Soc. Am. J. 56*:261–266.

Beaton, J. D., and Fox, R. L. (1971). Production, marketing, and use of sulfur products. *Fertilizer Technology and Use* (R. A. Olson, T. J. Army, J. J. Hanway, and V. J. Kilmer, eds.). Soil Science Society of America, Madison, Wisconsin.

Bower, C. A., and Fireman, M. (1957). Saline and alkali soils. In *Year Book of Agriculture 1957*. (A. Stettrud, ed) USDA, Washington, D.C. pp. 282–290.

Burns, G. R. (1967). Oxidation of sulphur in soils. Sulphur Institute of Technological Bulletin 13. Washington, D.C.

Chand, M., Abrol, I. P., and Bhumbla, D. R. (1977). A comparison of the effect of eight amendments on soil properties and crop growth in a highly sodic soil. *Indian J. Agric. Sci. 47*:348–354.

Doering, E. J., and Willis, W. O. (1975). Chemical reclamation for sodic strip-mine spoils. USDA ARS ARS-NC-20.

Dudal, R., and Purnell, M. F. (1986). Land resources: salt-affected soils. *Recl. Reveg. Res. 5*:1–9.

Eaton, F. M. (1950). Significance of carbonates in irrigation water. *Soil Sci. 69*:123–133.

Fenn, L. B., and Miyamoto, S. (1981). Ammonia loss and associated reactions of urea in calcareous soils. *Soil Sci. Soc. Am. J. 45*:537–540.

Fenn, L. B., Malstrom, H. L., Riley, T., and Horst, G. L. (1990). Acidification of calcareous soils improves zinc absorption of pecan trees. *J. Am. Soc. Hort. Sci. 115*:741–744.

Gilbert, R. G., Nakayama, S., and Bucks, D. A. (1979). Trickle irrigation: prevention of clogging. *Trans. ASAE 22*(3):514–519.

Gumaa, G. S., Prather, R. J., and Miyamoto, S. (1976). Effect of sulfuric acid on sodium-hazard of irrigation water. *Plant Soil 44*:715–721.

Gupta, R. K., and Abrol, I. P. (1990). Salt-affected soils: their reclamation and management for crop production. *Ach. Soil Sci. 11*:223–288.

Gupta, R. K., Bhumbla, D. K., and Abrol, I. P. (1984). Effect of sodicity, pH, organic matter and calcium carbonate on the dispersion behavior of soils. *Soil Sci. 137*:245–251.

Haynes, J. D. (1928). Studies with sulfur for improving permeability of alkali soil. *Soil Sci. 25*:443–446.

Helmers, S. G., and Miyamoto, S. (1990). Mechanical and physical practices for reducing salinity in pecan orchards. *Proc. 3rd Nat. Irrig. Sympos.*, ASAE, St. Joseph, Michigan, pp. 374–377.

Hilgard, E. W. (1889). The rise of the alkali in the San Joaquin Valley. *Univ. Calif. Exp. Sta. Bull. 83*.

Kelley, W. A., and Brown, S. M. (1934). Principles governing the reclamation of alkali soils. *Hilgardia 8*:149–177.

Keren, R., and O'Connor, G. A. (1982). Gypsum dissolution and sodic soil reclamation as affected by water flow velocity. *Soil Sci. Soc. Am. J. 46*:726–732.

Keren, R., Kreite, J. F., and Shainberg, I. (1980). Influence of size of gypsum particles on the hydraulic conductivity of soils. *Soil Sci. 130*:113–117.

Khorsandi, F. (1994). Phosphorus fractions in two calcareous soils as affected by sulfuric acid application. *J. Plant Nutri. 17*:1599–1609.

Kumar Ashok, and Abrol, I. P. (1986). Grasses in alkali soils. Bull. 2, Central Soil Salinity Research Institute, Karnal, India.

Leclercq, R., Bixby, D. W., and Fike, H. L. (1972)). Potential uses for sulphur dioxide. Tech. Bull. 19, Sulphur Institute, Washington, D.C.

Mathers, A. C. (1970). Effect of ferrous sulfate and sulfuric acid on grain sorghum yields. *Agron. J. 62*:555–556.

McCaskill, M. R., and Blair, G. J. (1987). Particle size and soil texture effects on elemental sulfur oxidation. *Agron J. 79*:1079–1083.

McFarland, M. L., Ueckert, D. N., and Hartmann, S. (1987). Revegetation of oil well reserve pits in west Texas. *J. Range Mgt. 28*:411–414.

McGeorge, W. T. (1939). Some problems connected with fertilization of alkali soils. *Calif. Citrogr. 24*:389.

McGeorge, W. T. M., and Green, R. A. (1935). Oxidation of sulphur in Arizona soils and its effect on soil properties. *Ariz. Agr. Exp. Sta. Tech. Bull. 59*:297–325.

McGeorge, W. T., Breazeale, E. L., and Abbott, J. L. (1956). Polysulfides as soil conditioners. Arizona Agr. Exp. Sta. Bull. 131, University of Arizona, Tucson.

McKee, A. G., and Co. (1969). Systems study for control of emissions by the primary nonferrous smelting industry. *U.S. Federal Clearing House.* P.B. 184. Washington, D.C.

Miyamoto, S., and Cruz, I. (1987). Spatial variability of soil salinity in furrow irrigated Torrifluvents. *Soil Sci. Soc. Am. J. 51*:1019–1025.

Miyamoto, S., and Enriquez, C. (1990). Comparative effects of chemical amendments on salt and Na leaching. *Irrig. Sci. 11*:83–92.

Miyamoto, S., and Ryan, J. (1976). Sulfuric acid for the treatment of ammoniated irrigation water II. Reducing Ca precipitation and sodium hazards. *Soil Sci. Soc. Am. J. 40*:305–309.

Miyamoto S., and Stroehlein, J. L. (1975). For improving water penetration into some Arizona soils—sulfuric acid. *Prog. Agr. 27*(2):13–16.

Miyamoto S., Ryan, J., and Bohn, H. L. (1973). Penetrability and hydraulic conductivity of dilute sulfuric acid solutions in selected Arizona soils. *Proc. Amer. Water Res. Assn. Arizona Section*, pp. 291–298.

Miyamoto, S., and Stroehlein J. L. (1986). Sulfuric acid effects on water infiltration and chemical properties of alkaline soils and water. *Trans. ASAE 29*:1288–1296.

Miyamoto, S., Ryan, J., and Stroehlein, J. L. (1975a). Potentially beneficial uses of sulfuric acid in southwestern agriculture. *J. Environ. Qual. 4*:431–437.

Miyamoto S., Prather, R. J., and Stroehlein, J. L. (1975b). Sulfuric acid for controlling calcite precipitation. *Soil Sci. 120*:264–271.

Miyamoto S., Prather, R. J., and Stroehlein, J. L. (1975c). Sulfuric acid and leaching requirements for reclaiming sodium-affected soils. *Plant Soil 43*:573–585.

Miyamoto, S., Stroehlein, J. L., and Ryan, J. (1975d). Sulfuric acid for the treatment of ammoniated irrigation water. I. Reducing ammonia volatilization. *Soil Sci. Soc. Am. J. 39*:544–548.

Miyamoto, S., Glenn, E. P., and Singh, N. T. (1994). Utilization of halophytic plants for fodder production with brackish water in subtropic deserts. In *Halophytes as a Resource for Livestock and for Rehabilitation of Degraded Lands* (V. R. Squires and A. T. Ayoub, Eds). pp. 43–75. New York: Kluwer Academic Publishing.

Mohammed, E. T. Y. (1972). The effect of various sulphur compounds on infiltration of selected soils. Ph.D. diss., University of California, Riverside.

Morril, L. G., and Dawson, J. E. (1967). Patterns observed for the oxidation of ammonium to nitrate by soil organisms. *Soil Sci. Soc. Am. Proc. 31*:757–760.

Moser, U. S., and Olson, R. V. (1953). Sulphur oxidation in four soils as influnced by soil moisture tension and sulphur bacteria. *Soil Sci. 76*:251–257.

Oron, G., and DeMalach, Y. (1989). Effect of dikes and sulfuric acid on cotton under effluent irrigation. *J. Irrig. Drainage Engr. 115*:463–473.

Oster, J. D., and Schroer, F. W. (1979). Infiltration as influenced by irrigation water quality. *Soil Sci. Soc. Am. J. 43*:444–447.

Overstreet, R, Martin, J. C., and King, H. M. (1951). Gypsum, sulfur and sulfuric acid for reclaiming an alkali soil of the Fresno series. *Hilgardia 21*:113–127.

Paulina, Y. W., Li. (1951). The oxidation of elemental sulfur in soils. M.S. thesis, University of Minnesota.

Phillips, A. B., and Scott, W. C. (1966). Preparation and properties of ammonia–sulfur solutions. *Agr. Ammonia News 16*(3):40–46.

Phillips, A. B., and Webb, J. R. (1971). Production, marketing, and use of phosphorus fertilizers. *Fertilizer Technology and Use* (R. A. Olson, T. J. Army, J. J. Hanway, and V. J. Kilmer, eds). Soil Science Society of America, Madison, Wisconsin.

Robbins, C. W. (1986). Sodic calcareous soil reclamation as affected by different amendments and crops. *Agron. J. 78*:916–920.

Roberts, S., and Koehler, F. E. (1965). Sulfur dioxide as a source of sulfur dioxide for wheat. *Soil Sci. Soc. Am. Proc. 29*:696–698.

Robinson, F. E., Cudney, D. W., and Jones, J. P. (1968). Evaluation of soil amendments in the Imperial Valley. *Calif. Agr.* Dec. issue: 10–11.

Ryan, J., Miyamoto, S., and Stroehlein, J. L. (1974). Solubility of manganese, iron and zinc as affected by application of sulfuric acid to calcareous soils. *Plant Soil 40*:421–427.

Ryan, J., Miyamoto, S., and Stroehlein, J. L. (1975a). Effect of acidity on germination of some grasses and alfalfa. *J. Range Mgt. 28*:154–155.

Ryan, J., Stroehlein, J. L., and Miyamoto, S. (1975b). Sulfuric acid applications to calcareous soils. Effects on growth and chlorophyll content of common bermudagrass. *Agron. J. 67*:633–637.

Ryan, J., Stroehlein, J. L., and Miyamoto, S. (1975c). Effect of surface-applied sulfuric acid on growth and nutrient availability of five grasses in calcareous soils. *J. Range. Mgt. 28*:411–414.

Sandoval, F. M., and Gould, W. L. (1978). Improvement of saline- and sodium-affected disturbed lands. In: *Reclamation of Drastically Disturbed Lands*. American Society Agronomy, Madison, Wisconsin, pp. 485–504.

Scholl, D. G., and Miyamoto, S. (1984). Response of alkali sacaton and fourwing salt bush to various amendments on coal mine spoils from northwestern New Mexico. II. Sodic Spoil. *Reclam Reveg. Res. 2*:243–252.

Scofield, C. S., and Headly, F. B. (1921). Quality of irrigation water in relation to land reclamation. *J. Agric. Res. 21*:265–278.

Shainberg, J., and Letey, J. (1984). Response of soils to sodic and saline conditions. *Hilgardia 52*(2):1–57.

Singh, G., Abrol, I. P., and Cheema, S. S. (1990). Effects of irrigation on *Prosopis juliflora* and soil properties of an alkali soil. *Inter: Tree Crop J. 6*:81–99.

Stroehlein, J. L., Miyamoto, S., and Ryan, J. (1978). Sulfuric acid for improving irrigation waters and reclaiming sodic soils. *Agri. Engr. and Soil Sci. Bull. 78*:5. University of Arizona, Tucson, AZ.

Stromberg, L. K., and Tisdale, S. L. (1979). Treating irrigated arid-land soils with acid-forming sulfur compounds. Tech Bull. 24. Sulphur Institute, Washington, D.C.

Suarez, D. L., Rhoades, J. D., Lavado, R., and Crieve, C. M. (1984). Effect of pH on saturated hydraulic conductivity and soil dispersion. *Soil Sci. Soc. Am. J. 48*:50–55.

Thorup, J. T. (1969). pH effect on root growth and water uptake by plant. *Agron J. 61*:225–227.

Tisdale, S. L. (1968). Sulphur as fertilizer. *New Fertilizer Materials* (Y. Araten, ed.). Centre International Engrais Chimiques. Noyes Development Corp., Rark Ridge, N.J.

Tisdale, S. L. (1970). The use of sulfur compounds in irrigated arid land agriculture. *Sulphur Inst. J. 6*(1):2–7.

Tobia, K., and Pollard, A. G. (1958). Effects of acidification of alkali and calcareous soils. *J. Sci. Food Agric. 9*:705–713.

U.S. Salinity Laboratory Staff (1954). Diagnosis and improvement of saline and alkali soils. USDA Handbook 60 (L. A. Richards, ed.).

Vodyanitskiy, Y. N., and Dokuchayev, V. V. (1985). Use of iron compounds for soil aggregation. *Poch vovedeniye 12*:49–54 (English translation).

Wallace, A., and Mueller R. T. (1978). Complete neutralization of a portion of calcareous soil as a means of preventing chlorosis. *Agron. J. 70*:888–890.

Weiss, F. T., Jungnickel, J. L., and Peters, E. D. (1953). Analysis of sulfuric acid and acid sludges from petroleum processes. *Anal. chem. 25*:277–283.

Wildman, W. E., Pecock, W. L., Wildman, A. M., Goble, G. G., Pehrson, J. E., and O'Connell, N. V. (1988). Soluble calcium compounds may aid low volume water application. *Calif. Agric.*, Nov–Dec. issue: 7–9.

Yahia, T. A., Miyamoto, S., and Stroehlein, J. L., (1975). Effect of surface-applied sulfuric acid on water penetration into dry calcareous and sodic soils. *Soil Sci. Soc. Am. Proc. 39*:1201–1204.

9
Mined and Industrial Waste Products Capable of Generating Gypsum in Soil

L. L. Somani and K. L. Totawat *Rajasthan College of Agriculture, Udaipur, Rajasthan, India*

I. INTRODUCTION

Alkali soils are widespread in several countries of the world and present a serious problem of immense practical importance for crop production in view of their inhospitable characteristics for plant growth despite inherent potentialities. The nature of the problem is such that it restricts economic utilization of land resources, thus adversely affecting crop production especially in semiarid and arid regions. For a long time mankind considered such soils as uncultivable wastelands and confined their energies to better soils. Because of the food shortage besetting mankind in various parts of the world, man is now forced to find ways of cultivating alkali soils that have highly deteriorated physical conditions due to the dispersive action of excessive sodium on their exchange complex. Realizing that geographical areas cannot be expanded and that enhancement of productivity of good soils is possible only to a limited extent, there is worldwide demographic pressure to attempt to cultivate all hitherto barren alkali lands.

Gypsum has been used for a long time and is in fact the most commonly recommended inorganic amendment for reclamation of alkali soils. But its supplies are being increasingly directed more and more towards other industrial uses of higher economic demand. The increasing cost of gypsum transportation and high cost involved in grinding and bagging are other factors that are restricting its wider use. This has diverted the attention of scientists towards making use of products capable of generating gypsum in the soil itself.

II. SULFUR

Elemental S can be a source of S for plants and a means of reducing alkalinity by generating gypsum in soil containing alkaline earth carbonates. However, it cannot neutralize soil alkalinity until it converts to sulfate. It is therefore important to know something about its reserves, mechanisms of oxidation, the factors that affect the oxidation, and the chemistry involved in the amelioration of soil alkalinity.

A. Sulfur Reserves

Sulfur is perhaps the most important chemical mineral and one of the most widespread native elements. It occurs as native S and in sulfides and sulfates. Native S is the chief commercial source.

Native S originates by several different means. Deposits associated with volcanism may be formed by condensation of S vapors, and by bacterial production of H_2S and oxidation to $S^{\cdot}$. Sulfur may be deposited by oxidation and by S bacteria from thermal springs containing H_2S. One or the other of these processes has probably given rise to most of the S of volcanic regions and also contributed the S of sedimentary beds. The chief S-producing countries of the world are the USA, Italy, and Japan. Of the total world output of S about 90% comes from the USA.

B. Oxidation

For many years, there were two schools of thought as to the process by which S is oxidized in the soil. One emphasized chemical (auto or nonbiological) oxidation; the other biological oxidation.

1. Nonbiological Oxidation

Abiotic oxidation of S to SO_4 takes place as shown here (Wiklander et al., 1950):

$$2S + 2H_2O + 3O_2 = 2H_2SO_4$$

S oxidation can occur even in autoclaved soil (Nor and Tabatabi, 1977). Evidences now indicate that S can be oxidized in the soil by purely chemical processes, but these are usually much slower and therefore of less importance than oxidation involving microorganisms (Burns, 1968).

2. Biological Oxidation

Many autotrophic and heterotrophic microorganisms usually are available to oxidize S. In general, organisms that oxidize S are present wherever reduced forms of S are found under favorable oxidizing conditions. A rapid increase in numbers of S-oxidizing organisms occurs when reduced forms of S are added to the soil.

The best known, and usually considered to be the most important group of S-oxidizing microorganisms, are autotrophic bacteria belonging to the genus *Thiobacillus*. Inoculation with Thiobacilli appears to be a logical choice for enhanced S oxidation, since bacteria like *T. thiooxidans* gain energy solely by the oxidation of S. There is considerable evidence that S is oxidized more rapidly in soils containing an abundance of these bacteria then in soils from which the bacteria are absent (Wainwright et al., 1986). Some better known species of Thiobacilli are *T. thiooxidans, T. thioparus, T. denitrificans, T. ferrooxidans* and *T. intermedius*.

Stimulation of S oxidation in soil inoculated with *T. thiooxidans* (an obligate chemolithotroph) in the presence of organic matter indicate some kind of associative S oxidation between these bacteria and heterotrophs (Li and Caldwell, 1966). Soils devoid of Thiobacilli have been found capable of oxidizing S. Vitolins and Swaby (1969) found that heterotrophic yeasts and several genera of heterotrophic and facultative autotrophic bacteria were far more numerous than the strict autotroph and could play an important role in S oxidation in many soils. Autotrophs were, however, more efficient than the heterotrophs where conditions suited them. Heterotrophs, with their dependency on organic materials, give variable response in terms of S-oxidation. Furthermore, the oxidation of S in their life cycle is not crucial to their existence as it is with certain of the autotrophs. *Fusarium solani, Aureobasidium pullulans, Aspergillus niger, Trichoderma harzianum, and Pseudomonas* sp are considered potential S-oxidizing heterotrophs (Wainwright et al., 1986).

3. Factors Affecting Sulfur Oxidation

Temperature. Microbial oxidation of S will occur at temperatures as low as 4°C and up to about 55°C (Breed et al., 1948). The process of oxidation is very slow at or near these extremes. There is marked increase in the activity of S-oxidizing microorganisms from 23°C to 40°C (Fox et al., 1964, Li and Caldwell, 1966). At temperatures above 55°C to 60°C the organisms are killed (Breed et al., 1948). Optimum temperatures for S oxidation are not the same for all organisms, but temperatures between 27°C and 40°C include most (Rudolfs, 1922).

Soil Moisture and Aeration. The general equation expressing the oxidation of elemental S by microorganisms is

$$S + 1\tfrac{1}{2}O_2 + H_2O \rightarrow H_2SO_4$$

This reaction requires molecular oxygen and water. In soils with high moisture levels, lack of oxygen may be the limiting factor (Lee, 1955), while lack of water may limit S oxidation in soils with low levels of moisture.

Sulfur oxidizing bacteria are mostly aerobic, and their activity will decline if oxygen is lacking due to water logging (Jones et al., 1971). Generally, S is oxidized quite rapidly in the soil moisture range (50–75% WHC) (WHC = water holding capacity) that is optimum for growth of plants. Oxidation of elemental S is most rapid at soil moisture levels near field capacity (Kittams and Attoe, 1965). The oxidation of S is also associated with deep placement. Even in soils that are permanently flooded (e.g., paddy soils) aerobic and anaerobic zones exist together. An oxidized layer develops in the upper part of the flooded horizon where oxygen supply from algae and weeds exceeds consumption. Rice plants also have the ability to transmit oxygen, which is absorbed by the stoma or produced during photosynthesis, down through the roots into the neighboring soil. As rice plants can occupy a large volume of the planted soil, oxidized zones can occur that allow oxygen, even under these flooded conditions, for the growth and metabolism of aerobic microorganisms (Wainright, 1984). As a result, S can exist in these soils in all its oxidation states. Both autotrophic *Thiobacilli* and heterotrophic *Beggiatoa*, which are known to oxidize S, have been isolated from flooded soils (Somani, 1994).

Soil Reaction. Although *Thiobacilli* are able to function in a pH range of less than 2.0 to 9.6, both *Thiobacillus ferrooxidans* and *Thiobacillus thiooxidans* have an optimum pH below 5.0 for their activity (Quispel et al., 1952). It therefore raises doubts on the effectiveness of these organisms under alkaline conditions. Under such conditions S oxidation is mostly by chemical process, which proceeds at a very slow rate (Bloomfield, 1967). Accordingly, addition of S as an amendment in alkali soil cannot be expected to bring changes in soil pH to a range optimum or even desirable for the activity of *Thiobacillus* spp. However, Rudolfs (1922) and Waksman et al. (1923) observed fairly rapid oxidation of S in alkali soils. According to them, soil pH is not critical for the activity of *Thiobacillus* spp. Within the genus, *Thiobacillus thiooxidans* covers the pH range of 2.0 to 9.0. Rupela and Tauro (1973a,b) also reported *Thiobacillus novellus* to be capable of tolerating a very high concentration of Na in its environment. The possibility of existence of alkali tolerant species of *Thiobacilli* in salt affected soil has been indicated by several workers (Rupela and Tauro, 1973a; Rupela and Tauro 1973b; Chandra and Boolen, 1960). Moreover, the pH within the microenvironment surrounding an oxidizing S particle may be considerably more acid than is measured in the bulk of soil sample (Hart, 1962). Such a low pH at the microsite accelerates the oxidation of S (even under an alkali soil situation), and the whole process once started is speeded up more and more. Therefore under natural conditions a fairly rapid oxidation of S is always guaranteed, even in alkali soil, provided it is well aerated. The role of *Thiobacilli* is in accelerating the otherwise very slow oxidation of S. Even relatively sparse populations of *Thiobacilli* contained in the soil usually multiply rapidly after S containing materials are applied (Moser and Olson, 1953). It is, however, advisable to enrich the soil with effective *Thiobacillus* culture for quick reclamation with S. Many other autotrophs and heterotrophs capable of oxidizing S are also active within this range (Weir, 1975).

Effect of Salts. A number of studies have been reported on the effects of NaCl on S oxidation, mainly in relation to high concentrations of salt found in sodic soils that can be reclaimed by adding S. Fawzi (1976) showed that NaCl (9%) reduced the rate of S oxidation in soil and completely inhibited the process when added at a concentration of 11%. Similarly, S oxidation in Terra Rosa and Rendzina soils was still observed following the addition of 8% NaCl but was completely inhibited by 10% NaCl (Keller, 1969). The relatively high tolerance of S oxidizers to high levels of free salts and exchangeable sodium is an important fact (Escolor and Lugo Lopez, 1969).

Nutrient Supply. The nutrient requirements of S-oxidizing microorganisms have not been elucidated in detail, but these microorganisms require most of the nutrients that are essential for higher plants plus some others such as Co and V (Nicolas, 1963). The nutrient needs of the S-oxidizing microorganisms appear to be met by most soils capable of growing crops. Stimulated activity of S-oxidizing organisms may lead to temporary nitrogen shortage for crop plants following incorporation of S in soils (Weir, 1975).

Generally, ammonium salts are necessary for the development of *Thiobacilli* (Starkey, 1925). Urea and amino acids do not appear to be sources of available nitrogen for *Thiobacillus* (Starkey, 1925). Nitrates exert a distinctly injurious action on both growth and respiration of the S oxidizers (Waksman and Starkey, 1923), but S bacteria are relatively tolerant of other anions, and in particular sulfates (Keller, 1969). This is fortunate, as S oxidation, particularly in alkaline soils, results in considerable increase in soluble calcium, sulfates, potassium, phosphorus, and other nutrients (McGeorge and Greene, 1935). Similarly, elemental S is not toxic to *Thiobacilli* (Waksman and Starkey, 1923). High concentrations of ammonia in banded anhydrous ammonia did not depress S oxidation (Parr and Papendick, 1966). To some extent, chloride retards S oxidation (Keller, 1969).

Although heavy metals are highly toxic to several microorganisms, some strains of S bacteria show extraordinary tolerance in the oxidation of S compounds from metal ores (Bryner and Jameson, 1968).

Organic Matter. Organic matter is not essential for the activity of autotrophic bacteria, but heterotrophs require a source of organic carbon. Organic materials serve as the energy source for heterotrophs and in many cases supply certain growth factors (Lee, 1955). Gibberellic acid has been found to increase the rate of S oxidation in certain soils by as much as 50% (Burns, 1968) concomitant with increased numbers of soil microorganisms. However, in general, S is readily oxidized in soils with or without the addition of organic materials (Li and Caldwell, 1966; Jones et al., 1971).

Microbiological Population and Inoculation. Most aerable soils contain S-oxidizing organisms (Starkey, 1966), but the rate of S oxidation may be quite variable among soils (Joffe and McLean, 1922); this is commonly attributed to differences in microbial populations. Moser and Olson (1953) in comparing two soils found that a high count of *T. thiooxidans* coincides with the greater ability of one soil to oxidize S.

Inoculation, usually with *T. thiooxidans* (Kittams and Attoe, 1965; Li and Caldwell, 1966), but also with heterotrophs (Vitolins and Swaby, 1969), usually speeds the oxidation process, but such responses probably give only a short-term advantage during the first month. However, under certain conditions uninoculated S has been observed to be more rapidly oxidized than inoculated S. The addition of S to the soil normally results in a rapid buildup of S-oxidizing microorginisms and improved oxidizing power of the soil (Vitolins and Swaby, 1969). Moreover, organisms normally will not survive very long in dry S, thus making inoculation on a field scale difficult (Burns, 1968). This and the often overriding effects of environmental factors could account for the general lack of response to inoculation in field trials (McGeorge and Greene, 1935; Rudolfs, 1922).

Rupela and Tauro (1973a) showed that alkali soils in India possess low populations of *Thiobacilli*. They suggested that the rapid reclamation of these soils could be brought about by enrichment with efficient S-oxidizing bacteria (Rupela and Tauro, 1973b). Despite these studies, the inoculation of soils with *Thiobacilli* has not become standard practice.

An alternative to inoculation with S-oxidizers is to add soil that is rich in these organisms to soils that lack them, i.e., a form of indirect inoculation. Kittams and Attoe (1965) treated ten different soils in this way. S-oxidation in the previously oxidation-limited soil was in all cases increased—in one soil as much as tenfold.

The foregoing account on factors affecting the oxidation of S shows that under the arid and semiarid environments of alkali areas, S may oxidize fairly rapidly provided inoculation is used. S-oxidation in soil to some extent can be hastened by manipulating soil conditions besides the particle size, rate, method, and time of application of S.

Rate of Sulfur Application. At moderate rates of application, the percentage of the S oxidized usually is not affected by the amount applied (Waksman, 1922). However, at higher rates, as are common when S is used as an amendment in alkali amelioration, the percentage of S oxidized generally decreases with increasing rates of application (McGeorge and Greene, 1935). In spite of this, increasing the rate of application of S increases the S surface available for oxidation, which results in a linear increase in the release of total SO_4-S (Li and Caldwell, 1966; Chopra and Kanwar, 1968). Some control over the rate of oxidation can be accomplished, however, by adjusting the particle size range (Weir et al., 1963).

Placement of Sulfur. Mixing elemental S with the soil usually results in the most efficient use of the applied S. Nevertheless, band placement and top dressing can be quite effective under certain conditions (Fox et al., 1964). Mixing the S with the soil maximizes the soil S contact, reduces possible effects of the buildup of S oxidation products, and often improves S moisture relationships, all of which enhance S oxidation.

Where S is topdressed, soil moisture relations could favor complete mixing of the S with the soil, particularly in the field. However, Conrad (1950) observed poor results with topdressing of S, mainly because of inadequate moisture conditions on soil surface.

Band placement increases the concentration of S within the oxidation zone and can reduce oxidation (Parr and Papendick, 1966). This may be an effect of reduced contact between the S and the microorganisms or less contact between S and soil with its buffering effects.

Particle Size and Form of Sulfur. Elemental S can occur in a great variety of physical forms, but such differences within the solid forms of S appear to have little effect on oxidation rate. Simon and Schollenberger (1925) compared ground S with sublimed forms and concluded that surface area, i.e., particle size, was more important than surface shape.

As the particle size increases, the surface area per unit weight of S decreases, and generally the rate of sulfate production per unit weight also decreases (Simon and Schollenberger, 1925; Fox et al., 1964). Fox et al. (1964) showed that the relationship between surface area of S and the rate of its release as SO_4-S is linear.

Although the rate of oxidation of elemental S is related to particle size, there appears to be a limit beyond which further reduction in particle size is not desirable.

C. As an Amendment for Alkali Amelioration

The main object of applying chemical amendments is to furnish soluble calcium to replace adsorbed sodium. Chemical amendments used in the reclamation could be put under two categories, first, those supplying soluble calcium salts, i.e., gypsum, and second, acids and acid formers like sulfuric acid, S, and pyrite etc. The choice of amendment is governed by cost, availability, and soil characteristics.

Elemental S can be a source of S for plants and a means of reducing alkalinity in soils. Elemental S has several valuable properties such as resistance to leaching that can be advantageous in certain situations, but the major factor in the increased interest shown in it is its high analysis of S.

1. Mechanism

The effectiveness of S as an amendment depends on its oxidation to H_2SO_4 and on the presence of $CaCO_3$, which could react with H_2SO_4 to provide calcium in soil solution. The oxidation of S is largely a biological process and is brought about by microorganisms of the group *Thiobacillus* (and some heterotrophs) resulting in the formation of H_2SO_4 as shown in the formula given earlier.

Under moist closed conditions the reaction proceeds as follows:

$$H_2SO_4 + 2CaCO_3 + 4Na\text{-}Clay \rightarrow 2Ca\text{-}Clay + Na_2SO_4 + 2NaHCO_3$$

Thus oxidation of one atom of S theoretically results in the replacement of $4Na^+$ ions by Ca^{2+} ions. Under dry soil conditions the reaction proceeds as:

$$H_2SO_4 + CaCO_3 + 2Na\text{-}Clay \rightarrow Ca\text{-}clay + Na_2SO_4 + CO_2 + H_2O.$$

In this case only $2Na^+$ ions per S atom could be replaced. Free O_2 is essential for S oxidation so that CO_2 will escape into the atmosphere under aerated conditions (McGeorge and Greene, 1935).

However, if the soil is free of $CaCO_3$ the reaction proceeds as:

$$H_2SO_4 + 2Na\text{-}Clay \rightarrow 2\ H\text{-}Clay + Na_2SO_4$$

The soil is thus converted into an acid soil. It is thus clear that the value of S as an amendment depends on the presence of $CaCO_3$ in the soil.

2. Sulfur Requirement

At present there is no direct method to evaluate the S requirement of soil. It is indirectly worked out from the gypsum requirement of soils. Readers are advised to see the pyrite requirement of soil.

3. Gypsum vis-à-vis Sulfur

The use of S may not be superior to gypsum, so far as its immediate reclamative efficiency on a chemical equivalence basis is concerned, mainly because S is effective only after it is oxidized to sulfuric acid by microorganisms. Somani and Saxena (1981) reported that S takes a period of 2 to 3 months for its oxidation and for reaching chemical equilibrium comparable to that of gypsum. This suggests that the slowness of S as an alkali ameliorant could be compensated for by applying it in advance to permit its oxidation. This led Somani (1981) and Somani and Saxena (1982) to record better ameliorative influence of S as compared to gypsum. The ameliorative efficiency of S is considerably improved if used in conjunction with organic materials, possibly because organic matter hastens the activity of heterotrophic S oxidizers in soil, besides improving soil physical properties (Somani and Saxena, 1981).

4. Prospects of Sulfur Use

Encouraging results were obtained with S in the reclamation of extreme types of black alkali soils (Wursten and Powers, 1934). Gidnavar et al. (1972) found application of 12.5 cartloads of FYM (farmyard manure) along with 300 kg/ha

S to improve the rainfed saline-alkali soil. Somani (1972) obtained a yield of 1884 kg/ha of barley from calcareous saline alkali soil following green manuring with dhaincha plus incorporation of 1250 kg/ha of elemental S, whereas the control provided a yield of 421 kg/ha only. The soil pH also dropped to 8.3 as compared to 9.0 in the control plot.

III. PYRITE

Pyrite is a mineral containing iron and S, and generally it has the chemical composition FeS_2. Pyrite is found all over the world in igneous and metamorphic rocks and at some places as sedimentary deposits as well. Pyrite is highly pyrophoric in nature and produces sulfuric acid and iron sulfate on coming in contact with air and water.

A. Origin

Pyrite (FeS_2) and ferrous sulfide (FeS) are quite insoluble and can be considered the end phases of the reductive transformation of S in muds and soils. Iron sulfides are thus expected in all the anaerobic sediments where adequate amounts of S and reactive iron are present along with some oxygen acceptor to permit reduction of S compounds. Adequacy of reactive iron, even for the high accumulation of sulfides, could be easily assumed, since the very common transformation of ferrous into ferric and *vice versa* can maintain optimum supply of reactive iron. The reactive S may be either derived from brackish water or from sea water (sea water is known to posses more than 800 mg/l of SO_4-S) or from organic compounds (Clark, 1925). According to Starkey and Wright (1945), anaerobic reduction of sulfates in the presence of oxidizable organic substances or elementary iron leads to the formation of sulfides in marine sediments. Correns (1935), however, stressed that formation of sulphide was restricted neither to marine sediments nor to the availability of sulfates.

Regardless of the source of hydrogenated S, the formation of sulfide can be represented by one of the following reactions:

$$4Fe(OH)_3 + H_2O + 4H_2S \xrightarrow{OM} 4FeS + 11H_2O + CO_2 \quad (1)$$

$$FeCO_3 + H_2S \rightarrow FeS + H_2O + CO_2 \quad (2)$$

$$2Fe(OH)_3 + 3H_2S \rightarrow 2FeS + S + 6H_2O \quad (3)$$

Reaction 1 can proceed only when an oxygen acceptor like organic matter (OM) is present.

The presence of the elemental form of S along with FeS is considered essential for the formation of pyrite (Verhoop, 1940) as shown here:

$$FeS + S \rightarrow FeS_2 \quad (4)$$

The possibilities of pyrite formation directly from FeS and H_2S is shown here:

$$2H_2S + 2FeS + O_2 \rightarrow 2FeS_2 + 2H_2O \tag{5}$$

or from ferric hydroxide and H_2S:

$$4Fe(OH)_3 + 8H_2S + O_2 \rightarrow 4FeS_2 + 14H_2O \tag{6}$$

B. Occurrence and Distribution

In general, pyrite, along with minerals of other metallic sulfides, occurs all over the world, mostly in igneous and metamorphic rocks and very rarely in sedimentary form. The later is more reactive than the former two forms.

Pyrite is the major S source in developing countries of Asia. China is the largest pyrite producer and user in the world. Pyrite is also exploited on a significant scale in the Philippines, Korea, and India (Bain and Service, 1987). Pyrite provides around 70% of China's S requirements, with an output of two million MG of S per year. Total reserves are estimated at 3.3 billion MG. Pyrite is mined per se and is also produced as a by-product in nonferrous metal operations. Some pyrite also arises in association with coal production. Korea is the second largest producer of pyrite in the region. Output is estimated at around 400,000 MG of S per year. Philippines pyrite production is currently around 150,000 MG of S per year. There is potential for more. Pyrite is also produced as a by-product of copper mining activities in the Philippines. Thailand has pyrite deposits around one million MG, but there is no commercial exploitation. Indian pyrite deposits of high grade (40% S) are 350 million MG while pyrite shales (10% S) are about 1500 million MG.

C. Chemical Composition

Pyrite, as said, contains both iron and S.; it also has some micronutrients. Its geological structure has a chemical composition similar to FeS_2, i.e., ferrous sulfide. Although iron pyrite minerals have various physical forms such as crystalline, massive, or powdery, the powdery form of the sedimentary type is chemically the most reactive. A detailed chemical composition of agricultural grade pyrite and overlying pyritic ferrous shale given (Table 1) shows that it contains appreciable quantities of zinc, copper, manganese, and iron, which are likely to have significance in plant nutrition. The overlying pyritic shale is about 5 to 6 m thick and contains on an average about 10% S. The underlying shale below the pyrite bed does not contain any S. As the pyrite is highly reactive and pyrophoric, the chemical reaction is accelerated in the presence of air and water. As a consequence of this reaction, the mine water and the atmosphere both become highly acidic in nature.

Table 1 Chemical Composition (%) of Low Grade Pyrite

Constituent	Pyrite (shales)	Pyrite (powder)
Iron	11.7	20–22
S	11.7	22–24
MgO	0.9	0.1–0.5
CaO	4.3	0.1–0.3
Al_2O_3	11.3	8–12
SiO_2	46.9	45–50
Carbon	3.2	2–5
Zinc	—	0.02
Copper	—	0.05
Manganese	—	0.01

Source: Jaggi (1977); Tiwari (1989).

Pyrite containing more than 30% S has industrial uses of high economic value, particularly for the manufacture of sulfuric acid. Agricultural grade pyrite is a by-product of the mining of acid grade pyrite.

The composition of pyrite given above shows that the low grade iron pyrite is not of any industrial use, indeed it pollutes the nearby mining environment, but it has great potential for restoring the productivity of alkali wastelands and improving the fertility of the soil. It has unique qualities of serving as an amendment for reclamation of alkali soils besides being a source of plant nutrients S and iron in addition to some other micronutrients like Zn, Cu, Mn, etc. Like gypsum and elemental S, pyrites also have the dual role of amendment conditioner and fertilizer (Tandon, 1987).

D. Oxidation of Pyrite in Soil

The usefulness of pyrite as an amendment or as a source of plant nutrient is predominantly governed by its oxidation to sulfate.

Like S, pyrite can be oxidized yielding sulfuric acid both abiologically and biologically. The same microorganisms that bring about S oxidation play important roles in biological oxidation of pyrite as well.

1. Autooxidation

Freshly mined and crushed pyrite particles become oxidized when exposed to humidity and air. This oxidation results in the formation of acid, which remains on the site of reaction in the form of layers. The pyrite particles surrounded by acid layers cannot react further until the acid layers around them are removed. This restriction to further oxidation of pyrite is removed to some extent in sodic soil, since a continuous acid base reaction occurs in them due to the presence of carbonates of sodium and calcium. Thus the sodic environment of soils helps pyrite to continue its autooxidation process. Since it is an acid base reaction, the role of pH by itself in pyrite autooxidation hardly needs any special consideration. However, the rate and degree of oxidation can depend on various morphological, climatic, and soil factors. These aspects need attention in developing suitable technology for optimum autooxidation of pyrite in sodic soils. What is needed is agronomical manipulation to create a congenial environment for pyrite to become oxidized to its full capacity for securing maximum benefit in the reclamation of alkali soils (Sinha and Jha, 1984).

Autooxidation of pyrite in soil has been reported to be a sluggish process, and consequently its usefulness as an acidogenous amendment for reclamation of calcareous sodic soil or as an S-carrier was never rated high (Gupta and Abrol, 1989; Gupta and Abrol, 1990). The process become faster under the catalytic influence of microorganisms (Stumm and Morgan, 1979). Inoculation with effective *Thiobacilli* is thus a prerequisite to ensure fast oxidation of pyrite.

2. Suggested Oxidation and Metabolic Pathway

Under aerobic conditions and in the presence of moisture, pyrite is oxidized to sulfuric acid. The oxidation is partly chemical and partly microbial in nature (Weir, 1975; Somani et al., 1981; Somani, 1984; Tiwari, 1989; Tiwari et al., 1984). The whole sequence of reactions is shown here:

$$2FeS_2 + 2H_2O + 7O_2 \rightarrow 2FeSO_4 + 2H_2SO_4 \text{ (chemical)} \quad (7)$$

$$4FeSO_4 + O_2 + 2H_2SO_4 \rightarrow 2Fe_2(SO_4)_3 + 2H_2O \text{ (}\textit{T. ferrooxidans}\text{)} \quad (8)$$

$$Fe_2(SO_4)_3 + FeS_2 \rightarrow 3FeSO_4 + 2S \text{ (chemical)} \quad (9)$$

$$2S + 3O_2 + 2H_2O \rightarrow 2H_2SO_4 \text{ (}\textit{T. thiooxidans}\text{)} \quad (10)$$

It is thus evident that pyrite is first oxidized to ferrous sulfate by a chemical process. The ferrous sulfate so produced is then oxidized to the ferric salt under the influence of *Thiobacillus ferrooxidans*. Pyrite may also react with this ferric salt and become chemically oxidized, producing elemental S and ferrous sulfate (Tupatkar and Sonar, 1995). The S thus produced is subjected to oxidation by *Thiobacillus thiooxidans* leading to the formation of sulfuric acid. The acidity so produced encourages the activity of microorganisms involved in the oxidation and helps to promote the further process of oxidation (Harmsen et al., 1954). Rudolfs (1922) suggested that this oxidation is accelerated by bacteria and that elemental S acts as a catalyst. Jensen (1927) also demonstrated the stimulus of this oxidation by bacteria (*Thiobacillus* spp.), although the process also proved to be possible under sterile conditions.

Through the oxidation of iron, S is released either as elemental S or as sulfate ions. Thus endothermic substances are formed that can be utilized by S oxidizing microorganisms. The oxidation of S may be considered a final reaction, from which the microbes derive free energy. S oxidation starts only when the pyrite is disintegrated by oxidation of the iron. Part of the released S may then be absorbed directly by the bacterial cells, but the bulk of the S first accumulates in the elemental form. The mechanism of the penetration of this elemental S into cells is still obscure. Umbreit et al. (1942) suggested solution of the elemental S in the fat droplets near the poles of the cells. These fat droplets are characteristic for the most active S oxidizing microbes *Thiobacillus* spp.

3. Factors Affecting Pyrite Oxidation

The basic requirements and conditions required for oxidation of pyrite are the same as for S, so that all those factors that affect S oxidation also govern the oxidation of pyrite in the soil. In order to avoid duplication of information, only points that require specific/additional mention from the viewpoint of pyrite oxidation are described hereunder.

Particle Size. Like S, the rate of oxidation of pyrite also increases as particle size is reduced (Fig. 1). But unlike with S, grinding beyond a certain fineness has no effect on its rate of oxidation. This has been frequently demonstrated in incubation studies and pot and field trials (Vlek and Lindsay, 1978).

The results of Singh and Mishra (1986) also show that with increase in the fineness of the material, the oxidation of pyrite was enhanced. About twice as much pyrite of –200 mesh was oxidized on the very first day as compared to the material of –50 + 100 mesh. The influence of particle size on rate of ox-

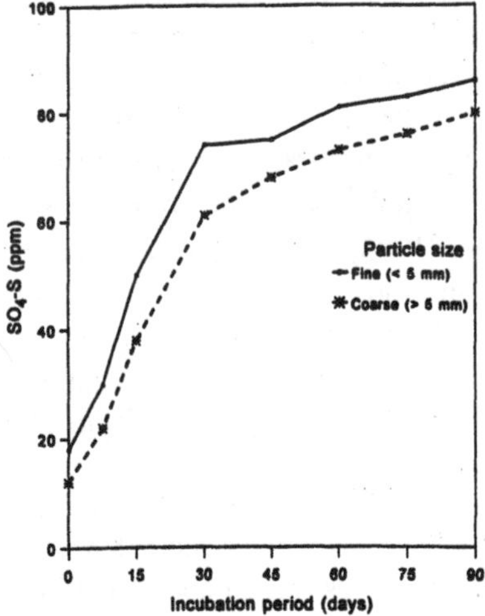

Figure 1 Effect of particle size of pyrites on oxidation of sulfur in a non-calcareone alkaline soil. (Source: Tiwari et al., 1984.)

idation of pyrite narrowed down after one week. This indicates that the rate of oxidation of pyrite in the beginning was related to the surface area per unit mass, which would increase with the fineness. Narrowing down of the particle-size effect during the later period may be because of buildup of the population of pyrite oxidizing bacteria in the soil in every case.

The relative efficiency of pyrite increased from 31 to 65% with decrease in their particle size from −32 + 72 mesh to −150 +300 mesh.

Very fine grinding does not provided faster oxidation. This may be because the particles of naturally occurring pyrite on exposure to air under moist conditions undergo self-disintegration due to autooxidation, which is a time-consuming process and demands a certain degree of fineness before application. The larger-sized particles thus disintegrate into smaller ones, exposing fresh mineral surfaces for such oxidative process. This continues, and the bigger particles of pyrite ultimately shatter into a powder form (Tiwari et al., 1984).

Very fine grinding of pyrite is not required for pyrite oxidation in order to make the soil treatment economic.

Chemical Composition and Purity. Harmsen et al. (1954) stressed the role of the S : Fe ratio in the oxidation of pyrite. A ratio of 2 : 1 is quite favorable for the oxidation, but an increasing proportion of S slows the oxidation process. Quispel et al. (1952) also reported a high S : Fe ratio in the stable phase of pyrite-containing old soils to exceed 3 : 1 and was as high as 4 : 1 in some of the soils. Since the low-grade pyrite contains S : Fe in a ratio nearing 1 : 1 (Jaggi, 1977; Jaggi, 1982; PPCL, 1982, 1986), there are good possibilities of its oxidation.

Most natural pyrite contains appreciable amounts of heavy metals like copper, manganese, and zinc, which are toxic to many microorganisms. However, the pyrite and S oxidizing *Thiobacilli* have shown remarkable tolerance to appreciably high concentrations of such heavy metals (Bryner and Jameson, 1958). Quispel et al. (1952) failed to observe the oxidation of purified pyrite even in the presence of *Thiobacilli*. Only in crude pyrite preparations was sulfuric acid formed. It thus appears that pyrite oxidation requires some impurities.

Moisture Regime and Aeration. Bhatt et al. (1978) noticed that when pyrite is surface dressed on the soil, there is faster and more autooxidation of S to sulfuric acid than when it is incorporated with the soil and mixed in the upper 10 cm layer. It show that easy availability of oxygen helped the reaction and made it proceed faster. This effect was reflected in the drop of soil pH, depth of the wetted soil front, and cumulative water intake.

Oxidation of pyrite cannot easily be stopped. Although only minimal oxidation is possible in dried soil, air drying of initially wet soil caused extensive oxidation, while complete flooding inhibited its kinetics (Sinha and Jha, 1984).

The researches conducted show that for maximum oxidation, pyrite needs to be surface applied and the soil kept moist at field capacity. Under such conditions there is ample availability of both oxygen (from the air) and water (from the soil). These are the two main ingredients basically needed for the oxidation process whereby S in pyrite is converted into sulfuric acid.

Organic Matter. Prasad et al. (1982) were of the view that if pyrite is applied along with organic manure, metal complexes are formed that oxidize faster than simple substances. Presumably owing to microbial action, the organisms can utilize the organic part of these complexes and precipitate iron, thus liberating the S part of the pyrite. The decomposition of organic materials and increased root and microbial metabolism releases CO_2 to build higher partial pressure of CO_2, which lowers soil pH and thus ultimately enhances the activity of *Thiobacillus* spp. and the rate of pyrite oxidation in alkali soils (Somani and Saxena, 1981; Somani, 1984, 1986). Organic matter and its decomposition products play a vital role in improving the physical conditions including porosity and aeration, ensuring better oxygen supply for pyrite oxidation.

Temperature. Singh and Mishra (1986) investigated the effect of temperature on the oxidation of pyrite in a Mollisol. Their results (Table 2) show that there was 12.3% more sulfate production at 35°C than at 25°C. The difference between the two incubation temperatures widened with progress in time. Assuming that all the sulfate produced from oxidation of pyrite in the soil was recovered by extraction

Table 2 Effect of Temperature and Particle Size of Pyrite on Sulfate Production (mg/kg) in Soil

Incubation period (days)	Temperature (°C)		Particle size (mesh)			
	25	35	−50+100	−100+200	−200	Mean
0.25	170	175	164	175	179	173
1	307	336	261	321	384	322
7	365	428	333	391	466	397
14	425	497	393	470	520	461
21	487	561	446	536	590	524
28	542	631	514	590	656	587
42	617	697	582	665	725	657
56	696	757	648	739	793	727
70	753	814	709	783	830	774
112	777	853	753	824	870	816
Mean	512	575	480	549	601	—

	Temp.	Particle size	Incubation period	Temp. × particle size	Particle size × period	Temp. × period
CD (5%)	5.0	6.7	8.3	NS	22.9	16.7

Source: Singh and Mishra (1986).

with 0.5 mol $NaHCO_3$, they reported that 50% of the added pyrite was oxidized in 30 and 26 days at 25 and 35°C, respectively. The oxidation rate was highest on the first day accounting for 21.6 and 24.1% at 25 and 35°C, respectively. The oxidation rate decreased drastically within a week, and it progressed slowly during the later period. After the 70th day, the rate of oxidation was around one mg/kg per day only. This is conceivable in view of chemical oxidation (autooxidation) of pyrite on the first day and inaccessibility of the pyrite surfaces for microbial attack during the later period owing to deposition of ferric hydroxide formed by the oxidation. This is why only 66.7 and 74.3% of the added pyrite was seemingly oxidized in 112 days of incubation at 25°C and 35°C, respectively.

E. As an Amendment-Conditioner for Alkali Soils Amelioration

The use of pyrite as an amendment is a recent development in the chemical amelioration of alkali soil (Somani, 1986). The use of pyrite has been found to be comparable to that of gypsum and has opened fresh avenues in the reclamation of alkali soils (Somani, 1984; Sinha and Jha, 1984; Tiwari et al., 1984; Somani and Totawat, 1993).

1. Mechanism

Effect on Soil Chemical Properties. The weight of evidence being overwhelmingly in favor of pyrite as an ameliorant of high promise for the reclamation of alkali soil, it is of interest to see the mechanism by which this reversal, from exchangeable sodium rich infertile alkali soil to exchangeable calcium rich fertile soil, is brought about. It has already been shown that the ultimate products of the oxidation of S in pyrite are sulfuric acid and iron sulfate. Both of these are of value in the reclamation of an alkali soil, as can be seen from the sequence of chemical reactions shown in Fig. 2.

Sulfuric acid is obtained on the oxidation of pyrite, and iron sulfate, which is also simultaneously formed, hydrolyzes in the presence of water to produce sulfuric acid (Stromberg and Tisdale, 1979). The other product, namely iron hydroxide, being insoluble, is precipitated. Sulfuric acid, formed from both sources, reacts with the free calcium carbonate of the soil to produce soluble calcium sulfate and carbon dioxide. Carbon dioxide acts further on calcium carbonate to produce calcium bicarbonate, which is also a source of soluble calcium as shown here:

$$CO_2 + H_2O \rightarrow H_2CO_3$$

$$CaCO_3 + H_2CO_3 \rightarrow Ca(HCO_3)_2$$

$$Ca(HCO_3)_2 \rightarrow Ca^{2+} + 2HCO_3^-$$

$$\text{Na-Soil} + Ca^{2+} \rightarrow \text{Ca-Soil} + 2Na^+$$

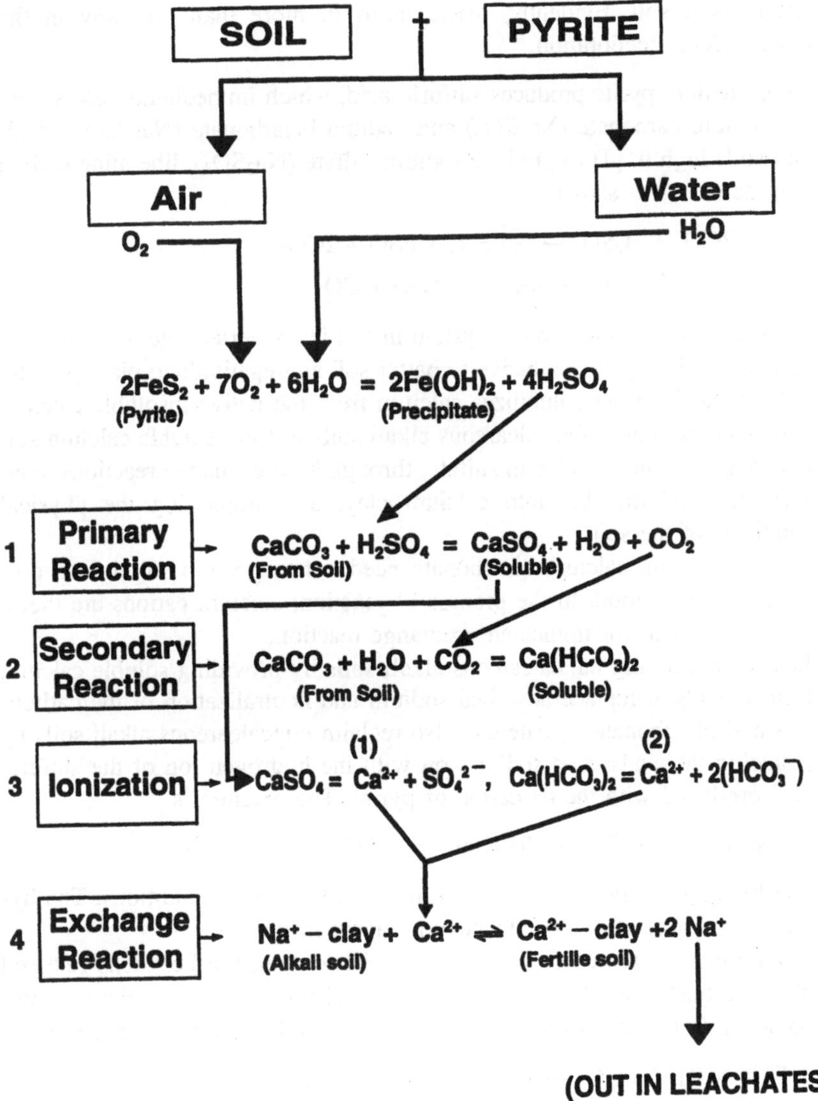

Figure 2 Sequence of chemical reactions of iron pyrite in soils in the course of the reclamation of alkali soils.

The end products of the reaction of sulfuric acid (both primary and secondary) with calcium carbonate are soluble calcium sulfate and calcium bicarbonate. Soluble calcium ion, from both these products (calcium sulfate and calcium bicarbonate), in turn reacts with the sodium clay of the alkali soil to produce calcium clay, which ultimately improves the physical properties of the highly dispersed alkali soil. Evidently, pyrite helps in more than one way in the process of alkali amelioration:

1. On oxidation, pyrite produces sulfuric acid, which immediately reacts with the sodium carbonate (Na_2CO_3) and sodium bicarbonate ($NaHCO_3$) of alkali soils high in pH to produce sodium sulfate (Na_2SO_4), liberating carbon dioxide (CO_2) as shown:

   $$2NaHCO_3 + H_2SO_4 \rightarrow Na_2SO_4 + H_2O + 2CO_2$$

   $$Na_2CO_3 + H_2SO_4 \rightarrow Na_2SO_4 + H_2O + CO_2$$

   Thus a carbonate–bicarbonate system in soil is converted into a sulfate system, which is highly conducive to better soil aggregation and plant growth.
2. Sulfuric acid further solubilizes calcium from the native insoluble calcium carbonate present in the calcareous alkali soils to form soluble calcium sulfate. The calcium of calcium sulfate, through base exchange reactions, converts the sodium clay into calcium clay, thus improving the physical condition of the soil.
3. Unlike gypsum, calcium bicarbonate need not to go into solution to produce calcium cation. In the process of reaction, calcium cations are themselves produced for immediate exchange reaction.
4. Besides improving the calcareous alkali soils by providing soluble calcium from $CaCO_3$ to replace adsorbed sodium and neutralization of the carbonates and bicarbonates, pyrite can also reclaim noncalcareous alkali soils by replacing their adsorbed sodium ion with the hydrogen ion of the sulfuric acid produced with the oxidation of pyrite. The reaction is

   $$2\text{Na-Clay} + H_2SO_4 \rightarrow 2\text{H-Clay} + Na_2SO_4$$

   The hydrogen clay is known to exhibit good physical conditions. The hydrogen ion acts as a good flocculating agent in the soil.
5. In addition to the beneficial effect of sulfuric acid, the Fe^{2+} ions released with the oxidation of pyrite also act as a proton donor and plays an important role in reducing the pH of the soil as indicated by the reaction

   $$Fe^{2+} + 3H_2O + \tfrac{1}{4}O_2 \rightarrow Fe(OH)_3 + 2H^+ + \tfrac{1}{2}H_2O$$

6. Iron sulfate, formed in the process of oxidation of pyrite, hydrolyses to produce sulfuric acid in soil; thus the quantity of acid is further augmented from this second source.
7. Besides proper root action stimulated by improved aeration and consequent availability of oxygen in soil air, metabolism of soil bacteria and fungi also starts. Both these processes augment the CO_2 supply in the root zone of the soil. Large volumes of CO_2 are also simultaneously produced when H_2SO_4 (from the oxidation of pyrite) reacts with free $CaCO_3$ of the soil or when it neutralise Na_2CO_3 and $NaHCO_3$ present in alkali soils.

8. A large volume of CO_2 produced in the above reactions increases the CO_2 partial pressure in the soil, which liberates calcium from native insoluble $CaCO_3$ (Table 3) for further exchange reaction with sodium in the soil and regulates the pH of the medium.

   The harmful effects of the high pH of calcareous soils result in the great difficulty experienced by plants in absorbing enough iron, manganese, boron, and phosphorus. As these plant nutrients are in an insoluble form in the soil at this high pH, the roots cannot always absorb enough of them for the requirements of the plant. However, the pH of the soil solution in a calcareous soil, which is more important than the pH of the soil itself, is greatly dependent on the carbon dioxide content of the soil air. If the carbon dioxide content of the soil solution can be increased, there will be lowering of its pH under some conditions; and the crop will be able to take up enough of some of these plant nutrients.
9. Pyrite itself also directly adds iron, S, and some micronutrients like Zn, Cu, Mn. Its acidic character prevents/reduces volatilization losses of nitrogen from added fertilizer nitrogen in the prevalent alkali soil media. The H_2SO_4 produced at the microsite environment around pyrite particles lowers the pH and not only improves the solubility of less soluble soil phosphates but also helps to increase the proportion of $H_2PO_4^-$ to HPO_4^{2-}. It is an established fact that $H_2PO_4^-$ ion is absorbed by plant roots much more readily than HPO_4^{2-} ion. It is beyond the scope of this chapter to discuss the ameliorative role of plant nutrients and the judicious application of these nutrients through fertilizers for sustaining remunerative crop production in alkali soils. Readers are advised to see these details elsewhere (Somani, 1996). It is however important to mention here that if the essential plant nutrients are made easily available to plant roots by the application of fertilizers, a better crop growth can be expected.

Effect on Soil Physical Properties. The soil as a medium for plant growth provides the environment for the germination and emergence of seedlings and for

Table 3 Solubility of $CaCO_3$ as Affected by the CO_2 Content and pH of the Soil

Content of CO_2 in air (% by volume)	pH	Solubility of $CaCO_3$ (mg/L)
0.00	10.23	13.1
0.03	8.48	63.4
0.30	7.81	133.4
1.00	7.47	202.9
10.00	6.80	470.0
100.00	6.13	1098.6

Source: Kovda (1973).

the development and functioning of the plant root systems. Existing literature (Baver et al., 1972; Pahwa and Gupta, 1982; Somani, 1987) shows that application of lime, gypsum, or organic manures improves the structure of the soils. Hydrated iron has also been reported to provide stable aggregates (Baver et al., 1972). One of the main objectives of applying amendments to alkali soil is restoration of soil structure that has been adversely affected by excessive accumulation of sodium on the exchange complex.

Somani (1981) and Ram and Prasad (1982) reported improved physical conditions of alkali soils treated with pyrite (Table 4). The water holding capacity and pore space rose from 29.2 and 37.2 to 32.5 and 40.7% respectively, after three months. These physical constants improved progressively with the dose of pyrite ranging from 12 to 24 MG/ha mainly because of lowering of exchangeable sodium from the exchange complex, which creates a dispersed condition in alkali soils. Verma and Gupta (1984) observed a decline of ESP (exchangeable sodium percentage) on leaching a pyrite-treated alkali soil with demineralized water as a consequence of the valence effect, according to which a dilution of the outside solution favors adsorption of the divalent cation Ca^{2+} over the monovalent Na^+ (Bear, 1964). The dilution effect is more prominent in soils or colloidal materials with high CEC (cation exchange capacity). A decrease in soil ESP was followed by a corresponding increase in a saturated hydraulic conductivity (Ks) of pyrite-treated alkali soil. Highest values of Ks were recorded in soil maintained at 45% moisture content during incubation. The Ks data suggest that at least a 120h incubation period followed by leaching of a moist soil should be allowed for efficient use of pyrite (Verma and Gupta, 1984). Pyrite has been reported to have a more beneficial

Table 4 Effect of Amendments on Soil Physical Properties and pH of Alkali Soils

Treatments	pH	Water holding capacity (%)	Pore space (%)	Reference
Control	10.5	29.2	37.2	Ram and Prasad (1982)
Pyrite (12 MG/ha)	9.9	30.8	39.3	
Pyrite (18 MG/ha)	9.8	30.9	38.7	
Pyrite (24 MG/ha)	9.6	32.5	40.7	
Control	9.11	33.5	32.1	Somani (1981)
Gypsum (50% GR)	8.67	36.5	35.5	
S (50% GR)	8.62	41.0	37.7	
Dhaincha (DA-GM)	8.67	38.8	38.9	
DA-GM + gypsum	8.53	42.2	43.4	
DA-GM + S	8.23	46.3	45.3	

GR = gypsum requirement; DA-GM = organic amendment.

effect over gypsum or rock-phosphate in combination with FYM or paddy straw in increasing the number of soil aggregates of larger size (>0.5 mm) (Singh and Srivastava, 1982).

The factors enumerated above that are responsible for the beneficial action of pyrite in the building up of crop productivity in calcareous alkali soils are shown in Fig. 3. All the different steps illustrated are mutually interdependent.

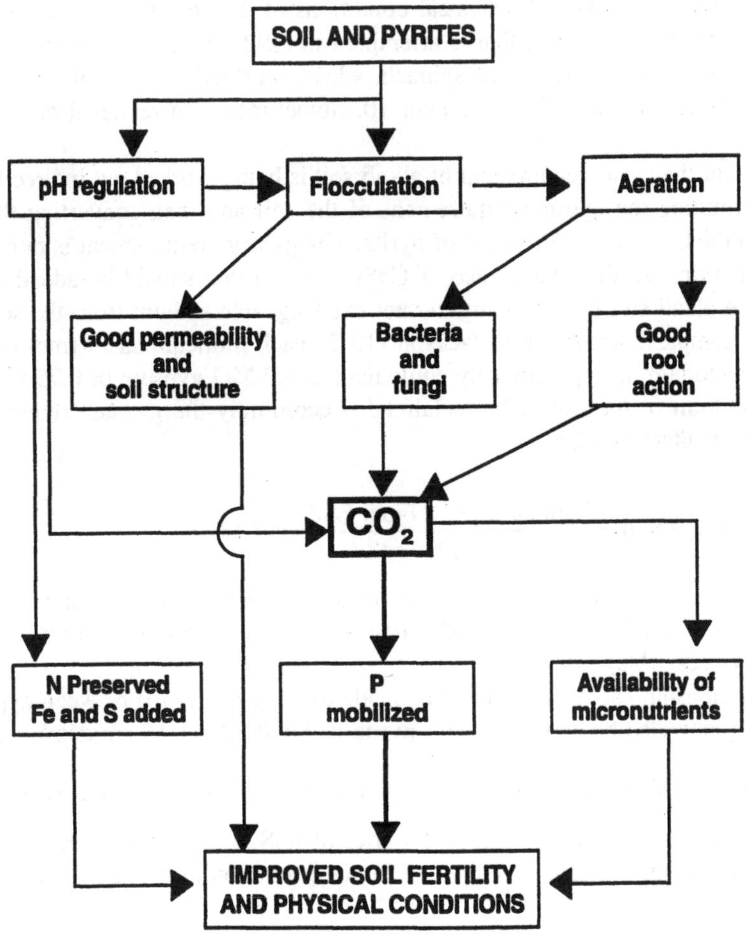

Figure 3 Sequence of changes and factors responsible for ameliorative action of pyrite in calcareous alkali soil.

But carbon dioxide plays the central role in producing the ultimate results. Optimum pH in the soil–water system is almost entirely controlled by CO_2 partial pressure, which regulates the ambient pH in the feeding zone of roots for maximum nutrient absorption and healthy growth of plants.

2. Pyrite Requirement of Alkali Soils

Amelioration of alkali soil involves the displacement of sodium from the exchange complex. Since an ESP of 10 or below is generally considered to be safe for maintaining favorable physical conditions of the soil, the replacement by calcium to this level is all that is attempted in field, though for certain tolerant crops like paddy, sugar beet, spinach, wheat, and barley, the replacement may be safe at a higher ESP level than 10. Sometimes a lower level can be injurious.

At present the pyrite requirement of alkali soil is being worked out indirectly after determining the gypsum requirement of the soil and then converting the figures on the basis of the S content of pyrite. The gypsum requirement is rather specific; it refers to the equivalence of $CaSO_4$. $2H_2O$ that should be added to reclaim the alkali soil by displacing the net exchangeable sodium from the soil exchange complex. According to Jackson (1973) each milliequivalent (me) exchangeable sodium is approximately equivalent to 4.0 MG gypsum or 0.75 MG S per ha-30 cm of the soil to be reclaimed. Accordingly the gypsum requirement (GR) is determined as

$$\text{GR me/100 g Soil} = \frac{\text{Initial ESP} - \text{Final ESP}}{100} \times \text{CEC}$$

The initial ESP is determined by analysis of soil before reclamation and final ESP usually kept 10 (except in cases of heavy soils like vartisols where a final ESP of 5 is desirable).

The relationship based on calculations made by Jackson (1973) between gypsum and S may also be correlated with pyrite in MG/hectare (15 cm depth) as

$$1 \text{ GR} = 2.25 \text{ MG gypsum} = 0.4 \text{ MG S} = 1.82 \text{ MG pyrite} = 1.24 \text{ MG } H_2SO_4$$

This assumes 22% S in pyrite and 93% purity of H_2SO_4. It is, however, advisable to develop some suitable method for direct estimation of pyrite need of alkali soils.

3. Agrotechniques to Improve the Efficiency of Pyrite

Keeping in view the factors that affect the oxidation of pyrite in soils, the following agrotechniques should be kept in mind for improving the efficiency of pyrite.

1. Properly crushed pyrite-should be used as an amendment–conditioner. It is not necessary that pyrite be very finely ground for this purpose.

2. The field should be properly prepared and given a preamendment treatment of light irrigation to bring the soil to its near field capacity soil moisture.
3. Application of pyrite should be done preferably in summer to take advantage of higher temperatures for more effective oxidation (Sharma and Gupta, 1986).
4. On moist soil surfaces pyrite should be dressed as thinly as possible, to ensure maximum availability of air to facilitate oxidation. Sufficient time should be allowed (3–6 weeks) for pyrite to oxidize on the surface of the soil, and then it should be incorporated into the soil.
5. A presubmergence aerobic period of 10–20 days should be allowed to permit enough oxygen for fast oxidation. The field should be subdivided into small compartments and filled with irrigation water to facilitate leaching of displaced sodium salts and the later transplanting of rice.
6. Rice should be preferred as the first crop to encourage a buildup of CO_2 partial pressure under submerged conditions and to permit leaching of replaced sodium salts. Rice culture should be done with a suitable alkali tolerant high-yielding rice variety, which should be adequately fertilized with N, P, K, and Zn. The number of seedlings per hill should be increased and distance between row to row should be decreased for more survival of seedlings and greater plant population per hectare.
7. Pyrite in quantity equivalent at least to 50% of the gypsum requirement of the soil to be reclaimed should be used. Higher quantities can also be used if quicker action is desired or the alkali soil is in advanced stage of deterioration.
8. Usually for alkali soil reclamation, a crop rotation of rice–wheat is generally recommended. Another useful rotation can be rice–berseem–rice–wheat. Introduction of berseem, which needs a greater amount of irrigation water, helps in still faster and larger oxidation in the cooler months of the *rabi* (spring) season than is possible with wheat. This can be done once, and, thereafter wheat can be taken in the crop sequence of rice–wheat–rice.
9. Transplanting of seedlings immediately after application of pyrite should be avoided, as it usually results in high mortality of rice seedlings, inferring that sufficient time should be provided between pyrite application and rice transplanting (Mehta, 1985).
10. The efficiency of pyrite could be considerably improved by making concomitant use of pyrite and organic residues, manures, and wastes like pressmud. The potentiality of sugar industry waste, sulfitation pressmud, in ameliorating alkali soil is well known (Singh et al., 1986; Somani, 1986). Singh et al. (1986) observed an additive effect when pyrite was used along with sugar industry waste pressmud. Rai et al. (1982) reported improved nodulation and yield of chickpea with pyrite-pressmud complex. Prasad et al. (1982) observed enhanced availability of iron when pyrite

was applied along with organic manure. Singh and Shrivastava (1982) concluded that application of FYM or straw in combination with pyrite is helpful in improving the physical conditions of soils. Chaudhary and Jha (1982) obtained remarkable yield improvements by applying pyrite in combination with organic matter (Fig. 4). It is thus advisable to make concomitant use of pyrite and organic materials for improving the ameliorative efficiency of pyrite.

4. Gympsum vis-à-vis Pyrite

Several experiments have been conducted in various parts of India to study the relative performance of pyrite and gypsum in reclaiming alkali soils. Singh and Sinha (1978) studied the effectiveness of gypsum, pyrite, and pyrite + gypsum (1:1) applied to meet 0, 50, 75, and 100% of gypsum requirements and worked out the relative effectiveness of the treatments using the equation

$$\text{Relative efficiency} = \frac{\text{ESP (pyrite or pyrite + gypsum)}}{\text{ESP (gypsum)}} \times 100$$

Their study revealed that gypsum had an edge over pyrite in all the soils. A similar conclusion was also drawn by others as well. (Anonymous, 1975; Somani et al., 1981). Sharma and Gupta (1986) reported pyrite to be 92.0 to

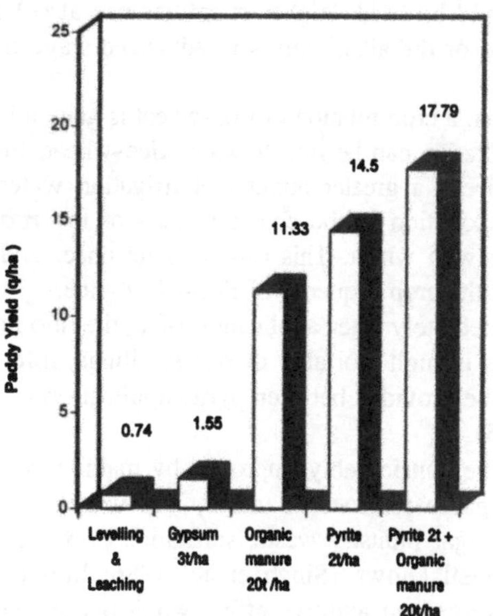

Figure 4 Relative effectiveness of different suggested ameliorative measures. (After Chaudhary and Jha, 1982.)

98.3% as effective as gypsum. They suggested that the dose of gypsum should be 1.3 times the gypsum requirement and of pyrite 1.5 times the gypsum requirement for achieving a soil ESP of about 10. Pathak et al. (1978) reported that gypsum proved to be more effective than pyrite in the initial stages but pyrite recovered with the progress of time and compared well with gypsum in reclaiming alkali soil and improving crop yields. According to them, the application of pyrite to meet 40% of the gypsum requirement proved as good as 60% of the gypsum requirement through gypsum application. Poor performance of pyrite in reclamation, particularly during early stages, could be attributed to the time required by pyrite to become oxidized in the soil before it is able to replace exchangeable sodium. Bhatt et al. (1978) stressed the need for undertaking long-term experiments with pyrite, as it has given better results over gypsum in many of the localities while it was just the reverse at other sites. Yadav (1977) and Somani (1984, 1986) have also pointed out the need for studying the relative effectiveness of pyrite against gypsum application to the main crop as well as subsequent crops on the same soil under similar climatic conditions.

Jaggi (1982) compared the relative performance of five different amendments–conditioners and found pyrite to be the most efficient one for improving both the physicochemical characteristics of the soil and the crop yield. Pyrite application proved superior to gypsum application. Rai et al. (1977) also reported yield response of several crops, such as sugarcane, paddy, wheat, and legumes, besides a favorable effect of pyrite in decreasing the content of exchangeable sodium from soil exchange complex (Ram and Prasad, 1982). Sinha and Singh (1982) found increased microbial activity, particularly of nitrogen fixing microorganisms, following application of pyrite.

5. Prospects for Pyrite Use

Most alkali soils have appreciable reserves of free calcium carbonate. The native reserve of calcium carbonate in most alkali soils is adequate to provide desired levels of soluble calcium, provided it is made to react with acid. The sulfuric acid produced by the chemical and microbiological oxidation of pyrite could thus bring in solution the desired amount of calcium to replace the sodium adsorbed on the exchange complex of the soil. There is thus a good possibility of utilizing the world's large reserves of low-grade pyrite in reclamation of alkali soils, particularly in areas where it is locally available and where transportation of gypsum involves heavy expenditure. Although the use of pyrite may not be superior to gypsum so far as its immediate reclamative efficiency on a chemical equivalence basis is concerned, yet it has remarkable residual effects and improves the fertility of the soil. Summing up the whole situation, there are good possibilities for utilizing pyrite in the reclamation of calcareous alkali soils.

IV. WASTE H_2SO_4 IN ASIA

The potential value of sulfuric acid for improving the productivity of sodic soils has long been recognized. However, its actual use has not been extensive, partly because of limited supplies, handling difficulties, and inadequate application criteria and partly because of high cost. But the availability of production or by-product H_2SO_4 has greatly increased in recent years, which expands the problem of environmental pollution (Miyamoto et al., 1975). This would eventually necessitate proper disposal and management of waste H_2SO_4. Agricultural use of by-product and waste H_2SO_4 for reclamation of sodic soils may offer a partial solution.

A. Potential Sources

The production of waste and by-product H_2SO_4 is likely to be higher and higher as air and water pollution abatements are implemented in smelters, coal burning power plants, S burning plants, and the petroleum and dairy industries. Copper, zinc, and lead smelters usually recover less than one-fourth of S, and the rest is mostly discharged as SO_2 except in places where laws mandate its trapping. Considerable H_2SO_4 will be produced if air pollution regulations are fully implemented. Most of these smelters are capable of producing over 1000 MG of concentrated H_2SO_4 every day (Miyamoto et al., 1975). Coal burning power plants and the petroleum industry can yield millions of MG of H_2SO_4 annually.

B. As an Amendment for Alkali Amelioration

1. Mechanism of Alkali Amelioration

The reclamation of alkali soil with H_2SO_4 is based upon the following chemical reactions.

If the soil is calcareous:

$$H_2SO_4 + CaCO_3 \rightarrow CaSO_4 + CO_2 + H_2O$$

$$2\ \text{Na-Clay} + CaSO_4 \rightarrow \text{Ca-Clay} + Na_2SO_4$$

In the zone of penetration, H_2SO_4 decomposes calcium and magnesium carbonates, giving rise to the formation of soluble calcium sulfate, which positively affects soil properties. The newly formed Ca and Mg salts participate in the soil colloidal chemical reactions, replacing active/toxic sodium from the exchange complex.

If the soil is non-calcareous:

$$2\ \text{Na-Clay} + H_2SO_4 \rightarrow \text{H-Clay} + Na_2SO_4$$

The H^+ from H_2SO_4 directly replaces Na^+ from the exchange complex, giving rise to H-Clay. Such soils are of course in better physical condition, but eventually they may develop into degraded alkali soils. It would be preferable to use waste H_2SO_4 in conjunction with lime in such soils.

2. Effect on Soil Properties

The ameliorative effect of H_2SO_4 is very quick, particularly if the soil is calcareous. It brings about instantaneous improvement in infiltration and other physical properties due to flocculation of highly dispersed clay. Tyulenina and Fomina (1978) observed increased activity of soil microflora, decreased peptization of clay, and increased available moisture reserve following application of H_2SO_4. Sulfuric acid application increased the rate of water penetration for most sodic soils, the effect being greater with increasing levels of exchangeable Na (Yahia et al., 1975). Kurbaton and Okorokov (1975) reported a drop in the electrokinetic potential of the solonetzes from 28 to 13 mV, thus resulting in aggregating the soil particles and increasing the average radius of the pores and the hydraulic conductivity of the soils.

Soil acidification with H_2SO_4 neutralized sodicity, decreased the ESP significantly, and improved the hydrophilic properties.

The high pH and lime content of sodic soil often limits solubility and availability of phosphorus and certain micronutrient like Fe, Mn, and Zn. In general, acidification of such soils increases the solubility of these elements originally present in the soil and those applied as fertilizers (Miyamoto and Stroehlein, 1975; Ryan et al., 1974).

3. Waste H_2SO_4 vis-à-vis Traditional Amendments–Conditioners

The relative performance of H_2SO_4, S, gypsum etc. has been evaluated by a number of workers under field conditions. Chena (1959) reported that H_2SO_4 removed the Na almost twice as rapidly as gypsum did. The results of a field study in western Siberia indicated that the ameliorative effects of HNO_3, $CaCl_2$, and $CaSO_4$ were approximately the same during prolonged leaching but was much weaker than the effect of H_2SO_4 on the permeability and Na replacement from a sodic soil (Dolozhenko, 1977). Gypsum needed a much extended period of leaching to increase crop yield to the same extent as the H_2SO_4 did (Branson and Fireman, 1960). H_2SO_4 obviously reacts very quickly with alkali soils. At times the use of H_2SO_4 on an equivalent basis is slightly costlier than gypsum. The quicker reclamation with H_2SO_4, however, pays off more than the higher initial cost by harvesting good yields following H_2SO_4 treatment (Kovda, 1973). Huge availability of waste and by-product H_2SO_4 offers an alternative to commercial H_2SO_4. Traditional amendments like gypsum, pyrite, and S bring about reclamation of the surface layer of alkali soils. As a result, the entire profile is not reclaimed, as with H_2SO_4. The farmer has to grow tolerant and semitolerant crops for several years after applying traditional amendments, but can go for normal cultivation to reap good profit with an H_2SO_4 treatment.

Corrosive and hazardous H_2SO_4 severely attacks the skin and clothes of the user upon contact. Its use requires proper equipment and safety training for handling and applications.

4. Prospects of H_2SO_4 Use

As early as 1915 Hunt (1916) was successful in reclaiming alkali soil by surface sprinkling of H_2SO_4. Concentrated H_2SO_4 sprinkled on the surface of sodic soils at a rate of 1.5 MG/ha increased water infiltration rates and seedling emergence and resulted in a good growth of crops (McGeorge et al., 1956). Shah (1975, 1977) made field scale utilization of waste H_2SO_4 for reclaiming alkali soils in Gujarat. His results showed that soil pH dropped from 9.2 to 7.2 and the hydraulic conductivity improved 20 times. Kurbaton and Okorokov (1975) reported increased yield of barley and oat hay on a soda solonetzes of Western Siberia following application of industrial waste H_2SO_4.

For efficient amelioration of sodic soils, application of H_2SO_4 should be followed by leaching to neutralize the alkalinity of the soil solution and to replace exchangeable sodium from the upper one meter of the soil. To increase the effectiveness of chemical improvement, the soil must be diked during acidification (Petrosian, 1978). Petrosian (1977) observed the ameliorative influence of waste H_2SO_4 in a very short time. High-yielding annuals and perennial crops can be successfully cultivated on treated sodic soils, securing very quick returns on capital investments.

Chauhan and Tripathi (1980) however, reported a harmful effect of H_2SO_4 in noncalcareous sodic soils. Bocskai (1969) stressed the use of waste H_2SO_4 in conjunction with lime dust for noncalcareous alkali soils of Hungary.

V. IRON AND ALUMINUM SULFATE

Iron and aluminum sulfate have been used in some regions for reclamation of alkali soils. When added to moist soils they hydrolyze rapidly, producing H_2SO_4 as shown here:

$$Al_2(SO_4)_3 + 6H_2O \rightarrow 2Al\,(OH)_3 + 3\,H_2SO_4$$

$$FeSO_4 + 2H_2O \rightarrow Fe(OH)_2 + H_2SO_4$$

$$FeSO_4 + 3H_2O + \tfrac{1}{4}\,O_2 \rightarrow Fe(OH)_3 + H_2SO_4 + \tfrac{1}{2}H_2O$$

$$Fe_2(SO_4)_3 + 6H_2O \rightarrow 2Fe(OH)_3 + 3H_2SO_4$$

The sulfuric acid so produced reacts with native $CaCO_3$ invariably present in alkali soils and produces $CaSO_4$. The supply of both $Al_2(SO_4)_3$ and $FeSO_4$ has been limited mainly to low-cost byproducts. They are applied in the same manner as S and pyrites. Both these amendments react very quickly.

Application of $FeSO_4$ neutralizes alkaline reaction of soil besides increasing the infiltration rate and permeability (Petrosian, 1977). Flocculation is more vigorous and pH reduction is greater with $FeSO_4$ than with $CaSO_4$ (Bottini and Lisanti, 1951). Manukyan (1975) and Nurijanian and Khizantsian (1975) have successfully reclaimed sodic soils in the Ararat plain using $FeSO_4$ as an amendment. Iron sulfate at 50% of the full rate gave as good amelioration of sodic soil as gypsum at 100% of the full rate (Vasil Chikova, 1964). The improvement in permeability with $Al_2(SO_4)_3$ was much more effective than $FeSO_4$ in producing persistent acidity in calcareous alkali soils. However, Vakil and Ray (1971) found $Al_2(SO_4)_3$ to be an inferior source of amendment as compared to gypsum.

VI. CONCLUSIONS

1. S and pyrite, when incorporated into sodic soils, are oxidized to H_2SO_4, which then dissolves $CaCO_3$ and provides soluble calcium for the replacement of sodium from the exchange complex of sodic soils.
2. The desired rate of S/pyrite oxidation can be achieved by manipulating soil conditions, rate, method, time, and particle size of S/pyrite.
3. The ameliorative efficiency of S/pyrite is considerably improved when they are used in conjunction with organic materials.
4. Use of S is particularly recommended for areas where transportation of other amendments like gypsum and pyrites is expensive.
5. Sulfur/pyrite should preferably be applied 2–3 months before starting leaching and cropping to permit its oxidation.
6. The use of S/pyrite/H_2SO_4 should be restricted to calcareous sodic soils, else they may lead to the formation of degraded alkali soils.
7. Pyrite/S can be oxidized both abiologically (autooxidation) and biologically, but the process of abiological oxidation is very slow.
8. Biological oxidation of S/pyrite could be mediated by both autotrophic and heterotrophic microorganisms. The best known and usually considered to be the most important group of organisms are autotrophic bacteria belonging to the genus *Thiobacillus*.
9. Since many alkali soils lack *Thiobacillus* spp., enrichment with an effective strain of *Thiobacillus* spp. is a must for successful reclamation of these soils with S/pyrites.
10. Waste and by-product H_2SO_4 is particularly a cheap and fast ameliorant for the reclamation of highly deteriorated calcareous alkali soils. Its use, however, requires proper equipment and safety training for handling and application.

VII. FUTURE RESEARCH NEEDS

1. There is a lack of precise information about the cost of reclamation through S and pyrite as compared to other prevalent amendments.

2. Although S and pyrite are considered to be slow acting, there is a need to make long-term amelioration trials to elucidate their effect on soils and crops under varying agroclimatic situations.
3. Investigations need to be made to find ways to ensure faster oxidation of S and pyrite.
4. The rate of CO_2 evolved during S and pyrite oxidation and the subsequent reaction of H_2SO_4 with $CaCO_3$ in alkali amelioration needs careful investigation.
5. There is a need to find methods for the direct estimation of the S/pyrite/H_2SO_4 needs of alkali soil. At present they are indirectly calculated from the gypsum requirements of the soils.
6. Long term trials need to be made to investigate the toxicity if any due to presence of different metallic ions present in waste H_2SO_4.

REFERENCES

Anonymous (1975). Pyrites as a soil amendment. Ann. Rep. CSSRI, Karnal (*India*), p. 114.

Bain, B., and Service, I. (1987). Potential S resources of developing Asia and the Pacific. Proceedings Symp. Fertilizer S Requirements and Sources in Developing Countries of Asia and the Pacific, pp. 151–155.

Baver, L. D., Gardner, W. H., and Gardner, W. R. (1972). *Soil Physics*. John Wiley, New York, p. 498.

Bear, F. E. (1964). *Chemistry of the Soil*. Oxford and IBH, New Delhi, p. 515.

Bhatt, M. N., Goel, K. N., and Uprety, M. C. (1978). Use of pyrites in reclamation of alkali soils in Uttar Pradesh. *Proceedings FAI-PPCL-DAUP Sem. on Use of Sedimentary Pyrites in Reclamation of Alkali Soils, Lucknow (India)*, pp. 107–114.

Bloomfield, C. (1967). Effect of some phosphate fertilizers on the oxidation of elemental S in soil. *Soil Sci. 103*: 219–223.

Bocskai, J. (1969). The use of acid resins supplied by the oil industry in the chemical amelioration of alkali soils. *Agrokem. Talajt. 18* (Suppl.): 336–338.

Bottini, O., and Lisanti, L. (1951). Research on saline soils of the Tavoliere. *Ann. Sper. Agran. Roma 5*: 233–67.

Branson, R. L., and Fireman, M. (1960). Reclamation of an impossible alkali soil. *Trans 7th int. Congr. Soil Sci. 1*: 543–552.

Breed, R. S., Murray, E. G. D., and Parker, H. A. (1948). *Bergey's Manual of Determinative Bacteriology*. Williams & Williams, Baltimore.

Bryner, L. C., and Jameson, A. K. (1968). Microorganisms in leaching sulphide minerals. *Appl. Microbiol. 6*: 281–287.

Burns, G. R. (1968). Oxidation of S in soils. *S. Institute Tech. Bull. 13*: 1–41.

Chandra, P., and Boolen, W. B. (1960). Effect of Gibral on nullification and S oxidation in different Oregon soils. *Appl. Microbiol. 8*: 31–38.

Chaudhary, S. N., and Jha, S. N. (1982). Effect of various measures and amendments in crop yield and soil properties of salt affected calcareous soils. *Proceedings FAI-PPCL-GADA Sem. on Management of Salt-Affected Calcareous Soils, Pusa (India)*, pp. 107–114.

Chauhan, C. P. S., and Tripathi, B. R. (1980). Influence of continuous and intermittent leaching with and without amendments on reclamation of saline-sodic soils. *Madras Agric. J. 67*: 109–112.

Chena, G. (1959). Reclamation of a saline sodium soil. *Agric. Tech. Mexico 8*: 17–19.

Chopra, S. L., and Kanwar, J. S. (1968). Oxidation of S in soils. *J. Indian Soc. Soil Sci. 16*: 83–88.

Clark, F. W. (1925). Data of Geochemistry. *U. S. Geol. Survey Bull.*, p. 770.

Conrad, J. P. (1950). Sulfur fertilization in California and some related factors. *Soil Sci 70*: 43–54.

Correns, C. W. (1935). Wissenschaftliche Ergebnisse der deutschen atlantischen (1925–1927). Le Lieferung, Berlin.

Dolozhenko, I. B. (1977). Effectiveness of chemical amendments in the leaching of sodic solonchaks in West Siberia. *Pochvovedenic 2*: 64–72.

Escolar, R. P., and Lugo Lopez, M. A. (1969). Sulfur transformation in a saline sodic soil of the Lajas Valley. *J. Agric. Univ. Puerto Rico, 53*: 118–123.

Fawzi, A. M. A. H. (1976). Sulfate reduction in poorly drained soil as influencing the availability of sulfur fertilizers to alfalfa and corn. *Soil. Sci. Soc. Amer. Proc. 28*: 406–408.

Fox, R. L., Atesalp, H. M., Kampbell, D. H., and Rhoades, H. F. (1964). Factors influencing the availability of sulfur fertilizers to alfalfa and corn. *Soil Sci. Soc. Am. Proc. 28*: 406–408.

Gidnavar, V. S., Gumaste, S. K., and Krishnamurty, K. (1972). Effect of amendments on the properties of saline-alkali soils. *Proc. Sem. Drought.* Banglore, 1968, pp. 306–312.

Gupta, R. K., and Abrol, I. P. (1989). Salt-affected soils: their reclamation and management for crop production. *Adv. Soil Sci. 11*: 223–272.

Gupta, R. K., and Abrol, I. P. (1990). Reclamation and management of alkali soil. *Indian J. Agric. Sci. 60*: 1–16.

Harmsen, G. W., Quispel, A., and Otzen, D. (1954). Observations on the formation and oxidation of pyrite in the soil. *Pl. Soil 5*: 423–448.

Hart, M. G. R. (1962). Observation on the source of acid in empoldered managrove soils I-Formation of elemental S. *Pl. Soil. 17*: 87–98.

Hunt, T. F. (1916). *Ann. Rept. Calif. Agr. Expt. Sta. Californea*, p. 100.

Jackson, M. L. (1973). *Soil Chemical Analysis.* Prentice Hall of India, New Delhi p. 498.

Jaggi, T. N. (1977). Pyrites for reclamation of saline-alkali soils. *Proceedings Indo-Hungarian Sem. Management of Salt-Affected Soils, Karnal (India).*

Jaggi, T. N. (1982). Use of Amjhore pyrites as alkali soil amendment. *Proceedings FAI-PPCL-GADA Sem. on Management of Salt Affected Calcareous Soils, Pusa (India)*, pp. 17–40.

Jensen, H. J. (1927). Vorkommen von *Thiobacillus thiooxidans* in danischem Boden, *Zentr. Bakteriol. Parasitenk., Abt. II, 72*: 242–246.

Joffe, J. S., and McLean, H. C. (1922). A note on oxidation of sulfur in Oregon soils. *Soil Sci. 14*: 217–221.

Jones, M. B., Williams, W. A., and Martin, W. E. (1971). Effect of waterlogging and organic matter on the loss of applied S. *Soil Sci. Soc. Am. Proc. 35*: 542–546.

Keller, P. (1969). The effect of sodium chloride on sulfur oxidation in soil. *Pl. Soil 30*: 15–21.

Kittams, H. A., and Attoe, O. J. (1965). Availability of phosphous in rock phosphate sulfur fusion. *Agron. J. 57*: 331–334.

Kovda, V. A., ed. (1973). Chemistry of saline and alkali soils of arid zones. *Irrigation, Drainage and Salinity, An International Source Book*. FAO, UNESCO, Paris.

Kurbaton, A. I., and Okorokov, V. V. (1975). The effect of amendments on the electrokinetic properties of solonetzes of Western Siberia, *Izv. Timiryazev S kh Akad. 5*: 104–108.

Lee, H. (1955). *Biochemistry of Autotrophic Bacteria*. Butterworths Scientific Publications, London.

Li, P., and Caldwell, A. C. (1966). The oxidation of elemental S in soil. *Soil Sci. Soc. Am. Proc. 30*: 370–372.

Manukyan, R. R. (1976). Improvement of a sodic solonchak-solonetz on the Ararat plain with mineral acids and copperas (based on laboratory experiments). *Sov. Soil Sci. 8*: 202–212.

McGeorge, W. T., and Greene, R. A. (1935). Oxidation of sulfur in Arizona soils and its effect on soil properties. *Ariz. Agri. Exp. Sta. Tech. Bull.* 59.

McGeorge, W. T., Breazeale, E. I., and Abbott, J. L. (1956). Polysulfides as soil conditioners. *Ariz Agric. Exp. Sta. Tech. Bull.* 131.

Mehta, K. K. (1985). Long term studies on the effect of pyrites application on the reclamation of alkali soils and crop production. *Curr. Agric. 9*: 69–72.

Miyamoto, S., and Stroehlein, J. L. (1975). Sulfuric acid for increasing water penetration into some Arizona soils. *Prog. Agric. Ariz. 27*: 13–16.

Miyamoto, S., Ryan, J., and Stroehlein, J. L. (1975). Potentially beneficial uses of sulfuric acid in South-Western agriculture. *J. Environ. Qual. 4*: 431–437.

Moser, U. S., and Olson, R. V. (1953). Sulfur oxidation in four soils as influenced by soils moisture tension and sulfur bacteria. *Soil Sci. 76*: 251–257.

Nicholas, D. J. D. (1963). *Plant Physiology. III. Inorganic Nutrition of Plants*. Academic Press, New York.

Nor, Y. M., and Tabatabi, M. A. (1977). Oxidation of elemental sulfur in soils. *Soil Sci. Soc. Am. J. 41*: 736–741.

Nurijanian, V. N., and Khizantsian, S. M. (1975). Efficiency of chemical reclamation of soda solinized soils against the background of closed horizontal drainage. *Int. Symp. Develop. Field Salt Affected Soils, Cairo*, 1972, pp. 499–503.

Pahwa, K. N., and Gupta, I. C. (1982). *Reclamation and Management of Salt-Affected Soils (1950–1981)*. Associated Publishing Co., New Delhi, p. 352.

Parr, J. F., and Papendick, R. I. (1966). Agronomic effectiveness of anhydrous ammonia-sulfur solutions. *Soil Sci. 101*: 336–345.

Pathak, A. N., Sharma, D. N., and Sharma, M. L. (1978). Studies on the performance of gypsum and iron pyrites in amelioration of salt affected soil. *Proceedings FAI-PPCL-DAUP Sem. on Use of Sedimentary Pyrites in Reclamation of Alkali Soils, Lucknow (India)*, pp. 45–53.

Petrosian, G. P. (1977). Chemical reclamation of solonetz-solonchaks in the Ararat plain and the possibility of utilizing saline waters for leaching and irrigation. *Managing Saline Waters for Irrigation* (H. E. Dregne, ed.). Texas Tech. Univ., Lubbock, Texas, pp. 466–479.

Petrosian, G. P. (1978). Technology and economic indexes of the chemical improvement of the sodic solonetz-solochaks of the Ararat plain of the Armenian SSR, *Pochvovedenie 9*: 59–73.

PPCL (1982). *Pyrite in the Reclamation of Alkali Soils*. Pyrites, Phosphates and Chemicals, Rohtas (Bihar), India, p. 40.

PPCL (1986). *Pyrites in the Management of Alkaline Calcareous Soils of North Bihar*. Pyrites, Phosphates and Chemicals, Rohtas (Bihar), India, p. 66.

Prasad, S. S., Pandeya, S. B., and Sinha, M. K. (1982). Pyrites-organic manures complex as a source of iron to calcareous soils, Bihar. *Proceedings FAI-PPCL-GADA Sem. on Management of Salt-Affected Calcareous Soils, Pusa (India)*, pp. 199–210.

Quispel, A., Harmsen, G. W., and Otzen, D. (1952). Contribution to the chemical and bacteriological oxidation of pyrite in soil. *Pl. Soil 4*: 43–55.

Rai, R., Ashraf, M. H., and Choudhary, S. N. (1977). Pyrite as an amendment. *Fertil. Technol. 14*: 370–372.

Rai, R., Sinha, N. P., Singh, S. N., Prasad, V., Singh, and Sinha, R. B. (1982). Effect of iron amended pressmud, pyrite-pressmud, complex and ferrous sulfate on nodulation, yield and quality of chickpea (*Cicer arietinum* Linn) in calcareous soil. *Proceedings FAI-PPCL-GADA Sem. on Management of Salt-Affected Calcareous Soils, Pusa* (India), pp. 187–197.

Ram, H., and Prasad, R. (1982). Effect of pyrites on properties of alkali soil. *Proceedings FAI-PPCL-GADA Sem. on Management of Salt-Affected Calcareous Soils, Pusa* (India), pp. 91–95.

Rudolfs, W. (1922). Oxidation of iron pyrites by S-oxidizing organisms and their use for making mineral phosphates available. *Soil Sci. 14*: 135–146.

Rupela, O. P., and Tauro, P. (1973a). Isolation and characterization of *Thiobacillus* from alkali soils. *Soil Biol. Biochem. 5*: 891–897.

Rupela, O. P., and Tauro, P. (1973b). Utilization of *Thiobacillus* to reclaim alkali soils. *Soil Biol. Biochem. 5*: 899–901.

Ryan, J., Miyamoto, S., and Stroehlein, J. L. (1974). Solubility of manganese, iron and zinc as affected by application of sulfuric acid to calcarious soils. *Pl. Soil 40*: 421–27.

Shah, R. K. (1975). A note on quick and cheap reclamation method for saline-alkali soils using dairy waste sulfuric acid. *Agril. Res. Newsl. 3*: 4–6.

Shah, R. K. (1977). A quick method for saline-alkali soil reclamation using sulfuric acid. *Proceedings Indo-Hungarian Sem. on Management of Salt—Affected Soils. CSSRI, Karnal* (India), pp. 134–139.

Sharma, O. P., and Gupta, R.K. (1986). Comparative performance of gypsum and pyrites in sodic vertisols. *Indian J. Agric. Sci. 56*: 423–429.

Simon, R. H., and Schollenberger, C. J. (1925). The rate of oxidation of different forms of elemental sulfur. *Soil Sci. 20*: 6.

Singh, H., and Mishra, B. (1986). Kinetics of pyrite oxidation in relation to solubilization of rock phosphate in a neutral soil. *J., Indian Soc. Soil Sci. 34*: 52–55.

Singh, K. N., and Shrivastava, A. R., (1982). Relative efficiency of pyrites on gypsum in combination with organic amendments on soil physical properties and crop yield. *Proceedings FAI-PPCL-GADA Sem. on Management of Salt-Affected Calcareous Soils, Pusa* (India), pp. 155–160.

Singh, S., and Sinha, S. C. (1978). Note on pyrites and gypsum trials at cultivators' fields in Varanasi and Mirzapur. *Proceedings FAI-PPCL-DAUP Sem. on Use of Sedimentary Pyrites in Reclamation of Alkali Soils, Lucknow* (India), pp. 121–126.

Singh, K. D. N., Prasad, C. R., and Singh, Y. P. (1986). Comparative study of pyrite and sulphitation pressmid on soil properties, yield and quality of sugarcane in calcareous saline sodic soil in Bihar. *J. Indian Soc. Soil Sci. 34*: 152–154.

Sinha, N. P., and Jha, S. N. (1984). Studies on pyrites as amendments for sodic soil and as nutrients. *Indian J. Agric. Chem. 17*: 75–101.

Sinha, N. P., and Singh, S. N. (1982).Pyrite as a soil amendment and source of plant nutrients—review. *Proceedings FAI-PPCL-GADA Sem. on Management of Salt-Affected Calcareous Soils, Pusa* (India), pp. 67–74.

Somani, L. L. (1972). Sulfur as an amendment in the reclamention of calcareous saline-alkali soils of Rajasthan. *Agric. Agro. Ind. J. 15*: 16–18.

Somani, L. L. (1981). Reclamation of saline-alkali soil using organic materials and inorganic amendments. *Agrokem. Talajt. 30*: 333–350.

Somani, L. L. (1984). Use of low grade pyrites as an amendment for alkali soils and to improve soil fertility—review. *Fertil. News 29*(7): 13–27.

Somani, L. L. (1986). Use of pyrites as a soil amendment, *S. in Agric. 10*: 16–20.

Somani, L. L. (1987). *Soil Physical Research in India (1925–85)*. Associated Publishing, New Delhi, p. 482.

Somani L. L. (1994). *Use of Pyrites in Agriculture*. Agrotech Publishing Academy, Udaipur (India), p. 255.

Somani, L. L. (1996). *Efficient use of Fertilizers*. Agrotech Publishing Academy, Udaipur (India), p. 288.

Somani, L. L., and Saxena, S. N. (1981). The effect of organic and inorganic amendments on the microflora and crop growth in calcareous saline-alkali soil. *Pedobiologia 21*: 192–210.

Somani, L. L., and Saxena S. N. (1982). Role of some sources of organic materials alone and in conjunction with inorganic amendments in reclamation of a typical calcareous saline-alkali soil of Rajasthan. *Agrochimica 26*: 55–63.

Somani, L. L., and Totawat, K. L. (1993). *Management of Salt-Affected Soils and Waters*, Agrotech Publishing Academy, Udaipur (India), p. 384.

Somani, L. L., Joshi, R. S., and Mehta, H. C. (1981). Pyrites to reclaim alkali soils. *Indian Farmer's Digest 14*(7): 26–28.

Starkey, R. L. (1925). Concerning the carbon and nitrogen nutrition of *Thiobacillus thiooxidans*, an autotrophic bacterium oxidizing sulfur under acid conditions. *J. Bacteriol. 10*: 165–193.

Starkey, R. L. (1966). Oxidation of reduced S compounds in soils. *Soils Sci. 101*: 297–306.

Starkey, R. L. and Wright, K. M. (1945). Anaerobic corrosion of iron in soil. *Bull. Techn. Sect. Am. Gas—Assoc.*, New York.

Stromberg, L. K., and Tisdale, S. L. (1979). Treating irrigated arid land soils with acid forming sulfur compounds. Tech. Bull. 21. Sulfur Institute, Washington.

Stumm, W., and Morgan, J. J. (1979). *Aquatic Chemistry*. John Wiley, New York, p. 583.

Tandon, H. L. S. (1987). S containing fertilisers. *Proceedings Symp. Fertilizer S Requirements and Sources in Developing Countries of Asia and the Pacific*, pp. 94–100.

Tiwari, K. N. (1989). *Pyrites, Sulfur Fertilizers for Indian Agricultural*. FDCO, New Delhi, pp. 72–82

Tiwari, K. N., Dwivedi, B. S., Upadhyay, G. P., and Pathak, A. N. (1984). Sedimentary iron pyrites as amendment for sodic soils and carrier of fertiliser S and iron—review. *Fertil. News 29*(10): 31–41.

Tupatkar, P., and Sonar, K. R. (1995). Effect of pyrite on release of Fe, P and S in a calcareous Inceptisol and yield of rice. *J. Indian Soc. Soil Sci. 43*: 696–698.

Tyulenina, L., and Fomina, V. (1978). Chemical improvement of solonetzes by the method of aerosol acidification. *Nauka Tr. Omsk. S—kh Inst. 171*: 8–14.

Umbreit, W. W., Vogel, H. R., and Vogler, K. G. (1942). The significance of fat in sulfur oxidation by *Thiobacillus thiooxidans. J. Bacteriol. 43*: 141–148.

Vakil, P., and Ray, N. (1971). Effect of chemical amendments on reclamation of saline alkali soil of Barewaha tehsil. *Proceedings, All India Symp. Soil Salinity, Kanpur* (India), pp. 161–166.

Vasil Chikova, S. I. (1964). Comparative investigation of the reclaiming effect of chemical substance on soda solonchaks. *Izv. Nauk. 3*(17): 16–20.

Verhoop, A. (1940). Chemische on microbiologische omzettingen van igzersulphiden in den boden *Thesis*, Leiden-Wageningen.

Verma, S. K., and Gupta, R. K. (1984). Effectiveness of pyrites in reclaiming sodic clay soil under laboratory conditions. *Z. Pflanenernahr. Bodenk. 147*: 680–689.

Vitolins, M. I., and Swaby, R. J. (1969). Activity of S oxidixing microorganisms in some Australian soils. *Aust. J. Soil Res. 7*: 171–183.

Vlek, P. L. G., and Lindsay, W. L. (1978). Potential use of finely disintegrated iron pyrite in sodic and iorn-deficient soils. *J. Environ. Qual. 7*: 111–114.

Wainwright, M. (1984). Sulfur oxidation in soils. *Adv. Agron. 37*: 349–394.

Wainwright, M., Nevell, W, and Granyston, S. J. (1986). Potential use of heterotrophic sulfur-oxidizing microorganisms as soil inoculants. *S in Agric. 10*: 6–11.

Waksman, S. A. (1922). Microorganisms concerned in the oxidation of sulfur in the soil. *J. Bacteriol. 7*: 609–616.

Waksman, S. A., and Starkey, R. L. (1923). On the growth and respiration of sulfur oxidizing bacteria. *J. Gen. Phys. 5*: 285.

Waksman, S. A., Wark, C. H., Joffe, J., and Starkey, R. L. (1923). Oxidation of S by microorganisms in black alkali soils. *J. Agr. Res. 24*: 297–305.

Weir, R. G. (1975). The oxidation of elemental S and sulfides in soil. *S in Australasian Agriculture* (K. D. Mclachlan, ed.). Sydney University Press, Sydney, pp. 42–49.

Weir, R. G., Barkus, B., and Atenson, W. T. (1963). The effect of particle size on the availability of brimstone S to white clover. *Aust. J. Exp. Agric. Anim. Husb. 4*: 314–318.

Wiklander, L., Hallgren, G., and Jonsson, E. (1950). Studies on Gyttja soils. III. Rate of sulfur oxidation. *Ann. Royal Agr. Coll. Sweden 17*: 425–440.

Wursten, J. L., and Powers, W. L. (1934). Reclamation of virgin black alkali soils. *J. Am. Soc. Agron. 26*: 752–762.

Yadav, J. S. P. (1977). Reclamation of salt-affected soils. *Proceedings of Indo-Hungarian Seminar on Management of Salt-affected Soils, Karnal* (India), pp. 134–159.

Yahia, T. A., Miyamoto, S., and Stroehlein, J. L. (1975). Effect of surface applied sulfuric acid on water penetration into dry calcareous and sodic soils. *Soil Sci. Soc. Am. Proc. 39*: 1201–1204.

10
Testing Soils for Lime Requirement

Michael N. Quigley *Cornell University, Ithaca, New York*

I. INTRODUCTION

Lime requirement (LR) refers to the amount of lime* required to neutralize the acidic components present in soils of low pH. Raising the soil pH to a satisfactory level for a specific crop has long been recognized as an important way to improve yield, and historical accounts are instructive in this regard (Lathwell and Reid, 1984). Calculations can be performed to determine the amount of lime that needs to be added to a given depth of soil, and in this way LR is defined as the amount of calcium carbonate or equivalent required to adjust 1 acre (0.405 hectares) of soil of specified depth, usually 8 inches (20.3 cm), to a selected pH, typically pH 7.0. It is assumed that 1 acre of soil 8 inches deep weighs 2.40×10^6 pounds (1088.6 metric tons or tonnes, MG).

The optimum soil pH for maximum economic yield is dependent on the type of crop grown. For instance, alfalfa, soybeans, and general vegetable crops grow best in soils of pH 7.0; barley, birdsfoot trefoil, and wheat grow best in soils of pH 6.5; and corn, clovers, grasses, and oats grow best in soils of pH 6.2. Since lime is available in many different forms, it is important to remember that the total neutralizing value—the percentage of material capable of neutralizing acidic components of soil, in terms of an equivalent amount of calcium carbonate—will differ also (Barber, 1984). An in-depth discussion of these and related considerations has been written elsewhere (Reid, 1987).

---

*Lime is commercially available in the form of dolomitic limestone (54% calcium carbonate $CaCO_3$, 46% magnesium carbonate $MgCO_3$), calcitic limestone ($CaCO_3$), burned lime or quicklime (calcium oxide CaO), hydrated lime (calcium hydroxide $Ca(OH)_2$), and slag (calcium silicate $CaSiO_3$).

A. Determination of pH

Important to the calculation of LR is the initial determination of its pH (Eckert and Sims, 1995; Handbook on Reference Methods for Soil Analysis, 1992; Peech, 1965; Grewling and Peech 1960). An appraisal of LR methodologies would not be complete without first reviewing the different means of measurement and media used. For soils that are very slightly acidic, neutral, or basic, there is usually little or no need to add lime to achieve an optimum pH at or above 7.0. In the case of basic soils of course, an acidic substance, such as ammonium sulfate or elemental sulfur, needs to be added instead.

Solutions of pH indicator dyes (Peech, 1965a) such as bromothymol blue ($C_{27}H_{28}Br_2O_5S$, CAS: 76-59-5), chlorophenol red ($C_{19}H_{12}Cl_2O_5S$, CAS: 4430-20-0), bromocresol green ($C_{21}H_{14}Br_4O_5S$, CAS: 76-60-8), phenol red ($C_{19}H_{14}O_5S$, CAS: 143-74-8), and similar substances (Woodruff, 1961) are useful in providing a quick indication of soil acidity. Commercially available portable or pocket pH testers are also finding favor in this application.

In the soil test laboratory, electrodes are typically calibrated versus pH 4.0 and 7.0 buffer solutions for acidic soils, and the repeatability of the procedure is assessed by analyzing several reference soils along with the samples. Most laboratories use automatized or automated pH equipment to analyze the large number of samples that they are required to process, and some have investigated the use of laboratory robots to obtain fast and accurate pH readings (Quigley and Reid, 1995; Brenes et al., 1995; Torres et al., 1993). No matter what type of equipment is used to maneuver the electrodes from sample to sample, soil pH determinations are typically performed in the laboratory by mixing the soil with deionized water in a 1:1 or 1:2 (v/v) ratio; a 5 or 10 cm^3 scoop of soil is usually sufficient, and pH_w values are reported.

An alternative to the use of deionized water is 0.01 M calcium chloride solution. The soil pH_{Ca} values obtained with this medium are independent of dilution and also of the salts present. Providing a constant ionic strength is useful in analyzing a range of soils irrespective of their history or mineralogical content (Eckert and Sims, 1995; Handbook on Reference Methods for Soil Analysis, 1992; Conyers and Darey, 1988; Schofield and Taylor, 1995). Additional features behind the use of 0.01 M calcium chloride solution relate to the similarity in electrolyte concentration between it and a typical nonsaline soil of optimum moisture content, and reduced errors in pH value since flocculation of colloidal clay particles occurs; clogging of reference electrode junctions is always a possibility. Taking all this into account, it is important to note that pH_{Ca} values are approximately 0.7 pH units lower than pH_w values.

Classical soil pH determination in the laboratory relies on two important conditions (Peech et al., 1953): A freshly stirred suspension of soil should be allowed to stand for one hour. An electrode pair (reference and glass indicator) should be used, the bulb of the indicator electrode immersed in the sediment, and the junction of the reference electrode immersed in the supernatant. With regard to the latter, combination electrodes may be used, but those allowing for a suitable placement in the soil–water suspension are preferred. With so many different types of pH electrode available—whether they be variants of the standard electrode pairs or combination electrodes, or the newer variety of field effect transistor—most perform similarly when used only for soil pH determination.

B. Determination of Lime Requirement

Once a soil's acidic nature-has been established, a value for LR can be obtained by following one of a number of different methods. The simplest method is based upon determination of soil pH, with subsequent consultation of a table of lime applications. Cornell Nutrient Analysis Laboratories for instance markets a number of pH indicator soil test kits that come complete with an instruction sheet and tabulated values. These lime recommendations are often found useful by farmers in New York state, but it is recommended that a laboratory test for LR be conducted should the soil pH be found to be less than pH 6.1 (Reid, 1987).

Many articles concerning laboratory based LR determinations have been written (Eckert and Sims, 1995; Handbook or Reference Methods for Soil Analysis, 1992; McLean, 1982; Lathwell and Peech, 1964), and methods are tailored to suit the general types of soils under test. Most methods are chemically based, although organic matter determination has been successfully used as a measure (McLean, 1982; Clark and Nichol, 1966; Kaeney and Corey, 1963). Of the chemical methods, some depend on measuring the buffering capacity of the acidic components found in soil, others on determining the amount of exchangeable hydrogen ions held by the soil particles using a titration procedure, and still others on the determination of exchangeable aluminum (McLean, 1982; Reeve and Sumner, 1970, Kamprath, 1970). The propriety behind use of any of these methods is based upon the type of soil to be analyzed.

The early methods of Woodruff (Woodruff, 1948a, 1948b; McLean et al., 1966), and the later and more accurate methods of Adams and Evans (Eckert and Sims, 1995; Handbook on Reference Methods for Soil Analysis, 1992; McLean, 1982; Hajek, et al., 1972; Adams and Evans, 1962), Yuan (McLean, 1982; Yuan, 1974), and Shoemaker, McLean, and Pratt (Eckert and Sims, 1995; Handbook on Reference Methods for Soil Analysis, 1992; McLean, 1982; McLean et al., 1978; McLean 1975; McLean et al., 1977; McLean, 1978; Shoemaker et al., 1961) are all based on measurement of the change in pH of a buffer solution due to the presence of acidic components in soil. The Adams–Evans and Shoemaker, McLean, and Pratt (SMP) methods have become particularly popular as quick-test methods. Differences in the composition of their buffer solutions relate to the general type of soil they are used to analyze: Shoemaker, McLean, and Pratt's (SMP) original single-buffer method is used with soils containing appreciable amounts of exchangeable aluminum, and that are expected to have LR values greater than 4,480 kg/ha or 2 short tons/acre, 4000 lbs/acre (McLean, 1982; McLean et al., 1966). The premise behind use of the SMP double-buffer method is based on Yuan's method (McLean, 1982; Yuan, 1974) and is preferred for soils having lower anticipated LR values (McLean et al., 1978; McLean et al., 1977). The Adams-Evans method is also used with soils that are anticipated to result in low values for LR (McLean et al., 1978; McLean et al., 1977), cation-exchange capacity, and organic matter content.

One of the ingredients of SMP buffer solution is triethanolamine, and it is this substance that is thought to interfere with pH_{SMP} determination by effectively blocking the reference junction (Brenes et al., 1997; Joyce et al., 1995). Best results appear to be obtainable using either replaceable junction or flow-through junction type reference electrodes (Quigley, unpublished data).

Another way in which LR can be calculated is by determination of the exchange acidity of a given soil. Exchange acidity is a measure of the amount of hydrogen ions that are replaceable by another cation in solution. The classic methods of Mehlich (Handbook on Reference Methods for Soil Analysis, 1992; Mehlich, 1976; Mehlich et al., 1976; Mehlich, 1953; Mehlich, 1943; Mehlich, 1939) and Peech (McLean, 1982; Thomas, 1982; Clark and Nichols, 1966; Peech, 1965b; Peech et al., 1962; Peech and Bradfield, 1948; Dunn, 1943) rely on the determination of exchange acidity and offer an alternative to the Adams–Evans and SMP procedures. The calculation of exchange acidity is influenced by the degree to which the soil is saturated with basic species. This quantity, called base saturation (Lathwell and Peech, 1964; Peech, 1965c), is calculated in percentage terms with reference to Formula (1):

$$\%\ \text{Base saturation} = \frac{\text{Sum of exchangeable bases}}{\text{Cation exchange capacity}} \times 100 \quad (1)$$

One disadvantage of the Mehlich and Peech methods is the rather lengthy equilibration period required. Another disadvantage—common to several of the quick-test methods currently used—is the interefering effect of triethanolamine (see above).

Since most states and provinces in North America now use at least one of the aforementioned methods, very little has been written about LR determinations since the mid-1980s. Those articles and book chapters that have been written are useful, however, because either they offer a concise summary of methods (Eckert and Sims, 1995; Handbook on Reference Methods for Soils Analysis, 1992; Tran and van Lierop, 1993) or they relate to comparative studies that have been performed (Warman et al., 1996; Doerge and Gardner, 1988; Quaggio et al., 1986; Edmeades et al., 1985). The latter nicely complement similar studies of earlier date (Brenes et al., 1997; Tran and van Lierop, 1981).

II. METHODS

Four different methods are reviewed here: the SMP single-buffer and SMP double-buffer methods, the Adams-Evans method, and the Peech method. I defer to the excellent description of Mehlich's method to be found elsewhere (Handbook of Reference Methods for Soil Analysis, 1992).

It is assumed throughout that the soil samples to be analyzed have been dried overnight at 50°C, crushed, and finally passed through a 20-mesh sieve.

A. Shoemaker, McLean, and Pratt's Single-Buffer Method

1. Preparation of Reagents

Prepare SMP buffer solution as follows: Dissolve 16.2 g *p*-nitrophenol ($NO_2C_6H_4OH$, CAS: 100-02-7), 26.0 g potassium chromate (K_2CrO_4, CAS: 7789-00-6), 477.9 g calcium chloride dihydrate ($CaCl_2.2H_2O$, CAS: 10035-04-8) in 4.5 L of deionized water. This is solution 'A'. In another container, dissolve 18.0 g of calcium acetate ($Ca(CH_2COOH)_2$, CAS: 62-54-4) in 2.5 L of deionized water. This is solution 'B'. Add Solution 'B' to Solution 'A', and stir for 2 hours. This is solution 'C'. Add 22.5 mL triethanolamine ($(HOCH_2CH_3)_3N$, CAS: 102-71-6), stir overnight, and dilute to 9 L. This is solution 'D'. Adjust the pH to pH 7.50 ± 0.05 using 10% sodium hydroxide solution. Add four drops of toluene ($C_6H_5CH_3$, CAS: 108-88-3) as preservative, and attach two drying tubes in series through the lid of the container. One tube should be filled with Ascarite™,[2] (or similar substance), the other with Drierite™,[2] (or similar substance) to prevent contamination by CO_2 and moisture.

2. Procedure

Add 10 mL aliquots of deionized water to flasks containing 10 cm^3 scoops of dried, finely ground, and sieved soil. Stir the suspension and then allow to settle. After 1 hour, determine pH_w with a calibrated pH meter. Add 20 mL of SMP buffer solution to each flask and shake for 10 minutes. Allow the suspension to settle for 30–60 minutes and determine the $pH_{SMP(1)}$.

3. Calculation

LR is calculated using Formula (2):

$$\begin{aligned}&[-50.820+(13.378\times pH_d)-(1.928\times pH_d\times pH_{SMP(1)})\\&+(7.315\times pH_{SMP(1)})]\times\left(\frac{10}{M}-SMP\times 8/6\right)\end{aligned} \tag{2}$$

[2]Ascarite is a registered trademark of Thomas Scientific; Drierite is a registered trademark of W. A. Hammond.

where

pH_d = Desired pH of soil (typically pH 7.0)
$pH_{SMP(1)}$ = Soil/Single-buffer SMP pH determined
M-SMP = Mass of soil (g) used for determination

The plow depth is usually taken to be 8 inches (20.3 cm).

Table 1 is a compilation of LR values. The accuracy of Formula (2) can be assessed by substituting in $pH_{SMP(1)}$ data from the table (together with a 5 g sample size) and comparing the values obtained to those given in the second column. Since pure $CaCO_3$ is unlikely to be used by the average farmer, the additional columns in the table make adjustments for agricultural grade limestone (Reid, 1987).

Tabulated values of lime in tonnes/ha can be converted back to Shoemaker, McLean, and Pratt's original short ton/acre values for 2.00×10^6 lbs (907.2 tonnes) of soil using Formula (3):

$$\text{Short tons per acre of lime for } 2.0\times10^6 \text{ lbs of soil}$$

$$= \frac{[(\text{tonnes/ha of lime for } 2.4\times10^6 \text{ lbs of soil})\times 0.8333]}{(1.12\times10^{-3})\times 2000} \quad (3)$$

where

1 short ton = 2000 lbs (1 long ton = 2240 lbs).

B. Shoemaker, McLean, and Pratt's Double-Buffer Method

1. Preparation of Reagents

Reagents are prepared as in Shoemaker, McLean, and Pratt's Single Buffer Method detailed above.

2. Procedure

Perform the single-buffer SMP procedure (as described above) as far as the determination of $pH_{SMP(1)}$. Following the directions of McLean et al. (McLean, 1982; McLean et al., 1978; McLean et al., 1977), to each soil/water suspension, add sufficient hydrochloric acid solution to be equivalent to that required to decrease the pH of 10.0 mL of pH 7.5 SMP buffer solution to pH 6.0 (1.0 mL of 0.206 N(M) hydrochloric acid solution = 0.206 meq). Shake the mixture for a further 10 minutes, allow to settle again for at least 30 minutes, and then determine the pH of the supernatant, $pH_{SMP(2)}$.

Table 1 Amounts of Lime or Ag-ground Limestone in Metric Tons per Hectare[a] Required to Adjust Soil pH to a Desired Level, Based on $pH_{SMP(1)}$, and a 5 g Sample Size[b]

	Desired pH				
	7.0	7.0	6.5	6.0	5.2
$pH_{SMP(1)}$	Pure $CaCO_3$	Ag-ground limestone[c]			
6.8	2.4	3.2	2.7	2.3	1.5
6.7	4.1	5.3	4.7	3.8	2.9
6.6	5.3	7.6	6.5	5.3	4.0
6.5	7.0	10.1	8.5	7.0	5.3
6.4	9.0	12.3	10.5	8.5	6.5
6.3	10.5	14.6	12.3	10.1	7.8
6.2	12.1	16.8	14.3	11.6	9.0
6.1	13.4	19.2	16.1	13.2	10.3
6.0	15.2	21.5	18.1	14.8	11.4
5.9	17.2	23.8	20.1	16.3	12.8
5.8	18.6	26.2	21.9	17.9	13.9
5.7	20.1	28.5	23.9	19.5	15.0
5.6	21.8	30.6	26.0	21.0	16.3
5.5	23.3	33.2	28.0	22.8	17.5
5.4	25.3	35.4	30.0	24.4	18.8
5.3	26.7	37.8	31.8	26.0	19.9
5.2	28.5	40.1	33.8	27.6	21.0
5.1	30.2	42.5	35.8	29.1	22.4
5.0	31.8	44.8	37.8	30.6	23.5
4.9	33.6	47.2	39.9	32.3	24.7
4.8	34.9	49.5	41.6	33.8	26.0

[a]See text for factor to convert back to Shoemaker, McLean, and Pratt's values in short tons/acre.
[b]Adapted from Ref. 15.
[c]Ag-ground limestone of 90% $CaCO_3$ or equivalent, plus total neutralizing power, and fineness of 40% < 100-mesh, 50% < 60-mesh, 70% < 20-mesh, and 95% < 8-mesh.

3. Calculation

LR is calculated using the following formulas (adapted from Handbook of Reference Methods for Soil Analysis, 1992; McLean, 1982; McLean et al., 1978; McLean et al., 1977). Specifically, acidity is calculated using Formula (4), and LR is calculated using Formulas (5), (6), or (7).

$$\text{Acidity }(A) = \Delta pH_{SMP(2)} \times \frac{\Delta d^{\circ}_{SMP(2)}}{\Delta pH^{\circ}_{2}} + \left[\left(\Delta pH_{SMP(1)} \times \frac{\Delta d^{\circ}_{SMP(1)}}{\Delta pH^{\circ}_{SMP(1)}} - \Delta pH_{SMP(2)} \times \frac{\Delta d^{\circ}_{SMP(2)}}{\Delta pH^{\circ}_{SMP(2)}}\right) \times \left(\frac{pH_d - pH_{SMP(2)}}{pH_{SMP(1)} - pH_{SMP(2)}}\right)\right] \quad (4)$$

where

$pH_{SMP(1)}$ and $pH_{SMP(2)}$ = Soil-SMP buffer pH before and after addition of HCl solution
$\Delta H_{SMP(1)}$ = Difference between pH 7.5 and $pH_{SMP(1)}$
$\Delta H_{SMP(2)}$ = Difference between pH 6.0 and $pH_{SMP(2)}$
$\Delta d^{\circ}_{SMP(1)}/\Delta pH^{\circ}_{SMP(1)}$ = Change in acidity per unit change in pH of 10.0 mL of pH 7.5 SMP buffer solution
$\Delta d^{\circ}_{SMP(2)}/\Delta H^{\circ}_{SMP(1)}$ = Change in acidity per unit change in pH of 10.0 mL of pH 6.0 SMP buffer solution
pH_d = Desired pH of soil (typically pH 7.0)

$$\text{LR (tonnes/ha)} = 45.5 - A \quad (5)$$

$$\text{LR (short tons/}2.00\times10^{6}\text{ lb)}^{*} = 16.9A - 0.43 \quad (6)$$

$$\text{LR (short tons/}2.04\times10^{6}\text{ lb)}^{\dagger} = 20.3A - 0.52 \quad (7)$$

4. Alternative Suggested Method of Calculation

A recent report of the SMP double-buffer method (Handbook of Reference Materials for Soil Analysis, 1992) suggests a further method of calculation of LR. Based on a 5 g soil sample, the method calls for calculation of LR using Formula [8]:

$$\text{LR (meq/100 g soil)} = 1.69y - 0.86 \quad (8)$$

where y = 20 A (i.e. the acidity value in meq/5 g soil sample)

LR values in meq/100 g soil are converted to amount of $CaCO_3$ per unit area with reference to Table 2; 1 short ton/acre = 2.2418 tonnes/ha.

C. Adams-Evans' Method

1. Preparation of Reagents

Prepare Adams–Evans buffer solution as follows: Dissolve 20 g *p*-nitrophenol ($NO_2C_6H_4OH$, CAS: 100-02-7) 15 g boric acid (H_3BO_3, CAS: 10043-35-3), 74 g potassium chloride (KCl, CAS: 7447-40-7), and 10.5 g potassium hydroxide (KOH, CAS: 1310-58-3) in 2.0 L of deionized water. Adjust the pH to pH 8.0 ± 0.1, and attach two drying tubes in series through the lid of the container. One tube should be filled with Ascarite™,[2] (or similar substance), the other with Drierite™,[2] (or similar substance) to prevent contamination by CO_2 and moisture.

*Refers to an area of 1 acre (0.405 ha) and a plow depth of 6 two thirds inches (16.7 cm).
†Refers to an area of 1 acre (0.405 ha) and a plow depth of 8 inches (20.3 cm).

Table 2 SMP Conversion Factors for Use in Calculating Amount of Lime per Unit Area: Imperial and SI Units

Amount of $CaCO_3$ required	Multiply LR in meq/100 g
Short tons/acre (8″) for 2.4×10^6 lbs soil	0.6
Tonnes/ha (20 cm) for 1088.6 tonnes soil	1.344
Short tons/acre (6-2/3″) for 2.0×10^6 lbs soil	0.5
Tonnes/ha (16.7 cm) for 907.2 tonnes soil	1.120

Source: Partly drawn and adapted from Ref. 5.

2. Procedure

Add 20 mL aliquots of deionized water to flasks containing 20 cm³ scoops of dried, finely ground, and sieved soil. Stir the suspension and then allow to settle. After 1 hour, determine pH_w with a calibrated pH meter. Add 20 mL of Adams–Evans buffer solution to each flask and stir twice with intervals of 15 minutes; the original method (Adams and Evans, 1962) called for at least 10 minutes. Allow the suspension to settle, and determine the pH_{AE}.

Note that other reports of the Adams–Evans methodology (Eckert and Sims, 1995; McLean, 1982) call for preparation of double strength AE buffer solution (i.e., the same mass of ingredients diluted to 1 L with deionized water, instead of 2 L), with addition to the same volumes of soil and water.

3 Calculation

LR is calculated with reference to table 3 and is based on adjustment of the soil pH to 6.5. Tabulated data was drawn from Adams and Evans' original published paper and has been adjusted to take account of (a) SI units; (b) the difference in the authors' assumed weight of 2.00×10^6 lbs (907.2 tonnes) for 1 acre (0.405 ha) of soil plowed to a depth of 6 two third inches (16.7 cm), and a weight of 2.40×10^6 lbs (1088.6 tonnes) for 1 acre (0.405 ha) plowed to a depth of 8 inches (20 cm) assumed elsewhere (Confeus and Daley, 1988; McLean et al, 1977). The relation between depth of soil and weight of soil is apparent with reference to Formula (9). Since 8 inches (20 cm) of soil weighs 2.40×10^6 lbs (1088.6 tonnes), and x inches of soil weighs 2.00×10^6 lbs (907.2 tonnes), then

$$x \text{ inches soil} = \frac{8 \times 2.0 \times 10^6}{2.4 \times 10^6} \tag{9}$$

$$\text{i.e., } x = 6\tfrac{2}{3} \text{ inches}$$

Adams and Evans (Adams and Evans, 1962) drew attention to the fact that agricultural limestone is not completely effective in neutralizing soil acidity

Table 3 Amounts of Lime in Metric Tons per Hectare[a] Required to Adjust a 6-2/3 Inch (16.7 cm) Furrow Slice of Soil to pH 6.5, Based on pH_w, pH_{AE}, and a 10 g Sample Size

	Soil pH_{AE}									
Soil pH_w	7.0	7.1	7.2	7.3	7.4	7.5	7.6	7.7	7.8	7.9
6.3	2.46	2.21	1.97	1.72	1.48	1.23	0.98	0.74	0.49	0.25
6.1	4.35	3.92	3.48	3.05	2.61	2.18	1.74	1.31	0.87	0.44
5.9	5.86	5.28	4.69	4.1	3.52	2.93	2.34	1.76	1.17	0.59
5.7	7.10	6.39	5.68	4.97	4.26	3.55	2.84	2.13	1.42	0.71
5.5	8.14	6.11	6.51	5.7	4.88	4.07	3.26	2.44	1.63	0.81
5.3	9.03	8.13	7.23	6.32	5.42	4.52	3.61	2.71	1.81	0.90
5.1	9.82	8.84	7.86	6.88	5.89	4.91	3.93	2.95	1.96	0.98
4.9	10.54	9.49	8.43	7.38	6.33	5.27	4.22	3.16	2.11	1.06
4.7	11.23	10.11	8.99	7.86	6.74	5.62	4.49	3.37	2.25	1.12
4.5	11.98	10.78	9.58	8.39	7.19	5.99	4.79	3.59	2.4	1.32

[a]See text for factor to convert back to Adams and Evans' values in lbs/acre.
Source: Adapted from Ref. 25.

(Reid 1987; Handbook of Reference Methods for Soil Analysis, 1992) and cited the work of other researchers (Schollenberger and Salter, 1943; Pierre and Worley, 1928) in suggesting that a factor of 1.5 be used in assessing the figures in the Table.

Tabulated values of lime in tonnes/ha can be converted back to Adams and Evans' original values in lbs/acre for 2.00 × 10[6] lbs (907.2 tonnes) of soil using Formula (10):

$$\text{lbs per acre of lime for } 2.00\times10^6 \text{ lbs of soil} = \frac{(\text{tonnes/ha of lime for } 2.40\times10^6 \text{ lbs of soil})\times0.8333}{1.12\times10^{-3}} \tag{10}$$

4. Alternative Suggested Methods of Calculation

The Council on Soil Testing and Plant Analysis recommend that based on a 10 g soil sample, a 10 mL water addition, and a 10 mL AE buffer addition, LR be calculated using Formula (11) (simplified and adapted from Handbook of Reference Methods for Soil Analysis, 1992):

$$\text{LR CaCO}_3\left(\frac{\text{T}}{\text{A}}\right)=8\left[\frac{8.00-\text{pH}_{\text{AE}}}{\text{pH}_{\text{W}}}\right]\times(\text{pH}_{\text{W}}-\text{pH}_{\text{d}})\times1.5 \tag{11}$$

where

pH_{AE} = Soil/Adams–Evans buffer pH determined
pH_w = Soil/water pH determined
pH_d = Desired pH of soil (typically pH 7.0)

The authors of a recently published comparative study of LR determinations of Maritime Provincial soils recommend that LR values be calculated using Formula (12) (adapted from Warman et al., 1996):

LR (tonnes/ha) to bring soil pH to 6.5 =

$$\left\{\frac{(-0.00011\times pH_{AE})+0.00088}{z}\right\}\times\ (z-0.25)\times 18700 \tag{12}$$

where

$$z=5.55-\frac{\sqrt{30.80-[9.08\times(7.79-pH_W)]}}{4.54}$$

D. Peech's Method

1. Preparation of Reagents

Prepare a solution containing 0.5 M barium chloride and 0.055 M triethanolamine as follows. Dissolve 1100 g of barium chloride dihydrate ($BaCl_2H_2O$, CAS: 10361-37-2) in 10 L of deionized water. Add 148 g of triethanolamine ($(HOCH_2CH_3)_3N$, CAS: 102-71-6) and dilute to 18 L. Use concentrated hydrochloric acid solution to adjust the pH to 8.00 ± 0.02. Attach a drying tube through the lid of the container to prevent contamination by CO_2.

Prepare a mixed indicator solution by dissolving 0.22 g of bromocresol green ($C_{21}H_{14}Br_4O_5S$, CAS: 76-60-8) and 0.075 g methyl red ($(CH_3)_2NC_6H_4N$: NC_6H_4COOH, CAS: 493-52-7) in 100 mL of 95% ethanol (C_2H_5OH, CAS 64-17-5).

Prepare an approximate 0.03 M hydrochloric acid solution by diluting 5 mL of the concentrated acid to 2 L with deionized water. Standardize repeatedly until three concordant results are obtained: Accurately weigh approximately 0.06 g of predried sodium carbonate ($CaCO_3$, CAS: 471-34-1) as primary standard and dissolve in 40 mL of deionized water. If an automated titrimeter is available, titrate with the standardized hydrochloric acid solution to pH 5.1, otherwise titrate manually to a blue-purple endpoint using the mixed indicator.

2. Procedure

Neutralize the acidic compounds in a 1 cm^3 scoop of dried, finely ground, and sieved soil with 25.0 mL of the buffered solution of 0.5 M barium chloride and

0.055 M triethanolamine. Use a 125 mL Erlenmeyer flask with a rubber stopper. Shake the mixture for one hour and allow it to settle overnight. Add 60 mL of deionized water to 5.0 mL of the supernatant and titrate this solution to pH 5.10 with the standardized 0.03 M hydrochloric acid solution, using one drop of mixed indicator solution. Incorporate blank determinations into the analysis using 5.0 mL of buffer solution without added soil. Manual or automated equipment can be used to perform a potentiometric titration as opposed to a visual end-point titration, but in this event greater care must be taken to check on the performance of the pH electrodes used. Recent work has focused attention on the detrimental effect of triethanolamine on the porous junction of the reference electrode (Jocye et al., 1995).

Peech's method defines LR as the number of pounds of $CaCo_3$ required to adjust 1 acre (0.405 ha) of soil and 6 inches (15.2 cm) depth to 100% of base saturation. This corresponds to adjustment of approximately 2.00×10^6 pounds (907.2 tonnes) of soil to approximately pH 8.

3. Calculation

Exchange acidity is calculated using Formula (13). LR is calculated using Formulas (14) and (15).

$$\text{Exchange acidity (meq/100 g)} = \frac{[\text{Blank volume (mL)} - \text{Sample volume (mL)}] \times \text{N HCL soln.} \times 500}{\text{Mass of Soil (g)}} \quad (13)$$

$$\text{Lime requirement (short tons/acre) at pH 8} = \text{Exchange acidity} \times 1000 \quad (14)$$

The following formula corrects for LR at pH7:

$$\text{Lime requirement at pH 7} = \frac{\text{Lime requirement at pH 8} \times (80 - \%\text{BS})}{100 - \%\text{BS}} \quad (15)$$

where % BS refers to the % base saturation of the soil, as estimated from its pH value.

III. CONCLUSION

There are many methods available for the determination of soil lime requirement (LR), but no single method is suitable for all soils under all conditions. In some circumstances, the slower procedures developed by Mehlich and Peech can provide accurate and timely results, but the methods of Shoemaker. McLean, and Pratt and Adams and Evans have found increasing favor due to their being faster processes. In an era when the majority of farmers are cognizant of the importance of soil testing in general, many laboratories rely on the rapid test methods to facilitate and expedite their services. An important consideration in establishing any given LR method is a critical repeatability study. Recent evidence strongly suggests that results obtained from the rapid test methods mentioned can be negatively influenced by the mere choice of electrode used to determine buffer solution pH.

WORLD WIDE WEB

Further details regarding the determination of lime requirement can be found by consulting the home page of Cornell Nutrient Analysis Laboratories on the World Wide Web: the URL is

http://www.cals.cornell.edu/dept/cnal/

ACKNOWLEDGMENTS

A number of people have been involved with methods development into lime requirement determination at Cornell Nutrient Analysis Laboratories. My work has been built largely on the efforts of Dr. Gregory Ferguson, Dr. W. Shaw Reid, Ms. Anna Bunnell, and Ms. Kathy McCracken.

REFERENCES

F. Adams and C. E. Evans. A rapid method for measuring lime requirement of red-yellow podzolic soils. *Soil Sci. Soc. Am. Proc. 26*: 355 (1962).

S. A. Barber. Liming materials and practices. *Soil Acidity and Liming*, 2d ed. (F. Adams, ed.). Agronomy Monograph No. 12, ASA-CSSA-SSSA, Madison, WI, 1984, pp. 171–209.

N. Brenes, M. N. Quigley, and W. S. Reid. Determination of soil pH using a laboratory robot. *Analytica Chimica Acta 310*: 319 (1995).

N. Brenes, M. N. Quigley, and W. S. Reid. Performance characteristics in the determination of soil-water pH and soil-SMP buffer pH using a flow-through junction combination electrode. *Commun. Soil Sci. Plant Anal. 28:* 759 (1997).

J. S. Clark and W. E. Nichol. The lime potential, percent base saturation relations of acid surface horizons of mineral and organic soils. *Can. J. Soil Sci. 46*: 281 (1966).

M. K. Conyers and B. G. Darey. Observations in some routine methods for soil pH determination. *Soil Sci. 145*: 29 (1988).

T. A. Doerge and E. H. Gardner. Comparison of four methods for interpreting the Shoemaker–McLean–Pratt (SMP) lime requirement test. *Soil Sci. Soc. Am. J. 52*: 1054 (1988).

L. E. Dunn. Lime-requirement determination of soils by means of titration curves. *Soil Sci. 56*: 341 (1943).

D. Eckert and J. T. Sims, Recommended soil pH and lime requirement tests. *Recommended Soil Testing Procedures for the Northeastern United States* (J. T. Sims and A. Wolf, eds.). Northeastern Regional Publication No. 493, University of Delaware, Newark, 1995, pp. 16–21.

D. C. Edmeades, D. M. Wheeler, and J. E. Waller. Comparison of methods for determining lime requirement of New Zealand soils. *N. Z. J. Agr. Res. 28*: 93 (1985).

T. Grewling and M. Peech. *Chemical Soil Tests*, Cornell Univ. Agric. Exp. Stn. Bull. 960, 1960.

B. F. Hajek, F. Adams, and J. T. Cope. Rapid determination of exchangeable bases, acidity and base saturation for soil characterization. *Soil Sci. Soc. Am. Proc. 36*: 436 (1972).

Council on Soil Testing and Plant Analysis. *Handbook on Reference Methods for Soil Analysis*. Athens, GA, 1992, pp. 51–69.

M. F. Joyce, P. A. Langan, M. N. Quigley, and W. S. Reid. A robotic procedure for the determination of soil lime requirement. *Commun. Soil Sci. Plant Anal. 26*: 3385 (1995).

E. J. Kamprath. Exchangeable aluminum as a criterion for liming leached mineral soils. *Soil Sci. Soc. Am. Proc. 34*: 252 (1970).

D. R. Keeney and R. B. Corey. Factors affecting the lime requirements of Wisconsin soils. *Soil Sci. Soc. Am. Proc. 27*: 277 (1963).

D. J. Lathwell and M. Peech. *Interpretation of Chemical Soil Tests*. Bulletin 995, Cornell University, Ithaca, NY, 1964.

D. J. Lathwell and W. S. Reid. Crop response to lime in the Northeastern United States. *Soil Acidity and Liming*, 2d ed. (F. Adams, ed.). ASA-CSSA-SSSA, Agronomy Monograph No. 12, 1984, pp. 305–332.

E. O. McLean. Recommended pH and lime requirement tests. *Recommended Chemical Soil Test Procedures for the North Central Region*. North Central Regional Pub. 221: 6 (1975).

E. O. McLean. Principles underlying the practice of determining lime requirements of acid soils by use of buffer methods. *Commun. Soil Sci. Plant Anal. 9*: 699 (1978).

E. O. McLean. Soil pH and lime requirements. *Methods of Soil Analysis: Chemical and Microbiological Properties*, 2d ed. (A. L. Page, ed). ASA, SSS, Madison, WI, 1982, pp. 199–224.

E. O. McLean, S. W. Dumford, and F. Coronel. A comparison of several methods of determining lime requirement of soils. *Soil Sci. Soc. Am. Proc. 30:* 26 (1966).

E. O. McLean, D. J. Eckert, G. Y. Reddy, and J. F. Trierweiller. An improved SMP soil lime requirement method incorporating double-buffer and quick-test features. *Soil Sci. Soc. Am. J. 42*: 311 (1978).

E. O. McLean, J. F. Trierweiler, and D. J. Eckert. Improved SMP buffer method for determining lime requirement of acid soils. *Commun. Soil Sci. Plant Anal. 8*: 667 (1977).

A. Mehlich. Use of triethanolamine acetate-barium hydroxide buffer for the determination of some base exchange properties and lime requirement of soil. *Soil Sci. Soc. Am. Proc. 3*: 162 (1939).

A. Mehlich. The significance of percentage base saturation and pH in relation to soil differences. *Soil Sci. Soc. Am. Proc. 7*: 167 (1943).

A. Mehlich. Rapid determination of cation and anion exchange properties and pHe of soils. *J. Assoc. Off. Agric. Chem. 36*: 445 (1953).

A. Mehlich. New buffer pH method for rapid estimation of exchangeable acidity and lime requirement of soils. *Comm. Soil Sci. Plant Anal. 7*: 637 (1976).

A. Mehlich, S. S. Bowling, and A. L. Hatfield. Buffer pH acidity in relation to nature of soil acidity and expression of lime requirement. *Comm. Soil Sci. Plant Anal. 7*: 253 (1976).

M. Peech. Exchange acidity. *Methods of Soil Analysis Part 2. Chemical and Microbiological Properties* (C. A. Black, ed.). ASA, Madison, WI, 1965, pp. 905–913.

M. Peech, Hydrogen ion activity. *Methods of Soil Analysis, Part 2. Chemical and Microbiological Properties* (C. A. Black, ed.). ASA, Madison, WI, 1965, pp. 914–926.

M. Peech. Lime requirement. *Methods of Soil Analysis Part, 2. Chemical and Microbiological Properties* (C. A. Black, ed.). ASA, Madison, WI, 1965, pp. 927–932.

M. Peech, R. L. Cowan, and J. H. Baker. A critical study of the $BaCl_2$-triethanolamine and the ammonium acetate methods for determining the exchangeable hydrogen content of soils. *Soil Sci. Soc. Am. Proc. 26*: 37 (1962).

M. Peech and R. Bradfield. Chemical methods for estimating lime needs of soils. *Soil Sci. 65*: 35 (1948).

M. Peech, R. A. Olsen, and G. H. Bolt. The significance of potentiometric measurements involving liquid junction in clay and soil suspensions. *Soil Sci. Soc. Am. Proc. 17*: 214 (1953).

W. H. Pierre and S. L. Worley. The buffer method and the determination of exchangeable hydrogen for estimating the amounts of lime required to bring soils to definite pH values. *Soil Sci. 26*: 363 (1928).

M. N. Quigley and W. S. Reid. Comparison of a combination pH electrode and field effect transistor electrode for the determination of soil pH and lime requirement. *Commun. Soil Sci. Plant Anal. 26*: 3371 (1995).

J. A. Quaggio, B. Van Raij, and E. Malavolta. Alternative use of the SMP-buffer solution to determine lime requirement of soils. *Comm. Soil Sci. Plant Anal. 16*: 245 (1985).

N. G. Reeve and M. E. Sumner. Lime requirements of Natal Oxisols based on exchangeable aluminum. *Soil Sci. Soc. Am. Proc. 34*: 595 (1970).

W. S. Reid. Soil fertility management. *Cornell Field Crops and Soils Handbook*, 2d ed. (W. J. Cox and S. D. Klausner, eds.). New York State College of Agriculture and Life Sciences, Ithaca, NY, 1987, pp. 27–50.

R. K. Schofield and A. W. Taylor. The measurement of soil pH. *Soil Sci. Soc. Am. Proc. 19*: 164 (1955).

C. J. Schollenberger and R. M. Salter. A chart for evaluating agricultural limestone. *J. Am. Soc. Agron. 35*: 955 (1943).

H. E. Shoemaker, E. O. McLean, and P. F. Pratt. Buffer methods for determining lime requirement of soils with appreciable amounts of extractable aluminum. *Soil Sci. Soc. Am. Proc. 24*: 274 (1961).

G. W. Thomas. Exchangeable cations. *Methods of Soil Analysis, Part 2. Chemical and Microbiological Properties*, 2d ed. (A. Page, R. H. Miller, and D. R. Keaney, eds.). American Society of Agronomy, Madison, WI, 1982, pp. 159–165.

P. Torres, J. A. Garcia-Mesa, M. D. Luque de Castro, and M Valcarel. Determination of soil pH by use of a robotic station. *Fresenius' J. Anal. Chem. 346*: 704 (1993).

T. S. Tran and W. van Lierop. Evaluation and improvement of buffer-pH lime requirement methods. *Soil Sci. 131*: 178 (1981).

T. S. Tran and W. van Lierop. Lime requirement. *Soil Sampling and Methods of Analysis* (M. R. Carter, ed.) Lewis Pub., Boca Raton, FL, 1993, pp. 109–113.

P. R. Warman, B. Harnish, and T. Muizelaar. A lime requirement test for Maritime Canada, and response time and effect of liming source on soil pH. *Commun. Soil Sci. Plant Anal. 27*: 1427 (1996).

C. M. Woodruff. Determination of the exchangeable hydrogen and lime requirement of the soil by means of the glass electrode and a buffered solution. *Soil Sci. Am. Proc. 12*: 141 (1948).

C. M. Woodruff. Testing soils for lime requirement by means of a buffered solution and the glass electrode. *Soil Sci. 66*: 53 (1948).

C. M. Woodruff. Brom cresol purple as an indicator of soil pH. *Soil Sci. 91*: 272 (1961).

T. L. Yuan. A double buffer method for the determination of lime requirement of acid soils. *Soil Sci. Soc. Am. J. 38*: 437 (1974).

11

Liming to Improve Chemical and Physical Properties of Soil

L. Darrell Norton and X. C. (John) Zhang *Agricultural Research Service, U.S. Department of Agriculture, Purdue University, West Lafayette, Indiana*

I. INTRODUCTION

Soil acidification is a natural process in soils from both acidic and basic parent materials. Soils usually become more acidic with time due to weathering. More weathered soils are generally more acidic. This process is often exacerbated by human activities such as intensive use of acid fertilizers and acid rains. As a result, soil acidity, continues to be a problem and is increasingly becoming a limiting factor for agricultural production in many areas. Excessive soil acidity must be neutralized in order to attain a desired soil pH and soil environment for better nutrient availability and crop growth.

Agricultural liming materials typically are materials whose calcium and magnesium compounds are capable of neutralizing soil acidity, or their accompanying anions are capable of reducing the activities of hydrogen and aluminum in the soil solution. The term "lime" correctly refers to calcium oxide (CaO) or quicklime, but in general it has been used in reference to limestones or liming materials in agricultural literature. By convention, the latter sense will be used in this chapter. Materials commonly used in agricultural land include oxides, hydroxides, carbonates, silicates of calcium and/or magnesium, and industrial by-products. In recent years, by-product gypsum has been found promising in alleviating subsoil acidity.

Liming soil to improve crop yields has a long history in the United States. Early settlers used marl to amend acid soil from the beginning of 19th century, and the positive response of crop yield was reported (Ihde, 1952). Large-scale experimental research on lime was conducted between 1880 and 1910 at agricultural experiment stations in several states (Barber, 1984). Favorable response of crop yield to liming at these stations stimulated the wide use of lime on agricultural land from the beginning of 20th century, with the fastest increase between 1935 and 1945, in which a tenfold increase in lime use was recorded. This rapid increase in usage was partially attributed to federal subsidy programs developed to alleviate land deterioration due to soil acidification. Since the 1950s, lime use has experienced ups and downs caused by changes in federal subsidy funding and in cropping management, but the increasing trend has been steady.

Application of lime to acid soils has increased crop productivity considerably in many areas of the United States. The effects of liming on crop growth and soil properties, though complicated and sometimes situation dependent, can be generalized as 1) neutralizing soil acidity and reducing the toxicities of Al^{3+}, Mn^{2+}, and H^+; 2) increasing the supply of Ca and/or Mg if dolomite is used; 3) increasing CEC (cation exchange capacity) and decreasing AEC (anion exchange capacity) for variable charge components in soils; 4) changing the adsorbed and desorbed chemical compositions in both solid and solution phases; 5) influencing the availability of phosphorus and micronutrients; 6) increasing microbial activity, especially improving nitrogen mineralization and fixation; 7) influencing soil physical properties by improving soil structure and soil aggregation; 8) in general, promoting reactions that produce H^+ and retarding reactions that consume H^+; and 9) affecting the activity of certain plant pathogens through the change of soil pH. These effects can be either beneficial or detrimental to plant growth, depending upon the situation, and often occur simultaneously and interdependently.

The amount of lime required to create a favorable environment for plant growth is highly situation specific, depending largely on soil properties and crop characteristics. It should also be influenced by economic feasibility and profitability. Not all acid soils need lime, nor should all lime-responsive soils be limed to the same pH level. In general, since interactions and intercompensations exist among the factors affected by liming and vary with environmental conditions, target pH even for the same crop may differ from one scenario to another. For highly weathered Oxisols and Ultisols in which the hydrous oxides of Al and Fe are predominant, a soil pH of about 5.5 with exchangeable Al being <20% of effective CEC would produce satisfactory plant growth (Adams, 1984); while liming to pH 6.5 to 6.8 is desirable for Alfisols and Mollisols (McLean and Brown, 1984). It is also recommended that for highly weathered soils lime sufficient to neutralize exchangeable aluminum is all that is needed. Regardless which method is used, the general consensus is that lime recommendation must be made based on calibrated crop response curves, or differences between soils and among crops must be considered.

This chapter will briefly discuss the chemical and physical properties of acid soils and the consequential effects of liming on soil properties. Liming materials, lime requirements, and liming practices will be introduced. The use of by-product gypsum for alleviating subsoil acidity and for ameliorating soil physical and chemical properties will also be covered.

II. CHEMICAL PROPERTIES OF ACID SOILS

Acid soil in this chapter is defined as one having a pH below 7. Soil acidity is the dominating factor of acid soils, which influences almost every aspect of the soil system. Total soil acidity is divided into active acidity and potential acidity. Active acidity is a measure of H^+ concentration or activity in soil solution, which is well represented by soil pH. Soil pH is defined as the negative logarithm of the H^+ activity in soil solution. Active acidity, which is presumably in equilibrium with potential acidity, is also referred to as the intensity factor of soil acidity. Potential acidity is the fraction of total acidity that needs to be neutralized in order to achieve a selected soil pH. It is well reflected by the liming requirement and therefore is often referred to as the capacity factor. Potential acidity can be further separated into exchange acidity and titratable acidity. Exchange acidity is the acidity that can be replaced from exchange sites with neutral salts. Titratable acidity is the fraction that can be neutralized by titration at a selected pH such as that produced by Al hydrolysis and dissociation of H^+ from organic and clay mineral compounds.

A. Formation of Soil Acidity

Soil acidity is determined by soil components and the nature and extent of chemical and biological reactions in the soil systems, and is largely controlled by the degree of weathering. According to generally held views, several sources were found to be responsible for soil acidification. Weathering and dissolution of parent materials by hydrolysis of CO_2 followed by leaching of basic cations (Na, Ca, and Mg) with bicarbonate, is the dominant soil acidification process in nature. Weathering of acid parent materials or oxidation of acid minerals produces additional soil acidity. Organic matter or humus contains reactive carboxylic and phenolic groups that act as weak acids and dissociate H^+ through deprotonation. It is also a potential source of the stronger acid HNO_3, produced by nitrification of organic nitrogen. Inorganic soil components such as aluminosilicate clay minerals and sesquioxides produce titratable acidity through deprotonation from hydroxyl groups at the clay surfaces and edges. Aluminum and iron hydrolyses are often major sources of soil acidity for highly weathered soils. Acid chemical fertilizers can also produce a considerable amount of acid by nitrification of ammonium fertilizer materials. In addition, removal of basic cations by plants, redox reactions, and acid rain may also cause soil acidification to a certain degree.

B. Significance of Soil pH

The pH of acid soils is ordinarily between 4 and 7. It is seldom below 4 unless free acids are present due to oxidation of sulfides. As an intensity factor of soil acidity, soil pH is a master indicator of soil systems and influences many of the chemical and biological processes occurring in soils, though it produces no hint about the total soil acidity. The pH is a good indicator of Al toxicity, which is generally the most limiting factor for crop production in acid soils. Severe problems are normally encountered when pH is below 5, because the most toxic species of Al^{3+} predominates in soil solution. No serious problems exist at pH > 5.5, particularly in highly weathered soils. The pH also indicates the solubilities of most chemical elements in soils and their associated availability, deficiency, or toxicity to plant growth. At low pH, Al and Mn become more soluble and can be toxic to plants. Solubilities of most micronutrients except for Mo decrease as pH increases, and deficiency symptoms are often seen when pH is greater than 7. The degree of Al hydrolysis is controlled by soil pH, which further affects total soil acidity and the level of Al toxicity to plants. Soil pH also has a significant impact on soil microbial activity. It has been shown that the rates of mineralization and nitrification increase with the increase of pH up to 7 (Alexander, 1980).

The pH-dependent characteristics of element solubility and chemical reactions influence chemical compositions and concentrations of exchangeable and soil solution cations. As pH increases, the Al saturation percentage normally decreases, while basic metal cation saturation increases. This has significant effects on clay flocculation, aggregate stability, and soil surface sealing. A decrease in exchangeable Al^{3+} due to an increase in soil pH tends to induce clay dispersion when electrolyte concentration in solution is low. For variable charge soils, negative charges at the clay edges increase through deprotonation or dissociation of H^+ from hydroxyl groups as soil pH increases (charge reversal). This charge reversal reduces edge-to-face flocculation and therefore promotes clay dispersion (van Olphen, 1977). Enhanced clay dispersion after liming is more likely to occur in highly weathered, acid soils where Al^{3+} is the dominant exchangeable cation.

C. Cation Exchange Capacity of Soil

Cation exchange capacity (CEC) is largely determined by total negative charge of soils. Total negative charge is primarily composed of permanent charge and variable or pH-dependent charge. Permanent charge, which is independent of environmental conditions such as soil pH under which it is characterized (Fig. 1), is formed from isomorphous substitution of higher valency cations by lower valency cations within the clay lattice. This type of charge varies greatly with clay minerals and changes with degree of weathering, with 2:2 and 2:1 clay > 1:1

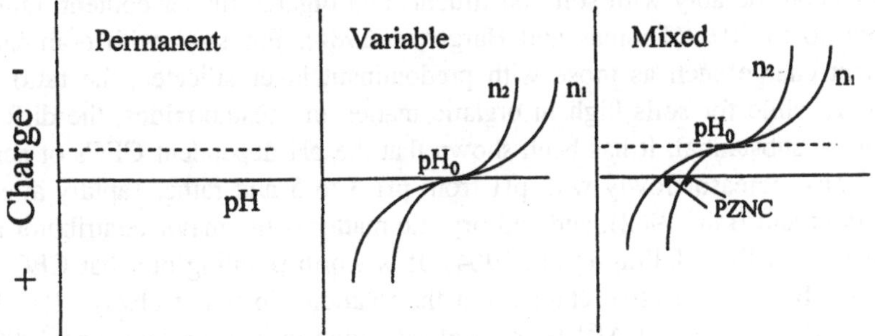

Figure 1 pH dependence of surface charge at two electrolyte concentrations for permanent, variable, and mixed charge systems. (After Sumner, 1992.)

clay > R_2O_3. Negative variable charge is caused by deprotonation from carboxylic and phenolic groups of organic matter or hydroxyl groups of clay and sesquioxides. This type of charge is dependent upon soil conditions such as soil pH, electrolyte concentration, and type of cation (Fig. 1). Soil pH at which net variable charge vanishes is called pH_0 (Uehara and Gillman, 1981). Net variable charge is negative above pH_0 and becomes positive below pH_0. Organic matter has a pH_0 around 3, and pH_0 of sesquioxides is between 8 and 9. In the pH range between 4 and 7, organic matter is negatively charged, while sesquioxides are positively charged. On a weight basis, organic matter by far has more charges than either layer silicates or sesquioxdes or complexes of both. In addition, negative charge produced by specific adsorption or exchange of ligands at the clay edges can be significant under certain circumstances. It also should be kept in mind that soil is a mixture of various types of clay minerals and organic matter, so that charge interactions and intercompensations may exist and should be considered when interpreting charge related experimental data.

Since CEC is largely dependent on total negative charge of soils, all factors that alter total negative charge affect CEC as well. Therefore CEC of a variable charge soil is also dependent on soil pH and electrolyte concentration under which it is determined. Due to the pH dependent nature of CEC, effective CEC, which is measured at soil pH by unbuffered neutral salt solutions such as KCl or $CaCl_2$, is preferred for soils with variable charges such as highly weathered acid soils, because it reflects more intrinsic cation holding capacity of the soil. In comparison, the other widely used methods such as $BaCl_2$-TEA (triethanolamine) at pH 8.2 and NH_4OAc at pH 7 tend to overestimate cation holding capacity. However, the CECs determined with these buffered methods can be used as standard references, because the experimental conditions are well controlled in these methods. The percent ratio of effective CEC to that at pH

8.2 varies considerably with soil constituents and organic matter content, ranging from 30 to 70% (Thomas and Hargrove, 1984). For soils with dominant permanent charge such as those with predominant layer silicates, the ratio is close to 1, while for soils high in organic matter and sesquioxides, the difference can be substantial. It has been shown that the pH dependent CECs of acid surface soils increase slowly with pH from pH 3 to 5 and rather rapidly afterwards (Pratt and Bair, 1962), and soil organic matter is the major contributor to pH dependent CEC (Helling et al., 1964). It is worth pointing out that CEC is determined by total negative charge, but the relationship is not always 1:1. At very low pH, exchange of Al^{3+} by neutral salt solution is often less than 100% because Al^{3+} is held tightly on adsorption sites due to large charge-to-size ratio. More importantly, complexation of Al^{3+} with organic matter, which makes it unexchangeable, counters some charges. As pH increases, Al^{3+} hydrolyzes, and the resultant hydroxyl-Al that is not exchangeable by neutral salt (Chernov, 1964) still counters some charges. However, more sites are expected to be liberated as Al hydrolysis continues. This may explain Pratt and Bair's (1962) observation that rapid increase of CEC occurred at pH > 5. Besides, the negative charge, which contributes to charge deficit in the electric double layer due to charge repelling (this will be covered later), is not accounted for by CEC measures.

Percent base saturation with Ca^{2+} and Mg^{2+} is another indicator of soil acidity and has significant implications to soil physical properties as well. Since the removal of these basic cations with bicarbonate from CO_2 hydrolysis is the major process causing soil acidification in nature, percent base saturation is inversely related to the degree of soil acidification. In contrast, percent saturation of Al^{3+}, which is the major source of exchangeable acidity, increases as acidity increases. On a basis of CEC at pH 8.2, percent base saturation at pH 7 was found to be around 80% for less weathered soils with predominant 2:1 and 2:2 clay minerals; and it was between 30 and 50% for organic matter as well as highly weathered soils with predominant kaolinite and oxides (Mehlich, 1942, 1943). Percent base saturation on an effective CEC basis may be more meaningful, but the problem of using effective CEC lies in the difficulties in obtaining reliable data due to large experimental variations.

The Ca^{2+}, Mg^{2+}, and Al^{3+} ions are the dominant cations exchangeable by neutral salts in acid soils. At pH < 5, Al^{3+} typically becomes predominant. Exchangeable H^+ is relatively small simply due to less attraction or affinity of adsorption sites to a lower valent cation (Coleman et al., 1959). This is supported by the fact that hydrogen-saturated montmorillonite is still dispersive and shows considerable swelling. Chemical composition of adsorbed and/or dissolved exchangeable cations influences not only chemical properties but also physical properties of an acid soil. It dictates clay behaviors (dispersion or flocculation), which further influence aggregate stability, formation of surface seals, water infiltration, and soil erosion. Chemical composition of adsorbed cations tended to affect clay behaviors more than that of soil solution (Miller et al., 1990). The flocculation power of a cation is positively related to its valence but negatively to its size. In general, the flocculation power of metal cations follows the order of $Al^{3+} > Ca^{2+} > Mg^{2+} > K^{+} > Na^{+}$. At pH < 5, predominant Al^{3+} keeps clay flocculated and aggregates stable. Shifting from Al^{3+} to Ca^{2+} domination can cause a variable degree of clay dispersion for some soils, depending on the ionic strength of soil solution. This will reduce hydraulic conductivity and increase erosion due to clogging of conductive pores by dispersed clay particles (Norton et al., 1993). On the other hand, substitution of monovalent cations and/or Mg^{2+} by Ca^{2+} promotes clay flocculation and therefore increases water infiltration. Clay flocculation can be further enhanced by elevated electrolyte concentration after liming. Contrarily, the deterioration of soil physical properties was observed when Ca^{2+} was substituted by Mg^{2+}. Therefore for achieving better soil physical properties, calcitic limestones are superior to dolomitic limestones.

D. Element Toxicity and Deficiency

Soil pH affects the solubilities and mobilities of most elements and therefore their toxicities or deficiencies to plants. Growth limiting factors that are commonly encountered in acid soils are toxicities of Al^{3+} and Mn^{2+} (H^{+} can be a problem at pH < 4) and deficiencies of Ca, Mg, P, and Mo. Aluminum toxicity, determined by its activity in the soil solution, is probably the most limiting factor for plant growth in acid soils. The toxicity becomes very severe at pH < 5 and can be a problem as high as pH 5.5 for kaolinite soils (Foy, 1984). The toxicity is often exacerbated by low levels of exchangeable Ca (Kamprath, 1984). Manganese toxicity is the second most limiting factor for plants in acid soils. It often occurs at pH ≤ 5.5. The degree of toxicity is also affected by total Mn and soil organic matter contents. Organic matter reduces Mn and Al toxicities by chelating excessive Al^{3+} and Mn^{2+} from soil solution.

Calcium deficiency is often associated with Al toxicity in acid soils due to excessive leaching of Ca^{2+} and higher exchange power of Al^{3+} relative to Ca^{2+}. Also, physiologically the antagonistic effect of Ca on Al alleviates Al^{3+} toxicity when Ca^{2+} levels are high, and the opposite is true when Ca^{2+} levels are low. This antagonistic effect further strengthens the link between Al toxicity and Ca deficiency. Similar to Ca^{2+}, excessive leaching loss of Mg^{2+} causes Mg deficiency to plant growth in some acid soils. This is often true in sandy soils due to the small cation holding capacity after liming soil with low Mg content materials. Replacement of Mg^{2+} by Ca^{2+}, followed by subsequent leaching, is the mechanism causing Mg deficiency. Phosphorus and Mo deficiencies are frequently encountered in acid soils. The formation of an insoluble Al and Fe phosphate at low pH and decreasing solubility of Mo with decreasing soil pH reduce their availability to plants.

III. LIME USE

A. Mechanisms of Neutralization

Carbonates, oxides, and hydroxides of Ca or Ca and Mg are commonly used as liming materials. The reactions of liming materials with acid soils are complex. Among them are carbonate and Al^{3+} hydrolyses, ion exchange, free H^+ neutralization, and ionization of hydrogen from functional groups of organic matter, layer silicates, and sesquioxides. Basic hydrolysis reactions of calcium and magnesium are

$$CaCO_3 + H_2O \rightarrow Ca^{2+} + HCO_3^- + OH^-$$

$$MgCO_3 + H_2O \rightarrow Mg^{2+} + HCO_3^- + OH^-$$

The hydration of calcium oxide is

$$CaO + H_2O \rightarrow Ca^{2+} + 2OH^-$$

The formed OH^- neutralizes free H^+ from soil solution as

$$OH^- + H^+ \rightarrow H_2O$$

The neutralization of soil solution H^+ will drive the sequential hydrolysis of Al^{3+} and its hydroxyl compounds. The generalized overall reaction may be written as

$$2Al^{3+} + 3CaCO_3 + 3H_2O \rightarrow 3Ca^{2+} + 2Al(OH)_3 + 3CO_2$$

The final products are precipitated $Al(OH)_3$ and exchangeable Ca^{2+} or/and Mg^{2+} if dolomite is used. Exchangeable sites are usually predominantly occupied by Ca^{2+} after the reaction of lime with acid soils. Abundance of Ca^{2+} on exchange sites and in soil solution, coupled with elevated ionic strength, keeps clay particles flocculated and soil structure stable. Under most circumstances, this will result in improvements in soil aggregation and hydraulic conductivity, and a reduction in soil erosion by water.

B. Liming Materials

Various liming materials have been used, but the most abundant are ground carbonates of Ca and Mg. The neutralization effectiveness of a liming material is evaluated by its $CaCO_3$ equivalent on a percent weight basis, called neutralizing value. It is defined as the amount of $CaCO_3$ required to equal the acid neutralization power of 100 g of the material.

Agricultural limestone is usually a mixture of calcite (pure $CaCO_3$) and dolomite (having equal molar ratio of $CaCO_3$ and $MgCO_3$). Usually, limestone is divided into calcitic and dolomitic based on the Mg content. Calcitic limestones are those containing little Mg, whereas dolomitic limestones include those containing a "substantial" amount of Mg, ranging from pure dolomite to those containing a few grams per kilogram of Mg (Barber, 1984). Normally, the neutralizing values of calcitic limestones range from 65% to a little more than 100%, depending on their impurities and Mg content. Pure dolomite has a neutralizing value of 108% due to the lower molecular weight of $MgCO_3$.

Calcium oxide (quicklime) is produced by burning calcitic limestone, and its quality depends on the raw materials. Pure CaO has a neutralizing value of 179%. It hydrates and reacts rapidly with soil and therefore may invoke temporary problems caused by high pH conditions. Calcium hydroxide [$Ca(OH)_2$] is produced rapidly by hydrating CaO. It is also called slaked lime or hydrated lime and has a neutralizing value of 136%. Marl is soft, unconsolidated calcium carbonate, usually mixed with clay. It is mostly low in Mg and has neutralizing values ranging between 70 and 90%, depending on clay content.

By-products of various slags have been used because of their liming and additional nutrient values. Three types of slags are commonly used. 1) Blast-furnace slags, which mainly consist of Ca and Mg silicates, are produced in the manufacture of pig iron. These slags usually have a high content of Mg. The neutralizing values range from 80 to 100% (Barber, 1984). Field tests have shown that these slags were as effective as ground limestones if applied on an equivalent basis. 2) Basic slag is produced in making steel from pig iron. It has a neutralizing value of about 60 to 70% (Tisdale et al., 1985) and is high in phosphorus. The phosphorus is from high-phosphorus iron ores and is readily available to plants. 3) Electric-furnace slag is produced in the preparation of elemental phosphorus from phosphorus rocks, or in the manufacture of iron and steel. It is mainly composed of calcium silicates. The slag from phosphorus production contains 0.9 to 2.3% of P_2O_5. Its neutralizing value ranges from 65 to 80% (Tisdale et al., 1985).

Recently there has been a return of interest in utilizing coal combustion by-products as soil amendments as an alternative use of these materials simply because the cost of landfill is rapidly increasing (Norton, 1995). Coal combustion by-products include fly ashes, bottom ashes, cyclone slags, spent bed materials from fluidized bed combustion (FBC), and flue gas desulfurization (FGD) sludges. Chemical and physical properties of these materials vary dramatically, depending on the type of coal and the burning technology. Fly ashes are divided into highly alkaline and acid-neutral. Alkaline fly ash has been tested for use as a liming material. A study by Phung et al. (1978) showed that fly ash had a neutralizing value of about 20%. The ash was effective in reducing soil acidity and supplying calcium to plants, but it may cause boron toxicity to plants if applied in large quantities. Many FBC and FGD materials contain considerable amounts of calcium sulfates, ranging from 75 to 100% by weight. Some materials contain calcium oxide or calcium carbonate as high as 35%. Some of these materials have been evaluated for potential use as liming agents. Korcak (1980) conducted a greenhouse study with FBC that consisted of 33% CaO and found that the neutralizing value of the material is about 20%. Another greenhouse study by Terman et al. (1988) with FBC waste that contained 33% CaO showed that fine waste materials were 47% as effective as fine $CaCO_3$ during a 5 week period of study for increasing pH.

By-product gypsum is the predominant mineral of FGD materials produced with forced air oxidation. It has an equilibrium pH value of around 6.7. Recent studies have shown that gypsum is effective in reducing Al^{3+} toxicity to plants, especially in alleviating subsoil acidity, owing to its higher solubility and mobility relative to limestones (Sumner, 1993). Besides the beneficial chemical effects, FBC and FGD materials have been widely evaluated and used for improving soil physical properties and preventing soil surface sealing. These materials have been shown to prevent clay dispersion, increase aggregate stability and water infiltration, and reduce soil surface sealing and erosion (Norton et al., 1993; Reichert and Norton, 1994; Zhang and Miller, 1996; Miller, 1987; Shainberg et al., 1989).

Other materials used as liming agents are flue kiln dust from cement plants, water treatment lime, sludge from pollution control systems, and refuse materials from sugar beet factories, paper mills, calcium carbide factories, and so on. Most of these materials have a high $CaCO_3$ equivalent and are usually used in local areas close to their sources. The neutralizing values and contents of trace elements may vary widely depending on the material. Research results regarding these materials are not well documented.

C. Kinetics of Neutralization

The final products of liming are exchangeable Ca^{2+}, precipitated $Al(OH)_3$ or/and $Fe(OH)_3$, and more favorable environments for plant growth. Therefore faster reaction rates are usually preferred for most acid soils. On sandy soils, however, slow reactions may be more beneficial because of the small cation holding capacity of the soils. In general, the rate of reaction is largely dependent on the dissolution rate of the limestone, which is influenced by a number of factors such as chemical composition, particle size, application method, rate of removal of OH^- ions, soil temperature and moisture content, and the partial pressure of CO_2.

Solubilities of limestones vary with chemical composition or impurities, and even with crystallinity for the same type of limestone. Solubility of calcite in water is slightly lower than that of magnesite ($MgCO_3$) under normal temperature and atmospheric pressure conditions, with dolomite is somewhere in between. However, the rate of reaction (reactivity) of calcitic limestones is much greater than that of dolomitic limestones. Apart from widespread availability and low cost of calcitic limestones, the fast kinetics is another reason for the widespread use of calcitic limestones. On the other hand, dolomitic limestones have higher basicity and provide an additional source of Mg. However, fields that have been limed with dolomitic lime and have high Mg/Ca ratios have been known to experience soil structural problems.

Particle size is an important factor influencing reactivities of limestones. Solubilities of limestones are generally very low (< 0.25 g/L), and the rates of dissolution in the soil are largely dependent on the surface area that is in contact with the soil solution, soil pH, and other factors affecting carbonate equilibria. The reactivities of limestones increase with an increase in particle fineness. In other words, finer particle sizes produce a faster upward pH adjustment, whereas larger sizes provide a slow and longer-lasting effect. Therefore desirable agricultural limestones should contain both coarse and fine materials. Many states require that 75 to 100% of the limestone pass an 8- to 10-mesh screen and that 20 to 80% pass anywhere from an 8- to a 100-mesh screen (Tisdale et al., 1985). As mentioned before, dolomitic limestones react with soil more slowly. For an equal reactivity, finer size distribution should be used when applying dolomite.

The rate of hydrolysis of limestone is dependent on the rate of removal of dissociated OH^- ions. The existence of strong acids or free H^+ will expedite the neutralization of OH^- and therefore enhance the dissolution rate. In general, reactivity of limestones decreases as pH increases. The dissolution rate of limestone in soil is also influenced by soil temperature and moisture content. An increase in either temperature or moisture level results in a faster dissolution rate, which of course will expedite the acid-neutralizing rate. Mixing limestone with soil greatly improves the reaction rate. The low solubility of limestone and the slow diffusion rate of Ca^{2+} in soil (Barber, 1984) necessitates a mixing application in order to achieve a faster reaction rate. This is not possible in no-tillage systems. Therefore a more soluble liming material is desirable for surface applications. In addition, the dissolution rate increases as the partial pressure of CO_2 in soil increases. Since partial pressure of CO_2 in soil is dependent on microbial activity, the difference in microbial activity can alter the dissolution rate of the limestones.

D. Lime Requirement

Lime requirement (LR) is the amount of lime needed to raise pH to a selected level or to reduce the content of toxic substance to a certain level. Many methods are available, and none of them is universally accepted for use under all circumstances. See Chapter 10. What method to use should be based on soil properties, accuracy of the method, and the time and labor required for the test. Generally, test methods fall into the following categories. 1) Soil-lime incubation. The procedure involves mixing incremental amounts of lime with appropriate amounts of soil and determining soil pH after equilibration. The lime requirement is determined based on the selected pH level. This method is normally reliable but takes months to complete. 2) Soil-base titration. This involves directly titrating an acid soil with a base, or mixing incremental amounts of base such as $Ca(OH)_2$ with predetermined aliquots of soil. Lime requirement is determined based on the measured pH after equilibration. It normally takes days to complete. 3) Soil-buffer equilibration. One method is leaching soil samples with a buffer solution, followed by titrating the leachate with a base. The lime requirement is determined based on the difference in amounts of base used for titrating the leachate and buffer solution. Another method is to measure the change of the pH of the buffer solution after equilibration instead of titration. The change in pH is translated into lime requirement based on a calibrated buffer titration curve. The first method takes hours to complete, the second only a few minutes. 4) Soil-neutral salt equilibration. This involves leaching or extracting soil samples with a neutral salt, followed by either titrating the leachate or determining base saturation by summing the basic metal cations in the leachate. This method is developed mainly to neutralize exchangeable acidity and to lower Al^{3+} toxicity. It is justified by the argument that Al^{3+} toxicity is the most limiting factor for plant growth in acid soils.

E. Amelioration of Subsoil Acidity with Gypsum

Crop growth on many subtropical and tropical soils is often restricted by subsoil acidity and the resultant Al phytotoxicity, which severely retards root growth. The resulting shallow rooting system is more vulnerable to drought stress. The ordinary liming practice of mixing lime in the tillage layer has very little effect on subsoil acidity owing to the low solubility and mobility of limestones. Direct mixing of lime with the subsoil is costly and mechanically infeasible. An alternative is to use a more soluble and mobile material such as gypsum to ameliorate subsoil acidity. Although the mechanisms are not yet definitely known, some studies during the past twenty years have shown that gypsum application can alleviate subsoil acidity and Al toxicity and promotes root growth and crop yield (Sumner et al., 1986; Noble et al., 1988; Pavan et al., 1984; Oates and Caldwell, 1985; Ritchey et al., 1980; Reeve and Sumner, 1972). See Chapter 7.

The direct effects of gypsum application on soil pH are small and inconsistent. Positive, negative, and indifferent effects have been reported (Ritchey et al., 1980; Oates and Caldwell, 1985; Pavan et al., 1984). However, when soil pH was corrected for salt and ionic strength effects, it usually increases with gypsum additions (Hue et al., 1985; Sumner et al., 1986). The pH increase is normally less than 0.3 pH unit, and such a small increase is not expected to have a significant impact on soil acidity through directly altering H^+ activity in soil solution. Many ideas have been proposed to explain the observed positive response to gypsum application. Different explanations result from the uncertainties about the real mechanisms involved and also reflect the complexity and interdependency of changes or reactions brought about by gypsum addition to acid soils. Since Al toxicity is the limiting factor for plant growth in acid soils, lowering Al^{3+} toxicity by reducing its activity through gypsum addition is conceived to be the primary mechanism that is responsible for the positive response of root growth and crop yields. The activities of Ca^{2+}, Al^{3+}, and other Al species, and the balance of these ions in solution, directly affect plant root growth (Noble et al., 1988), so the positive response of plant growth to gypsum should ultimately be attributed to both the elevated Ca^{2+} activity and the depressed Al^{3+} activity.

The Al^{3+} concentration in soil solution may be increased due to ion exchange caused by increased Ca^{2+} activity, but the Al^{3+} activity may also be reduced owing to increased ionic strength and formation of ion pair $AlSO_4^+$ (McCray and Sumner, 1990). Due to the physiologically antagonistic effects of Ca^{2+} on Al^{3+}, increased Ca^{2+} activity would compensate and alleviate the damage to root cells caused by Al^{3+} toxicity and suppress Al^{3+} uptake and further reduce physiological disorder due to the excessive Al accumulation in plants.

Pavan et al. (1982) have proposed that the formation of ion pair $AlSO_4^+$ may reduce Al toxicity. Noble et al. (1988) have shown that $AlSO_4^+$ was less toxic to soybean root growth in a nutrient solution study. Based on a chemical equilibrium model, they further showed that increases in SO_4^{2-} and Ca^{2+} concentrations reduced Al^{3+} activity through the formation of $AlSO_4^+$.

The proposed ligand exchange of OH^- by SO_4^{2-} on sesquioxide surfaces is conceived to increase SO_4^{2-} retention in soil and OH^- release to soil solution (Reeve and Sumner, 1972; Rajan, 1978). This adsorption type reaction is responsible for the small pH increase with gypsum addition and has been termed the self-liming effect by Sumner (1993). It has been postulated that the specific adsorption of SO_4^{2-} would increase negative surface charge density, which, coupled with preferential adsorption of Al^{3+}, would result in a lower Al^{3+} concentration in soil solution. Adams and Rawajfih (1977) proposed that $Al(OH)_3$ may react with SO_4^{2-} and produce $Al(OH)SO_4$ or $Al_4(OH)_{10}SO_4$ precipitation and production of free OH^-. This precipitation reaction also contributes to the self-liming effects of gypsum.

Increased downward movement of Al^{3+} with gypsum addition was reported in several column leaching studies (Oates and Caldwell, 1985; Kotze and Deist, 1975). This is substantiated by the increased ion exchange due to the elevated Ca^{2+} activity and the formation of ion pair $A1SO_4^+$ in soil solution and is supported by the observation that fluoride impurity in by-product gypsum greatly increased downward movement of Al^{3+} in soil columns compared with either pure or F-free gypsum (Oates and Caldwell, 1985). The enhanced movement of Al^{3+} was attributed to the formation of Al-F ion pair or complex. However, the leaching mechanism is believed to be important only for soils where hydraulic conductivity is usually high. In contrast, no substantial movement of Al^{3+} after gypsum application was observed in other studies, in which positive response of crop yield to gypsum was also observed (Pavan et al., 1984; Sumner et al., 1986). The discrepancy is probably due to the difference in soil hydraulic properties and suggests that Al^{3+} leaching is not the primary mechanism reducing Al^{3+} toxicity.

Most studies have shown positive crop yield responses to by-product gypsum application. Some may be due to the improved nutrient status of Ca and S, or P if phosphogypsum is used. Others may be due to the reduced Al^{3+} toxicity and the increased rooting depth, which is more resistant to drought stress, or a combination of both. The negative or lack of response on crop yield may result from K or Mg deficiency, which can be caused by excessive leaching of Mg and K, or the crop itself, which is not sensitive to Al^{3+} toxicity. In general, the nature and extent of crop response to gypsum application is dependent on the status of soil nutrients, the level of Al^{3+} toxicity, the crop sensitivity to Al^{3+} toxicity, the rooting depth of the crop, soil mineralogy, and the environmental conditions such as rainfall amount and distribution.

IV. EFFECTS OF LIMING ON PHYSICAL PROPERTIES OF ACID SOILS

A. Clay Flocculation and Soil Aggregation

Clay minerals are the most active components in a soil because of large surface area, small size, and charge phenomena. Soil physical and chemical properties are largely dependent on clay characteristics and behaviors (flocculation or dispersion). As a cementing agent, clay has a tremendous effect on soil physical properties such as aggregate and structural stabilities. Clay flocculation promotes soil aggregation and structural stability, while clay dispersion often leads to the deterioration of soil structure. The behaviors of clay particles can be well understood with the Gouy–Chapman electric double layer theory.

The electric double layer refers to a clay surface charge and a compensating counter ion charge (Fig. 2). The clay surface charge, normally negative, is composed of permanent charge and/or variable charge. The counter ion charge is a diffuse cloud of ions that are electrostatically attracted by the oppositely charged surface and have a tendency to break away from the surface to the bulk solution. Simultaneously, co-ions, which have the same sign as the surface charge, are repelled by the surface charge and tend to diffuse back into the counter ion cloud. Based on the electrostatic and diffusion theory (Boltzmann's theorem and Poisson's equation), concentration distributions of positive and negative ions in the double layer can be computed. A schematic illustration of the computed ion dis-

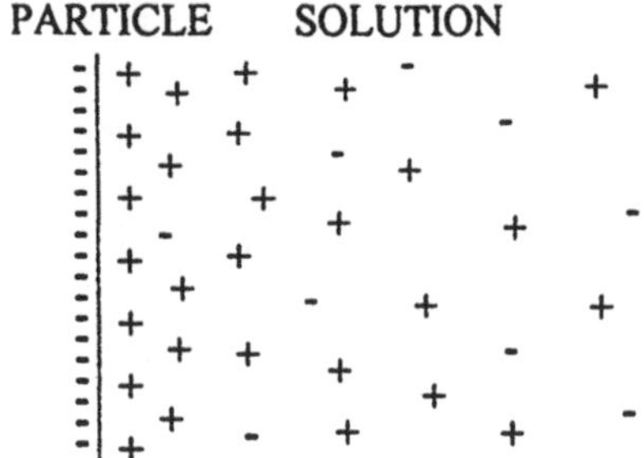

Figure 2 Schematic illustration of the diffuse electric double layer according to Gouy–Chapman theory. (After van Olphen, 1977.)

tributions at two electrolyte concentrations for both permanent and variable charge surfaces is presented in Fig. 3. The n_+ and n_- are cation and anion concentration distributions (number of ions per unit volume), and σ_+ and σ_- are total cation excess and anion deficit, respectively. Symbols with the prime sign stand for the corresponding values at a greater equilibrium concentration. The total surface charge ($\sigma = \sigma_+ + \sigma_-$) and the charge ratio of σ_-/σ_+ in response to the changes of electrolyte concentration are also given in Fig. 3. It clearly illustrates that the double layer is compressed towards the clay surface when electrolyte concentration is elevated. A similar response is also exhibited when a lower valency cation is replaced by a higher valency cation. The double layer compression or expansion dictates clay behaviors, as we shall see later.

When two particles approach each other, their diffuse counter ion clouds begin to interfere or repel each other. Based on the electric double layer theory, this repulsive potential or repulsive energy (V_r) decreases quasi-exponentially with increasing particle separation. As is shown in Fig. 4, curve V_r (namely the repulsive potential curve) is compressed towards the particle surface when the counter ion concentration is increased. An attractive energy (V_a), which works against the repulsive energy, is derived from the van der Waals attractive forces that are the sum of all the attractive forces between every atom of one particle and every atom of the other particle. The attractive potential curve decreases rapidly with particle separation and is inversely proportional to the second power of the distance from the particle surface.

The summation of the repulsive and attractive potential curves at each distance (termed net potential curve, V_n) determines the nature of the colloidal suspension. The V_n is suppressed as electrolyte is increased due to the reduction in the repulsive potential (Fig. 4). This results in a reduction in the energy barrier. When a critical concentration, at which the energy barrier disappears, is reached, rapid coagulation takes place. This concentration is termed the critical

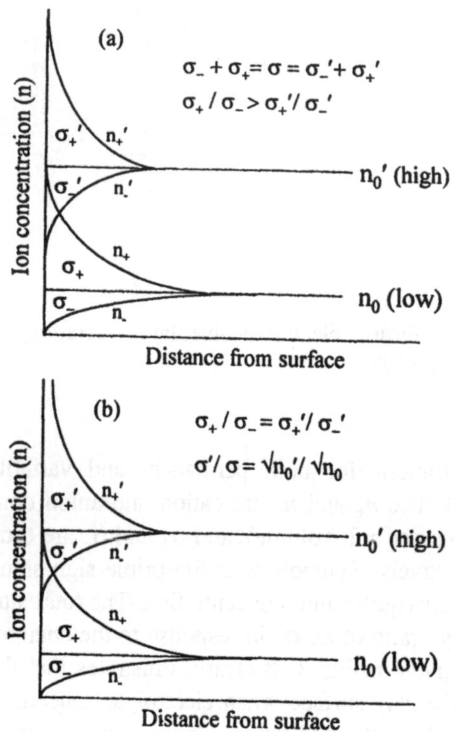

Figure 3 Ion or charge distribution in the diffuse electric double layer of a negatively charged particle surface at two electrolyte concentrations for (a) a constant surface charge and (b) a constant surface potential. (After van Olphen, 1977.)

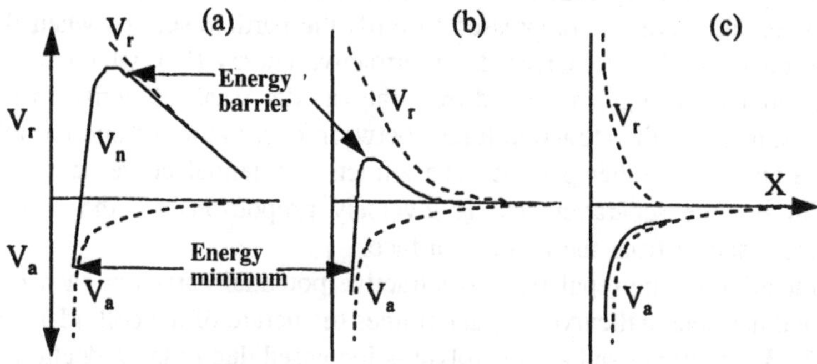

Figure 4 Repulsive, attractive, and net interactive energies as a function of particle separation for (a) low, (b) intermediate, and (c) high electrolyte concentrations. (Simplified after van Olphen, 1977.)

flocculation concentration (CFC). At the CFC, the flocculation rate depends entirely on the frequency of particle collisions. It should be pointed out that theoretically flocculation could occur at any concentration, but the flocculation rate becomes extremely slow when the energy barrier is high because the chances for clay particles to overcome the energy barrier and to be "trapped" in the energy minimum are very small, provided Brownian motion is the major means of causing particle encounters.

Under the conditions where the energy barrier vanishes, the CFC can be derived as

$$\mathrm{CFC} = \frac{\lambda[\tanh(ze\psi_0/(4kT))]^4}{A^2 z^6} \tag{1}$$

where

λ = a constant number
A = Hamaker constant
z = valence of the counter ions
e = electronic charge
ψ_0 = electrical potential at the clay surface
k = Boltzmann constant
T = absolute temperature

From Eq. (1), the dependence of CFC on $1/z^6$, known as the Schulze–Hardy rule, indicates that z has a dominant effect on flocculation. To obtain qualitative, comparative information on the effects of valence and electrolyte concentration on the double layer thickness, an "effective thickness" δ, which is defined as the center of gravity of the space charge in the double layer, can be computed by

$$\delta = \left(\frac{\varepsilon kT}{8\pi z^2 e^2 c N_A}\right)^{\frac{1}{2}} \tag{2}$$

where

ε = dielectric constant of the solvent
c = the molar concentration
N_A = Avogadro constant

The δ from Eq. (2) can only be used to represent the symbolic length of the double layer extension, since the double layer actually extends to infinity from the clay surface. However, Eq. (2) vividly depicts the impact of c and z on clay behaviors. Swelling pressure p_s, which is the repelling force between the two overlapped double layers, can be computed by

$$p_s = 2N_A ckT\left[\cosh\left(\frac{ze\psi_d}{kT}\right) - 1\right] \tag{3}$$

where ψ_d is the potential midway between the plates. The swelling pressure is considered to be identical to the osmotic pressure generated midway between the two plates. The dependence of p_s on z and c is not so obvious in Eq. (3) because the embedded function of ψ_d decreases exponentially with z and the square root of c. As a matter of fact, p_s decreases as z and c increase. This relationship has significant implications in soil mechanics and civil engineering applications.

The reverse process or clay dispersion is invoked when electrolyte concentration decreases to below CFC. This process is of great importance in preserving soil aggregate and structural stability. Spontaneous dispersion often takes place for soils with predominant swelling type clays or those high in exchangeable sodium percentage (ESP) when electrolyte concentration in the soil solution is below the CFC, while mechanically induced dispersion (with energy input, e.g., raindrop impact) is the dominant process for the soils with nonswelling clay minerals and low ESP. Gradual loss of clay particles deteriorates soil structure and sometimes leads to severe piping or tunnel erosion. Clay dispersion also plays a significant role in surface seal formation. The low electrolyte concentration of rainwater, coupled by mechanical raindrop impact, results in severe clay dispersion at the soil surface. Dispersed clay particles, which then clog conducting pores near the soil surface or settle on the soil surface during water infiltration, lead to the formation of a less permeable surface seal. This thin layer greatly increases runoff and erosion due to its extremely low hydraulic conductivity.

B. Effects of Chemical Composition and pH on Flocculation

Chemical composition and concentration on exchangeable sites and in soil solution greatly influence the clay behaviors in a soil. The flocculating power of cations depends on their valences and sizes. The Al^{3+} has the greatest flocculating power, while Na^+ often acts as a dispersing agent in soils. Therefore soil ESP is a good indicator of the degree of clay dispersion. Miller et al. (1990) reported that CFC increases linearly with ESP for three soil clays. This implies that electrolyte concentration necessary for flocculation increases as sodium concentration (ESP) increases. Soils that are high in ESP, such as sodic soils, often have poor aggregation and structure and very low hydraulic conductivity. Many studies have shown that ESP has a tremendous effects on surface seal and crust formation. The enhanced clay dispersion with Na^+ is because replacement of higher valency cations with Na^+ causes electric double layer expansion, which results in an increase in repulsive potential between clay particles.

Exchangeable Mg^{2+} and K^+ may also promote clay dispersion and cause soil structural deterioration compared with Ca^{2+} (Shainberg et al., 1988; Keren, 1989). Keren (1989) found that soils when treated with Ca salt showed a higher water infiltration rate than when treated with Mg salt. This is because the greater hydration radii of Mg (compared with Ca^{2+}) may shift the diffuse ion cloud away from the surface. Levy and van der Watt (1990) reported that an increase in K^+ exchangeable percentage in the K–Ca system enhanced clay dispersion and decreased hydraulic conductivity.

For variable charge soils, soil pH is an important factor influencing soil clay behaviors. Since negative pH dependent charge increases as soil pH increases, the increased total negative charge, of course, will result in an increase in repulsive potential. As a result, clay becomes more dispersive as pH increases. Reichert and Norton (1996) have shown that surface application of fluidized bed combustion bottom ash (containing 23% CaO and 73% anhydrite) on highly weathered soils caused enhanced clay dispersion due to elevated pH, which further resulted in greater runoff and soil loss rates. For acid soils when Al^{3+} is the dominant exchangeable cation (normally pH < 5), liming can cause severe clay dispersion and soil structural deterioration. As pH increases, Al^{3+} is replaced by exchangeable Ca^{2+} and then hydrolyzed. The lower flocculation power of Ca^{2+} compared with Al^{3+}, coupled by increased negative charges resulted from elevated soil pH, would enhance clay dispersion.

The beneficial effects of liming on soil physical properties also result from the elevated ionic strength in soil solution and the domination of Ca^{2+} on exchangeable sites. As discussed earlier, Ca^{2+} has the second highest flocculating power following Al^{3+}, which is the dominant species only at very low pH. Increasing exchangeable calcium percentage by replacing monovalent cations or Mg^{2+} undoubtedly inhibits clay dispersion and therefore promotes aggregation and soil structural stability. There is overwhelming evidence in this regard. Increased exchangeable calcium percentage and ionic strength increase soil resistance to aggregate breakdown (Zhang and Miller, 1996), preserve soil aggregates from disintegration by raindrop impact, reduce surface sealing and crusting (Miller and Radcliffe, 1992; Agassi et al., 1981; Shainberg, 1992), increase hydraulic conductivity (Chiang et al., 1987), and reduce surface runoff and erosion (Norton et al., 1993; Zhang and Miller, 1996; Miller, 1987; Shainberg et al., 1989). For highly weathered soils, which are often lower in cation holding capacity and electrolyte concentration, increased ionic strength usually plays a dominant role in flocculating clays; but for some soils, replacement of Al^{3+} by Ca^{2+} and/or increased pH dependent negative charge can enhance clay dispersion and deteriorate soil structure. A thorough knowledge of the soil chemistry and mineralogy are required for determining the optimum liming materials.

C. Soil Stabilization with Lime

Lime has been successfully used to control dam and canal embankment failures as well as piping erosion for soils with dispersive and expansive clays (McElroy, 1987; Garver, 1987). Clay becomes spontaneously dispersed in water for soils with swelling clays or those with high ESP. Gradual loss of clay particles causes deterioration of soil structure, which eventually leads to loss of matrix materials by water flowing through cracks or existing holes. This type of piping erosion often predominates in earth dams and on steep hill slopes where soils contain dispersive clays. Sloughing of embankments either of earth or lined with concrete is frequently caused by the swelling pressure of expansive clay, which is caused by expansion of the electric double layer due to high ESP or low ionic strength in ambient solution, as is presented in Eq. (3). Mixing lime with soil during construction successfully prevents clay dispersion and reduces swelling because of increased Ca^{2+} concentration in soil solution and on exchange sites. Field observation over 15 years has showed that 1 to 4% of calcium hydroxide has been successful in controlling piping erosion of highly dispersive clays (McElroy, 1987). Liming treatment has become a general practice in fixing failed embankments or in construction of canals on swelling soils with expansive clays (Garver, 1987).

V. CONCLUSIONS

Soil acidification, as a natural process, intensifies with the weathering of soil and parent materials, oxidation of organic matter, and increasing use of acid fertilizers. Therefore soil acidity will continue to be a problem for agricultural production in many agricultural soils. Excessive soil acidity must be neutralized in order to attain favorable soil environments for better plant growth. Liming materials commonly used in agricultural lands include oxides, hydroxide, carbonate of calcium and/or magnesium, and industrial by-products. In recent years, by-product gypsum has been found promising in alleviating subsoil acidity because of its greater mobility compared with limestones.

Application of lime to acid soils has a significant impact on soil chemical and physical properties. The direct and ultimate outcomes of lime application are the increased Ca^{2+} saturation percentage on the exchange sites and the elevated soil pH in soil solution. Soil pH is a master factor in acid soils that influences many chemical and biological processes occurring in soils. An increase in soil pH reduces the toxicities of Al^{3+}, Mn^{2+}, and H^+; increases the availability of phosphorus and most micronutrients; increases cation exchange capacity, especially for variable charge soils; and promotes microbial activity. Application of liming materials or by-product gypsum also substantially alters chemical compositions and increases chemical concentrations in both adsorbed and desorbed phases. This often leads to significant changes in soil physical properties.

The beneficial effects of liming on soil physical properties result from the elevated ionic strength in soil solution and the domination of Ca^{2+} on the exchange sites. Type of cations and ionic strength are the two major factors affecting clay behaviors (flocculation or dispersion), which further influence soil physical properties. Clay flocculation promotes soil aggregation and structural stability, while clay dispersion often leads to a deterioration of soil structure. Since Ca^{2+} has the greatest flocculating power among the common soil cations except for Al^{3+}, replacement of monovalent cations or Mg^{2+} by Ca^{2+} plus the elevated ionic strength in soil solution following liming promotes clay flocculation and therefore soil aggregation and structural stability. Many studies have shown that liming materials and by-product gypsum are effective in reducing aggregate breakdown by raindrop impact, surface sealing and crusting, soil swelling, surface runoff, and surface water erosion and piping erosion. However, liming an acid soil when Al^{3+} is the dominant exchangeable cation may cause clay dispersion and soil structural deterioration. Replacement of Al^{3+} by Ca^{2+}, coupled by the elevated repulsive forces due to the increased negative charges resulting from deprotonation, favors clay dispersion.

REFERENCES

Adams, F. (1984). Crop response to lime in the southern United States. *Soil Acidity and Liming* (F. Adams, ed.). Agron. Monogr. 12, ASA, CSSA and SSSA, Madison, Wisconsin, p. 380.

Adams, F., and Rawajfih, Z. (1977). Basaluminite and alunite: a possible cause of sulfate retention by acid soils. *Soil Sci. Soc. Am. J. 41*:686.

Agassi, M. J., Shainberg, I., and Morin, J. (1981). Effect of electrolyte concentration and soil sodicity on infiltration rate and crust formation. *Soil Sci. Soc. Am. J. 52*:1453.

Alexander, M. (1980). Effect of acidity on microorganisms and microbial processes in soil. *Effects of Acid Precipitation on Terrestrial Ecosystems* (T. Hutchinson, and M. Harvas, eds). Plenum, New York, pp. 363–364.

Barber, S. A. (1984). Liming materials and practices, *Soil Acidity and Liming* (F. Adams, ed.). Agron. Monogr. 12, ASA, CSSA and SSSA, Madison, Wisconsin, p. 380.

Chernov, V. A. (1964). *The Nature of Soil Acidity*. Soil Science Society of America, Madison, Wisconsin.

Chiang, S. C., Radcliffe, D. E., Miller, W. P., and Newman, K. D. (1987). Hydraulic conductivity of three southeastern soils as affected by sodium, electrolyte concentration and pH. *Soil Sci. Soc. Am. J. 51*:1293.

Coleman, N. T., Weed, S. B., and McCracken, R. J. (1959). Cation-exchange capacity and exchangeable cations in Piedmont soils of North Carolina. *Soil Sci. Soc. Am. Proc. 23*:146.

Foy, C. D. (1984). Physiological effects of hydrogen, aluminum, and manganese toxicities in acid soil. *Soil Acidity and Liming* (F. Adams, ed.). Agron. Monogr. 12, ASA, CSSA and SSSA, Madison, Wisconsin, p. 380.

Garver, L. L. (1987). Canal repair techniques using lime-stabilized soil. *Lime for Environmental Uses* (K. A. Gutschick, ed.). ASTM, Philadelphia, Pennsylvania, p. 147.

Helling, C. S., Chesters, G., and Corey, R. B. (1964). Contributions of organic matter and clay to soil cation-exchange capacity as affected by the pH of the saturating solution. *Soil Sci. Soc. Am. Proc. 28*:517.

Hue, N. V., Adams, F., and Evans, C. E. (1985). Sulfate retention by an acid BE horizon of an Ultisol. *Soil Sci. Soc. Am. J. 49*:1196.

Ihde, A. J. (1952). Edmund Ruffin, soil chemist of the Old South. *J. Chem. Education 29*:407.

Kamprath, E. J. (1984). Crop response to lime on soils in the tropics. *Soil Acidity and Liming* (F. Adams, ed.). Agron. Monogr. 12, ASA, CSSA and SSSA, Madison, Wisconsin, p. 380.

Keren, R. (1989). Water-drop kinetic energy effect on water infiltration in calcium and magnesium soil. *Soil Sci. Soc. Am. J. 53*:1624.

Korcak, R. F. (1980). Fluidized bed material as a lime substitute and calcium source for apple seedlings. *J. Environ. Qual. 9*:147.

Kotze, W. A. G., and Deist, J. (1975). Amelioration of subsurface acidity by leaching of surface amendments: a laboratory study. *Agrochemophysica 7*:39.

Levy, G. J., and van der Watt, H. v. H. (1990). Effect of exchangeable potassium on the hydraulic conductivity and infiltration rate of some South African soils. *Soil Sci. 149*:69.

McCray, J. M., and Sumner, M. E. (1990). Assessing and modifying Ca and Al levels in acid subsoils. *Adv. Soil Sci. 14*:45.

McElroy, C. H. (1987). Using hydrated lime to control erosion of dispersive clays. *Lime for Environmental Uses* (K. A. Gutschick, ed.), ASTM, Philadelphia, Pennsylvania, p. 147.

McLean, E. O., and Brown, J. R. (1984). Crop response to lime in the midwestern United States. *Soil Acidity and Liming* (F. Adams, ed.). Agron. Monogr. 12, ASA, CSSA and SSSA, Madison, Wisconsin, p. 380.

Mehlich, A. (1942). Base unsaturation and pH in relation to soil type. *Soil Sci. Soc. Am. Proc. 6*:150.

Mehlich, A. (1943). The significance of percentage base saturation and pH in relation to soil differences. *Soil Sci. Soc. Am. Proc.7*:167.

Miller, W. P. (1987). Infiltration and soil loss of three gypsum-amended Ultisols under simulated rainfall. *Soil Sci. Soc. Am. J. 51*:1314.

Miller, W. P., Frenkel, H., and Newman, K. D. (1990). Flocculation concentration and sodium/calcium exchange of kaolinitic soil clays. *Soil Sci. Soc. Am. J. 54*:346.

Miller, W. P., and Radcliffe, D. E. (1992). Soil crusting in the southeastern United States. *Soil Crusting: Chemical and Physical Processes* (M. E. Summer, and B. A. Stewart, ed.). Lewis, Boca Raton, Florida, pp. 233–266.

Noble, A. D., Fey, M. V., and Sumner, M. E. (1988). Calcium-aluminum balance and the growth of soybean roots in nutrient solutions. *Soil Sci. Soc. Am. J. 52*:1651.

Norton, L. D. (1995). Mineralogy of high calcium/sulfur-containing coal combustion by-products and their effect on soil surface sealing. *Agricultural Utilization of Urban and Industrial By-Products* (D. L. Karlen, R. J. Wright, and W. D. Kemper, eds.). Agron. Monogr. 58, ASA, CSSA and SSSA, Madison, Wisconsin.

Norton, L. D., Shainberg, I., and King, K. W. (1993). Utilization of gypsiferous amendments to reduce surface sealing in some humid soils in the eastern USA. *Soil Surface Sealing and Crusting* (J. W. A. Poesen, and M. A. Nearing, eds.), *Catena Suppl. 24*:79.

Oates, K. M., and Caldwell, A. G. (1985). Use of by-product gypsum to alleviate soil acidity. *Soil Sci. Soc. Am. J. 49*:915.

Pavan, M. A., Bingham, F. T., and Pratt, P. F. (1982). Toxicity of aluminum to coffee in Ultisols and Oxisols amended with $CaCO_3$, $MgCO_3$, and $CaSO_4{\cdot}2H_2O$. *Soil Sci. Soc. Am. J. 46*:1201.

Pavan, M. A., Bingham, F. T., and Pratt, P. F. (1984). Redistribution of exchangeable calcium, magnesium, and aluminum following lime or gypsum applications to a Brazilian Oxisol. *Soil Sci. Soc. Am. J. 48*:33.

Phung, H. T., Lund, L. J., and Page, A. L. (1978). Potential use of fly ash as a liming material, *Environmental Chemistry and Cycling Processes* (D. C. Adriano, and I. L. Brisbin, eds.). U. S. Dept. of Commerce, Springfield, Virginia, pp. 504–515.

Pratt, P. F., and Bair, F. L. (1962). Cation-exchange properties of some acid soils of California. *Hilgardia 33*:689.

Rajan, S. S. S. (1978). Sulfate adsorbed on hydrous alumina, ligands displaced, and changes in surface charge. *Soil Sci. Soc. Am. J. 42*:39.

Reeve, N. G., and Sumner, M. E. (1972). Amelioration of subsoil acidity in Natal Oxisols by leaching of surface-applied amendments. *Agrochemophysica 4*:1.

Reichert, J. M., and Norton, L. D. (1996). Fluidized bed combustion bottom-ash effects on infiltration and erosion of variable-charge soils. *Soil Sci. Soc. Am. J. 60*:275.

Ritchey, K. D., Souza, D. M. G., Lobato, E., and Correa, O. (1980). Calcium leaching to increase rooting depth in a Brazilian savannah Oxisol. *Agron. J. 72*:40.

Shainberg, I. (1992). Chemical and mineralogical components of crusting. *Soil Crusting: Chemical and Physical Processes* (M. E. Sumner, and B. A. Stewart, eds.). Lewis, Boca Raton, Florida, pp. 33–53.

Shainberg, I., Alperovitch, N., and Keren, R. (1988). Effect of magnesium on the hydraulic conductivity of sodic smectites-sand mixtures. *Clays and Clay Minerals 36*:432.

Shainberg, I., Sumner, M. E., Miller, W. P., Farina, M. P. W., Pava, M. A., and Fey, M. V. (1989). Use of gypsum on soils: a review. *Adv. Soil Sci. 9*:1.

Sumner, M. E. (1992). The electrical double layer and clay dispersion. *Soil Crusting: Chemical and Physical Processes* (M. E. Sumner, and B. A. Stewart, eds.), Lewis Boca Raton, Florida.

Sumner, M. E. (1993). Gypsum and acid soils: the world scene. *Advances in Agronomy 51*:1

Sumner, M. E., Shahandeh, H., Bouton, J., and Hammel, J. E. (1986). Amelioration of an acid soil profile through deep liming and surface application of gypsum. *Soil Sci. Soc. Am. J. 50*:1254.

Terman, G. L., Kilmer, V. J., Hunt, C. M., and Buchanan, W. (1978). Fluidized bed boiler waste, a source of nutrients and lime. *J. Environ. Qual. 7*:147.

Thomas, G. W., and Hargrove, W. L. (1984). The chemistry of soil acidity. *Soil Acidity and Liming* (F. Adams, ed.). Agron. Monogr. 12, ASA, CSSA and SSSA, Madison, Wisconsin, p. 380.

Tisdale, S. L., Nelson, W. L., and Beaton, J. D. (1985). *Soil Fertility and Fertilizers.* Macmillan, New York, p. 754.

Uehara, G., and Gillman, G. (1981). *The Mineralogy, Chemistry, and Physics of Tropical Soils with Variable Charge Clays.* Westview Press, Boulder, Col.

van Olphen, H. (1977). *An Introduction to Clay Colloid Chemistry.* John Wiley, New York, p. 318.

Zhang, X. C., and Miller, W. P. (1996). Physical and chemical crusting processes affecting runoff and erosion in furrows. *Soil Sci. Soc. Am. J. 60*:860.

12
Designing Synthetic Soil Conditioners Via Postpolymerization Reactions

D. L. Bouranis *Agricultural University of Athens, Athens, Greece*

I. DESIGNING POLYMERIC SOIL CONDITIONERS

Various polymers have been used as soil conditioners for the improvement of those properties of soil that limit water movement, root development, and aeration. Azzam (1980, 1983), Wallace and collaborators (1986), Cook and, Nelson (1986), Terry and Nelson (1986), Mitchell (1986), and many others have worked on such agricultural polymers. The extent of application of synthetic soil conditioning polymers is very wide, but their use in practice has not spread as expected owing to high cost. The importance of this objective is easily appreciated when one considers the tremendous capital investment required to apply synthetic conditioners to the large cultivated annually.

Not all synthetic polymers exhibit soil conditioning properties, but several of them are capable of improving and stabilizing soil. Several also seem to be stable against the attacks of soil microorganisms, and therefore amelioration endures much longer. From the point of view of agricultural applications, the actions of synthetic polymers applied to the soil are mostly located in one of the following, or combinations thereof: 1) Stabilization of the soil structure derived from aggregate formation. This action is also indirectly associated with properties such as porosity, the ability of water penetration, rhizosphere aeration, etc. 2) Increase of the water-holding capacity of the soil as a result of the addition of swellable hydrophilic polymers. 3) Prevention of erosion by the air or by water.

The action of the polymer can be accomplished by various mechanisms that include interaction of the immobilized groups with natural compounds in the soil. These materials usually contain groups similar to those of naturally polymerized organic compounds, that is, amino, hydroxyl, and carboxyl groups. Despite that, in the case of water-insoluble conditioners, their action comes from the availability of intramolecular space for water retention. The ability of polymers to form stable aggregates with soil particles is a function of molecule architecture, their degree of polymerization, the nature of attached groups, and the extent of the various ionic or other replacements that have taken place in the active groups. Any factor or soil condition that affects one or more of the characters of the macromolecule also affects the ability of this macromolecule to create stable aggregates. The concentration of synthetic polymer that must be added to the soil is also an important parameter.

For all these reasons, the design of a synthetic polymer soil conditioner, and its preparation as well, are a complicated processes, since several factors have to be taken into consideration (Bouranis et al., 1995). The classification of polymers that present soil conditioning properties is also somewhat complicated. Such a classification is depicted in Fig. 1. *Polymeric soil conditioners* (PSC) should be distinguished as either *natural* or *synthetic*. Natural PSCs are polyuronic acids, alginic acids, polysaccharides, and humus. The action of microorganisms on the natural organic material and the entailed biodegradation led to the development of synthetic, nonbiodegradable PSCs. These materials can be used for an extended time and under controlled conditions.

A way to achieve a synthetic PSC is the polymerization of the appropriate monomer(s) by following various conditions. The advantage of this procedure is the availability for selection from a variety of compounds and conditions. Acrylic polymers, styrene based polymers, polyethyleneoxide, and poly(vinyl acetate) are such examples. A second way involves *the modification of a preexisting macromolecule* regarding the expected properties of the final product. The advantage of this procedure is that it can be connected with polymer recycling, since modification is applied to already existing polymerized materials. Postpolymerization sulfonation of polystyrene and styrene based copolymers and cellulose derivatives, and the modification and hydrolysis of polyacrylonitrile, are typical examples.

The solubility of a PSC in water is a principal property. In this way, a PSC may be *water-soluble* or *water-insoluble*. Water-soluble synthetic soil conditioners, which are more intensively studied today, are hydrolyzed poly(acrylonitrile), poly(acrylic) acid (which coinsides with the previously mentioned polymer when its hydrolysis is complete), poly(methacrylic acid), poly(vinyl alcohol), poly(acrylamide), polysaccharides grafted on a polymer, various hydrophilic copolymers, etc. The hydrophilic character is ensured by the immobilization of polar groups on the polymer backbone (Fig. 2a). Most of the previously described polymers have shown an intensive hydrophilic character owing to the presence of polar groups within the polymer chains.

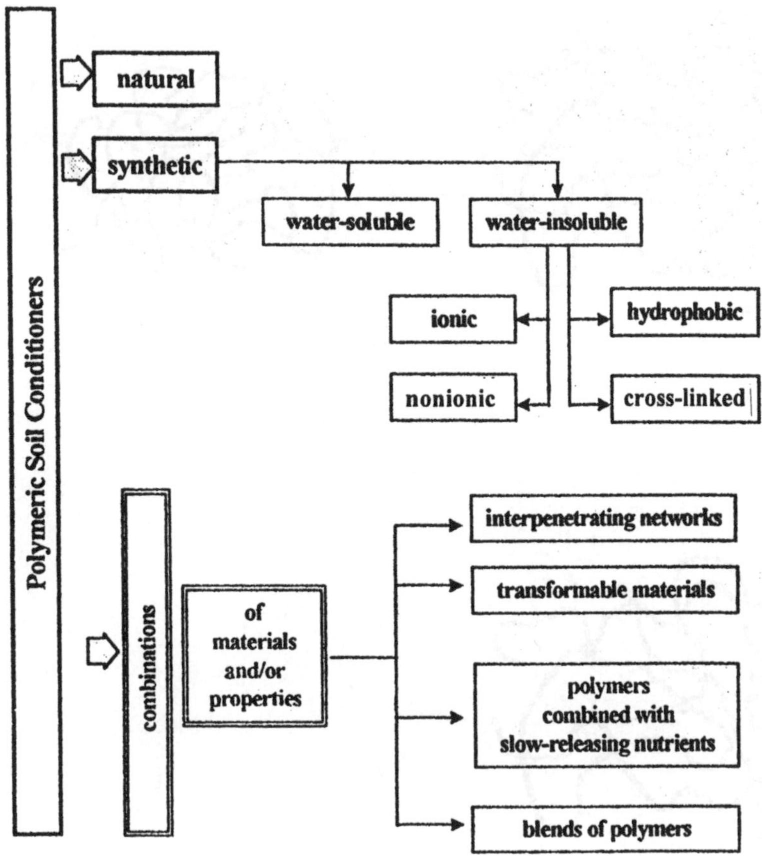

Figure 1 A classification of polymeric soil conditioners. (Modified from Bouranis et al, 1995, with permission.)

On the other hand, some examples, like bitumen emulsions and styrene-butadiene copolymer applications, indicate that a hydrophobic polymer may also be useful in some cases. Water-insoluble PSC can be either hydrophobic or cross-linked (Figs. 2b and 2c). A *hydrophobic* PSC is water-insoluble because of the lack or deficiency of a polar character. The styrene-butadiene copolymer is such an example. Polar groups are absent in this case to ensure the hydrophobic character of the material. A *cross-linked* PSC bears physical or chemical cross-links within macromolecular chains. Such a structure is responsible for the swelling ability. Examples are cross-linked sulfonated polystyrene and cross-linked polyacrylamide. Physical cross-links are usually domains incompatible with the other

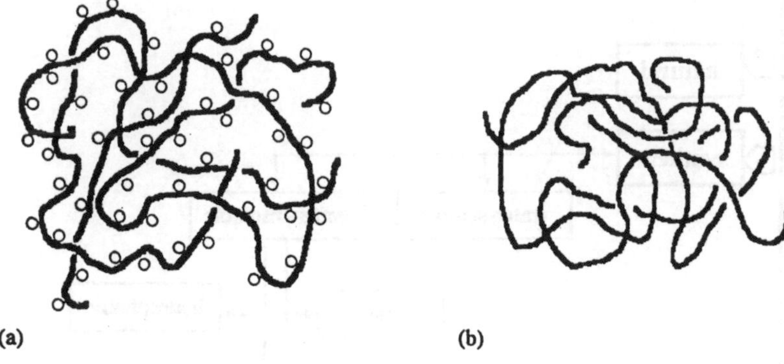

Figure 2 The shapes of various type of polymeric soil conditioners. (a) The shape of a water-soluble conditioner. Symbol ○ indicates a polar group. (b) The shape of a linear water-insoluble conditioner. (c) The shape of a cross-linked conditioner. Symbol ● indicates cross-links. (From Bouranis et al, 1995.)

polymer areas that tend to bind each to another. Chemical cross-links are usually obtained as a result of bifunctional cross-linking action during or following polymerization. The difference between cross-linked and linear (or grafted) substrates is an important issue in polymer application. When a cross-linked structure is desired, compounds that develop cross-links of physical or chemical nature are necessary. According to the first method of polymerization, a bifunctional reagent (like methylene-bis-acrylamide or divinyl benzene) is usually employed. In the case of postpolymerization, modification reagents (like dichloromethyl-benzene derivatives or ionomers) ensure conditioner swelling properties.

Polymeric soil conditioners may be *ionic* or *nonionic*. Most of the previously mentioned polymers are polyelectrolytes. In their hydrogen form, they may have strong or weak acidic groups, but they are usually applied in final form, their sodium form. This form is dissociated in an aqueous environment, leaving polyanions on the polymeric chain. Polystyrene sulfonate and polyacrylic acid are two typical examples of ionic hydrophilic polymers. Other materials, like polyvinyl alcohol, are hydrophilic ones, since they carry polar groups, but in general they are nonionic polymers.

The most complicated aspect is when the PSC is not pure but consists of combinations of ingredients or materials having combinations of desirable properties. Recent works point out the need to classify the synthetic soil conditioning polymers into specific categories, which derive from combinations of properties and/or materials. Such categories may be: 1) interpenetrating networks, 2) blends of soil conditioning polymers, 3) polymers combined with slow-releasing nutrients (slow release type), and 4) transformable materials, that is, slightly cross-linked polymers being in water-insoluble form, which are converted to a water-soluble form shortly after their incorporation into the soil. Both forms may exhibit soil conditioning properties (e. g., HSSAN).

In a blend of two conditioners (Fig. 3a), combined properties are expected. In an interpenetrating network (Fig. 3b), another macromolecule exists within a cross-linked product. In a transformable material (Fig. 3c), chain breaking forms a water-soluble product coming from a cross-linked network. It is also possible for a fertilizer to be attached on a cross-linked polymer substrate (Fig. 3d). (For further analysis see Bouranis et al., 1995.)

II. SYNTHESIS AND STRUCTURE OF CROSS-LINKED POLYMERIC SOIL CONDITIONERS

Cross-linked PSC may be formed from chloromethylated polystyrene or styrene copolymers, since these styrene units are accessible to electrophilic substitution of one of the hydrogen atoms by a chloromethyl group. According to their preparation conditions, cross-linked PSC can be divided into three major categories:

a. Cross-linked PSC Prepared by Copolymerization: This class of PSCs is based on monoolefin employment (styrene, acrylic acid, acrylamide, methacrylic acid, etc.) and reactive diolefins as cross-linking agents. Such cross-linking agent application includes bis-acrylamide derivatives and copolymers with dimethacrylic esters of 1, 12(*p*-oxyphenyl) dodecane. Regarding preparation technique, almost every copolymerization technique can be employed (Davankov et al., 1974).

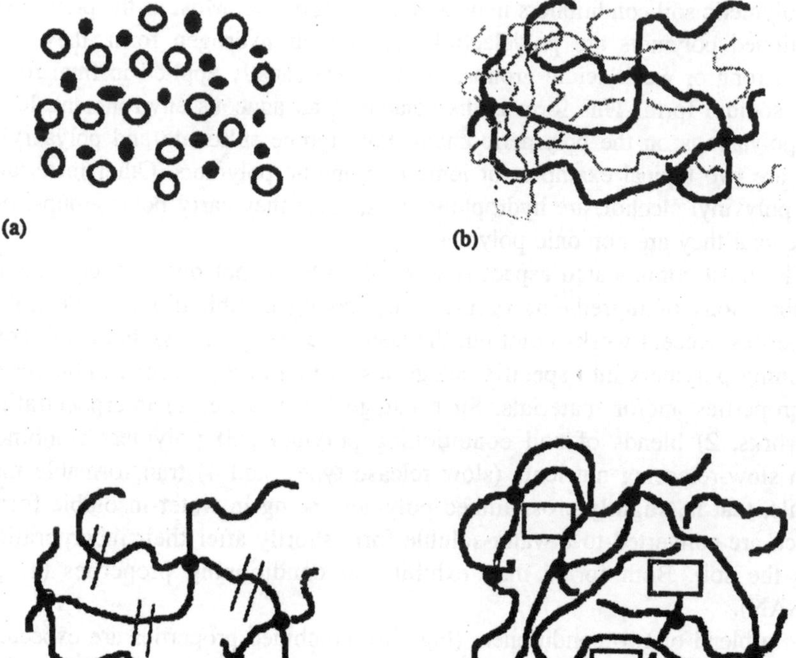

Figure 3 The shapes of various combinations of polymeric soil conditioners. (a) The shape of a blend of two conditioners. (b) The shape of an interpenetrating network of two conditioners. (c) The shape of a transformable conditioner. Symbol // indicates the positions of chain breaking. (d) The shape of a cross-linked conditioner containing an entrapped fertilizer. Symbol □ indicates the sites of the entrapped compound(s). (From Bouranis et al, 1995.)

b. Cross-linked PSC Prepared by Postpolymerization: Such PSCs are based on various modifications of linear macromolecules and grafted or block copolymers with multifunctional compounds as cross-linking agents. Uniformity of cross-linking, application of polymer recycling, and maintaining of polymer architecture characterize this category (Grassie and Gilks, 1973; Bussing and Peppas, 1983; Peppas and Valkanas, 1977; Barar et al., 1983a; Iovine and Ray-Chaudhuri, 1984).

c. Cross-linked PSC Derived as a Combination of Both Procedures: PSC obtained from a combination of the previously mentioned routes are networks that derive from additional cross-linking of a preformed standard copolymer. The advantage of the procedure is the preparation of more homogeneous cross-linked structures (Davankov et al., 1974; Belfer and Glozman, 1979; Goldstein and Schmuckler, 1972).

The recent trends and perspectives of cross-linked PSCs during the last decade are mostly connected with postpolymerization modifications of linear or grafted polymers. A typical solution procedure involves dissolution of the homopolymer or copolymers in a thermodynamically good solvent and addition of the cross-linker or a cross-linker solution to the reaction vessel. Various Friedel–Crafts catalysts have been used during macronets preparation (Theodoropoulos et al., 1994a,b; Peppas and Valkanas, 1977; Grassie et al., 1970; Theodoropoulos, 1993; Barar et al., 1983).

The cross-linking polymeric materials may be 1) *standard* or 2) *macronets*, depending on the type of bond formation within the polymer chains (Theodoropoulos and Konstantakopoulos, 1996). The dimension of divinylbenzene is usually used as a reference in order to distinguish between macronet and standard networks. Postpolymerization cross-linking of polymers has advantages compared to the common copolymerization techniques. A uniform macronet structure can be normally achieved, owing to the statistical distribution of the cross-links (Tsyurupa and Davankov, 1980). This structure indicates high swelling ability of the polymers in various media, for the same degree of cross-linking. By following a postpolymerization route it is possible to control polymer architecture and even obtain block-order networks, having cross-links in separate domains.

Macronet polymers constitute a new category of materials that has been studied recently for the purpose of soil conditioning. This category of PSC includes cross-linked polymer networks, bearing bulky molecules as cross-links. Cross-linking is responsible for the three-dimensional network structure that characterizes these materials. Their characteristic swelling and elastic properties are attributed to the presence of physical or chemical cross-links within polymer chains (Flory and Rechner, 1943; Flory, 1975). The nature of cross-links strongly influences the properties of the cross-linked polymer. Cross-links in macronets preserve polymer chains at a distance. In this way, some unusual structural characteristics and properties are also attributed to macronet networks, such as high swelling ability, permeability of large molecules, and specific network design parameters. The compatibility of macronets with soil and their hydrophilic character as well increase proportionally to the number of ionic groups that are introduced into their structure.

The term macronets was accompanied by the term "isoporous" for years, especially concerning macronets prepared via postpolymerization cross-linking. Actually, the term "isoporous" was introduced in order to describe networks in which cross-links are introduced during postpolymerization modifications (Davankov et al., 1974; Davankov and Tsyurupa, 1980). Following a postpolymerization modification, a statistical distribution of cross-links is expected, since no preferable domains are normally existing. Consequently, a rather narrow pore size distribution was detected for macronets prepared in solution (Tsyurupa and Davankov, 1980). A macronet structure cannot always be isoporous, since pore size distribution is influenced by various factors (Jerabek and Setinek, 1989).

Some cross-linked three-dimensional hydrophilic polymers may undergo changes in response to stimuli such as temperature, pH, and ionic composition of the external solution (Yoshida et al., 1995; Theodoropoulos et al., 1996). These changes may have a reversible or an irreversible nature. Also, a large number of phase transitions occur. The possible presence of strongly and weakly acidic groups on the same polymeric substrate demonstrates that the swollen network presents a behavior that is strongly affected by the pH of the external solution. This characteristic is desired for soil conditioning. In this case their action comes mainly from their high water-holding capacity and secondarily from the availability of ion-exchange sites. Except for the release of residual moisture on soil, the presence of carboxylic acid groups in the polymer indicates that properties like aeration and resistance to erosion can be improved as well (Azzam, 1980).

III. THE POSTPOLYMERIZATION REACTION PRODUCTS

A. Cross-Linked Sulfonated Polystyrene

Cross-linked sulfonated polystyrene (PSS) consists of benzene rings substituted by sulfonic groups. For this reason, an extremely vigorous relationship with water is obtained during swelling. This behavior is more intense as macronet sulfonated polystyrene attains water-holding capacities of several hundred times its weight in water and aqueous solutions. Sulfonated polystyrene resins can be used as PSC in acid forms or converted to final ion forms (such as Fe^{2+}, K^{+}, or Na^{+}) that can be released to the plant via ion-exchange process.

Such structures can be obtained by the introduction of bulky groups between the styrene units of linear polystyrene via Fridel–Crafts reactions. Processes like these lead to three-dimensional polymer networks, which appear to exhibit advantages compared to the conventional styrene-divinylbenzene copolymers (Tsyurupa et al., 1974; Davankov et al., 1974; Tsyurupa and Davankov, 1980). 1,4-dichloromethyl-2,5-dimethylbenzene has been shown to be a desirable cross-linking agent for polystyrene (PS) modification (Peppas et al., 1976; Peppas and Valkanas, 1977; Barar et al., 1983; Bussing and Peppas, 1983a,b; Peppas and Staller, 1982). Friedel–Crafts cross-linking of PS can be performed in solution, by using various solvents and catalysts until gelation occurs. However, it is afterward necessary to crush the products to the appropriate particle size distribution needed for the following sulfonation reaction, since rather small particles are required to achieve a high-yield final product in a short time. More uniform products are obtained by a suspension reaction in various media, but the PS particles have to be isolated before the introduction of the sulfonic group (Peppas et al., 1976; Peppas and Valkanas, 1977; Barar et al., 1983a,b; Bussing and Peppas, 1983; Peppas and Staller, 1982; Rabek and Lucki, 1988).

Novel efficient processes have been developed for the synthesis of cross-linked PSS, which are based on the use of concentrated sulfuric acid as a suspending medium and a Friedel–Crafts catalyst for the cross-linking reaction of PS and simultaneously as a sulfonating agent (Theodoropoulos et al., 1992). Most of the swollen particles were found to be in spherical form due to the action of sulfuric acid as the suspending medium. Aggregates of two or more spherical particles could also exist. The average molecular weight between cross-links of the products after cross-linking and sulfonation was determined using a modified Flory–Rehner equation. The cross-linked and sulfonated spherical particles also exhibit ion-exchange properties. High ion-exchange capacities (4–5.5 meq/g) are observed that depend upon the degree of sulfonation and the degree of cross-linking (Theodoropoulos et al., 1992).

B. Hydrolized Polyacrylonitrile

Hydrolized polyacrylonitrile (HPAN) was primarily used as an aggregating agent within soil species. Polyacrylonitrile (PAN) preparation is usually based on free radical polymerization using redox systems, azo compounds, or peroxides as free radical initiators. Hydrolysis of PAN is an easily performed high-yield reaction in the presence of acidic or alkaline catalysis. Thus HPAN can be considered as a postpolymerization soil conditioner. Following hydrolysis a purification and neutralization stage is required. HPAN is applied to soil in dry form or as an aqueous solution.

It should be mentioned here that poly(acrylic) acid that corresponds to the same formula is produced via a polymerization reaction, usually in water or in water/oil dispersions. Poly(acrylic) acid is a well-known conditioner in linear or cross-linked form, but it cannot be classified as a postpolymerization conditioner. In 1952 Martin et al. applied HPAN as a soil conditioner using Na^+ and Ca^{2+} ions as the final form of polar group. This product, under the commercial name Krilium (Monsanto Chemical Company), was applied and studied for many years with promising results. Allison (1952) also used HPAN in sodium form using low concentrations. A few years later, Chepil (1954) applied HPAN in the presence of vinyl acetate maleic anhydride (VAMA) in order to improve soil stabilization and resistance to erosion. Other authors have also reported positive effects on physical soil properties by the addition of HPAN in the presence of inorganic compounds (Jones and Martin, 1957; Mosolova, 1970). HPAN in these examples was used as a water-soluble conditioner in order to improve soil properties, while higher-yield crops were obtained in every case.

C. Hydrolyzed Sulfonate Styrene Acrylonitrile

Macronet PSS is a monofunctional network. Hydrolyzed sulfonated styrene acrylonitrile copolymer (HSSAN) resin contains both sulfonic and carboxylic groups on the same backbone, so it can be considered as a multifunctional HSSAN combining the known effect of HPAN on mechanical properties and soil erosion and simultaneously the ability of PSS to retain water. HPAN has been much studied as a soil conditioner. It has been demonstrated that water penetration and the physical properties of soils are improved by adding this conditioner as a polymer solution. On the other hand, cross-linked water-insoluble polymers have been shown to hold large water quantities, even when added at low concentrations, thus improving the soil's water capacity. PSS has been applied to soil mixtures owing to its swelling ability and ion-exchange properties. Thus the preparation and application of a novel material that could provide the combined properties of PSS and HPAN were a challenge.

This resin must have the structure of an insoluble network, exhibiting high swelling ability in water. The swelling ratio is a property that can be tailored for certain soil conditioning applications. Network density is the most significant property of swollen materials, and in this line resins with different concentrations of the cross-linking agent can be synthesized. According to this designing scheme, this macronet multifunctional network has also been prepared by developing single reactor processes. Hydrolysis of acrylonitrile is a well-known reaction that can be performed in the presence of protonic acids like hydrogen chloride or sulfuric acid (Janacek et al., 1975; Bevington et al., 1958). In addition, sulfonation of macronet polystyrene is a rather easily applied procedure (Peppas and Staller, 1982; Theodoropoulos et al., 1993a). In this case, sulfuric acid is a common medium for the introduction of sulfonic groups. Sulfuric acid is also used as a Friedel–Crafts catalyst for a postpolymerization cross-linking and as a secondary cross-linking agent itself (Theodoropoulos et al., 1992). Thus it is possible to obtain networks having both sulfonic and carboxylic groups. Products having different degrees of cross-linking are achieved using bifunctional cross-linking agents and a compound of sulfur trioxide/trialkyl phosphate (Iovine and Ray-Chaudhuri, 1984). An alternative technique to prepare highly swellable macronets with different degrees of cross-linking having two ionic groups has been described (Theodoropoulos et al., 1995). Using a random poly(styrene-co-acrylonitrile) copolymer as raw material, a series of HSSAN copolymer networks via postpolymerization modification has also been prepared. HSSAN has excellent water-absorption properties, and it was a challenge to use it as a PSC that improves water-retention by the soil substrate.

However, another interesting fact was observed as well. One of the products of the described process, which corresponds to zero nominal cross-linking agent concentration, has exhibited a transition between the gel and solution form under some specific conditions. Critical phenomena in hydrogels have attracted much interest owing to their scientific importance and applications (Ilmain et al., 1991). The transition of slightly cross-linked hydrogels to polyelectrolyte solutions has been known for years (Flory, 1975). In our case, due to the high swelling ability and the parallel presence of carboxylic groups, the transformable polymer combines properties of highly swollen hydrogels with properties of water-soluble soil aggregating agents. This *in situ* transformable soil conditioner (Bouranis et al., 1994) acts in addition to natural aggregating and water-retaining agents. Although pores exist in this structure, the porosity of the polymers remains at low values, and samples can be classified as gel-type resins.

D. Macronets from Recycling Polymers

As mentioned, an interesting possibility is that of using recycled polymers as raw materials for PSC preparation, since this modification can be applied for already polymerized materials. This direction in conditioner synthesis is expected to acquire great status in the coming years, owing to the degradation of the initial material properties incurred by recycling it for the same use. In many cases recycling is directed towards the retrieval of the initial properties of synthetic materials, but it often leads to undesired deterioration of their mechanical properties. An alternative approach to polymer recycling may help to lessen environmental pollution. Polymer recycling is usually associated with large-scale applications. The use of soil conditioners prepared from recycled hydrophobic materials is promising for polymer disposal. In this direction, recycled polymers produced from well-known commercial products have been achieved following postpolymerization procedures.

For a hydrophobic material to obtain such properties, the polymer has to be altered by the partial or full introduction of ionic groups such as —SO_3H, —COOH, —NH_2 (Vasheghani et al., 1990; Wang and Bloomglied, 1990; Belfer and Glozman, 1979; Andreopoulos, 1989a,b). However, a water-soluble linear hydrophilic polymer may become a pollutant. This problem can be faced by the introduction of stable chemical cross-links between the linear macromolecules (Barar et al., 1983a,b). The sequence of the two processes depends on the type of the recycled polymer; a combination of these two is also possible.

Modified ionic resins derived from recycled commercial polymers have been prepared according to postpolymerization processes. The addition of swollen sulfonated gels on soil substrate had positive effects on its properties and on plant cultivation (Theodoropoulos et al., 1993b). Since the compatibility of a synthetic hydrophobic material and natural substrate is low, the polymers were cross-linked so as to eliminate their solubility in water, which can cause pollution. Besides, the sulfonic group was introduced into the polymer chains to ensure that a hydrophilic character is achieved. Polystyrene-containing fountain pens (Bic) and shaving products (Gillette) have been used as raw materials (Theodoropoulos et al., 1994). They were cross-linked and sulfonated according to a modified version of a suspension technique (Theodoropoulos, 1993). Another interesting approach is a reaction of solid material in the absence of solvent that can easily be applied in polymer recovery as well.

E. Natural Polymers Grafted with Synthetic Compounds

Natural polymers, usually starch or cellulose modified with synthetic compounds, can be obtained via a postpolymerization route. In this case the natural polymer is mostly used as raw material. The advantage of these conditioners is the combination of soil–polymer compatibility, due to the natural polymer, and the mechanical and biological stability of the synthetic compound (usually polyacrylonitrile-polyacrylamide). Besides, since this category refers mainly to cross-linked and water-insoluble conditioners, high swelling ratios are obtained here, as a result of the variety of the grafting reactions. The common way to obtain natural polymers grafted with synthetic compounds is grafting of an unsaturated monomer (i.e., acrylamide, acrylic acid, acrylonitrile) onto starch or cellulose. If acrylamide or acrylonitrile is used, grafting is followed by a hydrolysis reaction. No specific cross-linking agent is normally employed. Cross-linking arises from the degree of hydrolysis either of a chemical nature or with the remaining nonionic domains (physical cross-links).

A typical modification in this area involves starch modification by polymerization of acrylonitrile. Since radical polymerization is usually employed, ceric salts are mainly used as initiators. Hydrolysis is usually performed in alkaline environments under pressure and heat. The products are purified and neutralized following preparation. Since starch is a natural product, extraction of fats and other pretreatments give upgraded properties of the final product (Weaver, 1976a,b,c, 1984). An excess of the synthetic compounds is usually used when acrylonitrile or acrylic acid is used, and similar reaction systems are employed, while the step of saponification is performed under mild conditions or even avoided. Water/methanol, water/toluene, and water/cyclohexane are also reported as solvents (Reid, 1977; Heidel, 1988). A variety of examples of starch-polyacrylamide conditioner application in soil systems is reported (Barry et al., 1991; Kizawa et al., 1990; Weaver, 1984) while blends of starch-grafted polyacrylonitrile or polyacrylamide with unmodified starch or polyvinyl alcohol

(PVA) are reported (Pó, 1994). Starch grafted, acrylic-acrylamide copolymer conditioners should be classified in the present category (Henderson, 1985).

Regarding cellulose as raw material, various types of products can be obtained by varying the hydrolysis reaction conditions and the kind of cellulose. The swelling ability, the mechanical gel strength, and the yield of the final product are influenced by these factors. When low mechanical strength products are obtained, the addition of a bifunctional cross-linking agent (e.g., bis(acrylamido)acetic acid) gives improved mechanical properties as a result of the increased degree of cross-linking (Pó, 1994). Carboxymethylcellulose (Kadyrov, 1989) and cellulose-methacrylic acid copolymer (Matsushima, 1986) are reported as soil conditioners, while an *in situ* production of modified poly(saccharide) is also reported (Barclay and Lewin, 1985). Humic acids grafted onto polymers (Amirkhanova et al., 1982; Dzhanpeisov et al., 1990) should be considered as natural substrates on synthetic compounds. As previously mentioned, the combination of the natural part's compatibility with soil and the stabilization as a result of the synthetic polymer is the main advantage of the products. Some additional advantages arise from the absence of organic solvents in the majority of the processes and the convenience of the purification stages.

F. Miscellaneous Modified Polymers

There is also a variety of polymers designed as soil conditioners via postpolymerization reactions that cannot be classified in the previously described categories. For example, Inomoto and Ito (1972) discuss PVA modification via acetylation by using aminaldehydes. A mixture of PVA and polyvinylpyrrolidone was subjected to cross-linking via irradiation (Orihara, 1988). The cross-linking of methacrylic acid–acrylonitrile copolymer in the presence of formaldehyde is also a related example (Takeuchi and Tanaka, 1978).

IV. PROPERTIES OF CROSS-LINKED POLYMERIC SOIL CONDITIONERS

A. Water Retention

As mentioned, the water-absorbing polymers are able to swell and absorb amounts of water or aqueous solutions as high as 10–1000 times their own weight. According to Pó (1944), water-absorbing polymers should have high water absorbency and retention (they should absorb and hold high amounts of fluid), even under load of other saturated materials ("suction power"); high water absorption rate (high amount of fluid absorbed per unit of time); good gel strength after absorption (the material should not become slimy and flow away); the fluid absorption should be reversible (the polymer should be able to absorb and release the fluid several times; they must be atoxic; they must not release traces of unreacted monomers; and monomers and production process costs should be low.

When one increases the cross-linking degree, the absorption rate and the gel strength increase as well, but the absorption capacity decreases. Water-absorbing polymer properties' performances are evaluated according to different methods. *Absorption capacity* is measured by putting a polymer sample in water for a time sufficient to saturate it and then weighing the gel obtained. "Absorption capacity," "gel capacity," "water absorbency," "ratio of absorption," "water retention value," and " swell index" are some of the commonly used definitions. They represent either 1) the weight ratio between swollen polymer and dry polymer or 2) the weight ratio between the absorbed water (swollen polymer weight minus dry polymer) and dry polymer. Because the dry polymer weight is much less than the swollen polymer weight, in practice the two figures coincide. *Absorption rate* values can be expressed in two ways, either 1) through the amount of fluid absorbed by a weighted amount of the polymer in a fixed short time or 2) through the time required by the polymer sample to absorb a given amount of fluid; the two parameters are in a contradictory relationship (Pó, 1994).

Macronets have the elasticity and swelling properties that derive from their three-dimensional network structure. A macronet polymer is able to swell more significantly than a standard network with the same degree of cross-linking (Davankov et al., 1973). The swelling ability of a macronet polymer depends on the nature and the nominal concentration of the cross-linking agent, the average molecular weight of the raw material, and the volume fraction of the initial polymer or monomer mixture. Regarding the swelling of ionic macronets in aqueous solutions, the degree of swelling depends also strongly on the ionic composition of the external solution. As the osmotic forces contribute to the swelling ability of the network, some rather high swelling degrees in the order of magnitude of several hundred times the volume of the dry polymer are attributed to the facility of extension of a macronet (Theodoropoulous et al., 1994; Theodoropoulos, 1993).

The kinetics of water-holding capacity (Kakoulides et al., 1993) achieved by high density PSS (PSS_{HD}) and low density PSS (PSS_{LD}) at 20°C, respectively, with water being the swelling medium, are illustrated in Figs. 4a and 4b. PPS_{LD} attained 4.6 times the maximum swelling attained by PPS_{HD}. This fact is due to the increased density of the net, which in turn is a result of differences in the concentration of the bifunctional cross-linking agent. The water retentions of the two polymers with respect to their maximum swelling versus time are shown in Fig. 4c. Water-holding capacity in tap water (400 μS/cm at 25°C) at saturation time was 13 and 58 g/g for PSS_{HD} and PSS_{LD}, respectively, while in distilled water (0.13 μS/cm at 25°C) the respective values were 19 and 74 g/g, which is much greater. This is attributed to osmotic pressure differences that are created by differences in the concentration of the electrolyte in the polymer and

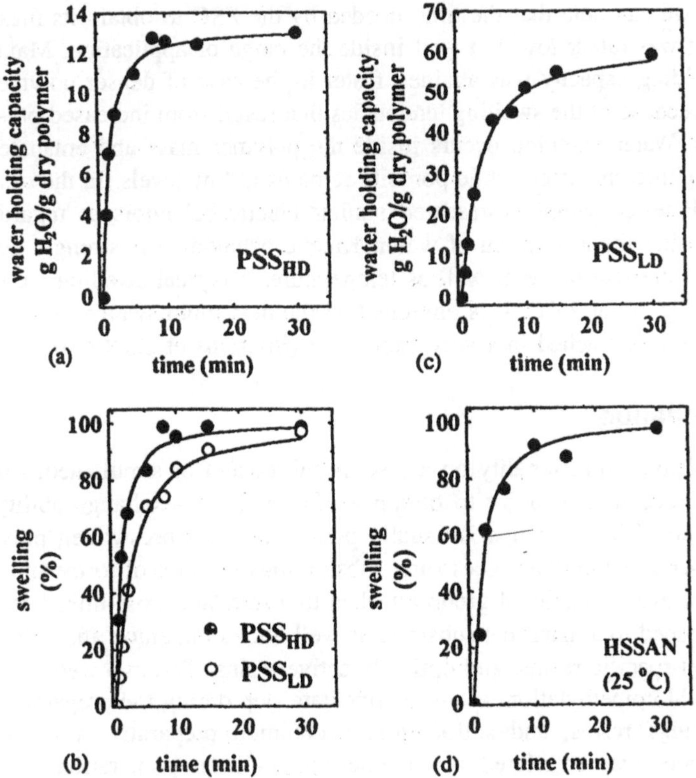

Figure 4 The kinetics of water-holding capacity achieved by a high-density cross-linked sulfonated polystyrene (PSS_{HD}, Fig. 4a) and a low-density cross-linked sulfonated polystyrene (PSS_{LD}, Fig. 4b). Typical swelling curves for PSS_{HD}, PSS_{LD}, (Fig. 4c), and hydrolyzed sulfonated styrene-acrylonitrile HSSAN (Fig. 4d).

the surrounding solution. As this difference increases (in the case of distilled water), so do the osmotic pressure and the swelling tendencies. Data were fitted using the equation

$$C_w = \frac{C_{w,max}(t/T)^n}{1+(t/T)^n}$$

where: C_W = the water capacity of the polymer at 20°C (g/g), $C_{w,max}$ = the water capacity of the polymer at swelling equilibrium state, t = time (min), T = the time necessary to obtain 50% swelling, and n = a constant that is dependent on temperature and the structure of the material. According for the predefined kinetic model, it was calculated that T = 1.1 min, n = 1.23, $C_{w,max}$ = 13.1 g/g for PSS_{HD} and T = 2.8 min, n = 1.20, and $C_{w,max}$ = 60.6 g/g for PSS_{LD}.

In general, we can note that the time needed by the PSC to obtain its maximum swelling was rather low and well inside the range of application. Maximum water-holding capacity was attained faster in the case of denser polymer nets (PSS_{HD}) because of the swelling tendencies that result from increased presence of bonds. Water retention occurs inside the polymer mass and emanates from its hydrophilic character, while porosity remains at low levels. In the case of HSSAN, all tested samples presented similar kinetic behaviors as regards their swelling after the absorption of water. Kinetic behavior was strongly dependent on particle size range as well as temperature. A typical swelling curve is depicted in Fig. 4d at 25°C. It is obvious that the maximum swelling capacity of the resins was reached in a very short time (Bouranis et al., 1994).

B. Ion Exchange

Introduction of ionic functionality on a PSC is subsequent or simultaneous to cross-linking procedure, in order to obtain a PSC with ion-exchange ability. The introduction of ionic groups is usually performed on a preswollen polymer. A macronet structure undergoes such substitutions because of to the inner surface. Almost every functional group attached to a standard cross-linked matrix can be attached to a macronet substrate as well, including anion and cation exchangers, ion-specific resins, and optically active resins. The two well-studied reactions, chloromethylation, as an intermediate step during the preparation of anion-exchange resins, and sulfonation, a common preparative route of cation-exchangers, may be related to macronet ion-exchange preparation without using the typical bifunctional cross-linking agents. In some cases sulfone cross-linking is undesirable during macronet cation-exchange preparation, especially when cross-linking should be attributed solely to typical more bulky cross-linkers, or when low and controlled degree of cross-linking is required. Sulfone cross-linking can be avoided by using sulfone inhibitors (Peppas et al., 1981; Peppas and Staller, 1982; Iovine and Ray-Chaudhuri, 1984). Usually high ion-exchange capacities are observed for macronet ion-exchangers (Theodoropoulos et al., 1992; Tsyurupa et al., 1974). The ion-exchange properties are associated with an increase of the hydrophilic character and result in elevated swelling degrees of the polymer in polar media, when ionic groups are introduced to cross-linked macronets to a certain degree, usually more than 25% of the available units.

The cross-linking density, the linear dimensions of the cross-linker, and the diameter of the sorbed ion have strong effects on the permeability of macronet ion-exchangers for ions, since this is determined by the size of the inner channels in the network as well as by the distances between the polymer chains. Usually, for gel-type networks, high permeability is obtained in low degrees of cross-linking (Tsyurupa et al., 1974). Besides, by increasing the chain length of the cross-linking agent larger pores are detected.

The availability of ion-exchange sites combined with the high swelling ability is promising for soil conditioning, an area where swellable polymers of uniform structure are required. Although the swelling behavior of the macronet is well studied, a lack of detailed information exists concerning their ion-exchange properties (Theodoropoulos et al., 1996). The general aspects of the kinetic behavior of ion-exchangers is understood for simple systems. However, in the case of macronets, due to their high swelling ability the boundary conditions are more complex than in standard ion-exchangers. Ion-exchange kinetics is primarily based on transport phenomena (Belfer and Glozman, 1979; Boyd and Soldano, 1953; Buijs and Wesselingh, 1980; Davankov and Semechkin, 1977; Flynn et al., 1974; Helfferich, 1995; Limberti et al., 1980). According to this theory, the rate controlling step was suggested to be diffusion: 1) in the external solution up to the surface of the polymer (film diffusion), and 2) in the polymer particle itself (particle diffusion). An intermediate range of both mechanisms may also exist while the chemical exchange in the vicinity of the exchange groups is in general extremely rapid. In such phenomena, rate equations based on physical theoretical background and mechanistic approaches known to occur provide analytical information. In fact the diffusion coefficient is difficult to calculate without the assumption in ion-exchangers for several reasons: 1) A part of the cross-sectional area in the polymer is occupied by the network chains and is not available for ion penetration. 2) Large ions may be impeded in mobility by the network. 3) The interaction of the charged ions with the fixed ionic groups has to be taken into account. 4) Concerning penetration from a glassy state, the diffusion paths can be changing during ion-exchange processes. The last reason is more important for highly swellable polymers like macronets.

V. THE SOIL CONDITIONING ACTION OF CROSS-LINKED POLYMERS

A. Physicochemical Characteristics of Sand–Polymer Mixtures

Data for the water-holding capacity and the apparent density of sand, PSS_{HD}, and PSS_{LD} are given in Table 1. The water capacity for sand–polymer mixtures of 0.1, 0.2, and 0.3% w/w is given in Table 2 (Kakoulides et al., 1993). The water-holding capacity of the polymer decreased (dramatically) in the presence of sand, as pressure caused by the sand layers that surround the polymer particles overwhelms the osmotic pressure that is responsible for swelling tendencies.

Table 1 Water-Holding Capacity and Apparent Density Values for Sand, PSS_{HD}, and PSS_{LD} Swelled in Water

	Water-holding capacity (g/g)	Polymer apparent density (g/cm^3)
Sand	0.109	1.386 (dry) 1.558 (wet)
PSS_{HD}	9.4 (±0.9)	0.367 (±0.068)
PSS_{LD}	44.1 (±4.2)	0.679 (±0.044)

Source: Adapted from Kakoulides et al. (1993), with permission.

Table 2 Water-Holding Capacity of PSS Swelled After Its Incorporation into Sand–Polymer Mixtures

Polymer–sand mixture (%)	0.1	0.2	0.3
Dry sand mass (g)	1000	1000	1000
Dry mixture mass (g)	1001	1002	1003
		PSS_{HD}	
Wet mixture mass (g)	1116	1121	1234
Polymer WHC (g/g)	6	5	7.3
		PSS_{LD}	
Wet mixture mass (g)	1135	1165	1187
Polymer WHC (g/g/)	25	27	25

WHC: water-holding capacity.
Source: Adapted from Kakoulides et al. (1993), with permission.

The effect of the incorporation of original and recycled macronets in soil systems is given in Fig. 5. This effect appeared more intense as polymer concentration increased. Figures 5a and 5b demonstrate the swelling of mixtures that consist of different polymer–sand ratios versus the volume of added water, for PSS_{HD} and PSS_{LD}, respectively. Figure 5c presents the increase of column volume versus polymer concentration. The water-holding capacity of the mixtures increased proportionally to the rate at which the polymer was applied into the substrate. The increase in volume is attributed to retention of greater amounts of water in the presence of polymers. These figures indicate that water retention in substrate mixtures increased proportionally to the swelling capability and the polymer–sand ratio. The reason for not studying high rates of polymer application in substrate is that they cannot be used in such amounts because of technical and economic factors.

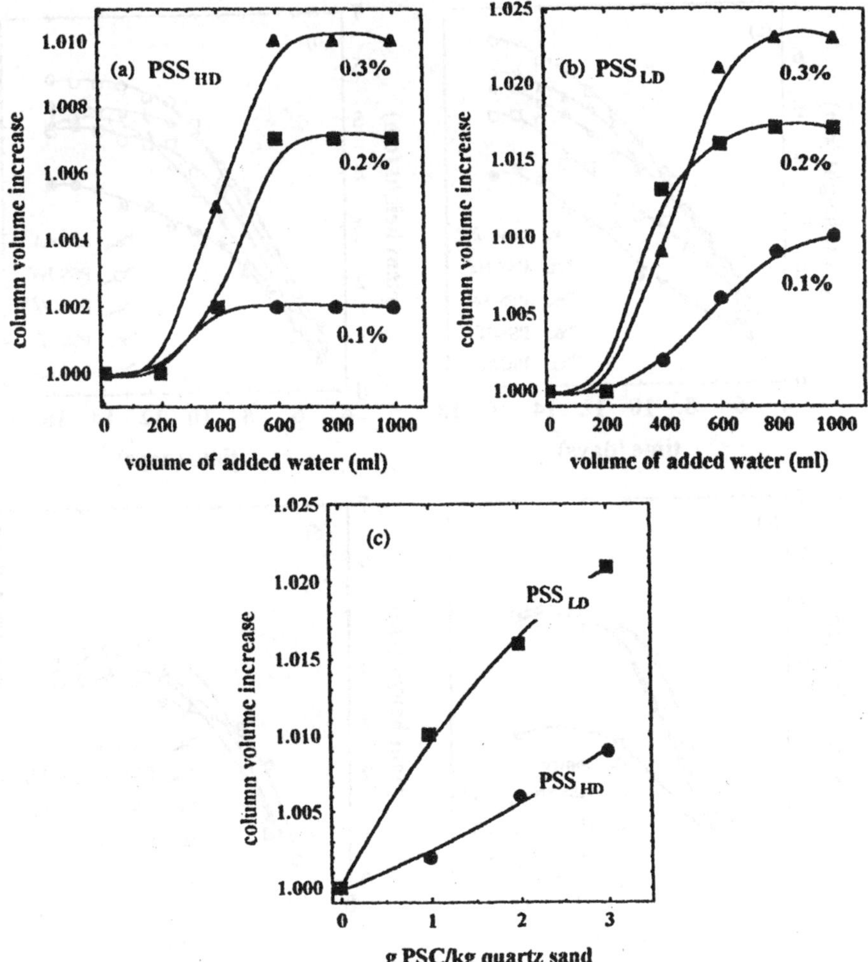

Figure 5 The swelling progress of high-density (a) and low-density (b) PSS–quartz sand mixtures versus the volume of added water, and (c) the maximum obtained swelling versus the concentration of the applied polymer. (Redrawn from Kakoulides et al., 1993, with permission.)

The above-mentioned data point out the short time that such cross-linked PSCs need to obtain maximum swelling in water. These products need less than 8 minutes to attain maximum swelling. Water capacity decreases in sand–polymer systems, while increased rates of sulfonated polystyrene increase the amount of water retained by substrate mixtures. Finally, the water capacity of mixtures is enhanced as the water capacity of the polymer itself increases.

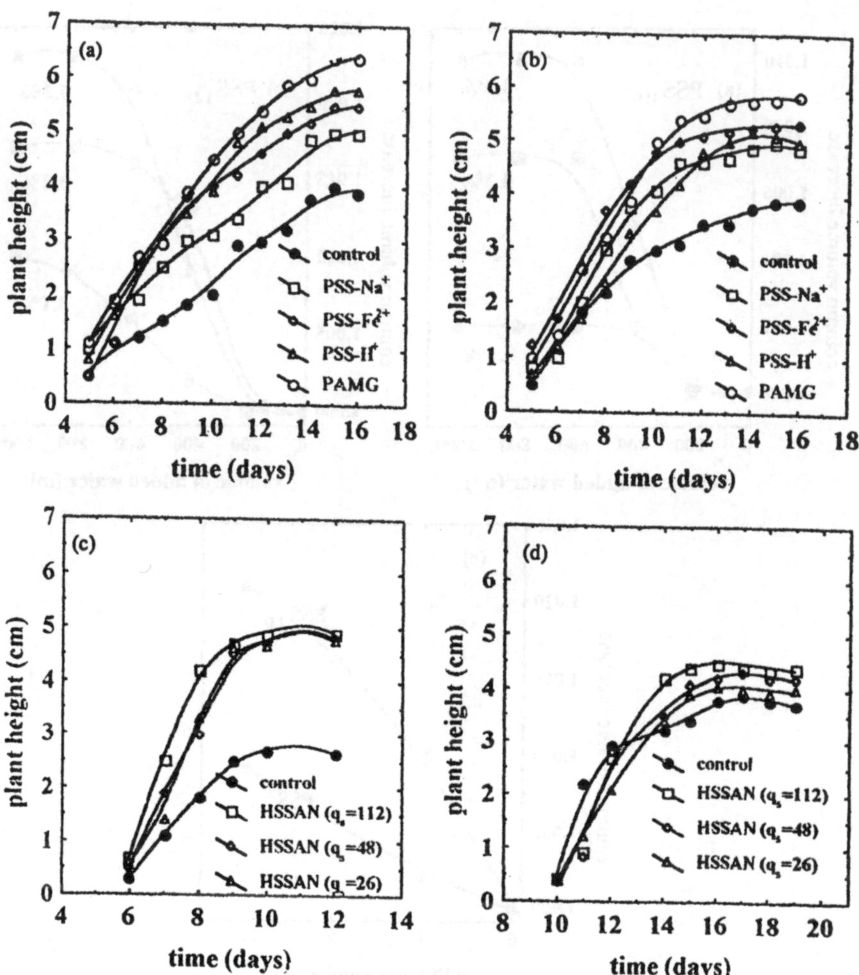

Figure 6 Plant height of tomato seedlings versus time in soil–polymer substrate or sand–polymer substrate, in the absence or presence of various polymeric soil conditioners prepared via postpolymerization reaction. (Adapted from Theodoropoulos et al., 1993, and Bouranis et al., 1994, with permission.)

B. Growth of Seedlings

The growth of tomato seedlings in a sandy loam soil and quartz sand in the presence of various PSCs is depicted in Figs. 6a and 6b, respectively. The obtained differences are attributed to positive effects of the resins as compared to plants cultivated in the absence of PSCs. Slightly improved effects were obtained in the case of the Fe^{2+} form in the sandy loam soil, and especially in the quartz sand, owing to the nutritional needs of the seedlings, but these deviations

Table 3 Dry Weight of Tomato Seedlings per Pot, Grown in Soil–Polymer and Sand–Polymer Mixtures as Substrates, in the Absence and Presence of Various Polymeric Soil Conditioners Prepared via Postpolymerization Reactions

	Dry weight (mg ±SE)	
Polymer	Soil	Sand
Control	3.1 (±1.2)	2.9 (±0.40)
PAMG	5.7	5.4
PSS-H^+	5.3 (±2.1)	5.0 (±1.8)
PSS-Na^+	5.3 (±0.6)	5.3 (±1.2)
PSS-Fe^{2+}	5.2 (±1.6)	5.4 (±1.8)

Source: Adapted from Theodoropoulos et al. (1993), with permission.

are not significant. The behavior of polyacrylamide gel (PAMG) was better, in comparison to various PSS forms. This fact can be attributed to the high holding capacity of polyacrylamide gels in water. Dry organic matter of tomato seedlings per pot is illustrated in Table 3, where similar results were obtained (Theodoropoulos et al., 1993b).

These results lead to several considerations: 1) The addition of cross-linked PSS has positive effects on soil–plant systems compared to reference experiments without conditioner. On the other hand, PAMG has given better growth results in every case, due to the enhanced swelling ratio of polyacrylamide gels. These facts also indicate that the swelling ratio and the water holding capacity as well are determining factors for the application of such insoluble soil conditioners in substrate mixtures. 2) The final ion form of the conditioner can have a positive effect on plant development, but the possible differences observed are not large. From the various ion species studied, the Fe^{2+} form gave slightly better results. 3) A very interesting issue is the ability of cross-linked PSS to be applied as a soil amendment in acid form. This point is becoming more important as we take into consideration that this is the form in which the polymer is obtained under current preparation conditions. Besides, the neutralization of a swollen network is a rather demanding process and has negative effects on the cost of the final product. Additional evidence on these points is presented in Table 3, where similar differences are shown for dry organic matter content per pot. The influence of the PSS-H^+ form on the pH of soil and sand as substrate is shown in Fig. 7. This influence is insignificant owing to the low concentration of the applied polymer in the soil mixtures. Actually these differences are more intense in quartz sand, since the ion-exchange processes of soil are also combined. The pH of the soil in-

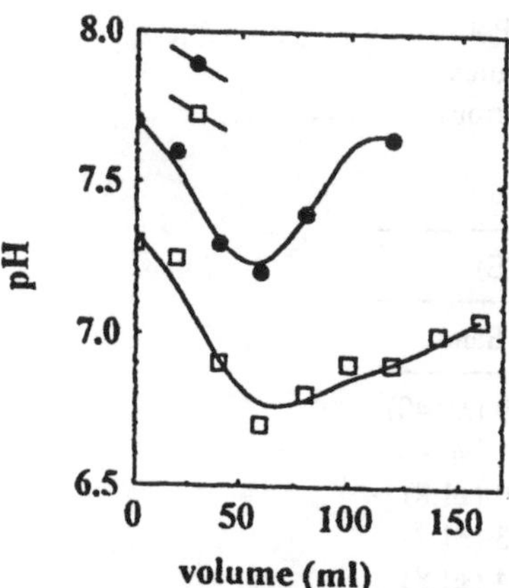

Figure 7 The influence of the PSS-H+ form on the pH of soil ■, and sand □, used as substrates versus the volume of the added water. (Redrawn from Theodoropoulos et al., 1993, with permission.)

creases to the initial level again, since water is replaced during the experiment and the resin becomes saturated. Thus ion-exchange resins having sulfonic groups attached on their backbones can be applied even in their acid form, since their effects on the pH of the substrate are insignificant.

The emergence and growth of tomato seedlings into the HSSAN-treated soil and the HSSAN-treated quartz sand mixtures are depicted in Figs. 6c and 6d, respectively. The presence of HSSAN increased significantly the height of seedlings. Differences in respect to swelling ratio were obtained from the first three days of seed germination in the HSSAN–soil mixtures. Then no further effects were detected. The height of the seedlings was about 84% higher (Table 4), and the polymer without cross-linking agent (sample A) presented the best action. Positive effects were also obtained when seeds were germinated in the HSSAN–quartz sand mixtures. The presence of HSSAN improved the height of seedlings by about 30% in the same manner as in the previous case. Sample A also presented the best action. Additional evidence concerning the value of HSSAN presented in Table 4 is given as the dry weight of seedling obtained per pot (mg) in the sandy-loam soil. HSSAN increased the biomass production, and the slightly cross-linked polymers appeared more effective than the denser networks. Polymer concentration was 0.15% (w/w). A significant decrease of agricultural polymer concentration used in soils was observed during recent years because of economic factors. In fact, a concentration in this range is adequate for plant growth. Similar positive effects were also obtained when lettuce seeds were germinated in the HSSAN–quartz sand mixtures (Bouranis et al., 1994).

Table 4 The Relative Increase of Tomato Seedling Heights and Their Dry Weight, Grown in Soil–Polymer or Sand–Polymer Mixtures as Substrates, in the Absence and Presence of HSSAN Samples Having Various Swelling Ratios

	Increase of seedling height (%)		Dry Weight (%)
Sample	soil	sand	soil
A	84.1	30.6	42.6
B	81.5	19.2	38.9
C	79.7	12.5	18.5

Sample	A	B	C
Cross-linking agent (% w/w)	0	1.5	4.5
Swelling ratio (q_s)	112	48	26

Source: Adapted from Bouranis et al., 1994, with permission.

These data indicate that the multifunctional resin HSSAN can be used as a soil conditioner. When this polymer is incorporated into substrate (soil or sand), it retains significant amounts of water, which contributes to the emergence and growth of seedlings. In this way, larger seedlings are obtained. It is obvious that the growth promotion of these polymers comes from their swelling ability. The network density, as a result of the bifunctional cross-linking agent concentration, is a critical factor for water-insoluble soil conditioners. Thus the swelling ratio of HSSAN on the emergence and growth of tomato and lettuce seedlings plays an important role. The polymer having a higher swelling ratio ($q_s = 112$, sample A) presented an interesting action. These results mean that no cross-linking agent or minor amounts are required for HSSAN best acting preparations. Even in the absence of a conventional cross-linking agent, a secondary cross-linking process can occur. This phenomenon may be due to the chemical nature attributed to the sulfone type cross-links or to the physical nature due to the orientation of chains in the network.

Nevertheless, these products seemed to be unstable compounds in aqueous environments, as a result of osmotic forces. Indeed, it was found that sample A was becoming a completely water-soluble material within a few months. This interesting property can be employed for the preparation of *in situ* transformable soil conditioners, affecting the water-holding capacity in the initial stage of plant growth, and the soil erosion afterwards, due to the action of the water-soluble hydrolyzed polyacrylonitrile on soil stability. In this way, slightly cross-linked products or products having low cross-linking agent concentrations are indicative of improved soil properties. During the following years, emphasis will be given to polymers having transition or limiting properties, such as water-insoluble materials slowly transformed to water-soluble materials.

C. Cross-Linked Polymers and Fertilizers

The action of a PSC and a controlled-release fertilizer was studied by Chatzoudis and Valkanas (1995a,b). The change in soil water-holding capacity and the release of potassium sulfate from conventional and controlled-release forms was evaluated using leaching experiments in soil columns. These experiments showed that it maintained increased water regimes in soil as well as the controlled release of nutrients from devices fabricated using conventional fertilizers and paraffin wax-modified rosin resins, which reduce fertilizer leaching into underground water and other aquifiers.

Tomato growth was followed by the comparison of biomass yield on a dry matter basis under different combinations of controlled-release or conventional fertilizer rates and soil conditioner applications. Tomato growth experiments showed that the combined application of PSS with controlled-release fertilizers resulted in an increase in plant yield on a dry mass basis and enhanced the plant nutritional status (Chatzoudis and Valkanas, 1995a).

Lettuce growth was also examined in pot experiments, under different soil–water conditions. In one experiment, the effect of PSS in combination with slow-release and water-soluble fertilizers was considered. It was shown that the use of soil conditioner with slow-release fertilizers gives greater yield than the use of water-soluble fertilizers alone. In another experiment, lettuce plant resistance to moisture stress with the use of PSS was evaluated. This experiment showed that there was no difference in the yield and quantity of lettuce plants between the trials until the irrigation interval of twenty days (Chatzoudis and Valkanas, 1995b).

D. The Agricultural Field Use of Polymeric Soil Conditioners

In a controlled-environment study (Volkmar and Chang, 1995) the capacity of two commercial hydrophilic PSC; based on sulfonated polystyrene, Grogel and Transorb (trade names of products), to mitigate the effects of recurring moderate water-deficit stress (dry-down to 50% field capacity before rewatering) on growth and yield of barley and canola was evaluated. Rates of 0.03, 0.12, 0.47, and 1.87 g polymer kg^{-1} sandy loam soil (1, 4, 16, and 64 times the recommended commercial application rate) were tested. Plants were grown at a soil moisture content of approximately 50% of field capacity. According to Volkmar and Chang (1995), neither polymer was effective at the commercially recommended rate. Barley and canola grain yields were unaffected at any Grogel rate, and Transorb had no effect on barley grain yield. Grogel at the highest rate enhanced early shoot mass, mature biomass production, and grain yield of barley and increased leaf RWC (relative water content). Canola had greater early and late vegetative biomass, but pod yield was not increased by Grogel at any rate. Transorb was most effective at four times the recommended rate, significantly increasing tiller and fertile spike number and mature biomass production at that rate. Leaf RWC were unaffected by Transorb treatment. Grogel stimulated root growth of barley but had no effect on roots of canola. Both polymers tended to increase consumptive water use. Spatial restriction was found to reduce drastically the water retention of both polymers and limit the absorbency of both polymers in this study. Volkmar and Chang (1995) concluded that the high rates of polymer required to elicit a crop yield response under relatively mild water-deficit conditions limit the value of these polymers for agricultural field use for the crop species tested.

VI. CONCLUDING REMARKS

The recent trends and perspective of cross-linked polymeric soil conditioners are particularly connected with postpolymerization processes. This way involves the modification of a preexisting macromolecule, concerning the anticipated properties of the final product. So cross-linked PSC, prepared via postpolymerization reactions are based on various modifications of linear macromolecules and grafted or block copolymers with multifunctional compounds as cross-linking agents. Especially, 1) recycled polymers produced from commerial products and 2) natural polymers, usually starch or cellulose modified with synthetic compounds, can be obtained via a postpolymerization route. An important point is that by following a postpolymerization process it is possible to control polymer architecture or even obtain block-order networks having cross-links in separate domains. Characteristics of this category of PSC are the uniformity of cross-linking, the application on polymer recycling, and the maintenance of polymer architecture. Such polymers have the structure of an insoluble network and are applied to soil as conditioners, because of their high water-holding capacity and ion-exchange properties. Most of the cross-linked hydrophilic polymers are used as water-retaining agents. The swelling ratio and the network density too are properties that can be tailored for certain soil conditioning applications. Novel techniques of polymer preparation, more complicated structures, materials with combined properties and lower cost, reflect the standards of future conditioners.

ACKNOWLEDGMENTS

I wish to express my appreciation to Prof. G. N. Valkanas, who introduced me to the subject of synthetic soil conditioners, and to Dr. A. G. Theodoropoulos for his excellent scientific cooperation and friendship as well.

ABBREVIATIONS

PSC: polymeric soil conditioners; PSS: sulfonated polystyrene; PS: polystyrene; PAN: polyacrylonitrile; HPAN: hydrolyzed polyacrylonitrile; HSSAN: hydrolyzed sulfonated styrene acrylonitrile; PVA: polyvinyl alcohol; PSSHD: high-density PSS; PSSLD: low-density PSS; PAMG: polyacrylamide gel.

REFERENCES

Allison, F. E. (1952). Effect of synthetic polyelectrolytes on the structure of saline and alkali soil. *Soil Sci 73*: 443–454.

Amirkhanova, A. K., Akkulova, Z. G., and Krichevskii, L. A. (1982). Study of electrochemical grafting of acrylamide to humic acids of coal. *Deposited Doc. Viniti* 6283–92 (in Russian).

Andreopoulos, A. G. (1989a). Hydrophilic polymer networks for agricultural uses. *Eur. Polym. J. 25*: 977–979.

Andreopoulos, A. G. (1989b). Preparation and swelling of polymeric hydrogels. *J. Appl. Polym. Sci. 37*: 2121–2129.

Azzam, R. A. I. (1980). Agricultural polymers; polyacrylamide preparation, application and prospects in soil conditioning. *Commun. Soil Sci. Plant Anal. 11*: 767–834.

Azzam, R. A. I. (1983). Polymeric conditioner gels for desert soils. *Commun. Soil Sci. Plant Anal. 14*: 739–760.

Barar, D. G., Staller, K. P., and Peppas, N. A. (1983a). Friedel–Crafts crosslinking methods for PS modification. 3. Preparation and swelling characteristics of cross-linked particles. *Ind. Eng. Chem. Prod. Res. Dev. 22*: 161–166.

Barar, D. G., Staller, K. P., and Peppas, N. A. (1983b). Friedel–Crafts crosslinking methods for PS modification. 4. Macromulecular structure of crosslinked particles. *J. Polym. Sci. Pol. Chem. 21*: 1013–1024.

Barclay, W. R., and Lewin, R. A. (1985). Microalgal polysaccharide production for the conditioning of agricultural soils. *Plant Soil 88*: 159–169.

Barry, P. V., Stott, D. E., Turco, R. F., and Bradford, J. M. (1991). Organic polymers effect on soil shear strength and detachment by single raindrops. *Soil Sci. Soc. Am. J. 55*: 799–804.

Bevington, J. C., Eaves, D. F., And Vale, R. L. (1958). Test on the hydrolysis of certain synthetic polymers. *J. Polym. Sci. 22*: 317–322.

Belfer, S., and Glozman, R. J. (1979). Anion exchange resins prepared from PS crosslinked via a Friedel–Crafts reaction. *J. Appl. Pol. Sci. 24*: 2147–2157.

Bouranis, D. L., Theodoropoulos, A. G., Theodosakis, G. I., and Valkanas, G. N. (1994). Effect of the swelling ability of a novel hydrolysed acrylonitrile-sulfonated polystyrene network on growth of tomato and lettuce seedlings. *Commun. Soil Sci. Plant Anal. 25*: 2273–2283.

Bouranis, D. L., Theodoropoulos, A. G., and Drossopoulos, J. B. (1995). Designing synthetic polymers as soil conditioners. *Commun. Soil Sci. Plant Anal. 26*: 1455–1480.

Boyd, G. E., and Soldano, B. A. (1953). Self-diffusion of cations in and through sulfonated polystyrene cation-exchange polymers. *J. Am. Chem. Soc. 75*: 6091–6099.

Buijs, A., and Wesselingh, J. A. (1980). Batch fluidized ion-exchange column for streams containing suspended particles. *J. Chromatography 201*: 319–327.

Bussing, W. R., and Peppas, N. A. (1983). Friedel–Crafts crosslinking methods for polystyrene modification. 1. Preparation and kinetics. *Polymer 24*: 209–216.

Chatzoudis, G. K., and Valkanas, G. N. (1995a). Monitoring the combined action of controlled-release fertilizers and a soil conditioner in soil. *Commun. Soil Sci. Plant Anal. 26*: 3099–3111.

Chatzoudis, G. K., and Valkanas, G. N. (1995b). Lettuce plant growth with the use of soil conditioner and slow-release fertilizers. *Commun. Soil Sci. Plant Anal. 26*: 2569–2576.

Chepil, W. S. (1954). The effect of synthetic conditioners on some phases, soil structure and erodibility by wind. *Soil Sci. Soc. Am. Proc. 18*: 386–391.

Cook, D. F., and Nelson, S. D. (1986). Effect of polyacrylamide on seedling emergence in crust-forming soils. *Soil Sci. 141*: 328–333.

Davankov, V. A., and Semechkin, A. V. (1977). Ligand-exchange chromatography. *J. Chromatography Chrom. Rev. 141*: 313–353.

Davankov, V. A., and Tsyurupa, M. P. (1980). Macronet isoporous styrene copolymers: unusual structure and properties. *Angew. Makromol. Chem. 91*: 127–142.

Davankov, V. A., Rogoshin, S. V., and Tsyurupa, M. P. (1973). Factors determining the swelling power of crosslinked polymers. *Angew. Makromol. Chem. 32*: 145–151.

Davankov, V. A., Rogoshin, S. V., and Tsyupura, M. P. (1974). Macronet isoporous gel through crosslinking of dissolved PS. *J. Polym. Sci. Symp. 47*: 95–101.

Dzhanpeisov, R. D., Akkulova, Z. G., Sokolova, T. M., Yakovleva, N. I., Krichevskii, L. A., Tkachenko, P. V., and Akanova, K. B. (1990). Research of semisynthetic soil structure-forms, based on the humic acids. *Izv. Akad. Nauk. Kaz. SSR, Ser. Biol. 5*: 71–75.

Flory, P. J. (1975). *Principles of Polymer Chemistry*. Cornell Univ. Press.

Flory, P. J., and Rechner, J. J. (1943). Statistical mechanics of cross-linked polymer networks. II. Swelling. *J. Chem. Phys. 11*: 521.

Flynn, G. L., Yalkowsky, S. H., and Roseman, T. J. (1974). Mass transport phenomena and models: theoretical concepts. *J. Pharmac. Sci. 63*: 479–510.

Goldstein, S., and Schmuckler, G. (1972). Sulfone formation during the sulfonation of crosslinked polystyrene. *Ion. Exch. Membranes 1*: 63–66.

Grassie, N., and Gilks, J. (1973). Friedel–Crafts crosslinking of polystyrene. *J. Polym. Sci. Pol. Chem. 11*: 1531–1552.

Grassie, N., Meldrum, I. G., and Gilks, J. (1970). Polystyrene with improved thermal properties, a new method of preparing crosslinked PS. *J. Pol. Sci. Pol. Let. 8*: 247–251.

Heidel, K. (1988). US 4,777,232. Starchem GmbH.

Helfferich, F. G. (1995). *Ion Exchange*. Dover, New York.

Henderson, J. C., and Hensley, D. L. (1985). Ammonium and nitrate retention by a hydrophilic gel. *HortScience, 20*: 667–668.

Ilmain, F., Kokufuta, E., and Tanaka, T. (1991). Volume phase transition of a gel driven by a hydrogen bonding. *Nature 349*: 400.

Inomoto, S., and Ito, T. Soil conditioner. Jpn. Kokai Tokkyo Koho, JP 72 21,477.

Iovine, C. P., and Ray-Chaudhuri, D. K. (1984). Process for the preparation of crosslinked sulfonated styrene polymers. U.S. Pat. 4,448,935.

Janacek, J., Stoy, A., and Stoy, V. (1975). Mechanical behaviour of partially hydrolyzed preswollen polyacrylonitrile networks. *J. Polym. Sci. Polym. Symp. 53*: 299–312.

Jerabek, K., and Setinek, K. (1989). Structure of macronet styrene polymer as studied by inverse steric exclusion chromatography and by selective sulfonation. *J. Polym. Sci. Pol. Chem. 27*: 1619–1623.

Jones, M. B., and Martin, W. P. (1957). Methods of evaluating aggregate stabilization by hydrolysed polysaccharide (HPAN) as it is affected by various inorganic salts. *Soil Sci. 83*: 475–479.

Kadyrov, N., Ayupova, V., and Daminova, M. (1989). Structure-forming capacity of polyelectrolytes (modified carboxymethylcelluloses) on soil dispersions. *Uzb. Khim. Zh. 3*: 76–78.

Kakoulides, E. P., Papoutsas, I. S., Bouranis, D. L., Theodoropoulos, A. G., and Valkanas, G. N. (1993). Synthetic macronet hydrophilic polymers as soil conditioners. 1. Kinetic characterization of macronet sulfonated polystyrene resins. *Commun. Soil Sci. Plant Anal. 24*: 1709–1720.

Kizawa, N., Yamanaka, Y., Tanaka, K., and Tanaka, T. (1990). Soil amendments acid regulators and water absorbing polymers. Jpn. Kokai Tokkyo Koho, JP 02 77,487.

Limberti, L., Madi J., Passino R., and Walis L. (1980). Ion-exchange kinetics in selective systems. *J. Chromatography 201*: 43–50.

Martin, W. P., Taylor, G. S., Engibous, J. C., and Burnett, E. (1952). Soil and crop responses from field applications of soil conditioners. *Soil Sci 73*: 455–471

Matsushima, M. (1986). Hydrophilic soil-improving agents. Jpn. Kokai Tokkyo Koho, JP 61 85,488.

Mitchell, A. R. (1986). Polyacrylamide application in irrigation water to increase infiltration. *Soil Sci. 141*: 353–358.

Mosolova, A. I. (1970). Effect of polymers (Krilium) on some physical and chemical properties of sod-podzolic and barley crops. *Biol. Nauki. 2*: 100–106.

Orihara, K. (1988). Synthetic high molecular-weight gel for water-holding in soil. Jpn. Kokai Tokkyo Koho, JP 63 15,881.

Peppas, N. A., and Staller, K. P. (1982). Friedel–Crafts crosslinking methods for polystyrene modification. 5. Sulfonated crosslinked polystyrene particles. *Polym. Bull. 8*: 233–237.

Peppas, N. A., and Valkanas, G. N. (1977). Crosslinking of PS by mono and bifunctional agents. *Angew. Makromol. Chem. 62*: 163–176.

Peppas, N. A., Valkanas, G. N., and Diamanti-Kotsida, E. T. (1976). Polycondensation of mono and bifunctional derivatives of *p*-xylene. *J. Polym. Sci. Polym. Chem. 14*: 1241–1247.

Peppas, N. A., Bussing, W. R., and Slight, K. A. (1981). Structure of Friedel–Crafts crosslinked polystyrene and sulfonated resins thereof. *Polym. Bull. 4*: 193–198.

Pó, R. (1994). Water-absorbent polymers: a patent survey. *J.M.S.-Rev. Macromol. Chem. Phys., C34*(4): 607–662.

Rabek, J., and Lucki, J. (1988). Crosslinking of PS under Friedel–Crafts conditions in DCE and CCl_4 solvents through the formation of strongly colored polymer $AlCl_3$ solvent complexes. *J. Polym. Sci. Polym. Chem. 26*: 2531–2551.

Reid, A. R. (1977). U.S. 4,028,290, Hercules Inc.

Takeuchi, S., and Tanaka, A. (1978). Soil amendments. Jpn. Kokai Tokkyo Koho, JP 78 18,646.

Tanaka, F. (1990). Thermodynamic theory of network-forming polymer solutions. 2. Equilibrium gelation by conterminous cross-linking. *Macromolecules 23*: 3790–3795.

Tanaka, K. (1991). Filler-incorporating hydrogels as soil conditioners. U.S. Pat. 5,013,349.

Terry, R. E., and Nelson, S. D. (1986). Effects of polyacrylamide and irrigation method on soil physical properties. *Soil Sci. 141*: 317–320.

Theodoropoulos, A. G., Bouranis, D. L., and Valkanas, G. N. (1992). Efficient, "One-pot" synthesis of suspension crosslinked sulfonated polystyrene via a Friedel–Crafts reaction. *J. Appl. Polym. Sci. 46*: 1461–1465.

Theodoropoulos, A. G. (1993). Preparation of strong acid network particles. The role of the catalyst in a multiple reaction environment. *J. Macrom. Sci. Macrom. Rep. A30*: 287–294.

Theodoropoulos, A. G., and Konstantakopoulos, I. C. (1996). Crosslinked polymers (macronets). *Polymeric Materials in Encyclopedia* (J. Salamone, ed.). CRC Press, Boca Raton, Florida (in press.)

Theodoropoulos, A. G., Tsakalos, V. T., and Valkanas, G. N. (1993a). Sulfone-type crosslinks in sulfonation of macronet polystyrene backbone. *Polymer 34*: 3905–3910.

Theodoropoulos, A. G., Bouranis D. L., Valkanas, G. N., and Kakoulides, E. P. (1993b). Synthetic macronet hydrophilic polymers as soil conditioners. II. The effect of the final ion form of crosslinked sulfonated polystyrene on seedling growth. *Commun. Soil Sci. Plant. Anal. 24*: 1721–1731.

Theodoropoulos, A. G., Valkanas, G. N., Stergiou, D. H., and Vlyssides, A. G. (1994a). Recycling of hydrophobic polymeric materials in the form of hydrogels and swelling promotion thereof. *J. Macrom. Sci. Macrom. Rep. A31*: 9–17.

Theodoropoulos, A. G., Konstantakopoulos, I. C., and Valkanas, G. N. (1994b). Synthesis of block ordered networks based on styrene ethylene butylene ABA triblock copolymer. *Eur. Polym. J. 30*: 1375–1380.

Theodoropoulos, A. G., Theodosakis, G. I., Valkanas, G. N., and Mamalis, A. D. (1996a). Preparation of bifunctional macronet polymers having both strongly and weakly acidic groups in one step: gel-solution transition of a transformable hydrogel. *Polymer 37*: 855–861.

Theodoropoulos, A. G., Bouranis, D. L., Vlyssides, A. G., and Kotsibou, E. D. (1996b). Ion-exchange kinetics in highly swellable macronets. *J. Macrom. Sci. Macr. Rep. A33*: 281–288.

Tsyurupa, M. P., and Davankov, V. A. (1980). The study of macronet isoporous styrene polymers by GPC. *J. Polym. Sci. Polym. Chem. 18*: 1399–1406.

Tsyurupa, M. P., and Davankov, V. A., and Rogozhin, S. V. (1974). Macronet isoporous ion-exchange resins. *J. Polym. Sci. Symp. 47*: 189–195.

Vasheghani-Farahani, E., Vera, J. H., Cooper, D. G., and Weber, M. E. (1990). Swelling of ionic gels in electrolyte solutions. *Ind. Eng. Chem. Res. 29*: 554–560.

Volkmar, K. M., and Chang, C. (1995). Influence of hydrophilic gel polymers on water relations and growth and yield of barley and canola. *Canad. J. Plant Sci. 75*: 605–611.

Wallace, A. (1986a). A polysaccharide (guar) as a soil conditioner. *Soil Sci. 141*: 371–373.

Wallace, A. (1986b). Effect of polymers in solution culture on growth and mineral composition of tomatoes. *Soil Sci. 141*: 395–396.

Wallace, A., and Abouzamzam, A. M. (1986). Interactions of soil conditioner with other limiting factors to achieve high crop yields. *Soil Sci. 141*: 343–345.

Wallace, A., and Wallace, G. (1986a). Effects of soil conditioners on emergence and growth of tomato, cotton and lettuce seedlings. *Soil Sci. 141*: 313–317.

Wallace, A., Wallace, G. A., (1986b). Effect of polymeric soil conditioners on emergence of tomato seedlings. *Soil Sci. 141*: 321–323.

Wallace, A., and Wallace, G. (1986c). Effects of very low rates of synthetic soil conditioners on soils. *Soil Sci. 141*: 324–327.

Wallace, A., and Wallace, G. A., (1986d). Additive and synergistic effects on plant growth from polymers and organic matter applied to soil simultaneously. *Soil Sci. 141*: 334–342.

Wallace, G. A., Wallace A. (1986e). Control of soil erosion by polymeric soil conditioners. *Soil Sci. 141*: 363–367.

Wallace, A., and Wallace, G. A. (1986f). Enhancement of the effect of coal fly ash by a polyacrylamide soil conditioner on growth of wheat. *Soil Sci. 141*: 387–389.

Wallace, A., and Wallace, G. A. (1989). Possible use of molecular weight polymers to flocculate soil in testing soils for available nutrients. *Soil Sci. 141*: 397.

Wallace, A., Wallace, G. A., and Abouzamzam, A. M. (1986a). Effects of soil conditioners on water relationships in soils. *Soil Sci. 141*: 346–352.

Wallace, A., Wallace G. A., and Abouzamzam, A. M. (1986b). Amelioration of sodic soils with polymers. *Soil Sci. 141*: 359–362.

Wallace, A., Wallace, G. A., Abouzamzam, A. M., and Cha, J. W. (1986c). Effects of polyacrylamide soil conditioner on the iron status of soybean plants. *Soil Sci. 141*: 368–370.

Wallace, A., Abouzamzam, A. M., and Cha, J. W. (1986d). Interactions between a polyacrylamide and a polysaccharide as soil conditioners when applied simultaneously. *Soil Sci. 141*: 374–376.

Wallace, A., Wallace, G. A., and Abouzamzam, A. M. (1986e). Effects of excess of a polymer as a soil conditioner on yields and mineral nutrition of plants. *Soil Sci. 141*: 377–380.

Wallace, A., Wallace, G. A., and Cha, J. W. (1986f). Mechanisms involved in soil conditioning by polymers. *Soil Sci. 141*: 381–386.

Wallace, A., Wallace, G. A., Abouzamzam A. M., and Cha, J. W. (1986g). Soil tests to determine application rates for polymeric soil conditioners. *Soil Sci. 141*: 390–394.

Wang, L., and Bloominglied, V. A. (1990). Osmotic pressure of polyelectrolytes without added salt. *Macromolecules 23*: 804–809.

Weaver, M. O., Bagley, E. B., Fanta, G. F., and Doane, W. M. (1976a). U.S. 3,935,099. USA Secretary of Agriculture.

Weaver, M. O., Bagley, E. B., Fanta, G. F., and Doane, W. M. (1976b). U.S. 3,985,616. USA Secretary of Agriculture.

Weaver, M. O., Bagley, E. B., Fanta, G. F., and Doane, W. M. (1976c). U.S. 3,997,484. USA Secretary of Agriculture.

Weaver, M. O., Otey, F. H., and Doane, W. M. (1984). Effect of some anionic starches on the stability of soil particles in water. *Starch/Staerke 36*: 56–60.

Yoshida, R., Uchida, K., Kaneko, Y., Sakai, K., Kikuchi, A., Sakurai, Y., and Okano, T. (1995). Comb-type grafted hydrogels with rapid de-swelling response to temperature changes. *Nature 374*: 240–242.

13

Improvement of Sandy Soils with Soil Conditioners

Abdulrasol M. Al-Omran and Abdulaziz R. Al-Harbi *College of Agriculture, King Saud University, Riyadh, Saudi Arabia*

I. INTRODUCTION

In this chapter, soil conditioners primarily are the cross-linked polymers that are sometimes called gel polymers. They absorb water and swell up to hundreds of times their dry weight. They are different from the water-soluble polymers discussed in Chapter 15 and elsewhere in this book. Gel polymers do much to improve sand soils. Water-soluble polymers do much to improve soils that contain clay. Some major differences between gel polymers and water-soluble polymers have been described (Wallace, 1994).

For the last 15 years there has been growing interest in using soil conditioners or superabsorbent materials to improve physical properties of coarse textured soils of arid and semiarid regions. The productivity of these soils is mostly limited by their low water holding capacity and excessive deep percolation losses, which result in low fertilizer and water use efficiencies. The application of soil conditioners has proved to be a means of alleviating some of these soil constraints. Numerous studies on the use of soil conditioners have shown their effectiveness in improving soil aggregates, increasing water holding capacity and water conservation, and reducing infiltration, saturated hydraulic conductivity, water diffusivity, and cumulative evaporation (Choudhary et al., 1995; El Shafei et al., 1994; Mustafa et al., 1988; Al-Omran et al., 1987). However, the use of soil conditioners on a large scale in agriculture was limited because of their high cost, insufficient longevity, reduction of water absorption capacity with salinity, and inconsistent results reported on their effects on soil chemical properties (Falatah et al., 1996; Falatah and Al-Omran, 1995). Recently, renewed attention has been focused on the use of soluble polymers, because of their new formulation, which are capable of improving soil physical properties at economical cost (Wallace and Wallace, 1994; Lentz et al., 1992; Levy et al., 1992).

II. EFFECT OF SOIL CONDITIONERS ON SOIL PROPERTIES

A. Physical Properties

The purpose of applying soil conditioners is to improve soil physical properties, thus to maintain soil productivity and reduce environmental problems. Over the past 20 years, many researchers have evaluated the effects of soil conditioners on physical and hydrological properties of soils. Research in the region has shown that soil conditioners when applied to coarse textured soils improve and stabilize soil aggregation (Al-Omran et al., 1987; Al-Omran and Shalaby, 1992), increase water holding capacity (Choudhary et al., 1995), suppress water evaporation from soil (Al-Omran et al., 1987; Choudhary et al., 1995; Shalaby, 1993), and control soil erosion (Levy et al., 1992).

1. Aggregation Index

Soil structure and its stability is an important property that influences water movement in soils. Maintenance of stable aggregates promotes infiltration in fine textured soils. Research in the region showed that applying soil conditioners was an effective means to stabilize soil aggregates (Levy et al., 1992; Al-Omran and Shalaby, 1992; Shainberg et al., 1990; and Helalia and Letey, 1988). Effect of a soil conditioner on aggregation was reported by Al-Omran et al. (1987). They found that addition of Jalma (a hydrophilic type polymer containing 24.5% humic acid and 3.8% polysaccharides) at the rate of 0.4% significantly increased aggregation indices of the three soils (Table 1). But when the Jalma rate was increased from 0.4% to 0.8%, the aggregation index was significantly increased by 10% in fine-textured soil, but there was no significant effect on coarse-textured soils. Further increase in the Jalma rate to 1.6% did not affect the aggregation index of all the soils tested. The improvement of aggregation in three soils at the low rate of 0.4% and in fine textured soils at the rate of 0.8% may be attributed predominantly to the chemical composition of Jalma, which contained 24.5% humic acid.

2. Bulk Density and Porosity

The bulk density of the soil is the ratio of soil mass to its volume and is usually expressed in g cm^{-3} and can be determined by the core technique. The porosity of the soil can be calculated from the bulk density estimation and particle density (assumed to be 2.65 g cm^{-3}). This is done by the well known formula

$$\text{Porosity } \% = \left(\frac{1 - \text{bulk density}}{\text{particle density}}\right) \times 100 \qquad (1)$$

Table 1 Aggregation Indices as Affected by Jalma Treatments

	Aggregation index (%)[a]		
Jalma rate (%)	Sandy loam	Loamy sand	Clay loam
0.0	60.5a[b]	59.2	50.4a
0.4	80.2b	81.2b	65.4b
0.8	82.9b	82.7	72.2c
1.6	83.3b	82.7	72.9c
LSD	6.24	7.91	4.93

[a]Aggregation index = *A-B*/*A* × 100, where *A* and *B* are silt plus clay of completely dispersed and not dispersed samples, respectively.
[b]Indices followed by the same letter for each soil are not significantly different at the 5% level according to Duncan's multiple range test.
Source: Al-Omran et al. (1987).

A sandy soil treated with a polyacrylamide soil conditioner, Broadleaf P4, resulted in an appreciable decrease in its bulk density. Compared to untreated soil, the percent decreases in bulk density were 44.3, 58.3, and 69.62 when soil was treated with the conditioner at 0.2, 0.4, and 0.6% concentrations. The decrease in bulk density subsequently affected porosity and void ratio of the soil. Increases in porosity of the soil when treated with 0.2, 0.4, and 0.6% conditioner were 67.5, 87.5, and 105.0%, respectively, compared to the control; while the increases of void ratio for the respective rates were 202.9, 347.7, and 580.6%. This remarkable effect was attributed to decrease in bulk density, which was the result of formation and stabilization of soil aggregates. Furthermore, the high swelling characteristics of such soil conditioners upon wetting resulted in numerous voids, decreased bulk density, and increased the porosity of the treated soil.

3. Swelling Index

Relative swelling index can be defined as the amount of distilled water retained at 1 kPa suction per gram of soil minus the amount of 1 M $CaCl_2$ solution similarly retained by nontreated soil. Soil swelling is an important property that influences infiltration and deep percolation especially in sandy soils. Al-Omran et al. (1987) reported that application of Jalma caused a significant swelling in the soils (Table 2). There was a general increase in relative swelling index with the increase in Jalma rate for the three soils. The increase in swelling of soils treated with soil conditioner may be attributed to the swelling characteristics of the soil conditioner.

In another study, El-Shafei et al. (1994) reported that under ponded and sprinkler irrigation there was a substantial increase in the expansion of soil treated with Acryhope conditioner (Fig. 1). This figure elucidates how relative expansion L_r, which is defined as L_e/L_i+L_e, where L_i is the initial soil length of

Table 2 Relative Swelling Indices as Affected by Jalma Treatments

	Relative swelling index (kg/kg)[a]		
Jalma rate (%)	Sandy loam	Loamy sand	Clay loam
0.0	0.002a[b]	0.000a	0.002a
0.4	0.112b	0.119b	0.050b
0.8	0.11b	0.114b	0.062b
1.6	0.254c	0.272c	0.234c
LSD	0.041	0.011	0.015

[a]Relative swelling index is the amount of distilled water retained at 1 kPa by each sample minus 1 M $CaCl_2$ solution retained at 1 kPa by the nontreated sample of the same soil.
[b]Indices followed by the same letter for each soil are not significantly different at the 5% level according to Duncan's multiple range test.
Source: Al-Omran et al. (1987).

the soil column and L_c is the absolute expansion above the initial soil length, was decreased with time during ponded infiltration as affected by soil conditioner rate. It can be concluded from the figure that most of the soil expansion had taken place during the first 10 minutes of irrigation. The results depicted in the figure illustrate that L_r was strongly affected by soil conditioner rate and that the higher the application rate the higher is L_r. Also the data show that the relationship between L_r and t is a power equation with $r > 0.972$. Another expression of relative swelling is COLE, coefficient of linear extensibility, expressed as

$$\text{COLE} = \left(\frac{V_m}{V_d}\right)^{1/3} - 1 \tag{2}$$

where V_m is the volume of the saturated soil sample (cm^3) and V_d is the volume of the air dried soil sample (cm^3). Sabrah (1994) calculated this term and reported a significant linear increase in swelling (COLE) with increasing application rate of soil conditioners. Recently, we evaluated the efficiency of a polyacrylamide (Broadleaf P4) on soil swelling when subjected to many wetting and drying cycles and concluded that swelling of sandy soil increased with increasing rates of soil conditioner. When Broadleaf P4 was applied at 0.2, 0.4, and 0.6% rates, the respective values for swelling at 0 wetting and drying cycle were 0.80, 1.40, and 2.33, and these values after 16 wetting and drying cycles were 0.66, 1.00, and 1.67 compared to the untreated soil.

4. Water Holding Capacity and Infiltration

The productivity of coarse textured soils, such as most of the agricultural soils in Saudi Arabia and the region, is limited due to their low water holding capacity, high infiltration rate, and deep percolation. Thus the management of

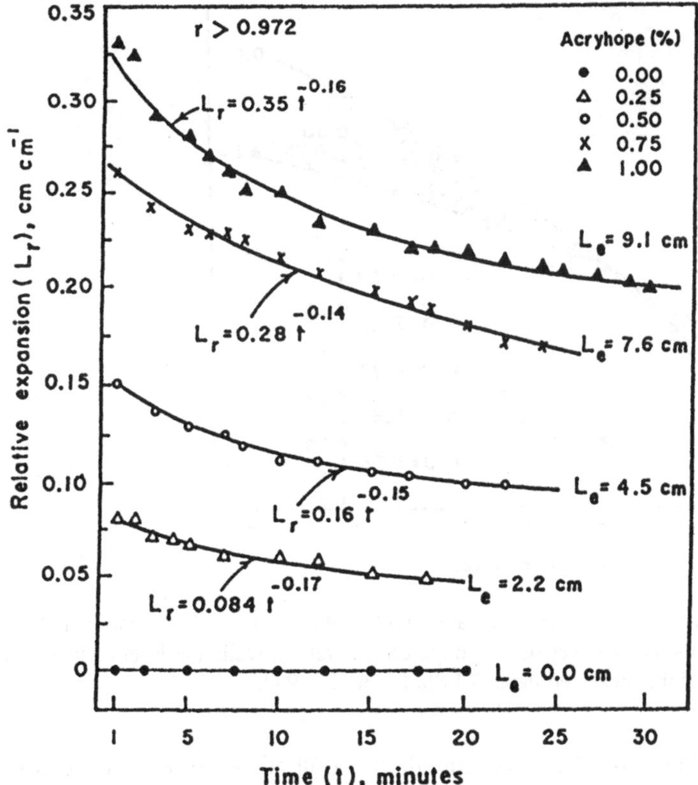

Figure 1 Relative expansion vs. time under ponded infiltration in sandy soil as affected by Acryhope application. L_c is the final value of the absolute expansion, while the inset equations are the regression of the measured data. (El-Shafei et al., 1994.)

these soils must aim at increasing water holding capacity (WHC) and reducing infiltration rate. It is well documented that water holding capacity increases with the addition of soil conditioners. Choudhary et al. (1995) studied the effect of four different synthetic polymers applied at different rates on WHC of sandy and loam soils. Their results showed that the effectiveness of soil conditioners to absorb water in sandy soil was higher than in loam soil.

Infiltration is usually defined as the entry of water into a soil profile, generally by downward flow through soil surface. This process is of great practical importance to irrigation and soil management. Thus many researchers have studied the effect of soil conditioners on infiltration and found that soil conditioners can be effective in reducing the infiltration rate of sandy soils (El-Shafei et al., 1994; Miller, 1979; Mustafa et al., 1988). An example study on the effect of a soil conditioner on infiltration is reported by El-Shafei et al. (1994). Figure 2 shows the

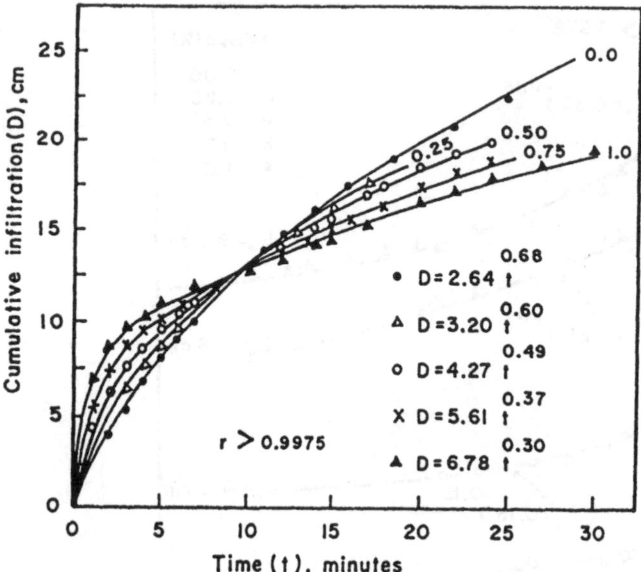

Figure 2 Cumulative infiltration in sandy soil treated with Acryhope. Numbers by the curves refer to the percentages of Acryhope concentration, while the inset equations are the regression of the measured data. (El-Shafei et al., 1994.)

effect of Acryhope conditioner on cumulative infiltration (D) under ponded irrigation. The results depicted in the figure illustrate that cumulative infiltration was strongly affected by soil conditioner rate. There were two stages of cumulative infiltration that produced opposite results. The first stage lasted for 8 minutes, during which infiltration increased with increasing soil conditioner rate. This stage was characterized by high water absorption capacity and high swelling. In the second stage, after approximately 8 minutes, cumulative infiltration decreased with increasing soil conditioner rate. The relationship between cumulative infiltration (D) at time (t) under ponded infiltration was presented by a power equation with $r > 0.997$. In general, the influence of soil conditioners on cumulative infiltration may be due to their effects on aggregation and swelling. In the early stage of applying irrigation, soil conditioners promote infiltration rate and limit capillary rise due to aggregation effect. However, the decrease in infiltration with increasing soil conditioner rate may be caused by swelling, which reduces the size of macropores and thus limits the infiltration.

5. Water Retention

Knowledge of the water retention curve of the soil is very important in land use for agricultural production and water management. Therefore the effects of soil conditioner rate on soil water retention and available water for plant growth have been evaluated by many researchers (El-Shafei et al., 1994; Al-Omran and Shalaby, 1992). The results presented in Fig. 3 show the relationship between soil pressure head (h) and volumetric soil water content (θ) as affected by Broadleaf P4. It is evident that added soil conditioner caused a considerable increase in soil water retention.

6. Saturated Hydraulic Conductivity

The relation between saturated hydraulic conductivity (K_s) and soil conditioner Broadleaf P4 rate is presented in Fig. 4. Data showed a significant reduction in K_s with the increase in this soil conditioner rate. This reduction may be caused by the swelling effect and the decreased size of macropores in the soil.

7. Unsaturated Hydraulic Conductivity

The relationship between unsaturated hydraulic conductivity $K(\theta)$ and soil water content (θ) calculated according to Campbell (1974) is

$$K(\theta) = K_s\left(\frac{\theta}{\theta_s}\right)^{2b+3} \tag{3}$$

where $K(\theta)$ is the unsaturated hydraulic conductivity in cm h^{-1} corresponding to (θ), K_s is the saturated hydraulic conductivity in cm h^{-1}, θ_s is the volumet-

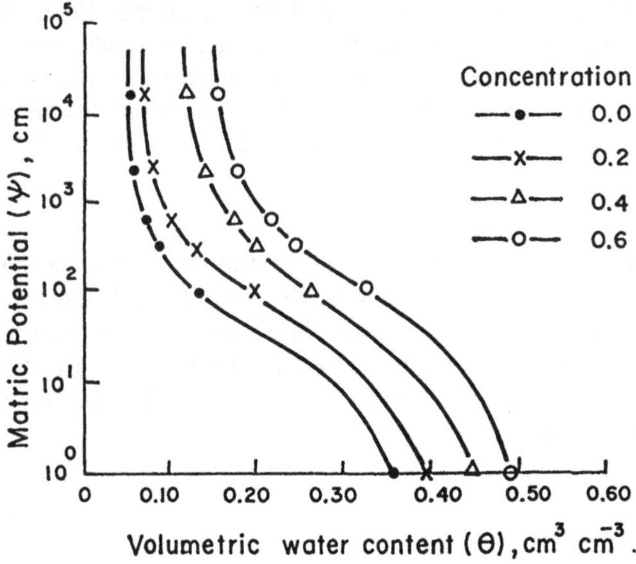

Figure 3 Relationship between matric potential (ψ) and water content (θ) of a sandy soil treated with Broadleaf P4.

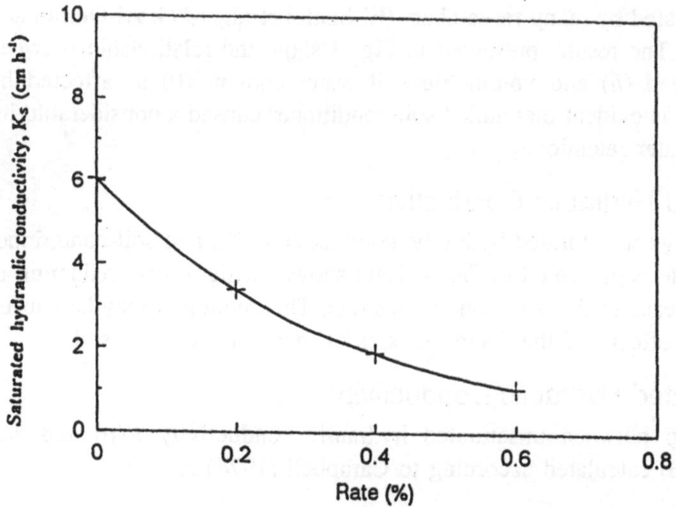

Figure 4 Saturated hydraulic conductivity (K_S) of a sandy soil as affected by four concentrations of Broadleaf P4.

ric saturated water content, and b is the soil parameter, which can be estimated from the soil water retention curve.

Data presented in Fig. 5 show the values of $K(\theta)$ as calculated by Eq. (2) for soil treated with Broadleaf P4 soil conditioner. Generally, the results indicate that values of $K(\theta)$ increased with increasing soil water content (θ). The results also showed that $K(\theta)$ tremendously decreased with increasing soil conditioner rate, which may be attributed to the swelling effect and the reduction in size of the macropores.

8. Diffusivity

The obtained soil water retention for sandy soil as affected by Broadleaf P4 rates can be used to calculate soil water diffusivity $D(\theta)$. If the logarithm of matric potential (ψ) values is plotted against the logarithm of the relative water content (θ/θ_s), one may expect a straight line within $10 \le \psi \le 1000$ cm range of matric potential, Fig. 6. The relationship between ψ and θ/θ_s can be expressed as the power function

$$\psi = \psi_e \left(\frac{\theta}{\theta_s} \right)^{-b} \tag{4}$$

where ψ is the soil matric potential, corresponding to the volumetric water content (θ), ψ_e is the matric potential, corresponding to the air entry value, θ_s is

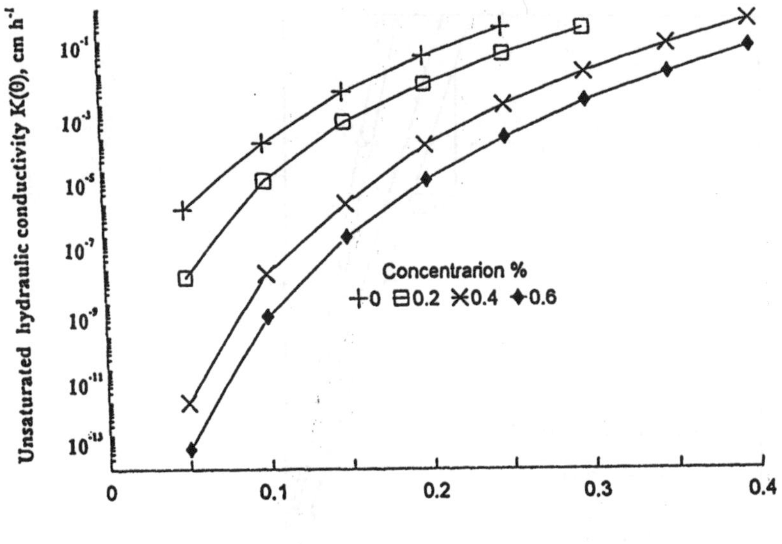

Figure 5 Relationship between soil moisture content (θ) and unsaturated hydraulic conductivity $K(\theta)$ of a sandy soil treated with Broadleaf P4.

the saturated water content, and b is the soil parameter. From Fig. 6, ψ_e and b can be evaluated. The relationship between $D(\theta)$ and θ is

$$D(\theta) = -K(\theta)\frac{d\psi}{d\theta} \tag{5}$$

$(d\psi/d\theta)$ can be calculated by taking the derivative of Eq. (4):

$$\frac{d\psi}{d\theta} = \left(\frac{-b\psi_e}{\theta_s{}^{-b}}\right) * \theta^{-b-1} \tag{6}$$

Combining Eqs. (3), (5), and (6) gives

$$D(\theta) = \frac{b\psi_e Ks \;\; \theta^{b+2}}{\theta s^{b+3}} \tag{7}$$

For $10 \leq \psi \leq 1000$ cm, data from Eq. (7) are presented in Fig. 7. The general shape of $D(\theta) - \theta$ curves is in accordance with that obtained by van Genuchten (1980) as predicted from the knowledge of the soil water retention and saturated hydraulic conductivity. Generally $D(\theta)$ increased with increasing soil water content (θ) and decreased with increasing soil conditioner rates.

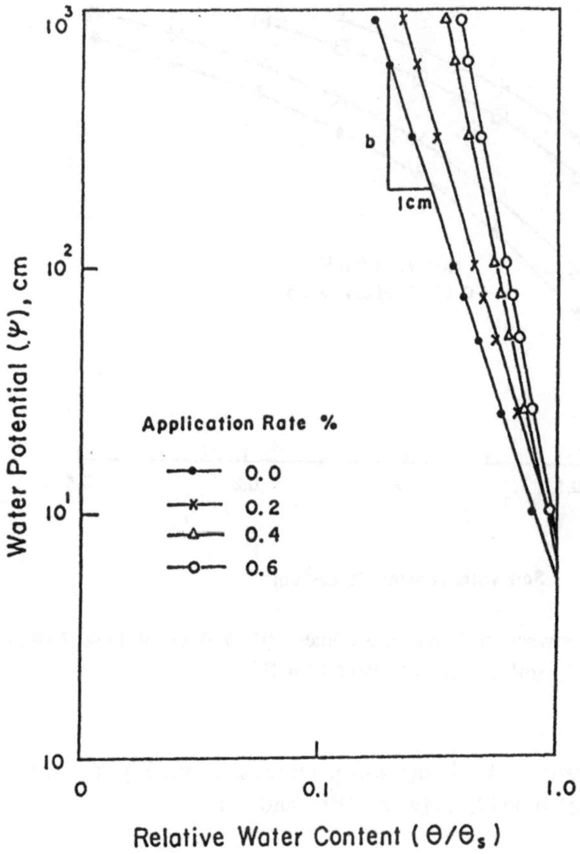

Figure 6 Relationship between matric potential (ψ) and relative water content (θ/θ_s) of a sandy soil treated with Broadleaf P4.

9. Cumulative Evaporation and Water Conservation

Soil conditioners can reduce evaporation and increase amount of water conserved (Al-Omran et al., 1987; Choudhary et al., 1995). Johnson (1984) found that polyacrylamide reduced evaporation, and effects varied with the type of commercial product used. In this respect Al-Omran et al. (1991) studied the effect of different commercial soil conditioners on the cumulative evaporation, and data are presented in Fig. 8. The results showed that soil conditioners significantly reduced cumulative evaporation. Results of another study by Al-Omran and Shalaby (1992), on the effect of soil conditioner rate on cumulative evaporation, are shown in Fig. 9. The results indicate that the soil conditioner significantly reduced cumulative evaporation of sandy soil, which resulted in a

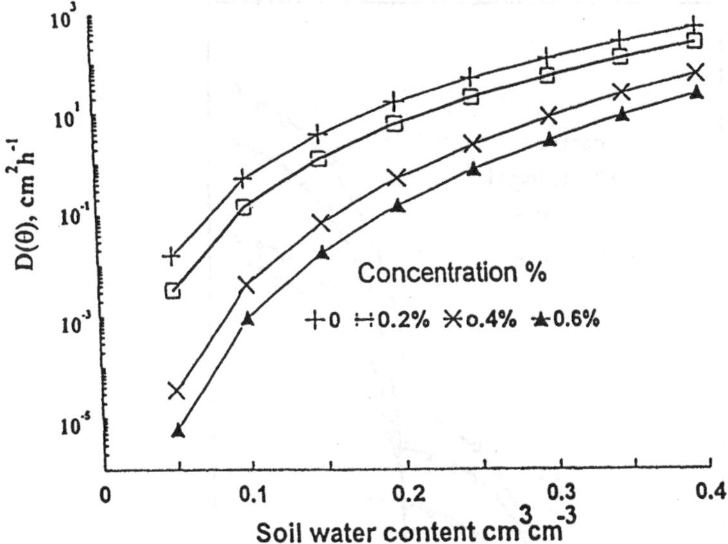

Figure 7 Relationship between soil water content (θ) and computed diffusivity *D* (θ) of a sandy soil treated with Broadleaf P4.

high amount of water conserved. The influence of soil conditioner on the cumulative evaporation may be due to its effect on aggregations which in return promote infiltration, limit capillary rise, and consequently suppress evaporation.

Most of the studies on soil conditioners were done on hydrophilic types that were mixed with surface (0–10 cm) soil. However, there were some other studies conducted to evaluate the effect of soil conditioner when positioned at different depths (Al-Omran et al., 1991), or when changing the depth in the treated layer (Shalaby, 1993). Figure 10 illustrates the effect of depth placement of 0.4% saturated layer barrier on evaporation. It is evident that the closer the soil conditioner is applied to the surface, the lower the amount of water evaporated, and thus the higher the amount of water retained below it, due to reduced capillary rise.

The effect of soil conditioner treated layer depth on evaporation is presented in Fig. 11. It is evident that Jalma treated at 0.05 m is most effective in reducing evaporation for each cycle. Increasing the depth of the treated layer to 0.10 or 0.15 m increased the cumulative evaporation.

B. Effect of Soil Conditioner on Some Chemical Properties

The coarse-textured soils of arid and semiarid regions are mostly characterized by their low CEC, negligible organic matter, alkaline pH, and high $CaCO_3$. For several years soil conditioners have been used to improve soil physical proper-

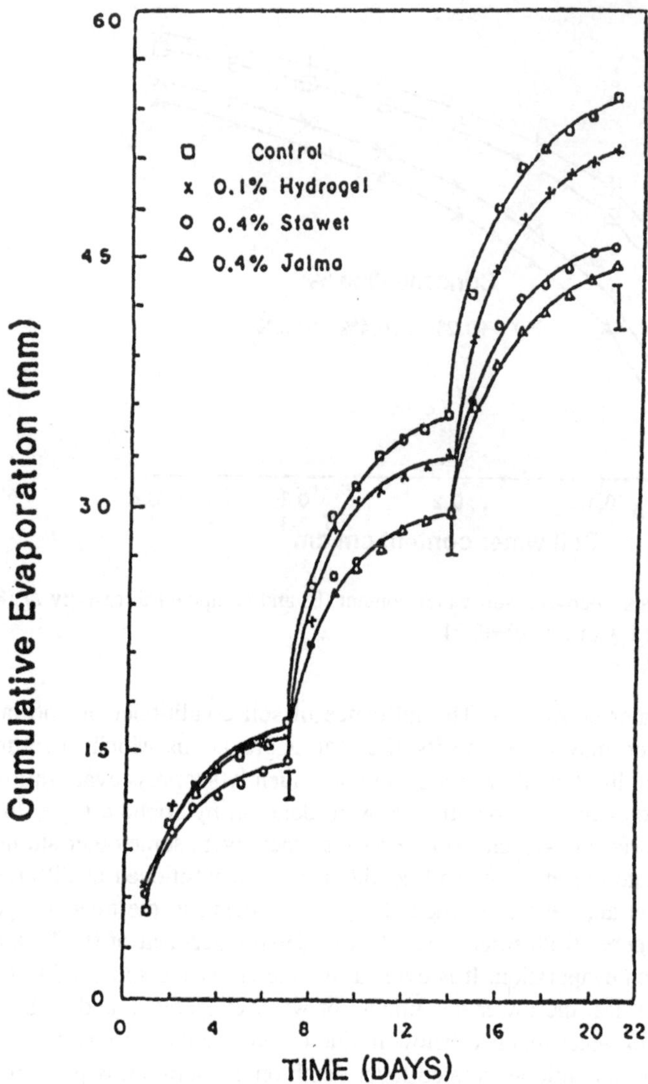

Figure 8 Intermittent cumulative evaporation of distilled water as affected by different gel conditioners. (Al-Omran et al., 1991.)

ties (Al-Omran et al., 1987). However, very little research has been conducted on the influence of soil conditioner on soil chemical properties. In the last two years, research has started to address the effect of these soil conditioners on pH, EC, and micronutrient availability (Falatah and Al-Mustafa, 1993; Falatah and Al-Omran, 1995; Falatah et al., 1996).

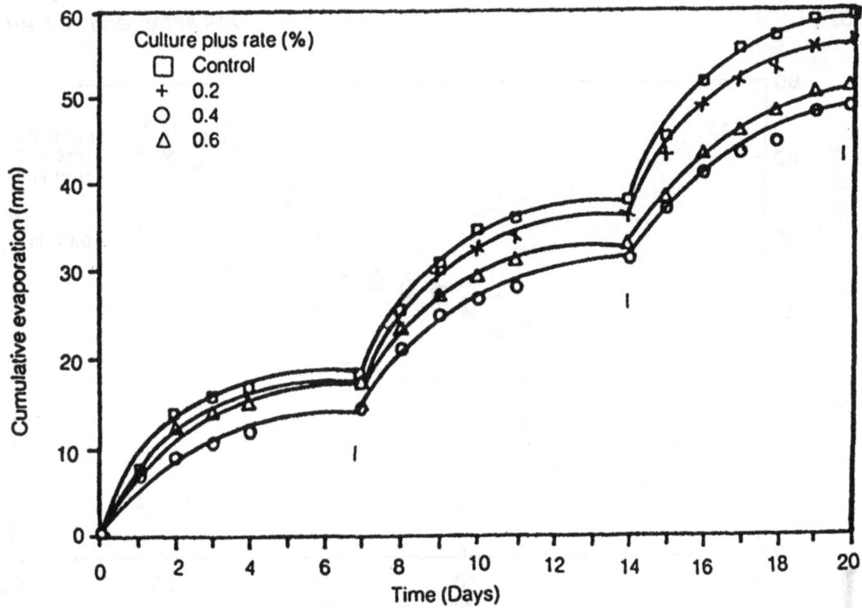

Figure 9 Cumulative evaporation from soil columns as affected by Culture Plus rate at EC of 1 dS m^{-1}. (Al-Omran and Shalaby, 1992.)

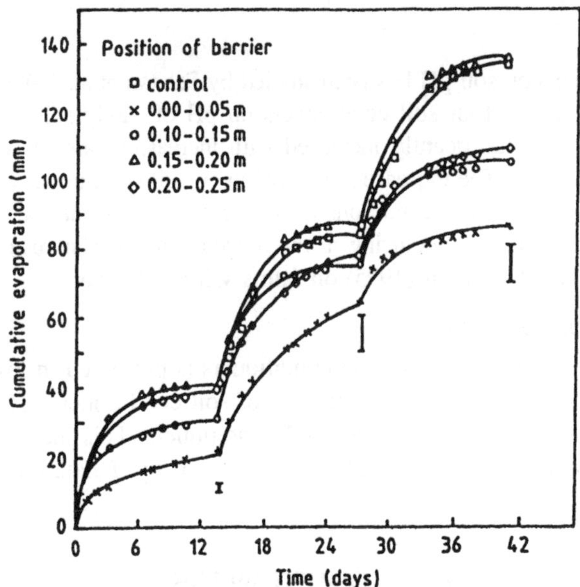

Figure 10 Intermittent cumulative evaporation from a Typic Toorripsamment as affected by position of 0.4% Jalma treated barriers. Vertical bars represent $LSD_{0.05}$. (Al-Omran et al., 1991.)

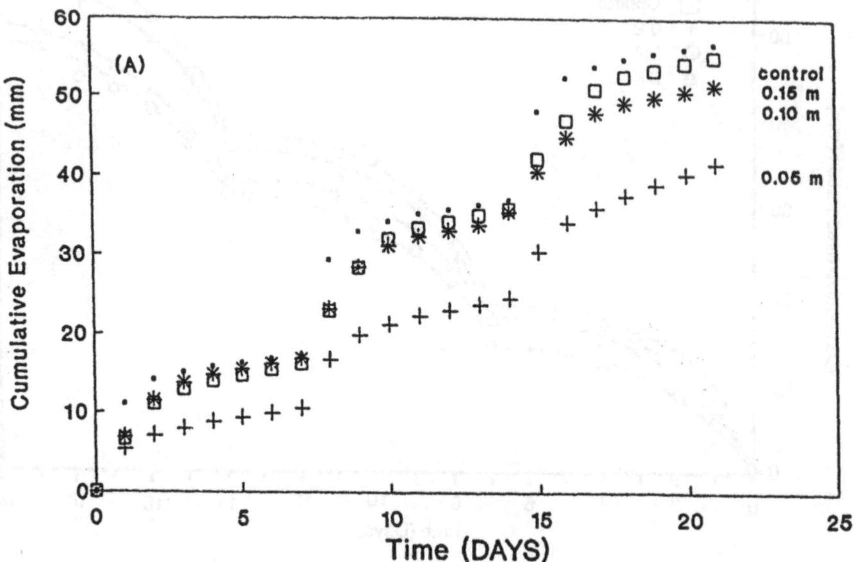

Figure 11 Intermittent evaporation at 0.4% Jalma as affected by depth of surface treated layer using tap water. (Shalaby, 1993.)

1. Soil pH

The effect of soil conditioners on soil pH has been studied by Falatah et al. (1996). Their results on the influence of four soil conditioners on pH of sandy soils are presented in Fig. 12. Soil pH significantly increased with increasing rate of soil conditioners relative to a control. The increase in soil pH upon soil conditioner application reflects both high pH of the conditioners used and the consumption of protons during the reduction process occurring in the treated soil. Similar results were reported by Falatah and Al-Omran (1995) on Jalma soil conditioner.

2. Soil Electrical Conductivity

The EC of the sandy soil treated with four soil conditioners is presented in Fig. 13. The EC significantly increased with the increase of soil conditioners. This increase might be due to the salt content of the soil conditioners. This increase is consistent with a previous study on Jalma conditioner by Falatah and Al-Omran (1995).

III. CROP RESPONSE TO SOIL CONDITIONERS

Greenhouse crops have the ability to benefit from advances in technology such as the addition of polymers to maximize productivity per unit of area. Recent developments in water absorbing synthetic polymers show that they improve seed germination, seedling establishment, and plant growth in greenhouses. Ini-

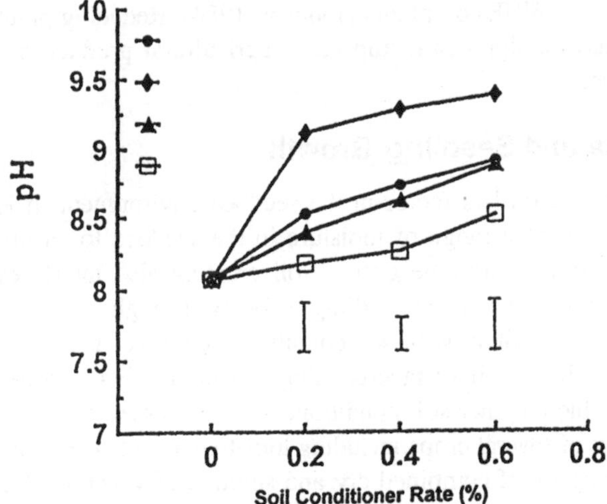

Figure 12 pH of a sandy soil as affected by various concentrations of four gel-forming conditioners. (Falatah et al., 1996.)

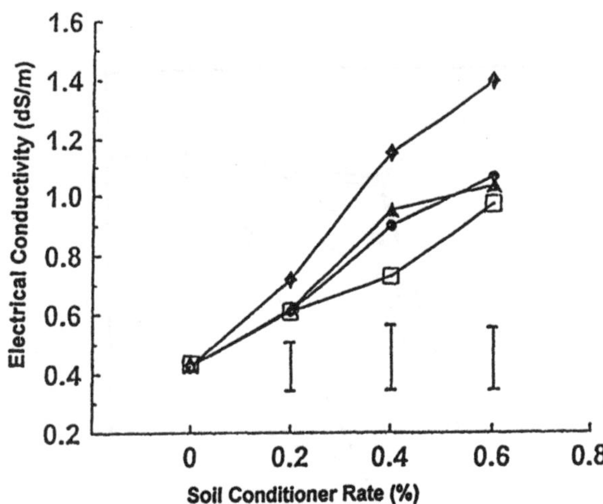

Figure 13 Electrical conductivity of a sandy soil as affected by various concentrations of four gel-forming conditioners. (Falatah et al., 1996.)

tial use of soil conditioner in horticulture was targeted to increase seed germination and seedling survival (Orzolek, 1993). This could result from the increased water retention and aeration of horticultural substrate used in crop production. Soil conditioners reduce the frequency and amount of irrigation by increasing water retention and suppressing evaporative losses, leading to an increase in water use efficiency (WUE) by plants (Johnson 1984). Reducing plant water application would have a significant impact on agricultural productivity in arid and semiarid regions.

A. Seed Emergence and Seedling Growth

Soil moisture conservation is a critical factor in the seedbed environment. It is necessary to maintain the desirable range of moisture in the seedbed to ensure faster germination and rapid seedling emergence. Soil crusting also interferes with seed germination by not allowing the seedling to break through.

Amendment of the growing medium with soil conditioner improves the moisture and aeration level and abates soil surface crusting (Seybold, 1994). Earlier results show that water-soluble polymer soil conditioner improves seed germination and seedling emergence of several crops including tomato, lettuce, beans, and peas (Quastel, 1954). Application of combined dry and solution of water-soluble polyacrylamide resulted in a marked increase and earlier seed emergence of lettuce and cotton (Wallace and Wallace, 1986). Research in the region also showed that cucumber seed cumulative germination rate increased with addition of a gel soil conditioner in sandy loam soil (Fig. 14, Al-Harbi et al., 1996).

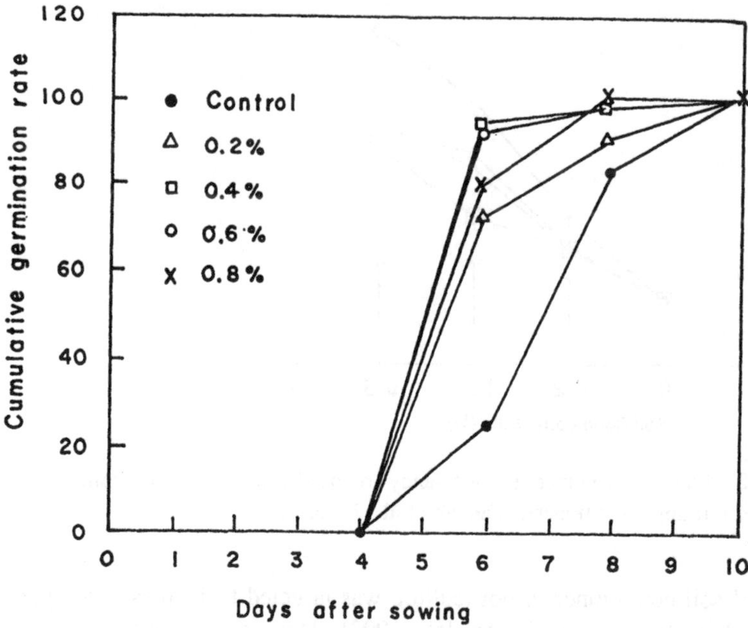

Figure 14 Effect of soil conditioner rate on cumulative seed germination percentage. (Al-Harbi et al., 1996.)

Soil crusting and moisture deficits following transplanting may result in a poor stand and stunted plants. Seedling development is a critical stage during crop establishment. Optimum seedling growth can ensure optimum plant vegetative growth and subsequently higher yield. Gel soil conditioners improve seedling establishment and growth. Part of this improvement is ascribed to better water supply (Gray et al., 1995). Leaf water potential of tomato seedlings increased with addition of a hydrophilic polymer to the growing medium (Henderson and Hensley, 1986). Al-Harbi et al. (1994) in their study on the effect of soil conditioner and irrigation frequency on tomato seedlings found that 0.6% of polymers added resulted in higher water potential compared to lower rates of polymers and in untreated soils (Fig. 15). Hydrophilic polymer addition reduced the variation in leaf water potential and maintained the leaf turgor pressure of sweet pepper and cabbage seedling grown under water stress, leading to greater growth (Chien and Woo, 1994). Cucumber seedlings growth also increased in a sandy loam soil amended with a hydrophilic polymer as shown in Table 3. Also the results indicate that the optimum rate for maximum effect of the polymers was at 0.3% (on a dry weight basis)(Al-Harbi et al., 1997).

The beneficial effect of soil conditioner on seedling growth may not be pronounced when water is not a limiting factor (Bres and Weston, 1993). Crop response to polymer addition could be affected by the presence of organic matter and fertilizer salts (Bowman, 1990; Orzolek, 1993).

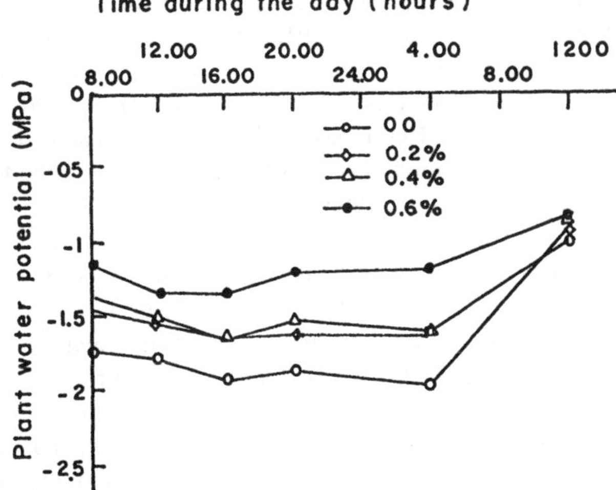

Figure 15 Effect of soil conditioner concentration (%) on plant water potential. (Al-Harbi et al., 1994.)

Table 3 Cucumber Plant Growth as Affected by Hydrophilic Concentration (on Dry Weight Basis)

Polymer concentration (%)	Plant height (cm)	Stem diameter (cm)	Leaf number (L/P)	Leaf area (cm^2)	Shoot fresh weight (g/p)	Shoot dry weight (g/p)
0[a]	24.4	0.62	7.1	386	17.3	2.26
0.1	35.4	0.77	8.9	671	31.2	4.32
0.2	43.8	0.73	9.6	765	36.0	4.79
0.3	49.8	0.82	10.0	816	42.9	6.10
0.4	43.6	0.79	9.6	734	37.3	5.32
$LSD_{0.05}$	5.2	0.11	0.9	114	5.2	0.78

[a]Control.
Source: Al-Harbi et al. (1997).

B. Plant Growth and Yield

The positive effect of soil conditioners may not be extended to the later stages of growth. It may be restricted to the early stages of seedling growth because root systems will grow out of the polymer zone (Callagan et al., 1989). Other investigators reported a significant impact on plant growth and yield with the addition of soil conditioner (Azzam, 1980; Orzolek, 1993). Yield of tomato plants grown in yellow podzolic soil in greenhouses was increased with addition of various types of soil conditioners (Sjamsudin et al., 1994). Flowering of tomato plants was advanced and fruit number was increased with the addition of hydrogel polymers (Szmidt and Graham, 1991; Ouchi et al., 1990). Improvement of vegetative growth and yield might be attributed to improved soil moisture content, reduced evaporation losses from the soil surface, and increased nutrient uptake (Ouchi et al., 1990). Reduced N leaching loss and increased uptake of nitrate-N and ammonium-N by the plant was also reported (Mikkelsen et al., 1993; Bres and Weston, 1993).

Growth and yield of lettuce were increased in clay loam soil amended with polystyrene soil conditioner at the rate of 0.2% under different water and nutrient treatments (Chatzoudis and Valkanas, 1995). Using different kinds of soil conditioner in different types of soils may contribute to the inconsistent results reported about the influence of soil conditioner on horticultural crops.

Salt accumulation in the soil is another limiting factor for agriculture in arid regions. In the greenhouse, soil salinity is caused by the heavy use of fertilizers and by saline irrigation water. Although efficiency of hydrophilic hydration is inhibited by the dissolved salts in the irrigation water and fertilizers, as previously reported, some reports indicate that the extent of growth reduction caused by salinity could be reduced by the addition of soil conditioners (Awad et al., 1986; Faraj, 1990). Hydrophilic polymers are found to be effective as soil conditioners in horticulture to improve salt tolerance in sand or light gravel substrate (El-Sayed et al., 1991), but this depends on the type of soil conditioner and the plant growth (Silberbush et al., 1993).

C. The Use of Soil Conditioner in Soilless Culture

Soil cultivation in greenhouses is still very common. Recent developments in soilless cultivation made it possible to overcome some problems associated with soil cultivation such as salt accumulation and soil-borne diseases. There is a wide variety of soilless media including natural and synthetic substrates such as peat, sand, gravel, perlite, vermiculite, and rock wool in the nutrient film technique (NFT).

Soil conditioners have potential application in soilless culture. Physical properties of inert substrates such as sand could be improved by a hydrophilic polymer amendment. Potential advantages include better water and nutrient retention. Hydrogel polymer can be used as an ecologically sound substrate to produce seedlings for soilless culture. Growth and subsequent yield of lettuce transplants produced in hydrogel substrate were similar to those produced using rock wool and peat substrate when grown in NFT (Paschold and Kleber, 1995).

IV. POTENTIAL USE OF SOIL CONDITIONERS IN THE KINGDOM OF SAUDI ARABIA AND CONCLUSIONS

As we mentioned earlier, the initial use of soil conditioners was projected to increase water holding capacity, reduce deep percolation, and suppress evaporation of sandy soils. Thus many recommendations were developed to use soil conditioners in sandy soils in the production of floral and nursery crops in greenhouses (Foster and Keerer 1990; Al-Omran et al., 1987; Wang, 1989). Most of the agricultural land in Saudi Arabia is sandy and characterized by low fertility, low organic matter, deep percolation, and low water retention. Because of these constraints and the scarcity of water in Saudi Arabia, many farmers were searching for methods that would decrease irrigation frequency as well as reduce the amount of irrigation water needed to grow their crops. One of these methods is the application of gel soil conditioners to sandy soils where the crop response to these polymers is great.

Soil conditioners are commercially available on the market in Saudi Arabia in solid or granular form and in liquid form as gel. Many of these solid forms (hydrophilic type) have been introduced to farmers under different commercial names such as Hydrogel, Acrihope, Sta Wet, Broadleaf P4, etc. to be utilized in many areas ranging from pollution control and agricultural production to the management of water movement.

In Saudi Arabia soil conditioners in different forms can be used to control sand and dust movement and can be put around trees in zones of diameter 1.5 m to supply water to the trees along highways. Also soil conditioners can be used in the production of flora and vegetables in the greenhouses. Soil conditioners can be used in most of the huge football stadiums around the kingdom to increase water holding capacity, reduce the hardness due to dry weather, maintain healthy ground cover after each game, and reduce the injuries sustained by players.

ACKNOWLEDGMENTS

The authors wish to acknowledge their thanks to Idrees M. Choudhary, Shalaby A. Sheashaa, and Mursi M. Mustafa, research associates in soil physics, for extending their help during the writing of this chapter.

REFERENCES

Al-Harbi, A. R., Al-Omran, A. M., Wahdan, H., and Shalaby A. A. (1994). Impact of irrigation regime and conditioner rate on tomato seedling growth. *Arid Soil Res. Rehab. 8*: 53–59.

Al-Harbi, A. R., Al-Omran, A. M., Choudhary, M. I., Wahdan, H., and Mursi, M. M. (1996). Influence of soil conditioner rate on seed germination and growth of cucumber plants (*Cucumis sativus L.*). *Arab Gulf J. Sci. Res. 14*: 129–142.

Al-Harbi, A. R., Al-Omran, A. M., Shalaby, A. A., Wahdan, H., and Choudhary, M. I. (1997). Growth response of cucumber to hydrophilic polymer application under different soil moisture level (in press).

Al-Omran, A. M., and Shalaby, A. A. (1992). Effect of water quality and gel-conditioner rate on intermittent evaporation. *J. King Saud Univ. Agri Sci. 4*: 273–286.

Al-Omran, A. M., Mustafa, M. A., and Shalaby, A. A. (1987). Intermittent evaporation from soil columns as affected by gel-forming conditioners. *Soil Sci. Soc. Am. J. 51*: 1593–1599.

Al-Omran, A. M., Al-Darby, A. M., Mustafa, M. A., and Shalaby, A. A. (1991). Impact of gel conditioners and water salinity on intermittent evaporation. *Egyptian J. Soil Sci. 31*: 575–588.

Awad, F., El-reman, M. A., and Assad, F. F. (1986). The combined effect of some soil conditioners and saline irrigation water. *Agrochimica 30*: 427–438.

Azzam, R. A. I. (1980). Agricultural polymers, polyacrylamide preparation, application and prospects. *Commun. Soil Sci. Plant Anal. 11*: 767–834.

Bowman, D. C. (1990). Fertilizer salts reduce hydration of polyacrylamide gels and affect physical properties of gel-amended container media. *J. Amer. Soc. Hort Sci 115*: 382–386.

Bres, W., and Weston, L. A. (1993). Influence of gel additive on nitrate, ammonium and water retention, and tomato growth in soilless medium. *HortSci. 28*: 1005–1007.

Callagan, T. V., Lindley, D. K., Ali, O. M., El Nour, H. A., and Bacon, P.J. (1989). The effect of water-absorbing synthetic polymers on the stomatal conductance, growth and survival of transplanted *Eucalyptus microtheca* seedling in Sudan. *J. Applied Ecol. 26*: 663–672.

Campbell, G. S. (1974). A simple method for determining unsaturated hydraulic conductivity from moisture retention data. *Soil Sci. 117*: 311.

Chatzoudis, G. K., and Valkanas, G. N. (1995). Lettuce plant growth with the use of soil conditioner and slow release fertilizers. *Commun. Soil Sci. Plant Anal. 26*: 2569–2576.

Chien, H. J. C., and Woo, N. C. (1994). The effect of hydrophilic polymer on leaf water potential in water-stressed seedlings of cabbage and sweet pepper. *J. Agric. Forestry 43*: 13–19.

Choudhary, M. I. Shalaby, A. A., and Al-Omran, A. M. (1995). Water holding capacity and evaporation of calcareous soils as affected by four synthetic polymers. *Commun. Soil Sci. Plant Anal. 26*: 2205–2215.

El-Sayed, H., Kirkwood, R. C., and Graham, N. B. (1991). The effect of hydrogel polymer on the growth of certain horticultural crops under saline conditions. *J. Exp. Botany. 42*: 891–899.

El-Shafei, Y. Z., Al-Darby, A. M., Shalaby, A. A., and Al-Omran, A. M. (1994). Impact of a highly swelling gel-forming conditioner (Acryhope) upon water movement in uniform sandy soils. *Arid Soil Res. Rehab. 8*: 33–50.

Falatah, A. M., and Al-Omran, A. M. (1995). Impact of a soil conditioner on some selected chemical properties of calcareous soil. *Arid Soil Res. Rehab. 9*: 91–96.

Falatah, A. M., and Al-Mustafa, W. A. (1993). The influence of gel-forming conditioner on pH, pe, and micronutrient availability of two torrifluvents. *Arid Soil Res. Rehab. 7*: 253–263.

Falatah, A. M., Choudhary, M. I., and Al-Omran, A. M. (1996). Changes in some chemical properties of arid soils as affected by synthetic polymers. *Arid Soil Res. Rehab. 10* (in press).

Faraj, M. A. (1990). The effect on soil and sugar beet plants of irrigating polyacrylamide treated soil with saline water. *Dissertation Abstracts International Sciences and Engineering 51*: 481B.

Foster, W. J., and Keever, G. J. (1990). Water absorption of hydrophilic polymers (hydrogels) reduced by media amendments. *J. Environ. Hort. 8*: 113–114.

Gray, D. J., Steckel, R. A., Miles, S., Reed, J., and Hiron, R. W. P. (1995). Improving establishment by a dibber drill. *J. Horti. Sci. 70*: 517–528.

Helaila, A. M, and Letey, J. (1988). Polymer type and water quality effects on soil dispersion. *Soil Sci. Soc. Am. J. 52*: 243–246.

Henderson, J. C., and Hensley, D. L. (1986). Efficacy of a hydrophilic gel as a transplant aid. *HortSci. 21*: 991–992.

Johnson, M. S. (1984). Effect of soluble salts on water absorption by gel-forming soil conditioners. *J. Sci. Food Agric. 35*: 1063–1066.

Lentz, R. D., Shainberg, I., Sojka, R. E., and Carter, D. L. (1992). Preventing irrigation furrow erosion with small applications of polymers. *Soil Sci. Soc. Amer. J. 56*: 1926–1932.

Levy, G. J., Levin, J., Gal, M., Ben-Hur, M., and Shainberg, I. (1992). Polymers effects on infiltration and soil erosion during consecutive simulated sprinkler irrigation. *Soil Sci. Soc. Amer. J. 56*: 902–907.

Mikkelsen, R. L., Behel, A. D., and Williams, H. M. (1993). Addition of gel-forming hydrophilic polymers to nitrogen fertilizer solutions. *Fertilizer Research 36*: 55–61.

Miller, D. E. (1979). Effect of H-SPAN on water retained by soils after irrigation. *Soil Sci. Soc. Amer. J. 43*: 628–629.

Mustafa, A. M., Al-Omran, A. M., Shalaby, A. A., and Al-Darby, A. M. (1988). Horizontal infiltration of water in soil columns as affected by gel-forming conditioner. *Soil Sci. 145*: 330–336.

Orzolek, M. D. (1993). Use of hydrophilic polymers in horticulture. *Hort. Technology 3*: 41–44.

Ouchi, S., Nishikawa, A., and Kamada, E. (1990). Soil-improving effects of a super-water-absorbent polymer (Part 2). Evaporation, leaching of salts and growth of vegetables. *Japanese J. Soil Sci. Plant Nutrition 61*: 606–613.

Paschold, P., and Kleber, J. (1995). Production of vegetable transplants for NFT in pure hydrogel. *Acta Horticultura 396*: 297–04.

Quastel, J. H. (1954). Soil conditioners. *Annu. Rev. Plant Physiol. 5*: 75–92.

Sabrah, R. E. A. (1994). Water movement in a conditioner-treated sandy soil in Saudi Arabia. *Journal of Arid Environments 27*: 363–373.

Seybold, C. A. (1994). Polyacrylamide review: soil conditioning and environmental fate. *Commun. Soil Sci. Plant Anal. 25*: 2171–2185.

Shainberg, I., Warrington, D. N., and Rengasamy, P. (1990). Water quality and PAM interactions in reducing surface sealing. *Soil Sci. 149*: 301–307.

Shalaby, A. A. (1993). Impact of gel-conditioned layer depths and water quality on water movement in sandy soils. *Arid Soil Res. Rehab. 7*: 281–291.

Silberbush, M., Adar, E., Malach, Y., and de-Malach, Y. (1993). Use of a hydrophilic polymer to improve water storage and availability to crops grown in sand dunes. II. Cabbage irrigated by sprinkling with different water salinities. *Agricultural Water Management 23*: 315–327.

Sjamsudin, E., Harjadi, S. S., Poerwanto, P., and Endang, S. (1994). Response of tomato to soil conditioners and NPK dosage on red yellow podzolic soil. *Acta Horticultura 369*: 344–351.

Szmidt, R. A. K., and Graham, N. S. (1991). The effect of polyethylene oxide hydrogel on crop growth under saline conditions. *Acta Horticultura 287*: 211–218.

van Genuchten, M. Th. (1980). A closed form equation for predicting the hydraulic conductivity of unsaturated soils. *Soil Sci. Soc. Am. J. 44*: 892–898.

Wallace, A. (1994). Gel polymers vs. water-soluble polymers. *HortSci. 29*: 606.

Wallace, A., and Wallace, G. A. (1986). Effect of soil conditioners on emergence and growth of tomato, cotton and lettuce seedlings. *Soil Sci. 141*: 313–316.

Wallace, A., Wallace, G. A. (1994). Water soluble polymers help protect the environment and correct soil problems. *Commun. Soil Sci. Plant Anal. 25*: 105–108.

Wang, Y. T. (1989). Medium and hydrogel affect production and wilting of tropical ornamental plants. *HortSci. 24*: 941–944.

14
Krilium: The Famous Soil Conditioner of the 1950s

Sheldon D. Nelson *Brigham Young University, Provo, Utah*

I. INTRODUCTION

Technological solutions to challenges facing the human family have been common in the recent past. Some of the most dramatic effects ever to occur on the biology of this planet have come from the application to soil of a broad range of agricultural chemicals. Crop yields and food quality have had dramatic increases in most agricultural areas during the past several decades where agricultural chemicals have been properly used. But modern agriculture continues to face the age-old problem of soil loss and soil degradation from improper agricultural practices. On many soils of the world, modern crop production practices as well as more ancient agricultural activities expose soil to the forces of erosion and destruction of the soil structure. Erosional losses and soil quality degradation contribute to decreased crop yields, increased sediment loads, and chemical pollution of surface waters. Historically this has lead to a declining economy and long term exposure and health risks, which all have significant political and social implications. Nonetheless, today's more environmentally conscious public often anticipates that technology will produce safer solutions to food production shortages and environmental concerns.

The enormous importance of the processes of stabilizing soil aggregates against the many natural and artificial forces that can reduce soil quality and potentially contaminate soil and water is little appreciated by people not working directly with the soil. A proper soil physical environment is one in which soil structure provides proper void space for the entrance and percolation of water and air. Soil porosity is associated with solubility of plant nutrients, the proper concentration of oxygen and carbon dioxide for root and microbial respiration, and the transfer of heat, all of which are of supreme importance to the aerobic biological life within the rhizosphere. And yet for many of the world's soils, aggregate instability and improper soil porosity are significant limitations to seedling emergence and plant growth, while soil erodibility continues to degrade soil and water quality.

For many years, soil scientists have investigated methods of soil stabilization, which have included the incorporation of soil amendments to improve soil tilth through maintaining or increasing the soil organic matter content. Historically the term soil amendment has been most frequently associated with organic additions to soils such as manures, composts, and organic waste products. At times the research involving soil amendments has provided important improvements in crop yields and soil water storage capacity while significantly reducing erosional sediments. The benefits of an adequate humus fraction in soils has long been associated with good soil physical properties, and the practice of manuring soils achieves the dual benefit of providing both essential plant nutrients and improved soil structure. However, as modern intensive agriculture has cultivated an ever increasing amount of land to feed an expanding world population, it has not been possible to produce sufficient quantities of manure or organic waste products that can be transported and applied in an economically efficient manner to meet the nutritional needs of high-yield crops or adequately to protect soils against the man-made and natural destruction of the physical conditions of these soils. Since World War II, chemical fertilizers have been used to meet crop nutrition demands, while the practice of manuring has decreased on many lands that are located far from a source of supply. This reduction in soil organic matter addition has had a negative effect on soil stability. In the light of the success of chemical fertilizers as substitutes for manures, it perhaps was only natural to hope that a chemical soil amendment could be found as a substitute for organic amendments.

II. A BRIEF HISTORY OF KRILIUM

During the 1950s agriculture appeared to be approaching the apex of a chemical solution to crop production barriers. Soil scientists had long been interested in the stabilization of soils for erosion control and improved plant growth. A wave of excitement swept over agricultural scientists shortly after the Symposium on the Improvement of Soil Structure was held by the Association for the Advancement of Science in 1951, at which a new group of water-soluble polymer (WSP) "soil amendments" was introduced that seemed to have a high probability of successfully stabilizing soils with very small application rates. This symposium was the impetus for many important investigations that were reported during that decade concerning the artificial stabilization of soil structure (Azzam, 1980; Bear, 1952; Quastel, 1954; Sherwood and Engibous, 1953).

The Monsanto Chemical Company had produced polymers of vinyl acetate maleic acid (VAMA or CRD-186) and hydrolyzed polyacrylonitrile (HPAN or CRD-189), which resembled some of the natural soil stabilizing agents found in organic matter. Monsanto applied the trade name Krilium to their soil stabilizing agent, which was formulated from vinyl acetate maleic acid and clay extenders. Monsanto's Krilium advertisements made broad claims for effectiveness of their "Year-round Soil Conditioner," claiming that when applied to "problem" soils the user could expect "improved soil workability, increased aeration, greater water holding capacity, faster germination, increased emergence, faster early growth, increased root formation, improved drainage, decreased crusting and ultimately, improved crop response." Their advertisements also noted that "most pronounced effects occur in hard-packed, crusty clay soils. In finer silt soils, the effect is less noticeable. In sandy soils, little appreciable improvement will be noted."

Various formulations of VAMA with clay extenders were marketed as Merloam, and pure VAMA was called Loamaker. The agricultural community and the gardening public were targeted with extensive advertising campaigns, and articles in major magazines included photographic instruction on proper application techniques. Many homeowners and some growers successfully used Krilium to amend soils, but too frequently users would not follow label directions, and failures were common. Some salesman were guilty of marketing the material without adequate instructions, leaving the user with the erroneous assumption that one could just spread the product on the surface of a poorly structured or compacted soil and Krilium would work its miracle. Complaints were made that it had an unpleasant odor, that it was a messy material to apply, and that soils were worse after treatment than before; these complaints did not help the public image of synthetic soil conditioners.

Other products were also made available to the scientific community, and some of these formulations were also sold through agricultural suppliers and gardening outlets. To most scientists and laymen, many of these synthetic soil conditioning products, regardless of formulation, were referred to as Krilium. Indeed, a soil conditioning product at that time was largely considered to be a synthetic chemical, and Krilium became almost a synonym for soil conditioner. However, besides Krilium (VAMA), the polymer HPAN (hydrolyzed polyacrylonitrile) was also sold in various formulations and by several companies under different names. Monsanto marketed HPAN Bondite. HPAN had several industrial uses at the time it was introduced to the agricultural community and was particularly successful as a drilling mud additive in the petroleum industry. It was observed that HPAN was more effective on some soils than was VAMA. During the Krilium period, isobutylene maleic acid (IBM) was also tested as a soil conditioner and found to have potential as a soil aggregate stabilizer.

These and a few other synthetic polymer materials usually produced dramatic results in laboratory experiments and demonstrations, and these results could frequently be duplicated in field trials, where quantifiable improvements in water infiltration, permeability, decreased surface crusting, improved seedling emergence, and improved yields were common. Some experimental results were so dramatic as to be almost unbelievable, yet failures with some soils and methods of using Krilium were also reported. Nonetheless, in most instances VAMA, HPAN, and IBM were highly successful in stabilizing soil aggregates and improving soil physical properties, particularly in soils with fine texture.

Rates of 100 mg (0.01%) to 20,000 mg (2.0%) of chemical conditioner per kg of dry soil made significant changes in aggregate stability, especially when dry conditioner powder or granular conditioner was thoroughly mixed with soil that approached air dryness. Claims were made in patents that rates as low as 0.001% were effective (Hendrick and Mowry, 1952b, 1953), but no published data have been reported to substantiate this. Successful techniques of applying conditioners in solutions to both dry and moist soils were also reported. Several studies were performed in which very high rates, several percent by dry weight, of conditioners were investigated. In some sandy textured soils, even very high rates were often not sufficient to improve soil aggregate stability measurably. Successful applications and favorable yield increases were reported on several alkaline and saline soils (Allison, 1952). Yield increases did not always occur even when Krilium treated soils were appropriately managed because soil structure was not the limiting growth factor.

The success stories, however, were for the most part sufficient to fuel the enthusiasm of the agricultural science community during the several years of research devoted to the Krilium breakthrough, even though economic realities doomed the widespread use of these products in mainstream agriculture from the outset. Krilium type products were marketed at \$4–5/kg in the 1950s, and the successful treatment of only a few centimeters of soil depth would have cost \$400–\$500/ha while the modification of soil to plow depth would have been several thousands of dollars per/ha. It was not surprising, then, that only soils used for very high return cash crops, container nurseries, landscapes, and hobby gardens could afford to use these products beyond a small trial basis.

When properly applied and incorporated into finer textured soils, the almost universal conclusion of the research community on the influence that these synthetic conditioners had on soil physical properties was the marked improvement of water infiltration rates. Improved infiltration was directly correlated with stabilized soil structure and an increased percentage of larger water conducting pores. These structurally improved soils often were less erodible, because the improved water intake and surface flow velocity greatly reduced runoff and sediment production. The more dramatic improvements were observed on soils of low surface stability that typically formed surface crusts following irrigation and rainfall that result in the disruption of soil aggregates. The improved soil structure was usually easily observable visually, and it was only natural to wonder what effect changes in porosity configuration would have on the water holding capacity of soils.

Most studies showed that the actual soil water holding capacity of treated soils did not show large differences over untreated soils when laboratory sorption/desorption data were compared. Soils that were slowly saturated and slowly dried, whether treated with conditioner or not, did not have large differences in available water holding capacity. However, several reports noted very significant increases in soil moisture availability to plants grown on conditioned soils as compared with checks. When the limiting condition for water infiltration rate was corrected by conditioner treatment, the effective water holding capacity of these structurally unstable soils did increase substantially. Increased soil water availability would often result in increased plant biomass production, with yield increases of up to 100% and occasionally much higher (Quastel, 1954). On soils subject to surface crusting, which influenced seedling emergence and water infiltration, very dramatic yield increases were observed in these Krilium-conditioned soils. However, many Krilium-treated soils sometimes produced only small plant growth increases, and occasionally a reduction in yield was reported. Obviously, Krilium was not the universal antidote for physical property deficiencies of all soil types, nor was improved plant growth a guaranteed result of conditioner application. The early observations and testimonials of dramatic results demonstrated on Krilium-treated soils, even when confirmed in many instances by a decade of careful and intensive research under a variety of field and laboratory conditions, did not prevent the economic reality that forced companies eventually to abandon the production of these soil conditioners.

III. LITERATURE RELATED TO KRILIUM

After six initial papers were presented at the Symposium on the Improvement of Soil Structure in 1951, studies on Krilium were published in scientific journal at an average of more than 20 per year during that decade, and numerous reports and non-journal articles were also published (Gardner, 1972). The peak publishing year was 1954 for journal articles (35 related references), but by the early 1960s research enthusiasm and industrial financial support for research programs investigating Krilium type conditioners had dwindled to insignificant levels. DeBoodt (1990) tabulated the number of scientific publications between 1952 and 1964 by years as 7, 22, 27, 32, 24, 22, 25, 13, 14, 12, 7, 1, and 3, respectively. Those articles and other pertinent literature are listed in the references herein.

The dearth of published articles after the mid-1960s does not mean that the scientific community had lost interest in methods of solving the soil aggregate stability issue. Quite the contrary, new chemical materials such as polyvinyl alcohols and acrylamide polymers were being investigated as promising soil stabilizers at lower concentrations than Krilium type materials (see Chapters 12, 13 and 15–17). Indeed today there appears to be a wellspring of scientific interest in water-soluble polyacrylamides similar to the Krilium frenzy of the 1950s. There are major economic and technical differences today. Are there lessons learned during that earlier era that are applicable to the use of more modern water-soluble polymers for soil stabilization in today's economic and environmentally conscious climate (Wallace and Nelson, 1986)? Can lower application rates of the much higher molecular weight water-soluble polymers effectively stabilize a wide range of soil types at economical rates? Can these materials be conveniently and safely applied with acceptable environmental consequences? Recent agricultural research provides favorable evidence for the safe and effective use of newer water-soluble polymers as summarized in this book.

IV. CONCLUSIONS

The 1950s were an era of excitement and optimism among soil science researchers concerned with solving one of agriculture's most challenging problems. Soil erodibility, water infiltration, and permeability had long been observed to be directly related to the stability of soil aggregates and suitable porosity. Agricultural tillage and production practices had been correlated with the destruction of favorable soil physical properties. In an era when chemical treatment of soils had dramatically improved crop yields through improved soil fertility and pest control, it was only natural to expect chemical technology to provide a solution to the problem of undesirable soil structure. The water-soluble polymers vinyl acetate maleic acid (VAMA) and hydrolyzed polyacrylonitrile (HPAN) in various formulations were found to stabilize soil peds, and in many experiments they produced dramatic improvements in soil physical properties and plant growth. Most of these soil stabilizing chemical formulations became known as Krilium after the name chosen by the Monsanto Chemical Company to represent their VAMA product. Scientific excitement soon turned to financial reality when it became clear in the 1960s that the per ha cost of Krilium was not economical for most agricultural crops. Field application techniques were also discouraging to many users and motivated the rejection of these products in mainstream agriculture. The need to improve soil quality is more acute today than it was in the 1950s and the search continues for production practices that can improve soil tilth and for economically and environmentally acceptable chemical means of soil stabilization.

REFERENCES

Abdel-Salam, M. A., and Pollard, A. G. (1959). Effect of synthetic polyelectrolytes on soil aggregation. *Sci. J. Poy. Coll. Sci. 27*:28.

Alderfer, R. B. (1954). Soil structure studies with synthetic conditioners. *Penn. Agric. Exp. Sta. Bull. 586.*

Aleksandrova, I. N. (1960). On the composition of humus substances and the nature of organo mineral colloids in soil. *Trans. 7th Internat'l Congr. Soil Sci. 2*:74–80.

Allison, L. E. (1952). Effect of synthetic polyelectrolytes on the structure of saline and alkali soils. *Soil Sci. 73*:443.

Allison, L. E. (1956). Soil and plant responses to VAMA and HPAN soil conditioners in the presence of high exchangeable sodium. *Soil Sci. Soc. Ami. Proc. 20*:147–151.

Allison, L. E. (1957). Effect of soil-conditioning polymers on the cation-exchange capacity. *Soil Sci. 83*:391.

Allison, L. E., and Moore, D. C. (1956). Effect of VAMA and HPAN soil conditioners on aggregation, surface crusting, and moisture retention in alkali soils. *Soil Sci. Soc. Am. Proc. 20*:143–146.

Archibald, J. A., and Erickson, A. E. (1955). Cation-exchange properties of a number of clay-conditioner systems. *Soil Sci. Soc. Am. Proc. 19*:444–446.

Baird, B. L., Bonnemann, J. J., and Richards, A. W. (1954). The use of chemical additives to control soil crusting and increase emergence of sugar beet seedlings. *Proc. Am. Soc. Sugar Beet Tech. 8*(1):136–142.

Bear, F. E. (1952). Synthetic soil conditioners. Proceedings of a symposium at Philadelphia, Pennsylvania, 1951. *Soil Sci 72*:419.

Bergmann, W., and Fiedler, H. J. (1954a). Synthetische polyelektrolyte als bodenverbesserungsmittel. 3. Der einfluss von Na-polyacrylat auf sedimentation und krümelstabilität von tonkolloiden. *Die Deutsche Landw.*, H. 4.

Bergmann, W., and Fiedler, H. J. (1954b). Synthetische boden verbesserungsmittel. *Der Deutsche Gartenbau, H. 12.*

Bergmann, W. (1959). Der einfluss von polyacrylät auf die krümelstabilität schwerer böden und auf das pflanzenwachstum. 24th Internat'l Symposium on Soil Structure, Ghent, Belgium, 28–31 May 1958, pp.112–116.

Bergmann, W., and Fiedler, H. J. (1954/1955). Synthetic soil conditioners, their action and use. *Wissenschaftliche Zeitschrift der Freidrich-Schiller-Universität Jena,* Reihe *4*:349–357.

Berryman, J. W. (1953). The effect of a soil conditioner in stabilizing the structure of a fallow on a degraded area of Cressy shaley clay loam. *Tasm. J. Agric. 24*:168.

Bolton, E. F., Fulton, J. M., and Aylesworth, J. W. (1955). The effects of two soil conditioners on some physical properties of a Brookston clay soil. *Can. J. Agric. Sci. 35*:51.

Bonnier, C., and Hely, F. W. (1953). The effect of certain synthetic resins which improve soil structure on the fungal population of a soil. *Bull. Inst. Agron. Gembloux 21*:14–28.

Bould, C., and Tolhurst, J. (1953). Soil conditioners. 1. The effect of the sodium salt of hydrolysed polyacrylonitrile (CRD 189) and of CRD 186 on nutrient availability and uptake by plants. *Long Ashton Res. Sta. Rpt. 1952*:49–54.

Brockman, F. J., and Allenby, O.C.W. (1953). Note on factors affecting the laboratory evaluation of soil conditioners. *Can. J. Agric. Sci. 33*:623.

Brockman, F. J., and Allenby, O. C. W. (1955). Factors affecting the laboratory evaluation of soil conditioners. *Can. J. Agric. Sci. 35*:27.

Chepil, W. S. (1954). The effect of synthetic soil conditioners on some phases of soil structure and erodibility by wind. *Soil Sci. Soc. Am. Proc. 18*:386–390.

DeBoodt, J. F. (1990). Applications of polymeric substances as physical soil conditioners. In: M. R. DeBoodt. ed. *Soil Colloids and their Associations in Aggregates.* Plenum Press, New York and London, pp. 517–556.

Dement, J. D., Martin, W. P., Taylor, G. S., and Alban, E. K. (1955). Effect of field applications of synthetic soil-aggregate stabilizers on plant emergence. *Soil Sci 79*:25.

Demortier, G., Droeven, G. (1953). Krilium. *Rev. Agric. Brux. 6*:1054–1098.

Demortier, G., and Droeven, G. (1954a). Does Krilium cause a lasting improvement in the physical properties of soils? *Rev. Agric. Brux. 7*:1089–1101.

Demortier, G., and Droeven, G. (1954b). The use of synthetic resins for structural improvement of loam and clay soils. *Rev. Agric. Brux. 7*:1198–1208.

Demortier, G., and Droeven, G. (1958). Synthetic resins and the improvement of soil structure: some complementary results. *Rev. Agric. Brux. 11*:347–353.

Doyle, J. J., and Hamlyn, F. G. (1960). Effects of different cropping systems and of a soil conditioner VAMA on some soil physical properties and on growth of tomatoes. *Can. J. Soil Sci. 40*:89.

Doyle, J. J., and MacLean, A. A. (1961). Use of a soil conditioner to increase the precision of soil fertility experiments. *Can. J. Soil Sci. 41*:86.

Duley, F. I. (1956). The effect of a synthetic soil conditioner (HPAN) on intake, runoff and erosion. *Soil Sci. Soc. Amer. Proc. 20*:420–422.

Emerson, W. W. (1955). Complex formation between montmorillonite and high polymers. *Nature 176*:461.

Emerson, W. W. (1956a). Synthetic soil conditioners. *J. Agric. Sci. 47*:117.

Emerson, W. W. (1956b). A comparison between the mode of action of organic matter and synthetic polymers in stabilizing soil crumbs. *J. Agric. Sci. 47*:350–353.

Emmerson, W. W. (1960). Complexes of calcium-montmorillonite with polymers. *Nature 186*:573–574.

Emerson, W. W. (1963). The effect of polymers on the swelling of montmorillonite. *J. Soil Sci. 14*:52.

Fiedler, H. J., and Bergmann, W. (1954a). Synthetische polyelektrolyte als bodenverbesserungsmittel. 2. Über die wechselwirkung zwischen polyelektrolyten und tonkolloiden im hinblick auf die ausbildung der krümelstruktur des bodens. *Die Deutsche Landw. H. 3.*

Fiedler, H. J., and Bergmann, W. (1954b) Synthetische polyelektrolyte als bodenverbesserungsmittel. 4. Der einfluss des polyacryläts auf die wasserstabilität von bodenkrümeln. *Die Deutsche Landw. H. 6*:313–318.

Fiedler, H. J., and Bergmann, W. (1955a). Wirkung verschiedener Bodenstruktur-Verbesserungsmittel. *Angew. Chem. 67*:699–704.

Fiedler, H. J., and Bergmann, W. (1955b). Synthetische Bodenverbesserungsmittel im Gartenbau. *Arch. Gartenbau 3*:(2)95–104.

Fiedler, H. J., and Bergmann, W. (1957a). Neue anwendungsmöglichkeiten synthetischer organischer substanzen in der technik. *Die Umschau in Wissenschaft und Technik. H. 4*:117–118.

Fiedler, H. J., and Bergmann, W. (1957b). Neue erfahrungen mit synthetisehen bodenverbesserungsmitteln. Auf dem wege zu fruchtbareren böden. *1957*(2): 44–46. Umschau Verlag, Frankfurt am Main, Germany.

Fischer, E. W., and Rentschler, W. (1957). The improvement of soil structure by synthetic high polymers. *Z. pflanzenernährung 76*:232–244.

Flaig, W. (1953). Clay Krilium. *Landb.-Forsch. 3*:88–91.

Frisch, H. L., and Simha, R. (1954). The adsorption of flexible macromolecules. *J. Phys. Chem. 58*:507.

Fuller, W. H., and Gairaud, C. (1954). The influence of soil aggregate stabilizers on the biological activity of soils. *Soil Sci. Soc. Am. Proc. 18*:35–40.

Gardner, W. H. (1952). Synthetic soil conditioners and some of their uses. *Proc. Wash. St. Hort. Assoc. 48*:115–122.

Gardner, W. H. (1972). Use of synthetic conditioners in the 1950s and some implications to their further development. In: M.F. DeBoodt, ed. Proceedings of a symposium on the Fundamentals of Soil Conditioning, Ghent, Belgium 17–21 April 1972, pp. 1046–1061.

Gardner, W. H., Hsieh, J. C., Ferguson, A. H., and Cannell, G. H. (1954). Alcohol–water method of measuring soil aggregate stability. Rpt. of Conf. on Soil Cond., Western Soc. Soil Sci. and Pac. Div. Amer. Assoc. Adv. Sci., Pullman, Wash.

Goodman, L. J. (1952). Erosion control in engineering works. *Agric. Eng. 33*:155.

Greenland, D. J., Lindstrom, G. R., and Quirk, J. P. (1962). Organic materials which stabilize material soil aggregates. *Soil Soc. Am. Proc. 26*:366.

Gussak, V. B. (1961). Experiment on the use of humic and polymerized preparations on scrozems to improve their structure and control erosion. *Pochvovedenie 8*:42.

Haise, H. R., Jensen, L. R., and Alessi, J. (1955). The effect of synthetic soil conditioners on soil structure and production of sugar beets. *Soil Sci. Soc. Amer. Proc. 19*:17.

Hanotiaux, G., and Manil, G. (1952). A note on synthetic resins applied for improving soil structure. *Bull. Inst. Agron. Gemblous 20*:66–71.

Haworth, F., and Nelder, J. A. (1955). The effect of Krilium on the yield of certain vegetable crops. *Brit. Soc. Promol. Res. Rep. 1954*:40–47.

Hayes, S. F., and Simpson, K. (1953). The effect of Krilium on a heavy clay soil. Edinburgh Scot. Coll. Agric. Misc. Publ. 116.

Healy, T. W., and LaMer, V. K. (1962). The absorption-flocculation reactions of a polymer with an aqueous colloidal dispersion. *J. Phys. Chem. 66*:1835.

Hedrick, R. M., and Mowry, D. T. (1952a). Effects of synthetic polyelectrolytes on aggregation, aeration and water relationships of soil. *Soil Sci. 73*:427–441.

Hedrick, R. M., and Mowry, D. T. (1952b). The improvement of soil structure by synthetic polyelectrolytes. *Chem. Industr.* 652–656.

Hedrick, R. M., and Mowry, D. T. (1953). Method of conditioning soils. U.S. Patent 2,625,529, 13 Jan 1953.

Hely, F. W., Bonnier, C., and Manil, P. (1954). Investigations concerning nodulation and growth of lucerne seedlings in a loess soil artifically aggregated to various levels. *Plant Soil 5*:121.

Hernando, V., and Jimeno, L. (1959). Study of the effect of Krilium on crops during three years. *Proc. Internat'l Symp. on Soil Struct. 1958*, pp. 81–88.

Holmes, R. M., and Toth, S. J. (1957a). Soil conditioners: their effect on coarse-textured soil, crop yields, and composition. *Can. J. Soil Sci. 37*:102.

Holmes, R. M., and Toth, S. J. (1957b). Physicochemical behavior of clay-conditioner complexes. *Soil Sci. 84*:479.

Homrighausen, E. 1957. Influence of the synthetic soil conditioner Rohagit S 7366 on soil structure. *Z. pflanzenernährung Bodenk. 78*:38.

Homrighausen, F. (1959). The moisture economy of soil as related to its structural conditions. *Proc. Internat'l Symp. on Soil Struct. 1958*, pp. 73–80.

Inden, T. (1960). The relationship of various conditions to the effects of soil conditioners on the growth of vegetables and on the uptake of nutrients. Bull. Fac. Agric. Mich. Univ., pp. 23–28.

Jacobson, H. G. M., and Swanson. C. S. W. (1958). Effect of soil type on duration of response to conditioner. *Soil Sci. 86*:216.

Jamison, V. C. (1953). Changes in air-water relationships due to structural improvement of soils. *Soil Sci. 76*:143.

Jamison, V. C. (1954). The effect of some soil conditioners on friability and compactibility of soils. *Soil Sci. Soc. Am. Proc. 18*:391.

Johnstone, F. E., Jr., Morris, H. D., Hanson, K. W., and H. W. Yuong. (1957). The effect of soil conditioners on the yields of sweet potatoes. *Proc. Am. Soc. Hort. Sci. 70*:403.

Jones, M. B., and Martin, W. P. (1957). Methods of evaluating aggregate stabilization by HPAN as it is affected by various inorganic salts. *Soil Sci. 83*:475.

Jones, M. B., Pratt, P. F., and Martin, W. P. (1957). The effect of HPAN and IBMA on the fixation and availability of potassium in several Ohio soils. *Soil Sci. Soc. Am. Proc. 21*:95.

Katchalsky, A., Lifson, S., and Eisenberg, H. (1952). Equation of swelling for polyelectrolyte gels. *J. Polymer Sci. 7*:571.

Kawai, K., and Okada, M. (1955). Effects of soil conditioner on the erosion of field soil and on the growth of sweet potato. *Proc. Crop Sci. Jpn. 23*:183–184.

Kita, D., and Kawaguchi, K. (1961). The mechanism of dispersion and flocculation induced in soil suspension by polyacrylic acid. 1. The bonding of soil particles by linear polyelectrolytes. *Jpn. Soil Sci. 32*:419.

Koral, J., Ullman, R., and Eirich, F. R. (1958). The adsorption of polyvinyl acetate. *J. Phys. Chem. 62*:541.

Kramer, M. (1955). Possibilities of the application of soil aggregating substances. *Agrokem. Talajt. 4*:94.

Kuipers, H., and Boekel, P. (1952). Results of a laboratory investigation on the action of Krilium. *Landw. Rijksuniv. 64*:731–735.

LaMer, V. K., and Healey, T. W. (1963). Adsorption and flocculation reactions of macromolecules at the solid–liquid interface. *Rev. Pure Appl. Chem. 13*:112–133.

LaMer, V. K., and Smellie, R. H. (1962). Theory of flocculation, subsidence and refiltration rates of colloidal dispersions flocculated by polyelectrolytes. *Clay Minerals 9*:295.

Laudelout, H. (1955). The effect of Krilium on soils of the Belgian Congo. *Bull. Inst. Stud. Agron. Congo Belge 4*:35–41.

Laws, W. D. (1954). The influence of soil properties on the effectiveness of synthetic soil conditioners. *Soil Sci. Soc. Am. Proc. 18*:378–381.

Leenheer, L., and DeBoodt, M. F. (1959). Practical importance of the use of soil conditioners such as Krilium in controlling the degradation of soil structure. *Proc. Int'l Symp. on Soil Struct. 1958*:89–96.

Lugo-Lopez, M. A., Bonnet, A., and Rico-Ballester, M. et al. (1957). Lack of response of some food crops to the application of synthetic soil conditioners to a clay soil in Lajas Valley. *J. Agric. Univ. Puerto Rico 41*:167.

MacIntire, W. H., Winterberg, S. H., Sterges, A. J. et al. (1954). Chemical effects of soil conditioner upon plant composition and uptake. *J. Agric. Food Chem. 2*:463.

Martin, W. P. (1953). Status report on soil conditioning chemicals. *Soil Sci. Soc. Am. Proc. 17*:1.

Martin, J., and Aldrich, D. G. (1955). Influence of soil exchangeable cation ratios on the aggregating effects of natural and synthetic soil conditioners. *Soil Sci. Am. Proc. 19*:50.

Martin, J. P., and Jones, W. W. (1954). Greenhouse plant response to vinyl acetate-maloic acid copolymer in natural soils and in prepared soils containing high percentages of sodium or potassium. *Soil Sci. 78*:317.

Martin, W. P., and Volk, G. W. (1952). Soil conditioners and fertilizers. *Fert. Rev. 27*(4):11.

Martin, W. P., Taylor, G. S., Engibous, J. C., and Burnett, E. (1952). Soil and crop responses from field applications of soil conditioners. *Soil Sci. 73*:455–471.

McCalla, T. M. (1959). Influence of soil conditioner HPAN on nitrification. *Trans. Can. Acad. Sci. 62*:53–57.

McNaught, K. J. (1955). Effect of synthetic soil conditioners on plant nutrient uptake. *New Zea. J. Sci. Tech. 36*:450.

Michaels, A. S. (1954). Aggregation of suspensions by polyelectrolytes. *Industr. Eng. Chem. 46*:1485.

Michaels, A. S., and Lamde, T. W. (1953). Soil flocculants and stabilizers. Laboratory evaluation of polyelectrolytes as soil focculants and aggregate stabilizers. *J. Agric. Food Chem. 1*:835.

Michaels, A. S., and Morelos, O. (1955). Polyelectrolyte adsorption by kaolinite. *Industr. Eng. Chem. 47*:1801.

Mitsui, S., Tensho, K., and Kurihara, K. (1956a). Effect of soil conditioners in increasing phosphate availability. 2. Effect of soil-conditioner treatment on growth, yield and composition of wheat in greenhouse experiments. *Jpn. Soil Sci. 26*:435

Mitsui, S., Tensho, K., and Kurihara, K. (1956b). Effect of soil conditioners in increasing phosphate availability. 1. Krilium treatment in relation to phosphate fixation by soil. *Jpn. Soil Sci. 26*:359–364.

Montgomery, R. S., and Hibbard, B. B. (1955). Theoretical aspects of the soil-conditioning activity of polymers. *Soil Sci. 79*:283.

Mortensen, J. L. (1957). Adsorption of hydrolyzed polyacrylonitrile on kaolinite. 1. Effect of exchange cation and anion. *Soil Sci. Soc. Am. Proc. 21*:385.

Mortensen, J. L. (1959). Adsorption of hydrolyzed polyacrylonitrile on kaolinite. 2. Effect of solution electrolytes. *Soil Sci. Soc. Am. Proc. 23*:199.

Mortensen, J. L., and Martin, W. P. (1954). Decomposition of the soil conditioning polyelectrolytes HPAN and VAMA. *Soil Sci. Soc. Am. Proc. 18*:395.

Mortensen, J. L., and Martin, W. P. (1956a). Effect of soil-conditioner–fertilizer interactions on soil structure, plant growth, and yield. *Soil Sci. 81*:33.

Mortensen, J. L., and Martin, W. P.. (1956b). Microbial decomposition and adsorption of synthetic polyelectrolytes in Ohio soils. U.S. Atomic Energy Comm. TID-7512:235–243.

Moss, G. R., Browning, H. A., and Southon, T. F. (1954). Synthetic soil conditioners. *New Zea. J. Agric. 88*:67.

Moustafa, A. H. I., and El-Shabassy, A. I. (1959). Effect of various soil conditioners on deteriorated soil. *Agric. Res. Rev. (Egypt) 37*:154.

Mowry, D. T., and Hedrick, R. M. (1953). Fertilizer and soil-conditioning compositions. U.S. Patent US 2,625,471, 13 Jan 1953.

Murdock, J. T., and Seay, W. A. (1954). The effect of a soil conditioner on uptake of superphosphate by greenhouse wheat. *Soil Sci. Soc. Am. Proc. 18*:97–98.

Murdock, J. T., Namise, N., and Anase, M. (1954). The effect of Krilium on the properties of soil. *Jpn. Agric. Eng. Soc.* 22:417.

Nath, T., and Nagar, B. R (1959). Effect of synthetic soil conditioners on the cation exchange capacity of soils and clays. *Curr. Sci. 28*:193.

Nath, T., and Nagar, B. R. (1960). Influence of cations on the aggregating effects of synthetic soil conditioners. *Curr. Sci. 29*:207.

Nicholls, R. L., and Davidson, D. T. (1957). Soil stabilization with large organic cations and polyacids. *Proc. Iowa Acad. Sci. 64*:349–381.

Nishikata, T., Mori, T. and Takeuchi, Y. et al. (1954). Studies on soil conditioners. 1. Influence of treatment with Krilium on the physical and chemical characteristics of soils. *Res. Bull. Hokkaido Nat. Agric. Exp. Sta. 66*:10–16.

Okuda, A., and Tokubo, K. (1956). The effect of soil conditioners. 4. *Jpn. Soil Sci. 26*:419.

Okuda, A., and Tokubo, K. (1957). Leak-preventing effect of synthetic soil conditioners in paddy field. *Soil Plant Food* 2:215–219.

Okuda, A., Hori, S., and Tokubo, K. (1953). Studies on the effect of soil conditioners. *Jpn. Soil Sci. 24*:89.

Okuda, A., Tokubo, K., and Kawasaki, T. (1956). The effect of soil conditioners. 5. *Jpn. Soil Sci. 27*:105.

Packter, A. (1957). Interaction of montmorillonite clays with polyelectrolyte. *Soil Sci. 83*:335–343.

Panabokke, C. R., and Quirk, J. P. (1975). Effect of initial water content on the stability of soil aggregates in water. *Soil Sci. 83*:185.

Pearson, R. W., and Jamison, V. C. (1953). Improving land conditions for conservation and production with chemical soil conditioners. *J. Soil Water Conserv. 8*:130.

Perez-Escolar, R., and Lugo-Lopez, M. A. (1957). The effect of synthetic soil conditioners on soil-aggregate stability and the production of potatoes and stringless beans. *J. Agric. Univ. Puerto Rico 41*:127.

Peters, D. B., Hagan, R. M., and Bodman, G. B. (1953). Available moisture capacities of soils as affected by additions of polyelectrolyte soil conditioners. *Soil Sci. 75*:467.

Peterson, C., and Kwei, T. K. (1961). The kinetics of polymer adsorption onto solid surfaces. *J. Phys. Chem. 65*:1330.

Pouwer, A. (1953). Laboratory tests with Krilium. *Meded. Direct. Tuinb. 16*:360–361.

Pringle, J., and Williamson, W. T. H. (1956). Some effects of a soil conditioner on a heavy and a light soil in Aberdeenshire. *J. Sci. Food Agric. 7*:540.

Pugh, A. L., Vomocil, J. A., and Nielsen, T. R. (1960). Modification of some physical characteristics of soils with VAMA, ferric sulfate, and triphenylsulfonium chloride. *Agron. J. 52*:399.

Quastel, J. H. (1953). Krilium and synthetic soil conditioners. *Nature 171*:7.

Quastel, J. H. (1954) Soil conditioners. *Ann. Rev. Plant Physiol. 5*:75.

Ramacharlu, P. T. (1956). Influence of synthetic soil conditioners on the evaporation of water from soil. *J. Indian Soc. Soil Sci. 4*:265.

Raney, W. A. (1953). Soil aggregate stabilizers. *Soil Sci. Soc. Am. Proc. 17*:76.

Rawlins, S. L., Kittrick, J. A., and Gardner, W. H. (1963). Electron microscope observations of freeze-dried soil conditioners. *Soil Sci. Soc. Am. Proc. 27*:354.

Rennie, D. A., Trough, E., and Allen, O. N. (1954). Soil aggregation is influenced by microbial gums, level of fertility, and kind of crop. *Soil Sci. Soc. Am. Proc. 18*:399.

Richard, F. (1959). The effect of Krilium in neutral and acid clay and loam soils. *Proc. Internat'l Symp. on Soil Struct. 1958*:97–102.

Richards, F., and Chausson, J. S. (1957). The artificial alternation of physical soil factors. *Mitt. Schweiz. Aust. Forstl. VersW 33*(1).

Ruehrwein, R. A., and Ward, D. W. (1952). Mechanism of clay aggregation by polyelectrolytes. *Soil Sci. 73*:485.

Ryaboklyach, V. A., Savitskaya, M. N., and Khomenko, A. D. (1963). Effect of polymer preparations on physical properties of soils and yield of agricultural plants. *Pochvovedenie 6*:97.

Schreiber, H. (1956). Investigations on the effect of synthetic soil conditioners on various physical properties of soil. *Z. Acker PflBau. 101*:361–394.

Sherwood, L. V., and Engibous, J. C. (1953). Status report on soil conditioning chemicals. *Soil Sci. Soc. Am. Proc. 17*:9.

Silberberg, A. (1962a). Adsorption of flexible macromolecule. 1. The isolated macromolecule at a plane interface. *J. Phys. Chem. 66*:1872.

Silberberg, A. (1962b). Adsorption of flexible macromolecules. 2. The shape of the adsorbed molecule.: the adsorption isotherm-surface tension. *J. Phys. Chem. 66*:1884.

Simpson, K., and Hayes, S. F. (1958). The effect of soil conditioners on plant growth and soil structure. *J. Sci. Food Agric. 9*:163–170.

Slater, C. S. (1953). Soil conditioners in soil conservation. *Agric. Eng. 34*:98–100, 102.

Smith, C. H. (1954) Influence of Krilium soil conditioner on sugar content of beets. *Proc. Amer. Soc. Sugar Beet Tech. 8*:143–146.

Smith, H. E., Schwatz, S. M., Gugliemelli, L. A., P. G. Freeman, and C. R. Russell. (1958). Soil-conditioning properties of modified agricultural residues and related materials. 1. Aggregate stabilization as a function of type and extend of chemical modification. *Soil Sci. Soc. Amer. Proc. 22*:405.

Strickling, E. (1957). Effect of cropping systems to VAMA on soil aggregation, organic matter, and on yields. *Soil Sci. 84*:489.

Swanson, C. L. W. (1952a). Soil conditioners awaken new interest in soils. *Frontiers Plant Sci. 5*(1):4.

Swanson, C. L. W. (1952b). A new concept: using chemicals for soil structure improvement. *J. Soil Water Conserv. 7*:61.

Swanson, C. L. W., and Jacobson, H. G. M. (1955). Effect of aqueous solutions of soil conditioner chemicals on corn seedlings grown in nutrient solutions. *Soil Sci. 79*:133.

Swanson, C. L. W., Hanna, R. M., and DeRoo, H. C. (1955). Effects of excessive cultivation and puddling on conditioner-treated soils in the laboratory. *Soil Sci. 79*:15.

Synthetic polyelectrolyte improves soil structure, prevents erosion (1951). *Chem. Eng. News 29*:5530.

Tamhane, R. V., and Chibber, R. K. (1955). Effect of synthetic soil conditioners on soil structure. *J. Indian Soc. Soil Sci. 3*:97.

Taylor, G. S., and Martin, W. P. (1953). Effect of soil-aggregating chemicals on soils. *Agric. Eng. 34*:550.

Taylor, H. M., and Henderson, D. W. (1959). Some effects of organic additives on compressibility of Yolo silt loam soil. *Soil Sci. 88*:101.

Taylor, H. M., and Vomocil, J. A. (1959). Changes in soil compressibility associated with polyelectrolyte treatment. *Soil Sci. Soc. Am. Proc. 23*:181.

Tokubo, K., and Okuda, A. (1958). Studies on the effect of soil conditioners. 7. Mechanism of the formation of water-stable aggregates in soils. *Jpn. Soil Sci. 29*:255.

Tokuoka, M., and Tokunaga, Y. (1954). The action of soil conditioner. *Jpn. Soil Sci. 25*:115.

Torstensson, G., Nilsson, N. M., and Fiedler, H. I. (1957). Investigations on soil conditioners for the improvement of structure. 1. Residual effects of different products. *K. Lantbr. Hogsk. Ann. 23*:323–341.

Toth, S. J. (1955). Report on performance of soil conditioning chemicals. *J. Ass. Off. Agric. Chem.* (Wash.) *38*:252.

Wahhab, A., Khabir, A., Azim, M. N., and F. Uddin. (1956). Effect of a synthetic polyelectrolyte on the chemical, physical, biological, and agronomic aspects of some Punjab soils. *Soil Sci. 81*:139.

Wallace, A., and Nelson, S. D. (1986). Special issue on soil conditioners. *Soil Sci. 141*:334–342.

Warrior, S. K., and Pollard, A. G. (1959). Action of synthetic soil conditioners on waterlogged soils. *J. Sci. Food Agric. 10*:565.

Weakly, H. E. (1960). The effect of HPAN soil conditioner on runoff, erosion and soil aggregation. *J. Soil Water Conserv. 15*:169.

Webber, L. R. (1959). Note on the aggregate-size distribution and aggregate stability in Haldimand clay treated with four soil additives. *Can. J. Soil Sci. 39*:252.

Weeks, L. E., and Colter, W. G. (1952). Effect of synthetic soil conditioners on erosion control. *Soil Sci. 73*:473.

Wester, R. E. (1953). Response of vegetable crops to soil conditioners. *Agric. Chem. 8*(7):48–50, 125–126.

Wester, R. E. (1955). Synthetic soil conditioner's effect on crop yield. *Agric. Chem. 10*(8):44–46, 99.

Willis, W. O. (1955). Freezing and thawing and wetting and drying in soils treated with organic chemicals. *Soil Sci. Soc. Am. Proc. 19*:263.

15

Some Uses of Water-Soluble Polymers in Soil

Guy J. Levy and M. Ben-Hur *Institute of Soils and Water, Agricultural Research Organization, The Volcani Center, Bet Dagan, Israel*

I. INTRODUCTION

Low infiltration rate, poor drainage, runoff, and soil erosion are among the principal factors responsible for soil degradation. Traditional soil conservation strategies for runoff and erosion control in cultivated lands are based mainly on measures that 1) protect the soil surface from raindrop impact using mulch, in order to prevent seal formation and soil detachment; 2) increase surface depression storage and soil roughness to reduce runoff volume and velocity; and 3) alter slope-length gradient and direction of runoff flow.

An alternative approach to soil conservation is one that advocates modification of some soil properties responsible for soil susceptibility to seal formation, runoff, and erosion. Increasing aggregate stability and preventing clay dispersion are known to control soil sealing, increase infiltration rate (IR), and reduce runoff in cultivated soils. Furthermore, stable aggregates at the soil surface are less sensitive to detachment by raindrop impact and to transportation by runoff water. Improving aggregate stability and preventing clay dispersion can be accomplished by applying additives (e.g., phosphogypsum, organic polymers) to the soil. The fact that seal formation, runoff, and soil erosion are surface phenomena allows the treatment of only the soil surface rather than mixing the soil conditioner with the entire cultivated layer. This, in turn, leads to a reduction in the amounts of additive needed and makes its use economically attractive.

In this chapter, soil conditioners (a group of soil additives) are defined as synthetic organic chemicals or chemically modified natural substances that stabilize soil aggregates and favorably affect soil structure and physical properties (Aslam, 1990). Interest in synthetic organic polymers as soil conditioners started in the early 1950s (see Chapter 14). Even though the conditioners were shown to improve soil physical properties, they failed commercially because of excessive cost, difficulty in use, and inconsistent results (Wallace and Wallace, 1990). The recent advances in the chemistry of synthetic polymers, coupled with the concept of treating the soil surface only, made the use of polymers more cost efficient.

A number of reviews have lately been published that discuss the use of organic polymers as soil conditioners in general (Azzam, 1980; Soil Science, 1986) and particularly in improving soil structure and physical properties (Wallace and Wallace, 1990; Soil Science, 1994; Seybold, 1994; Levy, 1995). This chapter discusses the use of commercial water-soluble polymers in controlling runoff and soil erosion, under arid and semiarid conditions where efficient water use and water conservation are of prime importance, with an emphasize on the Israeli experience.

II. POLYMERS: THEIR PROPERTIES AND INTERACTIONS IN THE SOIL

A. Polymer Properties

Polymers are small recognized repeating units (monomers) coupled together to form extended chains. Their chain lengths in solution range between a few thousand and 3×10^5 μm, and average diameters are 0.5–1.0 μm (Schamp et al., 1975). Besides being long, polymer chains are flexible, multisegmented, and polyfunctional. Polymers are characterized mainly by their molecular weight, molecular conformation (coiled or stretched), type of charge, and charge density.

Polyacrylamide (PAM) and polysaccharide (PSD) are two synthetic organic polymers that have recently been extensively investigated with respect to their efficacy as soil conditioners. PAM is a homopolymer compound formed by the polymerization of identical acrylamide and related monomers (Barvenik, 1994). Its molecular weight varies in the range of 10×10^4 Da (medium weight) to 15×10^6 Da (high weight). In the latter case, the PAM has $1–2 \times 10^5$ repeating units, each with a molecular weight of 71 Da (Barvenik, 1994). PAM can be cationic, nonionic, or anionic, with the anionic form being the most commonly used in soil conditioning. Anionic PAM can be produced by several processes. The commercial anionic PAMs are commonly produced by copolymerization of acrylamide and acrylic acid or a salt of that acid (Mortimer, 1991). The charge density of the anionic PAM can be manipulated by the degree of hydrolysis (Stutzmann and Siffert, 1977), or the pH of the system (Barvenik, 1994). Proportions of charged co-monomer in a PAM of <10%, 10–30%, and >30% are considered low, medium, and high charge density, respectively.

PSD is a guar derivative, guar being classified as a galactomannan consisting of mannose and galactose units (Aly and Letey, 1988). The PSDs have low to medium molecular weight (0.2–2.0 × 10^6 Da). Their charge density is determined by the number of substitutions of nonionic cis hydroxyl groups with quaternary ammonium, hydroxpropyl, and carboxyl groups in order to get a cationic, nonionic, and anionic PSD, respectively (Aly and Letey, 1988). Cationic low to medium charge density PSD is the one most commonly used.

B. Polymer-Solid Interactions

1. Polymer-Clay Interactions

The relationship between polymer treatment and soil physical and hydraulic properties is greatly affected by the adsorptive behavior of the polymer molecules on soil particles. Polymer adsorption on clay minerals was studied extensively and the data were summarized in several reviews (Greenland, 1972; Theng, 1979, 1982). Of the various polymer properties (molecular weight and conformation, type and density of charge, temperature, acidity, etc.), type of charge largely determines the mechanisms controlling polymer adsorption and thus received much attention (e.g., Theng, 1979, 1982).

The adsorption of nonionic flexible linear polymer onto a clay surface increases with the molecular weight of the polymer and generally leads to the desorption of numerous water molecules. The adsorption of such polymers is usually an entropy-driven process, and irreversible (Theng, 1982).

The adsorption of cationic polymers by clays occurs through electrostatic (Coulombic) interactions between the cationic groups on the polymer and the negatively charged sites on the clay surface. Polycations compete with exchangeable and electrolyte cations for exchange sites on the clay. Hence, adsorption of polycations by clay increases with a decrease in the valency of the exchangeable cation (Gu and Doner, 1992) and decreases with an increase in the electrolyte concentration of the solution (Aly and Letey, 1988). An increase in the solution pH, associated with an increase in negative edge sites on illite surfaces, increased the adsorption of a polycation by illite (Gu and Doner, 1992).

Negatively charged polymers tend to be repelled from the clay surface, and little adsorption occurs. In addition, anionic polymers do not tend to enter the interlayer space of expanding minerals (e.g., Ruehrwein and Ward, 1952). Adsorption is promoted by the presence of polyvalent cations that act as "bridges" between the anionic groups on the polymer and the negatively charged sites on the clay (Mortensen, 1962). Increasing the ionic strength of the solution reduces the electrostatic repulsion between the polyanion and the clay surfaces (Ruehrwein and Ward, 1952; Mortensen, 1959) and may also lead to decreased charge and size of the polymer (Mortensen, 1959); both of which enhance adsorption of polyanions (e.g., Aly and Letey, 1988; Gu and Doner, 1992). Acidic conditions, associated with an increase in positive edge sites on clays, also favor polyanion adsorption (Theng, 1982).

The amount of polymer adsorbed by clay depends not only on the type of polymer charge but also on the clay mineral. Gu and Doner (1992) reported that the amount of PSD adsorbed by Silver Hill Na-illite was 100, 35, and 8 g polymer per kg clay for cationic, nonionic, and anionic PSD, respectively (Fig. 1). Conversely, the amount of a nonionic PSD and a cationic PSD adsorbed by Wyoming Na-montmorillonite was similar, 80 g kg^{-1} (Aly and Letey, 1988). With respect to anionic PAM, only small quantities are adsorbed on Na-montmorillonite, 2–3 g kg^{-1} (Stutzmann and Siffert, 1977; Aly and Letey, 1988). By comparison, Wyoming Na-montmorillonite adsorbs 45 g kg^{-1} of nonionic PAM (Aly and Letey, 1988). Stutzmann and Siffert (1977) postulated that the small quantities of PAM adsorbed on the clay result from the fact that PAM is adsorbed only on the external surfaces of the clay particles, and hence that PAM adsorption depends on the external cation exchange capacity of the mineral.

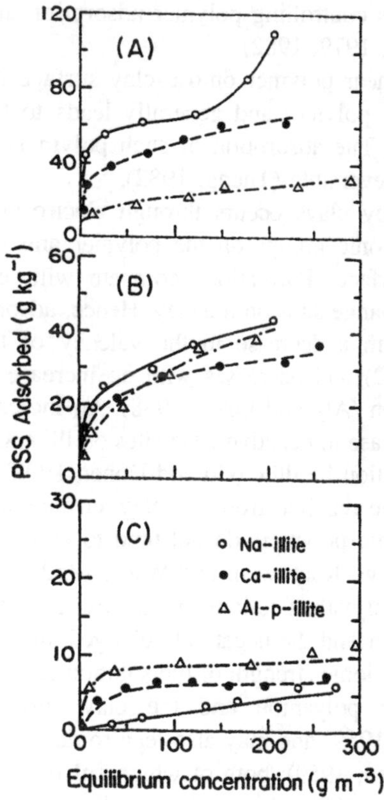

Figure 1 Adsorption isotherms of (A) cationic, (B) nonionic, and (C) anionic polysaccharide (PSD) by Na-, Ca-, and Al-*p*-illite. (From Gu and Doner, 1992.)

Ben-Hur et al. (1992a) suggested that polymer adsorption depends on the external charge density of the adsorbing clay rather than the adsorbing clay cation exchange capacity. Ben-Hur et al. (1992a) observed a considerably larger adsorption by illite than by montmorillonite, irrespective of the type of the polymer (i.e., cationic or anionic). The higher charge density of illite compared with montmorillonite (when external surfaces are considered) was regarded the reason for the higher adsorption of the polymers on illite (Ben-Hur et al., 1992a).

Polymer added to a colloidal suspension can act as a dispersant or flocculant, depending on the polymer properties and the electrolyte concentration in the suspension (Sato and Ruch, 1980). Significant interactions between polymer type and water quality with respect to their effect on flocculation of Na-montmorillonite were observed (Aly and Letey, 1988). Anionic PAM and PSD promoted flocculation in solutions of electrical conductivity (EC) = 0.7 ds m^{-1} for polymer concentrations >5 g m^{-3} but enhanced clay suspension stability in solutions of EC = 0.05 ds m^{-1}. Cationic and nonionic PSD promoted flocculation and the formation of large flocs at both water qualities. Flocculation of anionic polymers was explained by cation bridging and osmotic attraction, whereas in the case of cationic and nonionic polymers, flocculation was ascribed to charge neutralization (Aly and Letey, 1988). For Na-kaolinite, Nabzar et al., (1984) found that addition of nonionic high-molecular-weight PAM at concentrations $<0.2 \times 10^{-3}$ mg mL^{-1} or $>0.95 \times 10^{3}$ mg mL^{-1} stabilized the suspension, whereas at the intermediate concentration PAM enhanced flocculation. The results were explained by the correlation established between adsorption strength and the mode of particle association at the different PAM concentrations (Nabzar et al., 1984).

2. Polymer–Soil Interactions

Studies of polymer adsorption on clays were motivated by the understanding that clay minerals are the reactive fraction of the soil; thus polymer adsorption on clay material was considered as an accurate representation of polymer adsorption to soil. However, a review of studies on polymer adsorption on soils (Letey, 1994) showed that the aforementioned assumption was invalid.

Nadler and Letey (1989) studied the adsorption of three anionic polymers by a coarse loamy soil. Similar to anionic polymer–clay systems (Aly and Letey, 1988; Gu and Doner, 1992), Ca soils and initial high pH resulted in increased polymer adsorption; the increase, however, was very small. The adsorption levels were in the range of mg of polymer per kg of soil, being 2–3 orders of magnitude lower than those obtained by Aly and Letey (1988) for the same polymers on montmorillonite. It was postulated that 1) the higher specific surface area and the amount of charge available for adsorption associated with the clay-size fraction of the montmorillonite, and 2) the smaller accessible and active surfaces in the soil because of the presence of organic matter and aggregation are responsible for the lower adsorption of the polymers by the soil compared with clay (Nadler and Letey, 1989).

Malik and Letey (1991) extended the study of Nadler and Letey (1989) by including additional soils, washed quartz sand, and cationic polymers. Polymer adsorption on soil was in the following decreasing order: high-charge cationic PSD > low-charge anionic PAM > low-charge nonionic PAM. Adsorption on sand was only little less than on soil. Conversely, adsorption on clay was three orders of magnitude higher, and in the order low-charge cationic PSD > low-charge anionic PAM (Fig. 2). Thus Malik and Letey (1991) concluded the following: 1) adsorption by soils is mostly on external surfaces, and the polymers

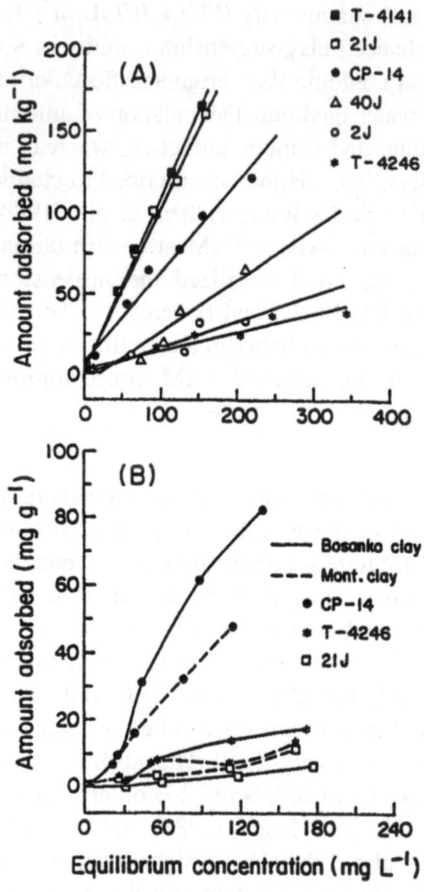

Figure 2 Adsorption isotherms of various polymers on (A) Arlington soil and on (B) clay extracted from Bosanko soil and specimen montmorillonite clay. (From Malik and Letey, 1991.)

did not penetrate the aggregates; 2) charge density of the polymer determined its adsorption on charged surfaces (e.g., clays); and 3) molecular size and conformation of the polymer determined its adsorption on weakly charged and noncharged surfaces (e.g., soils and sand); increasing molecular size and chain extension of the polymer lead to increased adsorption.

It can be concluded that solid properties that primarily affect polymer adsorption are

1. The surface area on which the polymer can be adsorbed onto; in soils, since the polymer hardly penetrates the aggregates, it is the external surface area of the aggregates.
2. Surface charge density and charge type, which determine the magnitude of the force by which the polymer is attracted to or repelled from the solid surfaces.

The polymer properties that affect its adsorption are mainly 1) its charge density and type, which influence electrostatic attraction or repulsion to the solid, and 2) its molecular weight and conformation, especially in the case of weakly charged or noncharged surfaces.

Desorption of polymers from soils rarely occurs. Nadler et al. (1992) measured desorption of PAM and PSD from soil material. Very little or no desorption occurred if the soil was kept wet. Upon drying, most or all of the polymer initially left in the solution became irreversibly bonded to the soil (Nadler et al., 1992). It is unlikely that all segments of the polymer can be simultaneously detached from the surface of the soil surface and remain detached long enough for the polymer to move away from the surface to the bulk solution (Nadler and Letey, 1989).

III. POLYMER USE IN DRY LAND FARMING

A. Introduction

Runoff and soil erosion problems are closely related to seal formation at the soil surface resulting from the impact energy of water drops, whether from rain or from overhead sprinkler irrigation. The phenomenon of seal formation and its effects on infiltration rate, runoff, and soil erosion are discussed in Chapter 7 of this book; hence just a brief synopsis is given here.

Runoff is directly related to seal development. The more developed the seal (i.e., the lower its permeability), the higher the runoff level. Runoff from cultivated fields is usually lost for crop production and accelerates soil erosion. Runoff leads to removal of fertilizers and pesticides from the field to the environment. Local runoff within the field accumulates in depressions and leads to poor water distribution, to local drainage and aeration problems, and subsequently to crop reduction (Letey et al., 1984).

Water erosion of soil is divided into 1) rill erosion, which is the product of soil detachment and transport by hydraulic shear stress in concentrated confined flow, and 2) interrill erosion, which results from detachment and transport by raindrop impact and shallow overland flow (Lane et al., 1987).

In arid and semiarid regions, the annual rainfall amount and the erosivity index are low (Agassi and Ben-Hur, 1991). Consequently, soil erosion in these areas is expected to be limited. However, when exposed to rain, soils from arid and semiarid regions develop a seal with low permeability at the soil surface (e.g., Chen et al., 1980; Ben-Hur et al., 1985), which leads to high amounts of runoff and soil loss. Moreover, Stern et al. (1991a) noted that for soils from semiarid regions, predominantly smectitic soils are very erodible compared to nonsmectitic ones. Hence, erosion is a severe problem in semiarid regions despite an alleged low erosivity index.

For prevention of seal formation, runoff, and soil erosion, it is preferable to add the polymer to the soil surface in solution form rather than as dry powder or granules (Wallace and Wallace, 1986). In dry land farming where surface sealing is formed under natural rain, polymer is commonly added to the soil surface, at a rate of <100 kg ha^{-1}, prior to the rainy season, by spraying-solution of low polymer concentration (500–1000 g m^{-3}) (e.g., Shainberg et al., 1989; Levy et al., 1991).

B. Controlling Runoff and Soil Erosion

The effect of polymers on stabilizing soil surface aggregates during rainstorms and hence on preventing runoff and erosion has been studied extensively under both laboratory and field conditions.

In the early 1970s, Gabriels et al. (1973) showed that surface application of a small amount (38 kg ha^{-1}) of an anionic polyacrylamide was highly successful in maintaining high IR values and preventing runoff.

Studies on the efficacy of surface application of polymers started in Israel in the mid 1980s. Shaviv et al. (1986) used a rainfall simulator to study the effect of an anionic polymer with a low molecular weight (70,000–150,000 Da) on the permeability of a sodic (exchangeable sodium percentage [ESP] 25.6%) silty loam from the northwestern Negev, Israel. They observed that in polymer application at a rate of 80 kg ha^{-1}, runoff started only after 50 mm of rain compared with almost an immediate initiation of runoff in the control. When the polymer application was supplemented by powdered gypsum at a rate of 2.4 Mg ha^{-1}, hardly any runoff was noted during the entire 75 mm storm (Shaviv et al., 1986).

Shainberg et al. (1990) investigated in the laboratory the effects of soil-surface application of a high-molecular-weight (2×10^7 Da), low-charge-density anionic PAM on seal formation and runoff in two nonsodic (ESP <5) Israeli soils, a loess (Calcic Haploxeralf) and a grumusol (Typic Chromoxerert). They applied 10, 20, and 40 kg ha^{-1} of PAM and found that the beneficial effect of rates above 20 kg ha^{-1} were insignificant in maintaining high IR (Table 1). Shainberg et al. (1990) also found that the effect of the polymer was dramatically enhanced when the polymer was added in combination with phosphogypsum (PG). For instance, percent runoff in the combined PAM and PG treatment was less than half than that when only PAM was used (Table 1). Shainberg et al. (1990) concluded that flocculation of the soil clay, achieved by the presence of PG, is apparently a precondition for efficient cementing and stabilization of aggregates at the soil surface by anionic polymers.

The amount of effective polymer reported by Shaviv et al. (1986) was four times that used by Shainberg et al. (1990), 80 and 20 kg ha^{-1}, respectively. This difference in the amounts of polymer needed to control runoff can be attributed to differences in the molecular weight of the polymers, suggesting that high-molecular-weight polymer is more effective in controlling seal formation and runoff, and hence a smaller application is required. The study of Levy and Agassi (1995) suggested that the efficacy of low-molecular-weight polymers depends on soil texture. They compared the effects of two negatively charged PAMs, a low-molecular-weight one (2×10^5 Da, PAM_L) and a high-molecular-weight one (2×10^7 Da, PAM_H), on infiltration of and erosion from a clay vertisol, a silty loam loess, and a sandy loam hamra, Israel. Both polymers were effective in maintaining higher IR and lower erosion levels than the control. In the medium- and coarse-textured soils (loess and hamra, respectively), the PAM_H was more effective in controlling seal formation and maintaining high IR than the PAM_L. In the fine-textured grumusol the effect of both polymers was comparable (Table 2). Levy and Agassi (1995) concluded that PAM_L is ef-

Table 1 Infiltration and Runoff Data

	Bet Qama Loess		Sede Yoav Grumusol	
Treatments[a]	FIR[b] mm h^{-1}	Runoff %	FIR mm h^{-1}	Runoff %
DW, PAM (0)	2.0	84.6	3.0	73.4
DW, PAM (10)	5.9	63.6		
DW, PAM (20)	6.5	52.1	7.4	47.6
DW, PAM (40)	6.5	51.3		
DW, PAM (0), PG	6.8	60.6	9.7	54.6
DW, PAM (10), PG	15.7	38.3		
DW, PAM (20), PG	23.5	19.8	20.5	22.1
DW, PAM (40), PG	23.9	18.6		

[a]DW = distilled water; PAM () = application in kg ha^{-1}; PG = phosphogypsum (5 Mg Ha^{-1}).
[b]FIR = final infiltration rate.
Source: Shainberg et al., 1990.

fective in fine-textured soils because the average distance between clay particles is short enough to enable effective binding between adjacent particles by the relatively short molecules of this polymer. In such soils the use of PAM_L is preferable due to solution viscosity consideration; the viscosity of the 1000 g m^{-3} PAM_L solution used was 1.26 mPa s^{-1}, compared with a viscosity of 3.89 mPa s^{-1} in the case of the PAM_H. In medium- and light-textured soils, where the distances between clay particles is longer than the "grappling distance" of the PAM_L molecules, the use of PAM_H is needed (Levy and Agassi, 1995).

Ben-Hur and Keren (1996) studied the effect of three different commercial polymers, a low-molecular-weight nonionic one (P-101), a cationic one with a medium molecular weight (CP-14), and a high-molecular-weight anionic one (CG), on aggregate formation, sealing, and infiltration rate in a sandy loam (Typic Rhodoxeralf) from Israel. All three polymers were effective in maintaining higher IR than that of the control; the P-101 was the most effective and the CG the least. The polymers significantly increased aggregate formation and the size of the aggregates formed. However, polymer efficacy on aggregation was in the order CG > CP-14 > P-101, opposite to that obtained in the IR experiment. The surface tension of the applied polymer solutions as well as the viscosity of these solutions were in the same order as that in the aggregate formation study (CG > CP-14 > P-101). Ben-Hur and Keren (1996) suggested that the ability of a polymer to maintain high IR is enhanced by the polymer's ability to penetrate into the surface aggregates; low molecular weight and low viscosity of the polymer solution contribute to this ability. Conversely, polymer effects on aggregate formation and the surface tension of its solution, which in

Table 2 Mean Final Infiltration Rates (FIR) in the Different Soils for the Two Different Polymers

Soil	Treatment[a]	Fir[b] mm h^{-1}
Grumusol—Bet Qama	Control	3.50 a
	PAM_L	8.25 b
	PAM_H	9.16 b
Loess—Nahal OZ	Control	4.72 a
	PAM_L	9.80 b
	PAM_H	14.65 c
Hamra—Morasha	Control	5.24 a
	PAM_L	8.17 b
	PAM_H	12.70 c

PAM_L = low-molecular-weight PAM; PAM_H = high-molecular-weight PAM.

[b]Within soils, means followed by the same letter do not differ significantly at the 0.05 level.

Source: Levy and Agassi, 1995.

turn, is expected to determine the polymer's capability to penetrate into the aggregates, are not dominant factors in determining soil susceptibility to sealing and low IR values (Ben-Hur and Keren, 1996).

The effects of sodic conditions, which are often found in soils from semiarid and arid regions, on the efficacy of high-molecular-weight PAM in controlling runoff was studied by Levy et al, (1995). They used samples from a silty loam loess and a clay grumusol, Israel, having ESP levels in the range of 3 to 25. Levy et al, (1995) noted that PAM was effective in controlling erosion at all ESP levels studied (Fig. 3). However, with respect to runoff, PAM was ineffective in significantly reducing runoff in soils with ESP >20 (Levy et al., 1995). Ben-Hur et al. (1992b) studied the effects of the addition of small amounts of a cationic PSD and two anionic PAMs varying in their charge den-

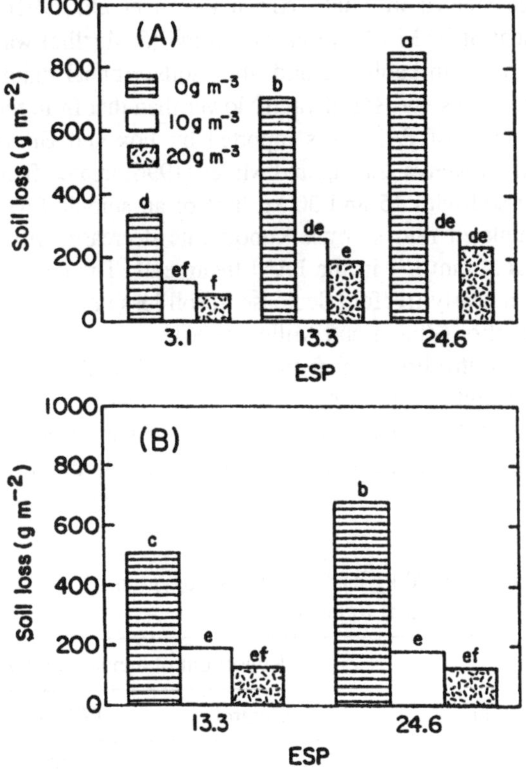

Figure 3 Soil loss as a function of amount of polymer added and soil ESP for a Grumusol irrigated with (A) tap water (EC = 0.8 dS m^{-1}) and (B) saline water (EC = 5.0 dS m^{-1}). Bars labelled with the same letter do not differ significantly at the 0.05 level. (From Levy et al., 1995.)

sity on samples of a sandy soil with ESP 8.5 and 30.6. The polymers did not affect soil final infiltration rate, but the two PAMs reduced soil loss at both ESP levels. Polymer application and soil ESP have conflicting effects on soil structure and aggregate stability. Polymer addition to the surface aggregates leads to 1) the stabilization of existing aggregates and improved bonding between adjacent aggregates (Schamp et al., 1975), and 2) clay flocculation in waters containing electrolytes (Aly and Letey, 1988). Conversely, increasing soil ESP increases the repulsion forces between clay particles, and hence the bonds between colloidal particles forming the aggregates weaken. Levy et al. (1995) postulated that since at high ESP (>20), repulsion forces exist not only between clay tactoids but also between single clay plates, leading to a significant weakening of the aggregates, the binding efficiency of polymers is not always sufficient to overcome these repulsions and control runoff and soil loss.

Soil conditioning with PAM was also found effective in predominantly kaolinitic and illitic soils. Field studies in South Africa (Stern et al., 1991b) showed that a combined treatment of PAM (20 kg ha^{-1}) and PG (5 $Mg^{-1}ha$) was very effective in controlling runoff from kaolinitic and illitic soils. Annual runoff percentage from PAM+PG-treated plots was significantly lower than that from the control (Table 3). Similar efficiency of PAM was reported by Fox and Bryan (1992) working with two soils from Kenya. Zhang and Miller (1996) treated field plots of a kaolinitic sandy loam soil with 15 and 30 kg ha^{-1} of an anionic PAM and exposed them to three simulated rain storms. Runoff rate decreased from 49.5 mm h^{-1} in the control to 16.4 mm h^{-1} in the PAM treatments. The considerable reduction in runoff led to a nearly fivefold decrease in soil loss in the PAM treatments compared with the control (Zhang and Miller, 1996).

The efficacy of polymers in controlling seal formation, runoff, and erosion depends on rain kinetic energy (Smith et al., 1990; Levin et al., 1991). The effect of combined application of PAM and PG on samples from hamra, loess, and grumusol from Israel under three rain kinetic energies (3.6, 8.0, and 12.4

Table 3 Effect of Polyacrylamide (PAM, 20 kg ha^{-1}) and Phosphogypsum (PG, 5 Mg ha^{-1}) on Percent Runoff from Natural Rainstorms

Soil classification	Annual rain (mm)	Runoff out of annual rain (%)	
		Control	PAM + PG
Paleudalf	214	40.2	20.7
Rhodustalf	107	72.5	27.9
Haplustalf	122	40.9	15.3

Source: Stern et al., 1991b.

kJ m^{-3}) was studied by Levin et al. (1991). The effect of PG and PAM+PG on decreasing total runoff from the entire rainstorm was generally similar at low-kinetic-energy rain 3.6 kJ m^{-3}). When high-kinetic-energy rain (12.4 kJ m^{-3}) was used, the effect of PAM+PG was greater than that of PG alone. Levin et al. (1991) hypothesized that at low-kinetic-energy rain clay dispersion was more important in the sealing process than aggregate disintegration. Hence the presence of the polymer that stabilizes the aggregates was less important at low kinetic energy than the presence of PG that prevents clay dispersion. Conversely, when high-kinetic-energy rain was applied, the use of polymer to prevent aggregate breakdown by the beating impact of the raindrops was essential. Similarly, a combined treatment of PAM and PG is significantly more effective than either of the two alone in preventing soil erosion. Soil loss in the PAM+PG was <5% of the loss observed in the control (Fig. 4). The efficiency of this treatment was ascribed to its favorable effect on reducing runoff, by reasoning that where there is no runoff there is no erosion (Smith et al., 1990). Treating the soil with a soil stabilizing agent (PAM) in addition to a source of electrolytes (e.g., PG) is imperative for successful control of runoff and soil loss under high-energy rainfall (Smith et al., 1990; Levin et al., 1991).

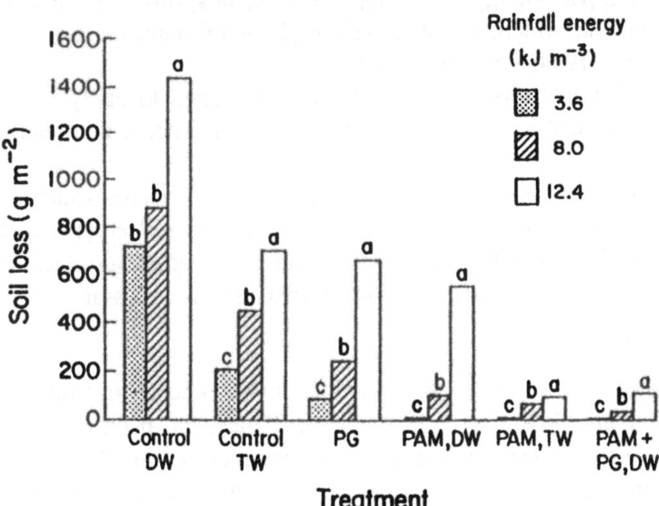

Figure 4 Total soil loss after an 80 mm storm for three kinetic energy levels of rain. Within treatments, bars labeled with the same letter do not differ significantly at the 0.05 level. Treatments: PAM = anionic polyacrylamide; PG = phosphogypsum; DW = distilled water; TW = tap water. (From Smith et al., 1990.)

IV. POLYMER USE IN IRRIGATED FIELDS

A. Introduction

In arid and semiarid regions, irrigation strategies must be designed to maximize or optimize production while conserving water and avoiding pollution of the environment. However, the increasing demand for food and irrigation development has brought about intensive use of marginal waters (e.g., wastewater, saline, and sodic water). The use of such water enhances soil degradation and its susceptibility to seal formation, and runoff and soil erosion initiation. These adverse phenomena depend not only on soil properties but also on the water application pattern (i.e., application rate and kinetic energy of the applied water) of the irrigation system. Irrigation systems can be divided according to their water application method into three main types, sprinkler, drip, and flooding. Flooding irrigation is not practiced in Israel, hence the discussion will concentrate on sprinkler and drip irrigation systems only.

B. Sprinkler Irrigation System

Sprinkler irrigation systems are characterized by a relative high impact energy of drops (Hudson, 1971). This causes the breakdown of surface aggregates and leads to seal formation and surface runoff (e.g., Ben-Hur et al., 1987, 1989). Moreover, the high impact energy of the applied water in sprinkler systems enhances soil detachment, which, coupled with high runoff rate, increases soil erosion (Agassi et al., 1994; Levy et al., 1991).

A moving sprinkler irrigation system (MSIS) is designed to apply a regulated amount of water within a relatively short period of time, which leads to a high application rate. Application rate of MSISs with spray nozzle emitters can be >100 mm h^{-1}, compared with ≤10 mm h^{-1} for fixed microsprinklers or impact sprinkler heads. Hence expected surface runoff and soil erosion under irrigation with a MSIS is substantially higher than under other sprinkler systems, and the use of soil conditioners for MSIS irrigation is necessary.

1. Laboratory Studies

Adding small amounts of polymers to the irrigation water to form dilute concentrations of polymers (5–30 g m^{-3}) in the applied water has been considered a logical alternative to surface application of the polymer commonly practiced in dry land farming. The amount of polymer added to the soil surface when applied via the irrigation water depends on irrigation depth. For common irrigations (<50 mm), the amount of polymer added will generally be 5–10 kg ha^{-1}, which is less than half of the amount of polymer sprayed on the soil surface in dry land farming. Furthermore, polymer application via the irrigation water ensures a constant presence of polymer at the soil surface. This concept has been tested in numerous laboratory studies using rainfall simulators for simulating sprinkler irrigation (Levy, 1995, and references cited therein).

Studies on the effect of dilute concentration of polymer in the irrigation water on seal formation, runoff, and erosion have indicated that the efficacy of the polymers depended mainly on polymer type and properties and water quality. At very low concentrations ($\leq$10 g m^{-3}), anionic and nonionic PSD had no beneficial effect of the infiltration rate of the soil (Ben-Hur and Letey, 1989). Conversely, 10 g m^{-3} of cationic PSD in the irrigation water significantly improved soil IR. The beneficial effect of the cationic PSD increased with an increase in PSD charge density (Ben-Hur and Letey, 1989). However, the relative beneficial effect of PSD tends to decrease with increasing sodium adsorption ratio (SAR) of the irrigation water (El-Morsy et al., 1991).

A comparison of the effects of low concentrations (10–20 g m^{-3}) of an anionic high-molecular-weight PAM and a cationic PSD on the IR of two Israeli soils showed that 1) a lower concentration of PAM (10 g m^{-3}) was needed for optimal effect on infiltration and runoff than that required if PSD was used (20 g m^{-3}), and 2) that the final IR was generally higher in the PAM than in the PSD treatments (Levy et al., 1992). It was postulated that the greater efficacy of PAM compared to PSD is related to PAM's longer molecules and limited adsorption, which make PAM more efficient in cementing aggregates and increasing their resistance to seal formation (Levy et al., 1992).

2. Field Studies

Beneficial effects of using polymers in irrigated fields on reducing runoff have also been established in field experiments. Runoff amounts measured from 3 m^2 cotton and peanut plots irrigated with a MSIS for various irrigation events are presented in Fig. 5. PAM was added to the soil surface, at a rate of 20 kg ha^{-1}, by spraying prior to the first irrigation. The cotton and peanut plots were irrigated by MSIS at an average water application rate of 160 mm h^{-1} and 100 mm h^{-1}, respectively. In general, runoff was significantly lower in the PAM treatment than in the control for both soils (Fig. 5). PAM was more effective in limiting runoff in the clay vertisol than in the silty loam loess. The differences in the runoff amounts between the control and PAM treatments resulted from PAM's activity that increased aggregate stability at the soil surface and hindered the development of seal formation.

The effect of PAM on soil erosion in fallow and cotton plots irrigated with a MSIS was studied by Levy et al. (1991). In fallow soil, soil loss in untreated plots was significantly higher than in PAM-treated ones. In the cotton plots, soil loss values in untreated plots were also higher than in the PAM plots, but the differences, in general, were not statistically significant (Table 4.). In the cotton plots, the plant canopy protected the soil surface from the impact of the waterdrops and consequently minimized the effect of the polymer (Levy et al.,

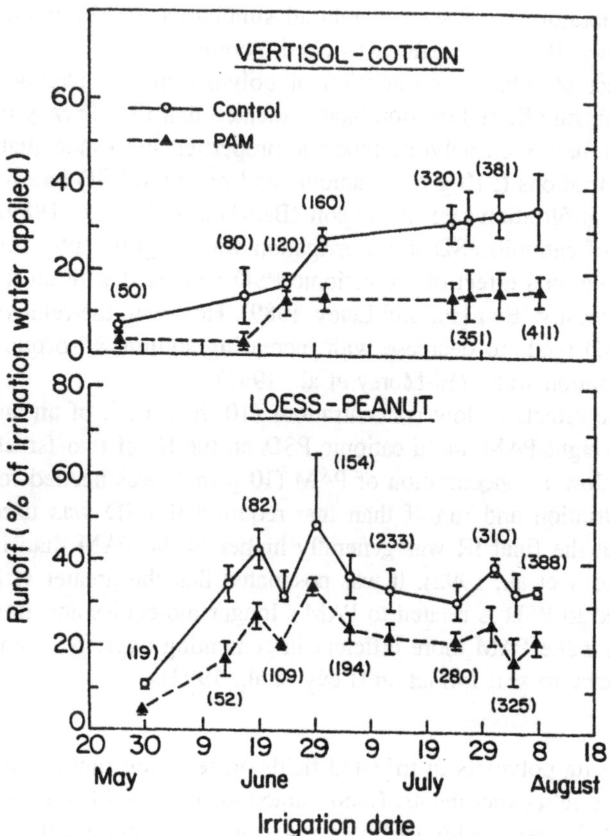

Figure 5 Percent runoff from PAM treated and control plots in two soils irrigated with a MSIS. Vertical lines indicate two standard deviations. Numbers in () indicate cumulative depth of irrigation mm. (From Levy et al., 1991.)

1991). The lower efficiency of the PAM in the cropped plots manifested itself also in the yield. Seed cotton yield was higher in the PAM (20 kg ha^{-1}) than in the control treatment; yet the difference was not statistically significant. Still, the results suggested that the higher yield can be attributed to the beneficial effect of PAM in improving the IR and reducing runoff and erosion.

The effect of a nonionic polymer P-101 on runoff and tuber yield of potato grown in a silt loam and irrigated with a linear MSIS was found to depend on the type of emitter used (M. Ben-Hur, unpublished data). In this study three emitters varying in their discharge rate were studied (Table 5). Addition of P-101 at a rate of 40 kg ha^{-1} significantly decreased runoff compared with the control. Runoff from the polymer-treated plots was 25%, 28%, and 37%

Table 4 Soil Loss from a Clay Loam Vertisol and a Silt Loam Loess Irrigated with MSIS for Control (Untreated Soil) and PAM (20 kg ha^{-1}) Treatments

Soil	Irrigation date Julian day	Irrigation depth mm	Soil loss gm^{-2} mm^{-1}	
			Control	PAM
		No crop		
Vertisol	205	40	12.03	2.06[a]
	214	31	22.04	8.57[a]
	221	30	32.76	10.76[a]
		With crop		
Vertisol	144	50	0.5	0.39
	166	30	0.56	0.32
	174	40	4.67	2.25[a]
	181	40	13.90	8.38
	205	40	8.70	4.32
	214	31	8.83	5.42
	221	30	7.91	4.29
Loess	170	30	2.13	0.43[a]
	180	45	13.06	8.54
	187	40	11.20	4.58
	195	39	8.32	4.60
	207	47	7.45	5.23
	214	30	8.36	5.76

[a]Significant differences between control and PAM treatments, for the same irrigation event at the 0.05 level.
Source: Levy et al., 1991.

of that from the control treatment for emitter No. 1, Spinner, and Super spray, respectively.

Average potato yield from P-101 plots was higher than from the control plots for the three emitters. The largest yield was in the Super spray emitter, 30 Mg ha^{-1}, which was ≈150% of the yield obtained in the control. The lowest yield was in emitter No. 1, ≈116% of the control. These differences in the P-101 efficiency resulted apparently from the differences in the application pattern of the emitters.

It can be concluded that polymer use in fields irrigated with a MSIS has a beneficial effect on runoff and erosion reduction and on yield increase in some crops. However, in order to reach eventually the use of polymers in commercial fields irrigated with a MSIS, more large-scale field experiments using different polymers and application methods are required to improve polymer handling and efficiency.

Table 5 Technical Data of the Various Emitters Used

Name	Manufacturer	Discharge of emitter L h^{-1}	Nozzle diameter mm	Space between emitters, m	Lateral discharge L m^{-1} h^{-1}
No. 1	Nelson	1254	5.9	2	650
Spinner	Nelson	2713	8.3	3	926
Super spray	Senninger	1777	6.7	2	925

C. Drip Irrigation

In drip irrigation water is applied to the soil with hardly any energy; consequently, seal development under this irrigation method is rare. However, in arid and semiarid regions, one of the most common soil problems is high levels of exchangeable sodium. Soils having exchangeable sodium percentage (ESP) >5% and a high percentage of expandable clay minerals may suffer from deterioration in HC and poor aeration and internal drainage, even during leaching, when water is applied to the soil with no energy (Shainberg and Levy, 1992). Sodicity related degradation of these properties can be attributed to clay swelling and dispersion, which in turn weakens aggregate stability in the entire soil profile. Treating the soil with polymers stabilizes aggregates and may prevent deterioration in soil HC (El-Morsy et al., 1991). However, it is required that the polymer reach the entire soil layer, in order for the polymer molecules to adsorb on the soil particles to prevent their disaggregation.

Shaviv et al. (1987a) tested the effects of an anionic polymer with a low molecular weight (60,000–150,000 Da), designated LIMA, on aggregate stability. They compared three methods of polymer application: 1) spraying a polymer solution on the soil surface and immediately incorporating the surface layer with the upper 12 cm soil layer, 2) mixing air-dried soil with a polymer solution in the treated soil, and 3) adding the polymer through a drip type irrigation system. Aggregate stability was determined on soil samples taken from 0–7 cm depth. Polymer efficacy was inferred from mean weight diameter (D) and mass fraction of aggregates >0.1 mm (F) (Table 6). Shaviv et al. (1987a) concluded that treating soil by drip irrigation increased the mean diameter of water-stable aggregates in the 0–7 cm layer. The other two methods were either less effective or more tedious and were thus considered inferior to polymer addition to the soil via the irrigation system.

However, under sodic conditions, soil degradation could exceed a depth of 0.9 m and more. Therefore the conditioning of the upper 7 cm in the field, as was found by Shaviv et al. (1987a), is insufficient. Furthermore, the applied amount of the polymer (100–130 kg ha^{-1}) to stabilize the aggregates only at the upper 7 cm of the soil is high and could thus not be considered economical.

Table 6 Mean Weight Diameter (*D*) and Mass Fraction of Aggregates >0.1 mm (*F*) for Soil Samples That Were Taken Before Planting (*BP*) and One Year After Planting (AP) for Different Polymer Treatments

	Spraying and rototilling Silt loam		Spraying and mixing Grumusol		Through the irrigation system Silt loam		Through the irrigation system Loessial alluvium Control		Through the irrigation system Loessial alluvium 100 kg ha^{-1}	
							Sampling depth, cm			
Parameter	Control	Polymer 1–1.6 g kg^{-1}	Control	Polymer 5 g kg^{-1}	Control	Polymer 130 kg ha^{-1}	0–3	3–7	0–3	3–7
D, mm BP	0.13 (0.02)[a]	0.18 (0.02)	0.38 (0.04)	0.82 (0.05)	0.12 (0.02)	0.26 (0.06)	0.15 (0.03)	0.19 (0.04)	0.58 (0.1)	0.46 (0.05)
D, mm AP	nd[b]	nd	0.25 (0.03)	0.65 (0.05)	nd	nd	nd	nd	nd	nd
F BP	0.22 (0.03)	0.28 (0.02)	0.91 (0.06)	0.98 (0.06)	0.22 (0.02)	0.52 (0.06)	0.36 (0.05)	0.44 (0.06)	0.81 (0.12)	0.68 (0.07)

[a]Standard deviation.
[b]nd = not determined.
Source: Shaviv et al., 1987a.

Sodic, fine-textured soils that contain expandable clay minerals tend to develop cracks when dry. Kamphorst (1989) suggested using cracks as pathways for water and salt transport as an alternative method for the commonly used expensive mechanical methods for soil reclamation. Kamphorst (1989) succeeded in a laboratory study in preventing complete closure of the cracks and maintaining high infiltration by applying gypsum and other calcium salt solutions directly into the open cracks.

Similarly, Malik et al. (1991) investigated the efficacy of dilute solutions of an anionic, medium charged (21% hydrolysis), high- molecular-weight ($10–15\times10^6$ Da) PAM, for prevention of rapid and complete closure of cracks in fine montmorillonitic calcareous California soils with ESP 8 and 25. Disturbed soil samples were packed in columns and irrigated with water containing 0, 25, 75, and 200 mg L^{-1} PAM. Prior to irrigation with the polymer solution, part of the columns were wetted first to obtain swelling and then were dried to create cracks. After irrigation, the soil columns in all the treatments were allowed to dry, and the HC of the individual soil columns was determined. In the noncracked samples, polymer application had only a small beneficial effect. Conversely, in the cracked samples in both ESP levels, the steady state HC increased linearly with an increase in the polymer concentration (Fig. 6). These results indicated that the polymer stabilized the cracks and prevented their complete closure upon rewetting (Fig. 7), which in turn increased soil HC. The beneficial effects of the polymer treatment in stabilizing cracks and maintaining high HC decreased with irrigation cycles but were still effective after 3 cycles (Malik et al., 1991).

It should be realized, however, that the use of soil cracks as pathways for polymer transport into the soil profile as suggested by Malik et al. (1991) is not useful in drip irrigation. In this irrigation system, the water movement is conducted mostly in small pores and not in cracks, particularly in the first stage of the irrigation when the cracks are still open.

Shaviv et al. (1987b) studied the effect of three anionic polymers, added to soil columns under simulated drip irrigation, on aggregate stability of a silt loam and clay soils. The effect of polymer molecular weight on the mean weight diameter (D, representing aggregate stability) along the soil column is presented in Fig. 8. Arrows indicate conditioning depth (H_d), defined as depth at which D equals two times the mean weight diameter (D_0) of the respective control soil column. The LIMA 87 (molecular weight of 75,000 Da) showed a significantly larger H_d value than the other two lower-molecular-weight polymers (<30,000 Da). Conversely, larger water-stable aggregates were obtained in the two upper soil layers with decreasing molecular weight. Shaviv et al. (1987b) suggested that higher-molecular-weight polymers should be used if the aim is to achieve deeper conditioning, whereas lower-molecular-weight polymers should be considered for obtaining larger water-stable aggregates at the soil surface.

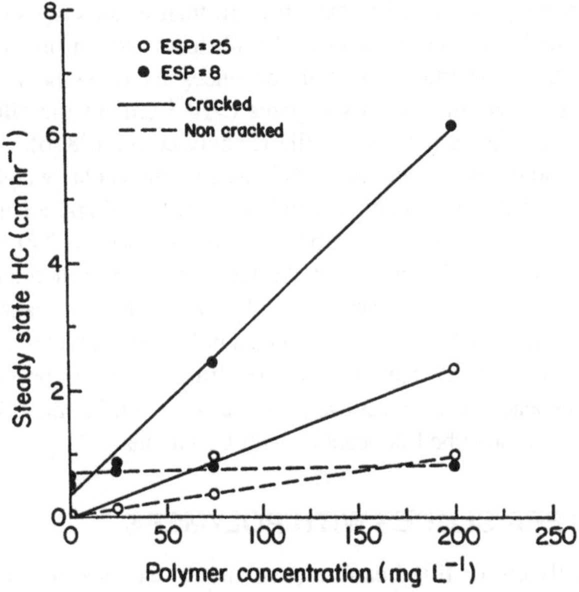

Figure 6 Steady-state hydraulic conductivity (HC) of clay soil at two ESP levels as a function of polymer concentration for cracked and noncracked samples. (From Malik et al., 1991.

Figure 7 Persistence of cracks at 0, 75, and 200 mg L^{-1} polymer treatment after 24 h of submergence under tap water. (From Malik et al., 1991.)

Polymer effects were more pronounced in the silt loam than in the clay soil. This was attributed to 1) the lower clay content in the silt loam, which poses a smaller demand for polymer adsorption and thus enhances its movement to deeper layers; and 2) a larger volume of macro pores (>10^{-4} cm) in the silty loam that eased polymer mobility in the soil profile (Shaviv et al., 1987b).

Ben-Hur et al. (1996) studied the effects of a nonionic low molecular weight ($10–20 \times 10^4$ Da) polymer on the bulk density and HC of a compacted and a non-compacted sodic clay soil. They found that polymer addition at a rate of 240 kg ha^{-1} under both drip and flood irrigation improved the HC of the compacted and noncompacted soils (Table 7). Polymer efficacy was higher under flood irrigation and for the noncompacted soil. Ben-Hur et al. (1996) concluded that polymer addition to clay soils in semi arid regions stabilizes the soil structure and improves its HC which thus enhances leaching of excess salts, and a reduction in soil ESP due to effective replacement of adsorbed Na cations with Ca cations.

V. STABILIZING STEEP SLOPES WITH POLYMERS

Steep bare slopes, especially cut or filled earth embankments, exposed to rainfall are extremely susceptible to erosion (e.g., Bryan, 1987; McIssac et al., 1987). Agassi and Ben-Hur (1991) found in a Typic Rhodoxeralf from the coastal plain, Israel, that annual erosion from 10 m long embankments with 48% slope exposed to annual rainfall of 520 mm was 299 Mg ha^{-1} for plots with a northern aspect and 416 Mg ha^{-1} for plots with a western aspect. These amounts of soil loss are more than one order of magnitude greater than soil loss from gentle (<10%) slopes (Warrington et al., 1989). The rill and gully erosion

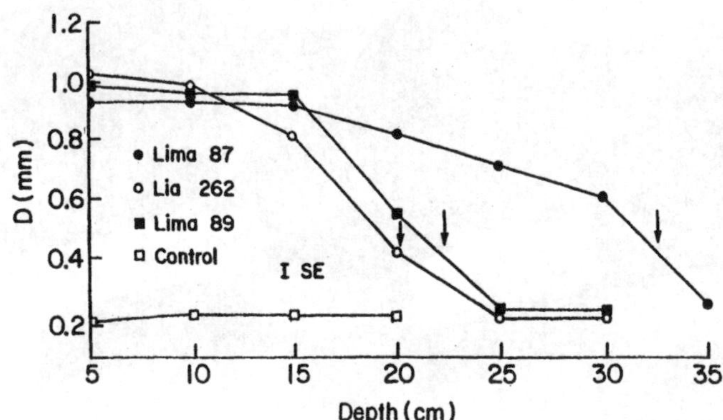

Figure 8 Mean weight diameter (D) vs. depth for various polymers applied to 2 to 84 mm Kissufim aggregates. Arrows indicate H_d (conditioning depth) values; SE stands for standard deviation. (From Shaviv et al., 1987b.)

Table 7 The Effect of Polymer Addition on Soil Bulk Density and Infiltration Rate

	Flood irrigation				Drip irrigation			
	Noncompacted		Compacted		Noncompacted		Compacted	
Parameter	Control	Polymer	Control	Polymer	Control	Polymer	Control	Polymer
Bulk density (mg m^{-3})	1.53 a[a]	1.53 a	1.62 a	1.56 a	1.55 a	1.54 a	1.56 a	nd
Final Infiltration rate (mm h^{-1})	5.10 c	33.30 a	5.60 c	25.60 ab	11.00 bc	22.80 ab	9.70 bc	12.60 bc

[a]Within a row, numbers followed by the same letter are not significant at the 5% level.
Source: Adapted from Ben-Hur et al., 1996.

that occur under such conditions can endanger the stability of the embankment and the nearby construction. Runoff water, from the embankment top or face, while flowing downslope, concentrates in natural longitudinal depressions that subsequently turn into rills and gullies. The gullies in the embankment can reach a depth of few meters, and when they reach the embankments or dike top, the consequences can be severe.

Runoff and soil erosion are closely related to the deterioration in the structure of the surface aggregates due to the impact energy of water drops (see Section II). Prevention of runoff and erosion processes in earth embankments can be achieved by either or both of the following: 1) protection of the soil surface against water drop impact energy, and 2) stabilization of aggregates at the soil surface by soil conditioners.

Numerous studies have proven the beneficial effect of soil surface mulching on runoff and erosion control (e.g., Meyer and Mannering, 1963; Barnet et al., 1967; Swanson et al., 1967; Meyer et al., 1972). Common materials for mulching are plant residues, gravel, rock fragments, and woodchips. In some regions, however, these materials are scarce, making their use costly and impractical.

The beneficial effects of the addition of small amounts of synthetic organic polymers, especially PAM and PSD, to the soil surface, on controlling seal formation, runoff, and erosion is well documented and has been discussed at length in the previous sections. However, the efficacy of using polymers in controlling erosion from steep slopes cannot be inferred from those studies because most data on the effect of polymers on erosion was collected in the laboratory in small pans (0.2 by 0.4 m) and/or on gentle slopes (<10%), conditions under which rill erosion can hardly develop (Meyer and Harmon, 1984). In smectitic soils, increasing the slope exponentially increases soil erosion (Warrington et al., 1989; Ben-Hur et al., 1992c). Furthermore, Agassi and Ben-Hur (1991) found in a steep slope (48%) that slope length affects erosion significantly. The annual amount of erosion per unit area in 10 m long plots was 6.4 times that in 1.5 m long plots. Agassi and Ben-Hur (1991) observed that in the small plots soil loss was mainly due to interrill erosion; conversely, in the large plots, rill erosion was the predominant one.

Agassi and Ben-Hur (1991) studied the effects of PAM and PSD on soil erosion from large plots with different lengths, slopes, and soils in different sites in Israel under natural rain conditions. A combined application of 70 kg ha^{-1} PSD together with 10 Mg ha^{-1} PG (PS+PG) treatment, or 20 kg ha^{-1} PAM together with 10 Mg ha^{-1} PG (PAM+PG) treatment, decreased soil loss by six- to tenfold, to <30 Mg ha^{-1}, in comparison with the untreated plots. Furthermore, visual observations of the soil surface in the untreated plots indicated that many rills were formed in these plots. Conversely, in the plots treated with 70 kg ha^{-1} PSD together with 10 Mg ha^{-1} PG (PS+PG), the soil surface was covered with stable aggregates, and no crust or rills were noted (Fig. 9). These re-

Figure 9 Photograph taken in April 1990 of the soil surface of control and PS+PG treatments at Bet-Yannay, Israel. (From Agassi and Ben-Hur, 1992.)

sults indicated that the combination of polymer+PG prevented the formation of rills along the steep slope. Apparently, prevention of seal formation in the polymer+PG treated plots decreased runoff amount and thus its erosivity to a level below the threshold value needed to initiate rill formation.

Planting plants is another way to control erosion from steep slopes. Plant canopy protects the soil surface from water drop impact energy, and in addition, plant roots stabilize the structure of the underlying soil. However, in arid and semiarid regions, the short rainy season and the long dry and hot season present challenging conditions for the successful establishment of plant cover. When the plants are young, their canopy and roots are incapable of properly protecting the soil surface (Ben-Hur et al., 1989) and stabilizing the soil structure. Thus during the rainy season seal formation, runoff, and soil erosion take place. Consequently, plants may be removed with the eroded soil, available water for the remaining plants dwindles, and plant development is severely hindered.

The effect of 70 kg ha^{-1} PSD together with 10 Mg ha^{-1} PG on the establishment and development of drought-resistant perennial plants growing on steep slopes with no irrigation under semiarid conditions was studied by Agassi and Ben-Hur (1992). Photographs of the plants at two different sites in Israel, after a dry summer, are presented in Fig. 10. Agassi and Ben-Hur (1992) noted that at least 90% of the plants became established, and that the plants looked fresh and vital at the end of the summer (dry season). The creeper canopy covered an area

Figure 10 Photographs of plants stabilizing steep slopes at Sharon and Soreq, Israel. (From Agassi and Ben-Hur, 1992.)

of 0.75–1.5 m^2, and the bushes were 0.8–1.2 m high. The soil surface between the plants showed no signs that rill erosion and seal formation had occurred during the rainy season (winter). The combined application of polymer and PG maintained high permeability and minimized surface runoff from the experimental plots. As a result, no plants were removed by runoff water, and more rainwater entered the soil profile and was thus available for plant use during the dry summer.

VI. CONCLUSION

The addition of small amounts of synthetic organic polymers to soils prevents degradation in soil hydraulic and physical properties. The polymers are especially effective when applied together with a source of electrolytes (e.g., phosphogypsum). Because polymers improve aggregate stability and soil structure, they reduce the tendency of soils to form seals, thereby preventing decline in soil HC and infiltration rate, reducing runoff and soil loss, and improving internal soil drainage. The beneficial effects of polymers extend over a wide range of conditions; dry land and irrigated cultivated soils, soils of varying mineralogy and sodicity, earth embankments with steep slopes, and rains with different kinetic energies.

The reviewed studies suggest that synthetic organic polymers can be considered as acceptable stabilizing agents that can replace conventional soil conservation practices successfuly. However, commercial use of polymers for the aforementioned applications is limited, which could in part be attributed to difficulty of handling of the currently available polymers and the uncertainty clouding their cost effectiveness.

REFERENCES

Aggasi, M., and Ben-Hur, M. (1991). Effect of slope length, aspect and phosphogypsum on runoff and erosion from steep slopes. *Aust. J. Soil Res. 29:* 197.

Aggasi, M., and Ben-Hur, M. (1992). Stabilizing steep slopes with soil conditioners and plants. *Soil Tech. 5:* 249.

Agassi, M. Bloem, D., and Ben-Hur, M. (1994). Effect of drop energy and soil and water chemistry on infiltration and erosion. *Water Resour. Res. 30:* 1187.

Aly, S. M., and Letey, J. (1988). Polymer and water quality effects on flocculation of montmorillonite. *Soil Sci. Soc. Am. J. 53:* 1453.

Aslam, M. (1990). Polymers as soil conditioners and sealing agents. *Pakistan Acad. Sci.* Islamabad.

Azzam, R. A. I. (1980). Agricultural polymers, polyacrylamide preparation, application and prospects in soil conditioning. *Commun. Soil Sci. Plant Anal. 11:* 767.

Barnet, A. P., Disker, E. G., and Richardson, E. C. (1967). Evaluation of mulching methods for erosion control on newly prepared and seeded highway back slopes. *Agron. J. 59*: 83.

Barvenik, F. W. (1994). Polyacrylamide characteristics related to soil applications. *Soil Sci. 158*: 235.

Ben-Hur, M., and Keren, R. (1996). Aggregate formation and prevention of soil surface sealing as affected by polymer properties. *Soil Sci. Soc. Am. J.* (in press).

Ben-Hur, M., and Letey, J. (1989). Effect of polysaccharides, clay dispersion and impact energy on water infiltration. *Soil Sci. Soc. Am. J. 53*: 233.

Ben-Hur, M., Shainberg, I., Bakker, D., and Keren, R. (1985). Effect of soil texture and $CaCO_3$ content on water infiltration in crusted soils as related to water salinity. *Irrig. Sci. 6*: 282.

Ben-Hur, M., Shainberg, I., and Morin, J. (1987). Variability of infiltration in a field with surface sealed soil. *Soil Sci. Soc. Am. J. 51*: 1299.

Ben-Hur, M., Plaut, Z., Shainberg, I., Meiri, A., and Agassi, M. (1989). Cotton canopy and drying effects on runoff during irrigation with moving sprinkler systems. *Agron. J. 81*: 752.

Ben-Hur, M., Malik, M., Letey, J., and Mingelgrin, U. (1992a). Adsorption of polymers on clays as affected by clay charge and structure, polymer properties and water quality. *Soil Sci. 153*: 349.

Ben-Hur, M., Clark, P., and Letey, J. (1992b). Exchangeable Na, polymer, and water quality effects on water infiltration and soil loss. *Arid Soil Res. Rehab. 6*: 311.

Ben-Hur, M., Stern, R., van der Merwe, A. J., and Shainberg, I. (1992c). Slope and gypsum effects on infiltration and erodibility of dispersive and nondispersive Soils. *Soil Sci. Soc. Am. J. 56*: 1571.

Ben-Hur, M., Keren, R., Grimberg R., and Chen, Y. (1996). Reclamation of saline sodic soils of Jezrael valley and improving their hydraulic properties by addition of a synthetic polymer. *Water and Irrig. 360*: 37 (in Hebrew).

Bryan, R. B. (1987). Processes and significance of rill development. *Catena 8*: 1.

Chen, Y., Tarchitzky, J., Morin, J., and Banin, A. (1980). Scanning electron microscope observations on soil crusts and their formation. *Soil Sci. 130*: 49.

El-Morsy, E. A., Malik, M., and Letey, J. (1991). Polymer effect on hydraulic conductivity of saline sodic soil conditions. *Soil Sci. 151*: 430.

Fox, D., and Bryan, R. B. (1992). Influence of a polyacrylamide soil conditioner on runoff generation and soil erosion: field test in Baringo district, Kenya. *Soil Technol. 5*: 101.

Gabriels, D. M., Moldenhauer, W. C., and Kirkham, D. (1973). Infiltration, hydraulic conductivity and resistance to water-drop impact of clod beds as affected by chemical treatment. *Soil Sci. Soc. Am. Proc. 37*: 634.

Greenland, D. J. (1972). Interactions between organic polymers and inorganic soil particles. Proceedings of a Symposium on Fundamentals of Soil Conditioning (M. De-Boodt, ed.). State Univ. Ghent, Ghent, Belgium, pp. 205–220.

Gu, B., and Doner, H. E. (1992). The interaction of polysaccharides with Silver Hill illite. *Clays Clay Miner. 40*: 151.

Hudson, N. (1971). *Soil Conservation.* Cornell University Press, Ithaca, New York, 1971.

Kamphorst, A. (1990). Amelioration of sodic clay soils by crack stabilization: an experimental laboratory simulation. *Soil Sci. 149*: 218.

Lane, L. J., Foster, G. R., and Nicks, A. D. (1987). Use of fundamental erosion mechanics in erosion prediction. *ASEA*, St. Joseph, Michigan, Paper No. 87-2540.

Letey, J. (1994). Adsorption and desorption of polymers on soil. *Soil. Sci. 158*: 244.

Letey, J., Vaux, H. J., Jr., and Feinerman, E. (1984). Optimum crop water application as affected by uniformity of water infiltration. *Agron. J. 76*: 435.

Levin, J., Ben-Hur, M., Gal, M., and Levy, G. J. (1991). Rain energy and soil amendments—effects on infiltration and erosion of three different soil types. *Aust. J. Soil Res. 29*: 455.

Levy, G. J. (1995). Soil stabilizers. *Soil Erosion and Rehabilitation* (M. Agassi, ed.). Marcel Dekker, New York.

Levy, G. J., and Agassi, M. (1995). Polymer molecular weight and degree of drying effects on infiltration and erosion of three different soils. *Aust. J. Soil Res. 33*: 1007.

Levy, G. J., Ben-Hur, M., and Agassi, M. (1991) The effect of polyacrylamide on runoff, erosion and cotton yield from fields irrigated with moving sprinkler systems. *Irrig. Sci. 12*: 55.

Levy, G. J., Levin, J., Gal, M., Ben-Hur, M., and Shainberg, I. (1992). Polymers' effects on infiltration and soil erosion during consecutive simulated sprinkler irrigations. *Soil Sci. Soc. Am. J. 56*: 902.

Levy, G. J., Levin, J., and Shainberg, I. (1995). Polymer effects on runoff and soil erosion from sodic soils. *Irrig. Sci. 16*: 9.

Malik, M., and Letey, M. (1991). Adsorption of polyacrylamide and polysaccharide polymers on soil material. *Soil Sci. Soc. Am. J. 55*: 380.

Malik, M., Amrhein, C., and Letey, J. (1991). Use of polyacrylamide polymers to improve water flow in high shrink-swell soils. *Soil Sci. Soc. Am. J. 55*: 1164.

McIsaac, G., Mitchell, F. J. K., and Hirschi, M. C. (1987). Slope steepness effects on soil loss from disturbed lands. *Trans. Am. Soc. Agric. Eng. 30*: 1087.

Meyer, L. D., and Harmon, W. C. (1984). Susceptibility of agricultural soils to interrill erosion. *Soil Sci. Soc. Am. J. 48*: 1152.

Meyer, L. D., and Mannering, J. V. (1963). Crop residues as surface mulches for controlling erosion on sloping land under intensive cropping. *Trans. ASAE 6*: 322.

Meyer, L. D., Johnson, C. B., and Foster, G. R. (1972). Stone and woodchip mulches for erosion control on construction sites. *J. Soil Water Conserv. 12*: 264.

Mortensen, J. L. (1959). Adsorption of hydrolyzed polysaccharide on kaolinite. II. Effect of solution electrolytes. *Soil Sci. Soc. Am. Proc. 23*: 109.

Mortensen, J. L. (1962). Adsorption of hydrolyzed polysaccharide on kaolinite. *Clays Clay Miner. Proc. 9th Nat. Conf.* (Ada Swinford, ed.). West Lafayette, Indiana Pergamon Press, New York, pp. 530–554.

Mortimer, D. A. (1991). Synthetic polyelectrolytes—a review. *Polymer Int. 25*: 29.

Nabzar, L., Pefferkron, E., and Varoqui, R. (1984). Polyacrylamide–sodium kaolinite interactions: flocculation behavior of polymer clay suspension. *J. Colloid Interface Sci. 102*: 380.

Nadler, A. and Letey, J. (1989). Adsorption isotherms of polyanions on soils using tritium labeled compounds. *Soil Sci. Soc. Am. J. 53*: 1375.

Nadler, A., Malik, M., and Letey, J. (1992). Desorption of polyacrylamide and polysaccharide polymers from soil material. *Soil Technol. 5*: 92.

Ruehrwein, R. A., and Ward, D. W. (1952). Mechanism of clay aggregation by polyelectrolytes. *Soil Sci. 73*: 485.

Sato, T., and Ruch, R. (1980). Stabilization of colloidal dispersions by polymeric adsorption. Marcel Dekker, New York.

Schamp, N., Huylebroeck, J., and Sadones, M. (1975). Adhesion and adsorption phenomena in soil conditioning. *Soil Conditioners.* Soil Science Society of America Special Publication No. 7 American Society of Agronomy, Madison, Wisconsin, 1973, pp. 13–22.

Seybold, C. A. (1994). Polyacrylamide review: soil conditioning and environmental fate. *Commun. Soil Sci. Plant Anal. 25*: 2171.

Shainberg, I., and Levy, G. J. (1992). Physico-chemical effects of salts upon infiltration and water movement on soils. *Interacting Processes in Soil Science* (R. J. Wagenet, P. Baveye, and B. A. Stewart, eds.). Advances in Soil Science, Lewis Publishers. p. 37.

Shainberg, I., Warrington, D. N., and Rengasamy, P. (1990). Water quality and PAM interactions in reducing surface sealing. *Soil Sci. 149*: 301.

Shaviv, A., Ravina, I., and Zaslavski, D. (1986). Surface application of anionic surface conditioners to reduce crust formation. *Assessment of Soil Surface Sealing and Crusting* (F. Callebaut, D. Gabreils, and M. De Boodt, eds.). Ghent, Belgium. p. 286.

Shaviv, A., Ravina, I., and Zaslavsky, D. (1987a). Field evaluation of methods of incorporating soil conditioners. *Soil and Tillage Res. 9*: 151.

Shaviv, A., Ravina, I., and Zaslavsky, D. (1987b). Application of soil conditioners solutions to soil columns to increase water stability of aggregates. *Soil Sci. Soc. Am. J. 51*: 431.

Smith, H. J. C., Levy, G. J., and Shainberg, I. (1990). Waterdrop energy and soil amendments: effects on infiltration and erosion. *Soil Sci. Soc. Am. J. 54*: 1084.

Soil Science (1986). Special Issue, Vol. 141: 311–397.

Soil Science (1994). Special Issue, Vol. 158: 233–305.

Stern, R., Ben-Hur, M., and Shainberg, I. (1991a). Clay mineralogy effect on rain infiltration, seal formation and soil losses. *Soil Sci. 152*: 455.

Stern, R., Laker, M. C, and van der Merwe, A. J. (1991b). Field studies on the effect of soil conditioners and mulch on runoff from kaolinitic and illitic soils. *Aust. J. Soil Res. 29*: 249.

Stutzmann, Th., and Siffert, B. (1977). Contribution to the adsorption mechanism of acetamide and polyacrylamide onto clays. *Clays Clay Miner. 25*: 392.

Swanson, N. P., Dedrick, A. R., and Dudeck, A. E. (1967). Protecting steep construction slopes against water erosion. *Highway Res. Record 206*: 46.

Theng, B. K. G. (1979). *Formation and Properties of Clay-Polymer Complexes*. Elsevier, Amsterdam.

Theng, B. K. G. (1982). Clay-polymer interactions: summary and prospectives. *Clays Clay Miner. 30*: 1.

Wallace, A., and Wallace, G. A. (1990). Soil and crop improvement with water-soluble polymers. *Soil Tech. 3*: 1.

Wallace, A., and Wallace, G. A. (1986). Control of soil erosion by polymer soil conditioners. *Soil Sci. 141*: 363.

Warrington, D., Shainberg, I., Agassi, M., and Morin, J. (1989). Slope and phosphogypsum effects on runoff and erosion. *Soil Sci. Soc. Am. J. 53*: 1201.

Zhang, X. C., and Miller, W. P. (1996). Polyacrylamide effect on infiltration and erosion in furrows. *Soil Sci. Soc. Am. J. 60*: 866.

16
Comparative Effectiveness of Polyacrylamide and Straw Mulch to Control Erosion and Enhance Water Infiltration

Clinton C. Shock *Oregon State University, Ontario, Oregon*

Byron M. Shock *Shock Computer Consulting, Ontario, Oregon*

I. INTRODUCTION

Irrigation induced erosion has long been a concern on furrow irrigated fields. Carter (1976) described an array of practices that growers can use to reduce sediment losses and runoff from furrow irrigated ground. When crops are grown on sloping soil, erosion rates can be high (Shock et al., 1988), and crop quality can suffer due to water stress.

Polyacrylamide, otherwise known as PAM, has recently been used to reduce soil erosion. The PAM discussed in this chapter is water-soluble PAM, with an anionic saturation of 18% and a molecular weight of 1.5×10^7 daltons. Using PAM, soil erosion has been reduced by 20–95% (Lentz et al., 1992). Directly applied to water, PAM binds the soil together so the water does not break off soil particles as easily when the water advances down the furrow. PAM also acts as an agent to settle particles of soil already suspended in the irrigation water. It gathers and carries sediment to the bottom of the furrow, instead of the sediment being carried off the field.

Straw mulch, when applied to the bottom of the furrow, acts as an agent to slow down water in the furrow. By slowing down the water in the furrow, the soil is not as easily eroded, and the larger wetted area in the furrow bottom improves infiltration (Berg, 1984; Brown, 1985; Brown and Kemper, 1987; Brown et al., 1988). Straw mulching by hand is very laborious and time consuming, but there are machines on the market pioneered by Joe Hobson Sr. that effectively apply wheat straw to the bottom of the furrow. The wheat straw has to be cut at about 25 cm length before baling to flow evenly through these machines. Mechanically applied straw mulch has been effective in reducing sediment loss in Malheur County, Oregon, under furrow irrigation in the last few years with a 95.0% reduction in season long sediment loss under onions (Shock et al., 1993a), a 91.5% reduction in season long sediment loss under sugar beets (Shock et al., 1993b), and a 95% reduction in season long sediment loss under wheat (Shock et al., 1994). For the purpose of this chapter, the term "straw mulch" will refer to the mechanical application of wheat straw 25 cm long to the bottom of the irrigation furrow followed by pressure on the straw to press it into place.

The use of PAM and straw mulch were compared season long for erosion control, net water infiltration, and soil water potential for two successive years at the same site. In 1994 the field was planted to potatoes and in 1995 the crop was onions.

II. FIELD MEASUREMENTS OF SOIL LOSS AND NET INFILTRATION

An essential problem for anyone wishing to estimate sediment losses, water inflow, water outflow, and net infiltration is to establish or adopt an objective measurement and calculation method. Onset times of water inflow and outflow, and interval measurements of water inflow rate, water outflow rate, and sediment yield, were recorded during each irrigation. Water inflow and outflow rates were recorded for the irrigated furrow next to the center hill in each plot.

For each water outflow reading, a 1 liter sample was placed in an Imhoff cone and allowed to settle for 15 minutes. Sediment content in the water, in g/l, was found to be related to the Imhoff cone reading (×) after 15 minutes by the equation

$$y = 1.015x \tag{1}$$

with $r^2 = 0.98$ and $P < 0.0001$.

Total inflow, outflow, infiltration, and sediment loss were integrated from field measurements using a Lotus Improv program, InfilCal (Shock and Shock, 1988) as publicly presented (Shock et al., 1992). The InfilCal program utilizes simple approximations of the integrals of inflow rates, outflow rates, and sediment content over time to estimate the inflow, outflow, and sediment loss. Each flow rate reading and Imhoff cone sediment sample is taken to be representative of the interval of time closest to the reading. In other words, if readings and sediment samples are taken at times t_{n-1}, t_n, and t_{n+1}, the flow rate reading and sediment sample at time t_n are taken to be representative of the interval of time from

$$\frac{t_n - t_{n-1}}{2}$$

to

$$\frac{t_{n+1} - t_n}{2}$$

The estimated inflow volume for the interval about t_n is therefore given by

$$InflowVolume\,(t_n) = \left(\frac{t_{n+1} - t_{n-1}}{2}\right) InflowRate\,(t_n) \tag{2}$$

where *InflowRate* (t_n) is the measured inflow rate at time t_n. The outflow volume is given by the same equation, replacing the inflow rate with the outflow rate. The estimated sediment loss for the interval about t_n is given by

$$SedLoss\,(t_n) = OutflowVolume(t_n)Content(t_n) \tag{3}$$

where *Content*(t_n) is the sediment content given by the above Imhoff cone method. Although measurements should be taken on an established schedule, the time intervals between measurements need not be constant.

Clearly, this method must account for the time between onset and the first measurement and, likewise, between the last measurement and shutoff. The inflow rate value at the time of onset is taken to be the same as the first inflow rate measurement after onset, and the value at the time of shutoff is taken to be the same as the last inflow rate measurement before shutoff. The outflow rate value at the time of onset is taken to be zero, as is the outflow rate at the end of the irrigation. Each of these values applies to the first half of the time interval between onset and the first measurement, or the last half of the time interval between the last measurement and shutoff.

Composite water samples were collected in 5-gallon (19 liter) buckets to obtain sediment samples for nutrient analysis during each irrigation. Sediment was analyzed for nitrate-N, ammonium-N, total N, phosphate-P, and total P.

III. EROSION AND INFILTRATION IN FURROW IRRIGATED POTATOES

Potatoes are usually grown using furrow irrigation in Malheur County and in southwest Idaho. When potatoes are grown with furrow irrigation on sloping soil, erosion rates can be high (Shock et al., 1988), and potato quality can suffer due to water stress where the water infiltration rate is so low that the crop's water needs are not met.

A. Procedures

The experiment was conducted in a Nyssa silt loam with a 3.0% slope at the Malheur Experiment Station, Oregon State University, Ontario, Oregon. Russet

Burbank potato seed was planted in hills 0.3 m tall with 0.91 m between rows. The experiment had a total of 27 plots, each plot 3 hills wide and 76 m long. The furrows of 12 plots were strawed at 0.9 Mg/ha before the first irrigation, and 15 plots were not strawed. Straw mulch was applied mechanically (Hobson Mulching System, Hobson Manufacturing Inc., Keiser, OR). Three of the unstrawed plots were treated with only PAM (Superflock® A-836 Flocculant, Cytec Co.) in the irrigation water. The rates of PAM reported are rates of the commercial product. All furrows were irrigated at approximately 12 l/min during every irrigation. Inflow and outflow readings were taken at approximate one-hour intervals for every irrigation. Imhoff cones were used to determine the amount of sediment loss at the same time as the outflow readings were taken.

Granular matrix sensors (GMS sold as Watermark Soil Moisture Sensor Model 200SS, Irrometer Co., Riverside, CA) were placed 21 m from both the top and the bottom of 10 plots. Sensors were installed 0.1 m from the middle of the top of the hill and placed so that the top of the sensor was 0.2 m deep, with four per plot. Two additional GMS were also placed at 0.4 m depth in two replicates of each treatment. Sensors were read daily at 8 A.M. starting June 21, 1994, using a 30 KTCD meter (Irrometer Co., Riverside, CA) that had previously been calibrated to soil water potential (Eldredge et al., 1993).

To apply PAM to the furrows, granular PAM was dissolved in water to a concentration of 0.02 to 0.1% depending on the irrigation. The PAM solution was applied directly into the irrigation water individually for each furrow just before the water advanced down the furrow. In such a small experiment, the PAM solution for each furrow was held in a 5 gallon nurse tank and was applied into the irrigation water. Gated pipe was arranged so that all plots were irrigated during each irrigation set, but the duration of irrigation in the nonmulched furrows was longer because of lower water infiltration rate. The crop was irrigated using alternating furrow irrigation. Wheel rows were irrigated during the first irrigation, then nonwheel rows were irrigated at the second irrigation, etc. On the first irrigation of each furrow, 1.12 kg/ha of PAM was applied. PAM was applied at 0.22 kg/ha to the wheel rows during the third irrigation. On the fourth (and all subsequent) irrigations, PAM was applied at 0.56 kg/ha.

Potato cultural practices were typical of those used in commercial fields (Shock et al., 1995). Shortly after planting, all tractor operations for the season were completed, so cultivation and tractor traffic during the irrigation season had no effect on soil loss or water infiltration.

B. Results

The potato crop emerged and grew normally. Irrigations began on June 1, 1994. The total water per hectare was greater than typical of commercial fields, because of furrow length, only 76 m (Table 1). Shorter irrigation durations were

Table 1 Effects of Mechanically Applied Straw Mulch and PAM on the Irrigation Duration, Total Applied Water, and Average 8 A.M. Soil Water Potential for Furrow Irrigated Potatoes, Malheur Experiment Station, Oregon State University, Ontario, Oregon, 1994

Erosion control treatment	Irrigation records			Average soil water potential, kPa[a]	
	Number	Duration, hr	Applied water, mm	0.2 m depth[b]	0.4 m depth
Check	16	587	2959	–47	–27
Straw	16	504	2517	–30	–17
PAM	16	540	2708	–32	–22
LSD (0.05)	—	—	61	–15	—

[a]From June 24 to August 23, 1994.
[b]Based on the average value of four sensors in each of four check plots, four plots with strawed furrows, and two plots receiving PAM.

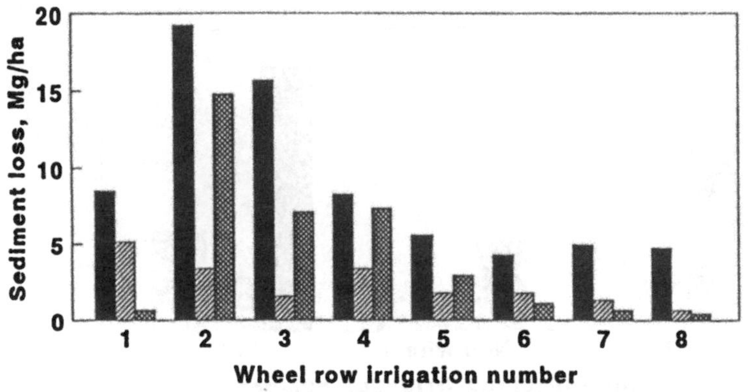

Figure 1 Effects of mechanical application of straw mulch or PAM treated irrigation water on sediment loss from wheel traffic furrows during eight successive irrigations on a Nyssa silt loam with a 3% slope. Malheur Experiment Station, Oregon State University, Ontario, Oregon, 1994, ■ check; ▨, straw; ▩, PAM.

needed to maintain adequate soil water potential for plots with straw mulch or PAM. Hence less total water was applied on the straw mulch and PAM plots.

Both PAM and straw mulch were effective at reducing sediment loss in both wheel and nonwheel traffic furrows (Figs. 1 and 2). Straw mulch was more effective than PAM in reducing total sediment loss in this trial, LSD(0.05) = 9.1 Mg/ha (Fig. 3).

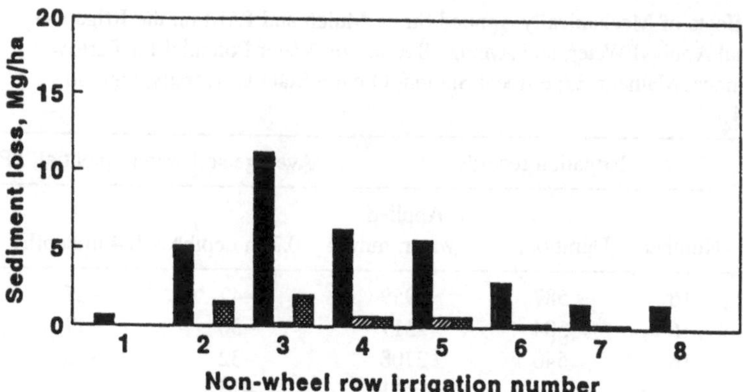

Figure 2 Effects of mechanical application of straw mulch or repeated use of PAM treated irrigation water on sediment loss from nonwheel traffic furrows during eight successive irrigations on a Nyssa silt loam with a 3% slope. Malheur Experiment Station, Oregon State University, Ontario, Oregon, 1994, ■ check; ▨, straw; ▩, PAM.

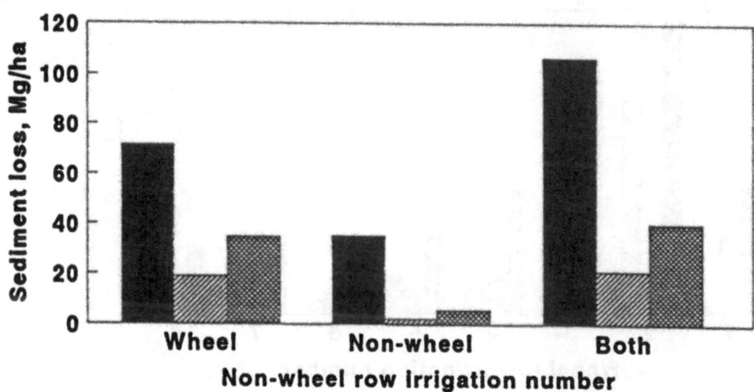

Figure 3 Reduction in season long sediment loss from furrow irrigations from mechanical application of straw mulch of PAM treated irrigation water. Season long sediment loss totals are reported for wheel traffic furrows, nonwheel traffic furrows, and for both together from a Nyssa silt loam with a 3% slope. Malheur Experiment Station, Oregon State University, Ontario, Oregon, 1994, ■ check; ▨, straw; ▩, PAM.

In the untreated check plots, a large proportion of the applied water was lost as runoff (Fig. 4). The use of PAM shifted the fate of water towards infiltration rather than runoff. Straw mulch was more effective in enhancing infiltration and reducing runoff than PAM.

Both straw mulch and PAM facilitated the maintenance of adequately wet soil water potential (Table 1, Fig. 5).

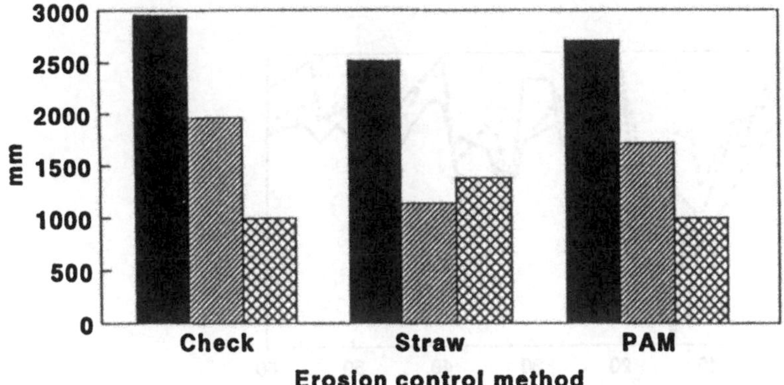

Figure 4 Effects of mechanical application of straw mulch or PAM treated irrigation water on the water applied, runoff, and infiltration in furrow irrigated potatoes from a Nyssa silt loam with a 3% slope. Malheur Experiment Station, Oregon State University, Ontario, Oregon, 1994. ■ applied; ▨ runoff; ▩ infiltration.

C. Discussion

The level of water stress suffered by the potatoes in the untreated check was similar to soil water potentials that have been proven to result in loss of tuber grade (Eldredge et al., 1992).

1. Limitations of This Trial

During the first irrigation of the wheel traffic furrows, more sediment was lost from the strawed furrows than was expected (Fig. 1). The straw mulching machine may have aggravated this soil loss, because too much weight was placed on sharply pointed press wheels during straw application. The pointed press wheels left a narrow crease in the bottom of the furrows. During the first irrigation, the water tended to follow the narrow crease, creating a channel below the straw and allowing soil loss. Straw mulching technique is evidently as important as the material and machinery.

During the second irrigation in the wheel traffic furrows, the PAM rate in the PAM treated plots was reduced to 0.22 kg/ha, and the product was applied evenly over the irrigation duration, rather than at the beginning during the water advance. During this irrigation PAM failed to control erosion (Fig. 1) and left a narrow channel that probably increased soil losses and decreased water infiltration during all subsequent irrigations of the wheel traffic furrows, since the field was never cultivated after irrigations began. The success of the PAM treatments in the nonwheel traffic furrows (Fig. 2) reveals their promise, and the partial failure in the nonwheel rows emphasizes the need for application technology. The PAM applied during the first irrigation of the wheel rows was in-

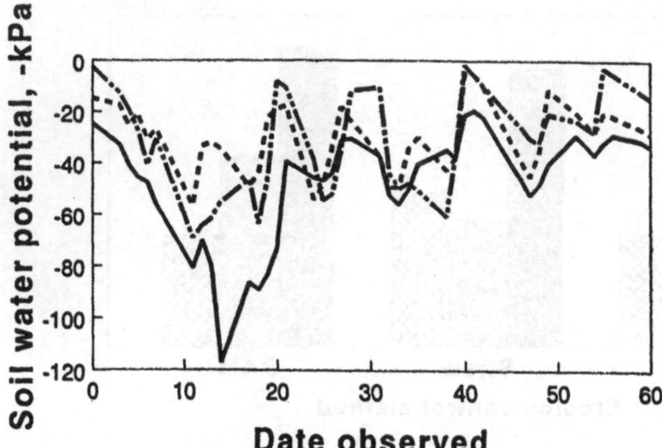

Figure 5 Soil water potential at 0.2 m depth in potato hills. Furrow irrigated potatoes, mechanical application of straw to the furrow bottoms, PAM in the irrigation water, or untreated for erosion control as a check. Malheur Experiment Station, Oregon State University, Ontario, Oregon, 1994. Unbroken line, check; dotted line, straw; dotted and dashed line, PAM.

adequate to protect these furrows during their second irrigation with the 0.22 kg/ha PAM rate distributed over the entire duration of the irrigation. The PAM rate should have been higher, or most of the PAM should have been applied during the advance of the irrigation water, or both.

2. Costs of application

The estimated cost of straw mulching is \$56.75 per acre or \$140.17 per hectare (Table 2). This includes the cost of buying and applying the straw at 0.9 Mg/ha, including tractor time, added labor, and renting a straw mulching machine. Added inconveniences of straw would be that there could be no cultivation in the field after the straw was applied, and the buyer would have to contract someone to furnish baled straw. Also, straw mulch could introduce volunteer wheat or weeds into the field.

The estimated cost of using PAM is \$68.25 per acre or \$168.58 per hectare (Table 2). This includes the cost of granular PAM, mixing it into a solution, and delivery. The inconveniences include setting up a system to apply the PAM, changing the rate of application, and monitoring the system for damage or clogging. Furrows treated with PAM have preserved furrow shape, which facilitates even water infiltration over the length of the field and facilitates cultivations. The PAM application is inconvenient when it needs to be repeated often on steep slopes.

Table 2 Estimated Costs of Mechanically Applied Straw Mulch or Using PAM Season Long to Reduce Erosion from Furrow Irrigated Potatoes, Malheur Experiment Station, Oregon State University, Ontario, Oregon, 1994

Estimated cost of straw mulching	$/acre	$/ha
80 hp tractor 0.4 h/ac	8.00	19.76
Mulcher use for one acre	20.00	49.40
2 workers salary plus benefits for 0.4 hr	6.00	14.82
800 lb straw, $1.00/50 lb bale (production cost by user)	16.00	39.52
Straw transport, $0.25/bale	4.00	9.88
Subtotal	54.00	133.38
5 month's interest 15 1%/month	2.75	6.79
Total	$56.75	$140.17
Polyacrylamide cost estimate	**$/acre**	**$/ha**
9 lb of raw material at $4.50/lb	42.75	105.59
Mixing, delivery, services	20.25	50.02
Equipment amortized	2.00	4.94
Subtotal	65.00	160.55
5 month interest at 1%/month	3.25	8.03
Total	$68.25	$168.58

Potato economic return may more than offset the cost of straw mulch or applying PAM. Potato yield and/or quality responses are expected from better crop irrigation because the crop is highly sensitive to water stress.

C. Summary

Polyacrylamide and straw mulch were used to reduce soil erosion on furrow irrigated potatoes. The PAM rate started out at 1.12 kg/ha per irrigation; then in the second irrigation of the wheel rows the rate was reduced to 0.22 kg/ha. The rate was increased to 0.56 kg/ha for all subsequent irrigations. Wheat straw was applied mechanically at 1.04 Mg/ha. Over 16 irrigations, the untreated furrows lost a total of 106.6 Mg/ha of soil washed out as sediment. The straw mulched furrows lost a total of 21.3 Mg/ha. The PAM treated furrows lost a total of 40.3 Mg/ha. Straw mulch decreased the amount of sediment loss by 80%, while PAM decreased the amount of sediment loss by 62%. Practical techniques are discussed that should improve the effectiveness of both PAM and straw mulch in reducing erosion.

During 16 irrigations, straw mulch increased water infiltration from 33.7 to 54.9% of applied water, while PAM increased the infiltration to 36.8%.

IV. EROSION AND WATER INFILTRATION IN FURROW IRRIGATED ONIONS

The purpose of this experiment was to evaluate and compare the effectiveness of mechanically applied straw mulch and PAM for erosion prevention and infiltration improvement in furrow irrigated onions.

A. Procedures

The experiment was conducted at the same Nyssa silt loam with a 3% slope as the 1994 trial using the same 27 plots with one more added for measurements with PAM (Shock et al., 1996). Harvested potato beds were left idle over the winter of 1994. Spring field work in 1995 prior to onion planting consisted of deep ripping in two directions, disking, and one groundhog operation. Four 0.56 m beds were made down the middle of each preexisting 2.73 m wide potato plot and then harrowed in preparation for planting. The onion variety Vision was planted on April 17. The field was cultivated on June 1. Cultural practices followed standard commercial practices for onion production.

The field consisted of 28 plots, each plot being 75 m long and 2.24 m wide. An unplanted strip 0.49 m was left between onion plots. There were twelve plots without straw or PAM, twelve plots treated with straw mulch, and four PAM plots. On June 13 straw mulch was mechanically applied to the furrow bottoms in twelve of the plots, at a rate of 630 kg/ha. There were no other cultivations after the straw mulch was applied. Gated pipe was arranged so that all plots were irrigated during each irrigation set. The crop was irrigated using alternate furrow irrigation, that is to say only every second furrow was irrigated all season, and all of these furrows were nonwheel traffic furrows. Furrows were irrigated at the rate of 8 l/min. Inflow and outflow measurements were taken hourly for every irrigation measured. Imhoff cones were used to measure the sediment content in the same outflow measurement samples.

The PAM solution was applied directly into the irrigation water individually for each furrow, starting just as the water advanced down the furrow. In such a small experiment, the solution for each furrow was held in a 5 gallon nurse tank and was applied into the irrigation water. During the first irrigation, 1.12 kg/ha of PAM was applied. During subsequent irrigations, PAM was applied at 0.56 kg/ha. The correct amount of PAM was applied to each furrow each irrigation by measuring out the precise amount of PAM stock solution and controlling the release rate of the stock solution with a valve at the head of each furrow. The PAM solution was metered into the water at a rate that would put approximately

Table 3 Irrigation Schedule for the Onion Field. The Field was a Nyssa Silt Loam with 3% Slope. Plots received Mechanically Applied Straw Mulch at 630 kg/ha, were Irrigated with PAM Treated Water, or Remained Untreated Check Plots, Malheur Experiment Station, Oregon State University, Ontario, Oregon, 1995.

Furrow irrigation #	Date	Duration	Method
	April 22	8 h	Sprinkler
	May 26	12 h	Sprinkler
	June 3	4 h	Sprinkler
	June 5	8 h	Sprinkler
1	June 22	24 h	Furrow
2	June 29	24 h	Furrow
3	July 5	24 h	Furrow
4	July 12	24 h	Furrow
5	July 18	24 h	Furrow
6	July 24	24 h	Furrow
7	July 28	24 h	Furrow
8	August 2	24 h	Furrow
9	August 15	24 h	Furrow
10	August 8	24 h	Furrow
11	August 21	24 h	Furrow

80% of the solution into the furrow during the time that the water initially advanced down the furrow length.

Granular matrix sensors were used to measure the soil water potential. Six sensors were placed in a PAM plot, six in a check plot, and six in a straw plot. Two of the six sensors were placed at 18.6 m, two were placed 37.4 m, and two were placed 55.7 m from the top of the field in each of the measured plots. The sensors were buried at a depth of 0.2 m and directly lined up under the onion row. Sensors were read daily at 8 A.M. starting July 18 using a 30 KTCD meter.

The onions were irrigated for emergence by sprinklers without PAM, at a rate of 2.5 mm per hour, four times over a period of six weeks, starting on April 22 (Table 3). A total of 81 mm of water was applied by sprinkler irrigation preceding the furrow irrigations. All subsequent irrigations were 24 h furrow irrigations. The field was furrow irrigated twelve times starting on June 22. On all but two of the furrow irrigations, inflow, outflow, and sediment loss data were collected. The data for the second and fourth irrigations were estimated by averaging the data from the irrigations immediately preceding and following the ones that were skipped.

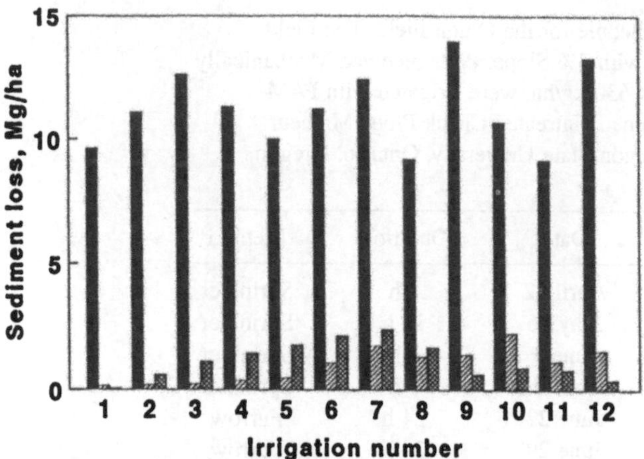

Figure 6 Average sediment loss during sequential furrow irrigations over a Nyssa silt loam with 3% slope. The field was planted to onions; then furrows received mechanically applied straw mulch at 630 kg/ha and were irrigated with PAM treated water during each irrigation or were left untreated. Malheur Experiment Station, Oregon State University, Ontario, Oregon, 1995, ■ check; ▨, straw; ▩, PAM.

B. Results

Sediment losses from the untreated check treatment ranged from 9 to more than 13 Mg/ha per individual irrigation, and losses were essentially undiminished all season (Fig. 6). Onion vegetation never grew to the extent necessary to contribute to a reduction in erosion potential. As the soil became more stable with time, soil stability was offset by erosion progressively narrowing the bottom of the irrigation furrow. Sediment losses from the PAM treated plots started very low, increased slightly mid-season, and then declined. Later in the season various spots in the furrows irrigated with PAM treated water began to fill with sediment. Furrows receiving straw mulch had very low amounts of sediment loss early in the irrigation season and proportionally more past mid-season. The poorer late season erosion control compared with the early season may have been caused by decomposition of the straw, burial of the straw, and the failure of the onions to provide cover to help reduce late season erosion.

Irrigation water treated with PAM significantly reduced sediment loss all season (Fig. 6). Seasonal total sediment loss averaged 134.6 Mg/ha from the check plots compared with 13.9 Mg/ha lost from the PAM plots and 11.9 Mg/ha lost from the straw mulch plots. Season total water infiltration increased from 335 mm in the check plots to 734 mm in the PAM plots and 1,021 mm in the straw mulch plots (Fig. 7). PAM and straw mulch did not differ significantly in

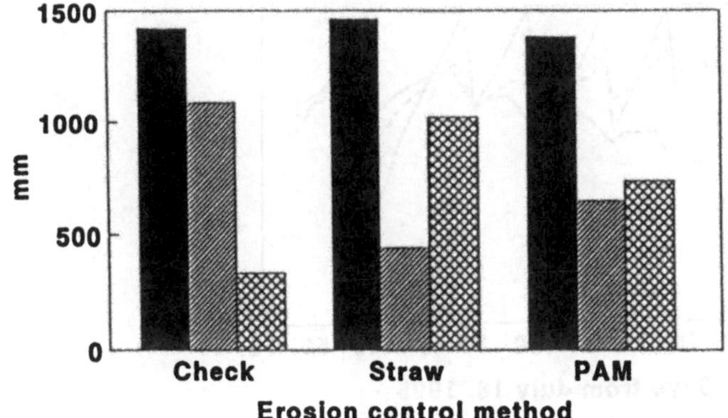

Figure 7 Season average effect of mechanically applied straw mulch at 630 kg/ha, or PAM, applied during each irrigation, on infiltration and runoff on furrow irrigated onions during 12 irrigations of a Nyssa silt loam with a 3% slope, on 75 m long runs, LSD (0.05) = 159 mm. Malheur Experiment Station, Oregon State University, Ontario, Oregon, 1995. ■ applied; ▨ runoff; ▩ infiltration.

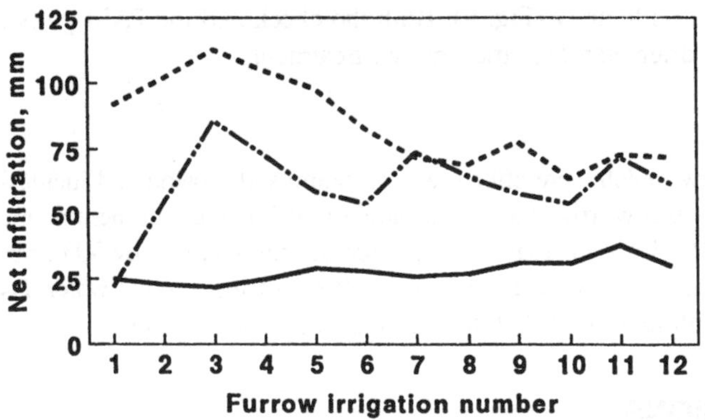

Figure 8 Average water infiltration in onions grown in a Nyssa silt loam with a 3% slope. Mechanical application of straw mulch at a rate of 630 kg/ha; PAM was applied during each irrigation. Malheur Experiment Station, Oregon State University, Ontario, Oregon, 1995. Unbroken line, check; dotted line, straw; dotted and dashed line, PAM.

preventing sediment loss, but straw mulch was significantly better than PAM at increasing infiltration (LSD (0.05) = 160 mm), particularly at the beginning of the season (Fig. 8). Runoff was reduced by both PAM and straw mulch, but there was significantly less runoff with a single straw mulch application at 0.63 Mg/ha than with 12 successive applications of PAM (LSD (0.05) = 159 mm).

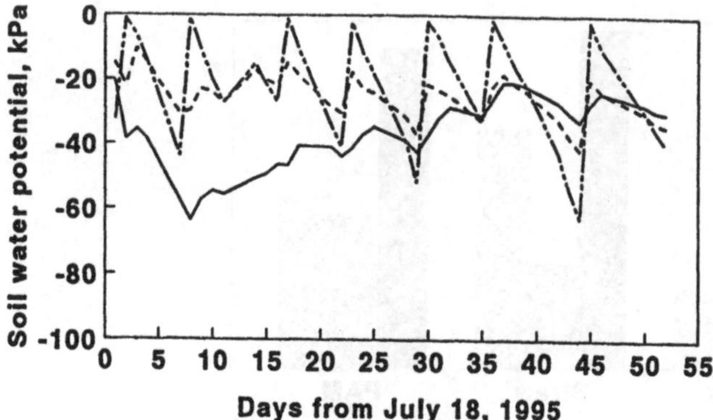

Figure 9 Soil water potential at 0.2 m depth in onion beds. Furrow irrigated onions received mechanically applied straw mulch at 630 kg/ha, PAM treated irrigation water, or an untreated check. The field was a Nyssa silt loam with a 3% slope. Malheur Experiment Station, Oregon State University, Ontario, Oregon, 1995. Unbroken line, check; dotted line, straw; dotted and dashed line, PAM.

The soil remained wetter when the water was treated with PAM or the furrows were mulched with straw (Fig. 9). Both the check and the PAM plots occasionally became drier than the straw mulch treatment.

C. Summary

Both PAM and straw mulch were effective in reducing soil erosion and increasing water infiltration in furrow irrigated onions. Straw mulch decreased the amount of sediment loss by 91%. PAM decreased the amount of sediment loss by 90%. Straw mulch increased water infiltration from 23.6 to 69.8% of the applied water. PAM increased water infiltration from 23.6 to 53.1% of the applied water.

V. CONCLUSIONS

Both PAM and mechanically applied straw mulch are effective methods for reducing sediment losses. Techniques of application and irrigation management are probably more important, for the erosion control success of either method, than any clear-cut advantage of one method over the other. The use of PAM requires less movement of materials and manpower than straw mulch.

While both PAM and straw mulch increased infiltration and reduced runoff, a single application of straw mulch increased infiltration and reduced runoff more than 16 repeated PAM applications in a potato trial and more than 12 repeated PAM applications in the onion trial. When the maintenance of wet soil is more important than other considerations, for example in the case of high onion yield and grade, straw mulch may achieve the desired soil water potential more easily than the use of PAM. Since PAM use is a recently developed technology, further advances in formulation or use may be possible to increase water infiltration.

ACKNOWLEDGMENTS

The trials reported here required many hours of careful measurements and crop care. We are grateful for the dedicated work of Daniel Burton, Erik Feibert, Kody Kantola, Lamont Saunders, Jan Trenkel, and Joanna Zattiero.

REFERENCES

Berg, Robert D. (1984). Straw residue to control furrow erosion on sloping, irrigated cropland. *J. Soil Water Conserv. 60*:58–60.

Brown, M. J. (1985). Effect of grain straw and furrow irrigation stream size on soil erosion and infiltration. *J. Soil Water Conserv. 40*:389–391.

Brown, M. J. and W. D. Kemper (1987). Using straw in steep furrows to reduce soil erosion and increase dry bean yields. *J. Soil Water Conserv. 42*:187–191.

Brown, M. J., W. D. Kemper, T. J. Trout, and A. S. Humpherys (1988). Sediment, erosion and water intake in furrows. *Irrig. Sci. 9*:45–55.

Carter, D. L. (1976). Guidelines for sediment control in irrigation return flow. *J. Environ. Qual. 5*:199–124.

Eldredge, E. P., C. C. Shock, and T. D. Stieber (1993). Calibration of granular matrix sensors for irrigation management. *Agron. J. 85*:1228–1232.

Eldredge, E. P., C. C. Shock, and T. D. Stieber (1992). Plot sprinklers for irrigation research. *Agron. J. 84*:1081–1084.

Lentz, R. D., I. Shainberg, R. E. Sojka, and D. L. Carter (1992). Preventing irrigation furrow erosion with small applications of polymers. *Soil Sci. Soc. Am. J. 56*:1926–1932.

Shock, B. M., and C. C. Shock (1988). InfilCal 5.0. Copyright 1988, 1989, 1993, 1995.

Shock, C. C., H. Futter, R. Perry, J. Swisher, and J. H. Hobson (1988) Effects of straw mulch and irrigation rate on soil loss and runoff. Oregon State University Agricultural Experiment Station Special Report 816, pp. 38–47.

Shock, B. M., C. C. Shock, and J. Hobson (1992). Calculations of surface irrigation inflow and outflow to estimate infiltration. First National Irrigation Induced Erosion Conference, Boise, Idaho, 1992.

Shock, C. C., J. H. Hobson, J. Banner, L. D. Saunders, and T. D. Stieber (1993a). An evaluation of mechanically applied straw mulch on furrow irrigated onions. Oregon State University Agricultural Experiment Station Special Report 924, pp. 71–77.

Shock, C. C., J. H. Hobson, J. Banner, L. D. Saunders, and B. Townley (1993b). Improved irrigation efficiency and erosion protection by mechanical furrow mulching sugar beets. Oregon State University Agricultural Experiment Station Special Report 924, pp. 172–177.

Shock, C. C., L. D. Saunders, B. M. Shock, and J. H. Hobson, M. J. English, and R. W. Mittelstadt (1994). Improved irrigation efficiency and reduction in sediment loss by mechanical furrow mulching wheat. Oregon State University Agricultural Experiment Station Special Report 936, pp. 187–190.

Shock, C. C., J. Zattiero, K. Kantola, and L. D. Saunders (1995). Comparative cost and effectiveness of polyacrylamide and straw mulch on sediment loss from furrow irrigated potatoes. Oregon State University Agricultural Experiment Station Special Report 947, pp. 128–137.

Shock, C. C., J. Trenkel, D. Burton, L. D. Saunders, and E. B. G. Feibert (1996). Season-long comparative effectiveness of polyacrylamide and furrow mulching to reduce sediment loss and improve water infiltration in furrow irrigated onions. Oregon State University Agricultural Experiment Station Special Report 964, pp. 178–185.

17

Use of Water-Soluble Polyacrylamide for Control of Furrow Irrigation-Induced Soil Erosion

Arthur Wallace *University of California–Los Angeles, Los Angeles, and Wallace Laboratories, El Segundo, California*

I. INTRODUCTION

Soil erosion caused by furrow and even by sprinkler irrigation systems is a serious problem on millions of hectares of irrigated cropland (Carter, 1990; Carter, 1993). The reports of Carter show that furrow irrigated land on 2% slopes can erode as much as 100 to 150 MG of soil per hectare per year. The problem of irrigation induced erosion has been recognized for many years, but only during the last few years has real progress has been made toward its control.

At least two groups have had phenomenal success in using water-soluble polyacrylamide (WS-PAM) (see Chapter 15) in furrow irrigation to control soil erosion. Soil erosion is very intense on sloping land and is often compounded as sediments containing pesticides, even pesticides used legally in former years, are moved into streams, lakes, and reservoirs. Erosion affects water quality as salts and nutrients also leave the land. Conversion of the irrigation procedure to sprinklers is one method of coping with the problem, but it is expensive at least for some types of agriculture and even it results in some erosion. The use of WS-PAM may be far more practical. An interim standard was approved in January 1995 (Spofford and Pfeiffer, 1996) and revised in July 1996 by the National Resources Conservation Service of the USDA that permits and guides the use of WS-PAM for erosion control in irrigation furrows in western states of the USA (see Table 1, which covers 5 pages). A conference was held in Idaho (Sojka and Lentz, 1996a) to explore in detail various aspects of the control of furrow irrigation induced erosion by WS-PAM and also the simultaneous effects on water infiltration into soil.

Table 1 Natural Resources Conservation Service Conservation Practice Standard

IRRIGATION EROSION CONTROL (POLYACRYLAMIDE)
(acres)
CODE 201 CA INTERIM

DEFINITION

The addition of polyacrylamide to irrigation water.

PURPOSES

To minimize or control irrigation induced soil erosion.

CONDITIONS WHERE PRACTICE APPLIES

On corrugation or furrow irrigation lands susceptible to irrigation induced erosion. This practice does not apply to peat soils or where irrigation waters exceed a sodium adsorption ratio (SAR) of 15.

CRITERIA

The polyacrylamide (PAM) will be of the anionic type meeting EPA and FDA acrylamide monomer limits, and shall be applied according to the labeling of the product for this use. Use shall conform to all federal, state, and local laws, rules, and regulations.

PAM will be used during the first irrigation after soil disturbance (Pre-irrigation is considered irrigation).

PAM will be added to irrigation water only during the advance phase of an irrigation. The advance phase will be considered to be from the time irrigation starts until water has advanced to the end of the furrows or corrugations.

The concentration of PAM in irrigation water applied shall not exceed 10 ppm. Premixed stock solutions are encouraged. Mixing of and/or application of materials shall be in accordance with the manufacturers recommendations.

CONSIDERATIONS

Other conservation treatments such as land leveling, irrigation water management, reduced tillage, crop rotations, etc. should be used in conjunction with this practice to control irrigation induce erosion.

Adjustment of the concentrations downward from 10 ppm may be used so long as no visible erosion occurs.

Secondary applications on untilled furrows may be needed but may not require as high a rate as the first application.

Where reasonably possible, the tailwater containing PAM should be used on other fields (or stored for a future irrigation).

PAM is a flocculating agent which can cause deposition in canals, laterals, head ditches, pipelines, furrows, or other locations where it comes in contact with sediment ladened waters. Downstream deposition from the use of PAM may require frequent cleaning to maintain normal functions.

The advance rate can vary greatly between hard rows (Wheel packed) and soft rows. Both PAM application and irrigation water management would benefit from treating these differences appropriately.

Consider the impacts of increases in infiltration of up to approximately 15 percent when PAM is applied.

Safety and Health

Consider proper health and safety precautions according to the label and industry guidelines. If inhaled in large quantities, PAM dust can cause choking and difficult breathing. A dust mask of a type recommended by the manufacturer should be used by persons handling and mixing PAM. PAM solutions can cause surfaces, tools, etc to become very slippery when wet.

Water Quantity

1. Effects on water budget components, especially relationships between runoff and infiltration.
2. The effect of changes in the water table on the rooting depth for anticipated land uses.

Water Quality

1. Downstream effects of erosion and yields of sediment and sediment-attached substances.
2. Effects on the salinity of the soil in the drained field.
3. Effects on the loadings of dissolved substances downstream.
4. Potential changes in downstream water temperature.

PLANS AND SPECIFICATIONS

Specification will be developed site specifically for each application. Specifications for this practice will be prepared for each field or treatment unit according to the criteria, considerations, and operation and maintenance described in this standard. Specifications shall consist of approved specifications, job sheets, and narrative statements on the Practice Requirements sheet.

Specification Guide

Follow sound irrigation water management principles.

- Generally infiltration increases (about 15 percent) when polyacrylamide is used. Stream sizes may need to be increased to keep the same balance between infiltration and runoff.
- Where stream sizes have been restricted due to excess soil movement in the past, it may be possible to increase the flows to provide a better infiltration / runoff balance.
- For the most uniform water application, advance streams containing polyacrylamide should reach the end of the field in the first 20 to 35 percent of the total set time.

Based upon soils, slope, and stream size, the necessary concentration of polyacrylamide may be reduced. For the best and most economic concentration, back off on the amount of polymer used until soil movement is noted, then increase slightly.

Lack of adequate turbulence is generally indicated by jellying and deposition of polymer material downstream of the application point.

OPERATION AND MAINTENANCE

Irrigations will be monitored and the PAM applications to irrigation waters will be discontinued when the advance phase has been completed.

All equipment will be operated and maintained to provide the uniform application rates as listed in Criteria (and as specifically stated in Practice Requirements sheet provided to the user). Rinse all equipment used to mix and apply PAM thoroughly with water to avoid formation of intractable PAM residues.

NATURAL RESOURCES CONSERVATION SERVICE
CONSERVATION PRACTICE SPECIFICATION

201 - IRRIGATION EROSION CONTROL (POLYACRYLAMIDE)

I. SCOPE

The work shall consist of adding polyacrylamide (PAM) to the irrigation water for the purpose to reduce soil movement (erosion).

II. MATERIAL

The polyacrylamide shall conform to the following requirements:

a. be anionic (negatively charged polyacrylamide) labeled as either, Acrylamide or Sodium Acrylate Copolymer.
b. Be charge density of 10 to 35 percent
c. Have a molecular weight of 6 to 15 Mg/mole
d. The monomer maximum concentration shall be 0.05 percent

III. POLYMER APPLICATION

The chemical (PAM) shall be added to the irrigation water at a rate to produce a concentration of 10 ppm, or as listed on the Practice Requirement sheet; and mixed by sound and turbulent mixing to yield a uniform distribution.

PAM shall be added to the first irrigation (Pre-irrigation) and on each irrigation that follows a soil disturbance. Re-apply the polymer to an irrigation when soil movement is noted or is predictable.

PAM will be added to the irrigation water only during the advance phase of irrigation. The advance phase will be considered to be from the time irrigation starts until water has advanced to the end of the furrows or corrugations.

The manufacturer's recommendations as listed on the product label shall be adhered to for mixing and application. Adjustment of the concentration downward may be used so long as no visible erosion occurs. Secondary applications on untilled furrows may be needed but may not require as high a rate as the first application.

Should tailwater contain PAM, the tailwater should be used on other fields, or stored for a future irrigation.

IV. BASIS OF ACCEPTANCE

This practice will be considered acceptable when the irrigation induced erosion has been reduced to the level as stated on the Practice Requirement sheet.

V. OPERATION AND MAINTENANCE

Irrigation will be monitored and the PAM applications will be discontinued when the advance phase has been completed with no sufficient soil movement.

Safety precautions as listed on the label, and industry guidelines shall be followed. PAM dust can cause choking and difficult breathing. A mask shall be used by persons handling and mixing PAM.

All equipment used to mix and to apply PAM shall be rinsed thoroughly with water to avoid formation of intractable PAM residues.

U.S. DEPARTMENT OF AGRICULTURE
NATURAL RESOURCES CONSERVATION SERVICE
CALIFORNIA

PRACTICE REQUIREMENTS
FOR
201 - IRRIGATION EROSION CONTROL (POLYACRYLAMIDE)

For: Business Name ______________________________

Job Location ______________________________

County__________ RCD__________ Farm/Tract No. __________

Referral No.________ Prepared By __________ Date __________

IT SHALL BE THE RESPONSIBILITY OF THE OWNER TO OBTAIN ALL NECESSARY PERMITS AND/OR RIGHTS, AND TO COMPLY WITH ALL ORDINANCES AND LAWS PERTAINING TO THIS INSTALLATION.

Installation shall be in accordance with the following drawings, specifications and special requirements. NO CHANGES ARE TO BE MADE IN THE DRAWINGS OR SPECIFICATIONS WITHOUT PRIOR APPROVAL OF THE NRCS TECHNICIAN.

1. Drawings, No. __________ __________ __________
2. Practice Specifications__________, __________, __________
3. Chemical agent to be used: ______________________________
4. Rate of Application: __________ lbs/1000 gals.
5. Special Requirements:______________________________

____

____

____

____

____

6. Special Maintenance Requirements: ______________________________

____

____

PRACTICE APPROVAL:

Job Classification: (Ref: Section 501 NEM)

Show the limiting elements for this job. This job is classified as, Class ___________

Limiting elements:	Units	
Area Treated	___________________	(acre)
___________________	___________________	
___________________	___________________	
___________________	___________________	

Approved by: ___________________________ Date: ___________________________

LANDOWNER'S/OPERATOR'S ACKNOWLEDGEMENT:

The landowner/operator acknowledges that:

a. He/she has received a copy of the drawings and specification, and that he/she has an understanding of the contents, and the requirements.

b. He/she has obtained all the necessary permits.

c. No changes will be made in the installation of the job without prior concurrence of the NRCS technician.

d. Maintenance of the installed work is necessary for proper performance during the project life.

Approved by: ___________________________ Date: ___________________________

PRACTICE COMPLETION:

I have made an on site inspection of the site (or I am accepting owner/contractor documentation), and have determined that the job as installed does conform to the drawings and practice specifications.

Completion Certification by:

/s/ ___________________________________ Date: ___________________________

Some of those using the polymer Krilium (see Chapter 14) 40 years ago did suggest that water-soluble polymers could be applied in solution to soil, but a Russian worker (Paganyas, 1975) was perhaps the first to suggest polymer use for control of irrigation induced soil erosion. Mitchell (1986), Terry and Nelson (1986), and Wallace et al. (1986) all reported that WS-PAM increased water infiltration into soil, as did Krilium earlier.

II. STANISLAUS COUNTY, CALIFORNIA

The West Stanislaus Hydrologic Unit Area Project (HUA), a Natural Resources Conservation Service project in Stanislaus County, California, is reducing agricultural nonpoint source pollution of the San Joaquin River (Hansen et al., 1995). HUA is a name created by the US federal government for a land area eligible for federal funding to handle pollution problems. Irrigation induced erosion in Stanislaus County carries sediment and pesticide residues associated with the sediment, including DDT pesticide isomers, into the San Joaquin River. The DDT was legally applied to the soil many years previously. Over 160 km of the San Joaquin and the entire span of the river in Stanislaus County are considered impaired according to the State Water Resources Control Board. The goal of the program is to reduce sediment in irrigation runoff to a level of 300 mg L^{-1} or less (opaque) compared to levels of 1,500 mg L^{-1} now common (chocolate brown).

WS-PAM, a synthetic polymer, is being tested and is providing effective in reducing the amount of sediment that leaves tailwater in irrigated fields (Hansen et al., 1995; McCutchan et al., 1993; and McElhiney and Osterli, 1996). Actually the program has gone beyond testing, as many thousand hectares have been treated commercially in Stanislaus County. Workers say they are getting in excess of 90% reduction of sediment leaving the field. Water infiltration rates in soil with polymer-treated water have increased as much as 30 to 40% in local projects. Treatments, according to those workers, can also increase crop growth and yield, owing to improved physical soil conditions, aeration, water movement, and decline in surface crusts that inhibit seed germination and emergence.

Research findings on the use of WS-PAM indicate from 95 to 98% reduction in total suspended solids (TSS) of tailwater, 55 to 76% reduction in bedload (TSS and bedload together represent a measure of total off-site movement of sediment), and increases in water infiltration.

The method of application used in 1993 was injecting a stock solution of WS-PAM into irrigation water. Application rates tested were 2.5, 5, and 10 mg L^{-1} applied continuously throughout the irrigation. Intermittent and advancing front techniques have also been used. Methods of application used in 1994 included using a stock solution of WS-PAM and a method of applying dry polymer directly to the irrigation water using a Gandy device to apply granules. This method appears to be effective, even at rates down to 1 mg L^{-1}. Although WS-PAM is nontoxic to humans and plants, its application has been investigated by California and other regulatory agencies for environmental effects. A review of environmental effects was made by Seybold (1994). No reasons have been found for not using WS-PAM.

One way to reduce the amounts of silt and clay going off farms is the use of sediment basins, which are large on-farm reservoirs where drainage water is held for two hours or more to give suspended clay particles time to settle out. The basins clean 90% of the silt from the water going into the river. However, they require periodic maintenance and can take up a substantial amount of land. WS-PAM is more effective and somewhat easier to use.

USDA conservation and education agencies in partnership with the Stanislaus HUA have successfully met water quality objectives the past five years through a comprehensive, integrated, locally managed watershed project. The West Stanislaus (HUA) project is one of 36 HUAs nationwide established by the USDA's 1991 Water Quality Initiative (McElhiney and Osterli, 1996).

Irrigation induced erosion has been studied in the West Stanislaus watershed area for 15 years, but the innovative evaluation and use of WS-PAM is a more recent practice that has great potential for reducing significant amounts of sediment and pesticide residues from entering the impaired San Joaquin River.

The Stanislaus team reports that approximately 24% (12,300 ha) of the total area in the HUA have adopted structural and managerial best management practices. Cumulative savings from runoff through 1995 as a direct result of HUA assistance were 436 kg of DDT isomers, 478,000 MG of sediment, 37.7×10^6 m^3 of irrigation water, and 19% average absolute improved irrigation efficiency. Much of this improvement is the result of use of WS-PAM (McElhiney and Osterli, 1996).

Nontreated furrows degrade without WS-PAM to the point where up to 75% or more of the applied irrigation water may run off the end of the field carrying significant loads of sediment and soil-adhered pesticide residues into manmade drains and intermittent streams and ultimately to the San Joaquin River. It can be essentially controlled with WS-PAM.

Water-applied PAM increases soil cohesion and strengthens aggregates in the irrigation streams by binding exposed soil particles together more securely (Sojka and Lentz, 1994c). This greatly reduces detachment and transport of sediments. Soil erodibility at the soil/water interface is reduced. WS-PAM also acts as a settling agent. It flocculates the fine clay particles dispersed by and carried in the flow, causing them to settle to the furrow bottom. Fewer dispersed fine particles remain in the water to block pores and reduce infiltration rates. Pore structure is maintained to prevent the usual reduction of water infiltration. This decreases runoff.

In a letter to the US Environmental Protection Agency, USDA-ARS scientists stated, "WS-PAM use for erosion control is at the heart of the concept of agricultural sustainability." The reasons given are that it provides a potent environmental benefit without negative impact on flora and fauna, it halts erosion (about 1 MG soil loss is prevented by as little as 100 g of WS-PAM used), and it increases infiltration, thus enabling conservation of water (McElhiney and Osterli, 1996). WS-PAM also allows changes in furrow irrigation management that provide more uniform water application. This reduces the potential for nitrate leaching. It eliminates substantial amounts of sediment, phosphorus, and pesticides from return-flows to rivers, and it greatly reduces biological oxygen demand. Because furrow reshaping and sediment pond or ditch cleaning are needed less frequently, WS-PAM also conserves fuel, therefore lessens air pollution, and reduces equipment wear and labor. WS-PAM has generated considerable interest and is part of an integrated approach for improved water quality in the West Stanislaus HUA Project.

In a study done in 1996 on a clay loam soil and discussed by the workers in Stanislaus County, WS-PAM at the rate of 1 mg L^{-1} in the first six hours of each 24 hours of irrigation was applied. The salt concentration of the irrigation water was only 128 parts per million (mg L^{-1}), so gypsum should have been used also (Wallace and Wallace, 1996). The crop was processing tomatoes. The rate of WS-PAM per hectare was 2.2 kg with an estimated cost of about $25 per hectare. Half of a 20 hectare field was treated and the other half used as a control. The treated half produced 14% more tomatoes than the control half for a gain of $728 per hectare. Since this was not a replicated experiment the investigators want it to be considered as a preliminary trial only. It does have some resemblance, however, to a test reported by Wallace and Wallace (1990) in which 3.3 kg per hectare of WS-PAM applied this time by sprinkler to a 25 hectare potato field increased yields by 25%. The WS-PAM was applied only once. The control 25 ha was half of the center pivot (Fig. 1). The sharp demarcation line between treatments indicates that the differences were valid.

Another grower in 1996 in Stanislaus County, California, says that his primary interest in WS-PAM is in its increased effect on water infiltration, which in his case was about 33% (reported by farm advisers). Decreasing sediment loss and saving water were valuable, he said, but getting adequate moisture into the soil, which means less stress on the crop and greater yields, was more important. This grower will continue evaluating the benefit of WS-PAM. In other field trials, water infiltration rates in furrows was increased by 15 to 45%. Another advantage that growers are finding is that use of WS-PAM in the irrigation water is less costly than installing sprinklers or even than using some microirrigation technologies as a means of water conservation. Some additional effects of WS-PAM on water utilization were suggested by Wallace and Wallace (1986).

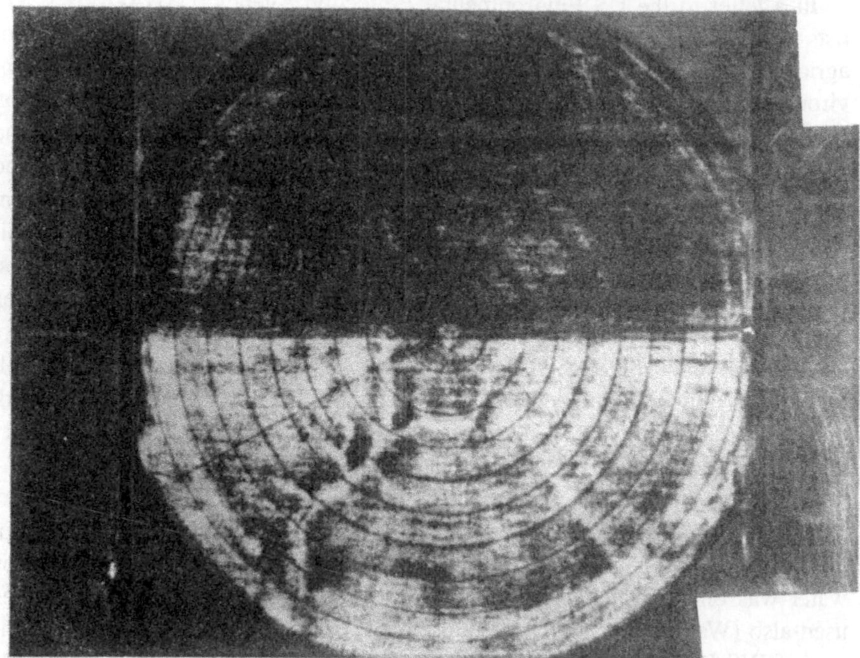

Figure 1 Infrared photograph (red shows as white) of a 50 ha potato field half of which received 3.3 kg/ha WS-PAM in the center pivot in 2.5 cm of water applied immediately after planting and before any rain or other irrigation. Gypsum was included. The treated half was predominately red, indicating increased longevity of the vines. Yield increase in the treated half was over 25%.

The Natural Resources Conservation Service and the University of California Cooperative Extension personnel in Stanislaus County say that the program including WS-PAM has had the following benefits:

Reduced irrigation induced soil erosion from 11–45 MG/ha to < 2.2 MG/ha
Reduced nonpoint source pollution to impaired water bodies
Reduced need for regulatory actions by state and federal agencies
Reduced impairment of rivers for endangered anadromous fish species
Reduced use of irrigation water
Reduced use of pesticides

III. IDAHO

WS-PAM has been given considerable attention in the past few yeas in Idaho for controlling irrigation induced erosion from furrows and also for increasing water infiltration (Lentz and Sojka, 1994a; Lentz and Sojka, 1994b; Lentz and Sojka,

1996a; Lentz et al., 1992a; Lentz et al., 1992b; Lentz et al., 1993a; Lentz et al., 1993b; Lentz et al., 1994; Lentz et al., 1995; Sojka and Lentz, 1993; Sojka and Lentz, 1994a; Sojka and Lentz, 1994b; Sojka and Lentz, 1995c; Sojka and Lentz, 1994d; Sojka and Lentz, 1996b; Sojka and Lentz, 1996c; Trout et al., 1994; Trout et al., 1993; Trout et al., 1995). The efforts in Idaho have been recently summarized (Sojka and Lentz, 1996b; Lentz and Sojka, 1996b). Over 50,000 ha have been treated to date. From 3 to 11 kg ha^{-1} of WS-PAM per year have been demonstrated as effective control with an average decrease of 94% of such erosion. An advancing front or phase technique has been used to maximize the effect of the small amount of WS-PAM used and to decreases the amount running off the field. Since WS-PAM used in irrigation water generally retards surface sealing of soil, both infiltration rate and lateral movement of the water entering soil are increased. The average increase for infiltration is about 15% and for lateral movement, about 25%. Both effects are related to the dilute solutions of WS-PAM and their effects on stability of soil aggregates; usually solutions of 10 mg L^{-1} are used. Higher concentrations of WS-PAM could decrease lateral movement due to viscosity (Letey, 1996). Dilute solutions favor movement of water into soil.

A certain amount of electrical conductivity (EC) or salt concentration in the irrigation water is needed for maximum effect of the WS-PAM (Ben Hur, 1994; Wallace and Wallace, 1996). Although calcium participates in binding of the anionic groups to clay, most of the effect of EC is probably due to ionic strength (Wallace and Wallace, 1996).

Idaho workers say that steep-slope, breaking-slope, and long-slope runs have great potential for benefit from use of WS-PAM. Research has shown that 10 mg L^{-1} of WS-PAM in the advance phase of irrigation will control erosion of slopes up to 3.5%. This use rate may be as little as 1 kg ha^{-1}. They recommend higher rates for higher slopes.

Shallow subsoil layers with poor infiltration properties will not hinder WS-PAM from controlling erosion but will have an adverse effect on total infiltration. Otherwise infiltration rates can be improved by WS-PAM with stabilization of only a few surface cm of soil. Benefits then can extend even below the zone of treatment.

For full benefits of both erosion control and water infiltration, it is important that the application of WS-PAM be started with the very first irrigation and also before any rain has fallen on the newly tilled or planted area (Wallace and Wallace, 1986). Surface damage to the soil can otherwise take place by slaking or sealing before the application of WS-PAM, which would minimize effect of the WS-PAM, especially for an improved infiltration rate. After any soil disturbance or recultivation, more WS-PAM should be applied. The use of WS-PAM cannot overcome the effects of wheels of instruments that destroy soil structure when they move through the furrows (see Chapter 16).

For repeat applications even after furrows have been remade, lower concentrations of WS-PAM, e.g., from 0.5 to 5 mg L^{-1}, may be useful. After furrows have been treated and irrigations are then made with water without WS-PAM, the Idaho researchers report a 50% loss of effect, and this cumulative with each irrigation, so that eventually virtually no effect remains. Perhaps some of this gradual loss can be avoided with curing or partial drying after the very first application (Wallace and Wallace, 1996) and by the use of very dilute WS-PAM solutions in follow-up irrigation.

Idaho workers report that infiltration rates decrease with time over the cropping season even if WS-PAM was used. WS-PAM tends to slow the rate of decline, however. During treatment, high concentrations of WS-PAM can inhibit water infiltration because of viscosity. Coapplication with gypsum would increase the EC and soluble calcium and also decrease somewhat the viscosity (Wallace and Wallace, 1996).

The range of reduced loss of sediment from fields in Idaho due to use of WS-PAM is 80 to 99% with an average of 94%. Furrow configuration is more stable and there is less need to reshape furrows during the season. Erosion control with WS-PAM also has other benefits. Off-field settling ponds can be smaller and require less frequent management. Less irrigation water can be used. Meeting water quality standards can be easier since sediment containing oxidizable organics, phosphates, and even pesticides (Agassi et al., 1995; Singh et al., 1996) are kept on the farms and do not move to streams.

Since WS-PAM can increase water infiltration rates and amounts, irrigation schedules may have to be adjusted to avoid over-irrigation or nonuniform irrigation. Without adjustments there could be a nonuniform distribution of water, since increased infiltration in the start of the run adds more water there. Use of WS-PAM in furrows of flat fields can add to this problem.

The WS-PAMs being used in furrow erosion control in Idaho are copolymers of polyacrylamide and acrylate, with 15 to 20% acrylates to give the negative charge. They have 12 to 15 million Dalton molecular weight and meet USEPA and FDA (Food and Drug Administration) limits of below 0.05% of monomer concentrations (Seybold, 1994). Label directions should be followed for full safety assurance, as there is itchiness caused by dust of WS-PAM for some persons.

Application of WS-PAM to irrigation water is not as easy as one would like and is still the subject of intensive research. Users can be discouraged if they have unhappy experiences, so those proposing and supporting such use must be careful and cautious. Some users prefer to make stock solutions of WS-PAM, and others apply dry WS-PAM granules to moving water with a Gandy apparatus (used both in Idaho and California, at least experimentally). It is important that turbulence be involved in putting WS-PAM into solution and that WS-PAM granules always go into water that does not already have WS-PAM. The limit of solution of the WS-PAM being used is around 2.5%, and that with difficulty. Heating water does help solution. Since partially dissolved WS-PAM particles do not pass through filters, it is important that filters in irrigation streams be avoided.

Idaho workers (Lentz et al., 1995) have outlined guidelines for choosing between dry and liquid procedures for application of WS-PAM to water going into furrows. Some of the advantages and disadvantages outlined by these workers are listed here. The final decision has not been made and is still the object of research and development.

Advantages of Liquid Application

It is easy to calculate and meter exact use rates.
It is easy to keep track of amounts applied, since volume can easily be recorded.
It requires minimal mixing in the water stream to work well.
It is slow to clog weed screens, filters, and narrow siphons.
It has low risk of exposure when the operator does not handle dry WS-PAM.
Applications can be accomplished without specialized mixing or metering equipment.

Disadvantages of Liquid Application

It may be more expensive than the granular method due to increased handling costs.
It requires bulk equipment that is not manually portable.
Large volumes of stock solution are needed for large fields.
Diluting the concentrate in the field takes considerable time and requires dedicated equipment.

Advantages of Dry Application

It uses portable equipment that can be moved manually.
A season's supply of dry WS-PAM can be purchased and stored at a farm.
It may be a low-cost form of WS-PAM.
There is less need to rely on suppliers or dealers to refill tanks on the farm.

Disadvantages of Dry Application

Application equipment tends to plug or clog.
It requires vigorous mixing for dissolution enough to give uniform application.
Solution time may be too long for ease in use.
It will rapidly plug weed screens and filters.
There is some danger of operators choking from inhalation of WS-PAM dust while filling units used to dispense the WS-PAM.
There is need to purchase or build application equipment.
There are greater WS-PAM losses from the field with runoff, since there is less control of the solution process. The rate of dissolving is too slow for full solution in short furrows.
There is less uniformity of distribution than with the liquid application procedures.

These workers recommend agitation of stock solution for at least 60 minutes after all WS-PAM has been introduced and, if possible, that the solution should stand overnight before use to insure full hydration of the WS-PAM. Part of this problem can be circumvented with smaller particle sizes of WS-PAM and careful mixing of WS-PAM into clean water or with continuous water without WS-PAM. Appropriate calculations are necessary to get proper concentrations of WS-PAM into the irrigation furrows (Lentz et al., 19950.

Another caution from the Idaho workers is that some irrigation water contains sediments, and if the WS-PAM is introduced before it enters delivery systems, some siltation may clog or cause other problems in the delivery system. This, too, is a subject of research. If the WS-PAM is introduced far ahead of the delivery system, the problem may be avoided due to removal of the sediment from the water by the WS-PAM.

The Idaho workers have addressed the problem of some of the WS-PAM escaping from the field and accumulating in streams and lakes. Lentz et al. (1996) reported a procedure for assay of WS-PAM in water to evaluate these effects. Wallace and Wallace (1996) say that sunlight does degrade and destroy the WS-PAM in water. This implies that WS-PAM will not accumulate in water bodies but will soon be degraded to carbon dioxide, water, and ammonia. Also the WS-PAM will settle out permanently with any clay in the water if the WS-PAM is not destroyed by sunlight.

New technology for injection of WS-PAM into irrigation systems may make these procedures obsolete. A procedure for making very high formulations of PAM with organic solvents could do away with the need to make large-volume stock solutions. These formulations are being tested and could help WS-PAM to be more easily used throughout the world.

IV. OTHER GEOGRAPHICAL USE AREAS

Idaho and California are not the only areas in the USA with vigorous ongoing programs to push the use of WS-PAM for control of erosion in irrigation furrows. Use is being tested in Oregon (see Chapter 16), Washington (Ley and Prest, 1996; Crose, 1996), and Colorado (Valliant, 1996). It is well that research and development be done in various irrigated areas for one important reason, namely, studies at Wallace Laboratories have strongly implied that very large differences exist in amounts of and kind of WS-PAM needed for various soils and soil conditions (Wallace and Wallace, 1996).

Zhang and Miller (1996) have used a spray technique experimentally to decrease crusting and furrow soil erosion with WS-PAM on a Cecil sandy loam soil in Georgia, USA. They used 1000 mg L^{-1} WS-PAM sprayed onto the soil to give 15 and 30 kg ha^{-1} of WS-PAM. The solutions contained 2.5×10^{-3} *M*

$CaSO_4$. Test pilots were sprayed with a pesticide sprayer and were allowed to dry for curing before being tested with simulated rainfall. The treatments were effective, and the authors concluded that the procedure could be satisfactory when high value crops were grown. The concentration of 1000 mg L^{-1} is high with perhaps too much viscosity to wet the soil easily. Higher water rates even if more water per hectare were needed should be more effective. Equal water and half as much WS-PAM could be satisfactory.

V. CONCLUSIONS

Water-soluble polyacrylamide has been used for over five years for control of irrigation furrow induced soil erosion in several USA locations. Not only has soil erosion been effectively controlled (usually well over 90%) but pesticide removal from fields has been decreased and also water infiltration has been increased. Although the procedure has been commercialized, research and development continue to improve application methods, as they have been a bit cumbersome. Organic solvents to produce high concentration stock solutions are being tested. It is expected that WS-PAM technology will increase many times to involve most furrow irrigated lands.

REFERENCES

Agassi, M., Letey, J., Farmer, W. J., and Clark, P. (1995). Soil erosion contribution to pesticide transport by furrow irrigation. *J. Environ. Qual. 24*:892.

Ben-Hur, M. (1994). Runoff, erosion, and polymers application in moving sprinkler irrigation. *Soil Sci. 158*:283.

Carter, D. L. (1990). *Soil Erosion on Irrigated Lands, Irrigation of Agricultural Crops* (B. A. Stewart, and D. R. Nielson, eds.). Agron. Monogr. 30. ASA, CSSA and SSSA, Madison, Wisconsin, p. 1143.

Carter, D. L. (1993). Furrow irrigation erosion lowers oil productivity. *J. Irrig. Drainage Eng. 119*:964.

Crose, H. C. (1996). Polymer use in the Columbia Basin. *Proceedings: Managing Irrigation-Induced Erosion and Infiltration with Polyacrylamide* (R. E. Sojka and R. D. Lentz, eds.). College of Southern Idaho, Twin Falls, Idaho, May 6–8, 1996, University of Idaho Miscellaneous Publication No. 101-96, p. 3.

Hansen, P. C., Osterli, P., and McElhiney, M. (1995). Polymers aid in erosion control. *Irrigation J. 45*(2):14.

Lentz, R. D., and Sojka, R. E. (1994a). Automated Imhoff cone calibration and soil loss infiltration analysis for furrow irrigation studies. Proc. 5th Int. Conf. Computers in Agriculture (D. G. Watson, F. S. Zazueta, and T. V. Harrison, eds.). *Amer. Assoc. Agri. Eng.*, St. Joseph, Michigan, pp. 858–863.

Lentz, R. D., and Sojka, R. E. (1994b). Field results using polyacrylamide to manage furrow erosion and infiltration. *Soil Sci. 158*:274.

Lentz, R. D., and Sojka, R. E. (1996a). Polyacrylamide application to control furrow irrigation-induced erosion. Proceedings of the 27th International Erosion Control Association meetings, Seattle, Washington, 27 Feb.–1 Mar., 1996, pp. 419–430.

Lentz, R. D., and Sojka, R. E. (1996b). Five-year research summary using PAM in furrow irrigation. *Proceedings: Managing Irrigation-Induced Erosion and Infiltration with Polyacrylamide* (R. D. Lentz, and R. E. Sojka, eds.). College of Southern Idaho, Twin Falls, ID, 6–8 May, 1996, pp. 20–27.

Lentz, R. D., Shainberg, I., Sojka, R. E., and Carter, D. L. (1992a). Preventing irrigation furrow erosion with small applications of polymers. *Soil Sci. Soc. Amer. J. 56*:1926.

Lentz, R. D., Shainberg, I., Sojka, R. E., and Carter, D. L. (1992b). Reducing furrow erosion with polymer-amended irrigation waters. *1992 Agron. Abs.*, p. 330.

Lentz, R. D., Sojka, R. E., and Carter, D. L. (1993a). Influence of irrigation water quality on sediment loss from furrows. Proc. SWSC Conf. on Agric. Res. to Protect Water Quality. Minneapolis, Minn., 21–24 Feb. 1993, pp. 274–278.

Lentz, R. D., Sojka, R. E., and Carter, D. L. (1993b). Influence of polymer charge type and density on polyacrylamide ameliorated irrigated furrow erosion. Preserving our environment: the race is on. Proc. 24th Int'l Erosion Control Assoc. Conf. Indianapolis, Ind., 23–26 Feb. 1993. Int'l Erosion Control Assoc., Steamboat Springs, Colorado, pp. 161–168.

Lentz, R. D., Foerster, J. A., and Sojka, R. E. (1994). Feasibility of a flocculation test to measure aqueous polyacrylamide activity in irrigated furrows. *1994 Agron. Abs.*, p. 358.

Lentz, R. D., Stieber, T. D., and Sojka, R. E. (1995). Applying polyacrylamide (PAM) to reduce erosion and increase infiltration under furrow irrigation. Proc. 1995 Winter Commodity Schools (L. D. Robertson, P. Nolte, B. Vodraska, B. King, T. Tindall, R. Romanko, and J. Gallian, eds.). University of Idaho Cooperative Extension, Moscow, Idaho, pp. 79–92.

Lentz, R. D., Sojka, R. E., and Foerster, J. A. (1996). Estimating polyacrylamide concentration in irrigation water. *J. Environ. Qual. 25*:1015.

Letey, J. (1996). Effective viscosity of PAM solutions through porous media. *Proceedings: Managing Irrigation-Induced Erosion and Infiltration with Polyacrylamide* (R. E. Sojka, and R. D. Lentz, eds.). College of Southern Idaho, Twin Falls, Idaho, May 6–8, 1996, pp. 94–96.

Ley, T. W., and Prest, V. I. (1996). Furrow flow rate modifications when using PAM. *Proceedings: Managing Irrigation-Induced Erosion and Infiltration with Polyacrylamide* (R. E. Sojka, and R. D. Lentz, eds.). College of Southern Idaho, Twin Falls, Idaho, May 6–8, 1996, p. 3.

McCutchan, H., Osterli, P., and Letey, J. (1993). Polymers check furrow erosion, help river life. *Calif. Agric. 47*(5):10–11.

McElhiney, M., and Osterli, P. (1996). An integrated approach for water quality: the PAM connection. ASA Calif. Plant Soil Conf. in Modesto, 18 Jan. 1996.

Mitchell, A. R. (1986). Polyacrylamide application in irrigation water to increase infiltration. *Soil Sci. 141*:353–358.

Paganyas, K. P. (1975). Results of the use of series K-compounds for the control of irrigational soil erosion. *Soviet Soil Sci. 7*:592.

Seybold, C. A. (1994). Polyacrylamide review: soil conditioning and environmental fate. *Comm. Soil Sci. Plant Anal. 17*:9.

Singh, G., Letey, J., Hanson, P., Osterli, P., and Spencer, W. F. (1996). Soil erosion and pesticide transport from an irrigated field. *J. Environ. Sci. Health. B31*(1):25–41.

Sojka, R. E., and Lentz, R. D. (1993). Improving water quality of return flows in furrow-irrigated systems using polymer-amended inflows. Proc. SWCS Conf. on Agric. Res. to Protect Water Quality. Minneapolis, Minn., 21–24 Feb. 1993, pp. 395–397.

Sojka, R. E., and Lentz, R. D. (1994a). Net infiltration and soil erosion effects of a few ppm polyacrylamide in furrow irrigation water. Proc. 2nd Int'l Sym. on Sealing, Crusting, Hardsetting Soils: Productivity and Conservation. Univ. of Greenland, Brisbane, Aust., 7–11 Feb. 1994, pp. 349–354.

Sojka, R. E., and Lentz, R. D. (1994b). Organic polymers and soil sealing in cultivated soils. *Soil Sci. 158*:267.

Sojka, R. E., and Lentz, R. D. (1994c). Polyacrylamide (PAM): a new weapon in the fight against irrigation-induced erosion. USDA-ARS Soil and Water Management Res. Unit. Station Note #01-94 (revised).

Sojka, R. E., and Lentz, R. D. (1994d). Time for yet another look at soil conditioners. *Soil Sci. 158*:233.

Sojka, R. E., and Lentz, R. D., eds. (1996a). *Managing Irrigation-Induced Erosion and Infiltration with Polyacrylamide*. College of Southern Idaho, Twin Falls, Idaho, May 6–8, 1996, University of Idaho Miscellaneous Publication No. 101-96.

Sojka, R. E., and Lentz, R. D. (1996b). A PAM primer: a brief history of PAM and PAM-related issues. *Proc. Managing Induced Erosion with Polyacrylamide*. Univ. of Idaho, Misc. Pub. pp. 11–20.

Sojka, R. E., and Lentz, R. D. (1996c). Polyacrylamide for furrow-irrigation control. *Irrig. J. 46*:8.

Spofford, T. L., and Pfeiffer, K. L. (1996). Agriculture irrigation polyacrylamide application standard. *Proc. Managing Irrigation-Induced Erosion with Polyacrylamide*. Univ. of Idaho, Misc. Pub. 101-96 (R. E. Sojka and R. D. Lentz, eds.). Moscow Idaho, pp. 49–51.

Terry, R. E., and Nelson, S. D. (1986). Effects of polyacrylamide and irrigation method on soil physical properties. *Soil Sci. 141*:317.

Trout, T. J., Lentz, R. D., and Asce, M. (1993). Polyacrylamide decreases furrow erosion. Management of Irrigation and Drainage Systems: Integrated Perspectives. Proc. 1993 Nat'l Conf. on Irrigation and Drainage Engineering (R. G. Allen, and C. M. U. Neale, eds.). Park City, Utah, 21–23 July 1993, pp. 191–197.

Trout, T. J., Carter, D. L., and Sojka, R. E. (1994). Irrigation-induced soil erosion reduces yields and muddies rivers. *Irrigation J. 44*:8.

Trout, T. J., Sojka, R. E., and Lentz, R. D. (1995). Polyacrylamide effect on furrow erosion and infiltration. *Trans. Amer. Assoc. Agric. Eng. 38*:761–765.

Valliant, J. C. (1996). Demonstration of PAM to reduce erosion on onions in the Arkansas River Valley of Colorado. *Managing Irrigation-Induced Erosion and Infiltration with Polyacrylamide*. Abstract. Twin Falls, Idaho, pp. 119–121.

Wallace, A., and Wallace, G. A. (1986). Effects of very low rates of synthetic soil conditioners on soils. *Soil Sci. 141*:324.

Wallace, A., and Wallace, G. A. (1990). Soil and crop improvements with water-soluble polymers. *Soil Tech. 3*:1.

Wallace, A., and Wallace, G. A. (1996). Need for solution or exchangeable calcium and/or critical EC levels for flocculation of clay by polyacrylamides. *Proceedings: Managing Irrigation-Induced Erosion and Infiltration with Polyacrylamide*. Twin Falls, ID (R. E. Sojka, and R. D. Lentz, eds.). Univ. of Idaho Misc Pub. No. 101-96, pp. 59–63.

Wallace, A., Wallace, G. A., and Abouzamzam, A. M. (1986). Effects of soil conditioners on water relationships in soils. *Soil Sci. 141*:346.

Zhang, X. C., and Miller, W. P. (1996). Polyacrylamide effect on infiltration and erosion in furrows. *Soil Sci. Soc. Am. J. 60*:866.

18

Some Living Plants and Some Additional Products Useful as Soil Conditioners and in Various Technologies

Arthur Wallace *University of California–Los Angeles, Los Angeles, and Wallace Laboratories, El Segundo, California*

I. SOME EFFECTS OF LIVING PLANTS ON PHYSICAL PROPERTIES OF SOIL (SOIL CONDITIONING EFFECTS)

A. Living Mulches

1. Historical

In a review paper titled "The Historical Roots of Living Mulch," Paine and Harrison (1993) begin with the following quotation: "Civilized man has marched across the face of the earth and left a desert in his footprints" (Carter and Dale, 1974), meaning that traditionally the soil gets poor care. They then proceed to give a brief history of caring for the soil, which is important to the philosophy of use of all soil conditioners as well as for living mulches. Part of that history (Paine and Harrison, 1993) is summarized here.

The most long-standing method of restoring soil productivity has been fallowing of the land (Fussell, 1965). While the use of manures and green manures was known in very early times, fallowing has, until very recently, been the most common practice for dealing with declining soil productivity. It has roots in Biblical history.

Europe lost between one-eight and one-third of its population to the plague between 1347 and 1351 A.D. This decreased the pressure on farmers to produce large amounts of grain. The large farm fields were then divided among individual farmers and there was a shift towards less labor-intensive methods and more pasture and livestock production to give an increased amount of manure to be applied to the land (Fussell, 1972). One result of these changes was the development of crop rotation. Crop rotation served satisfactorily to sustain the productivity of Europe's soils for several centuries. Numerous crop rotation schemes were developed in the 18th and 19th centuries involving the major crops of the period, wheat, barley, clover, peas, and pasture grasses. All systems included fallow seasons, and many involved green manure crops. Needed productivity for that time period was obtained.

Tull promoted the concept of tillage in the 18th century. "The finer land is made by tillage, the richer will it become, and the more plants it will maintain" (Tull, 1731). Tillage became an integral part of western agriculture but is now questioned; the major premise was erroneous. Today tillage, with its many drawbacks, is believed to be an obstacle to sustainability. Most goals of users of green manure and cover crops in recent years have been for the amelioration of the negative effects of tillage.

The humus theory was developed in the last decades of the 18th century (Fussell, 1971). The humus theory held, wrongly, that all necessary substances for plant growth came from the organic matter in the soil. Finally, in the 1830s, researchers accumulated evidence for proposing a mineral basis for plant nutrition. As a result, Liebig and his colleagues ushered in the modern era of crop nutrition and soil fertility. It became known that the value of manure was not only in its organic matter content but also in its mineral content as well. However, much attention was still given to green manures, but overemphasis of the need for minerals did lead to considerable loss of tilth properties of soil because soil organic matter levels generally decreased.

The abundance of cheap, fertile land on the North American continent 100 to 200 years ago discouraged most soil conservation practices. Many farmers moved west when the productivity of their old land declined.

Agricultural science grew slowly in the 1800s, during which an increasing awareness of factors involved in soil productivity developed. A strong belief that soil organic matter was a major factor in soil fertility continued for some time, however, in spite of increasing emphasis on mineral nutrients. The first production of synthetically formulated fertilizers was beginning in the late 1800s, but research was still largely focused on maintenance of organic matter levels in soil and the effect of crop production on their depletion. Fortunately many of these studies continue to document the long-term effects of different crop production systems (Paine and Harrison, 1993).

In 1925, the American Society of Agronomy sponsored a symposium on "soil deterioration" (Haskell, 1926). Conclusions were that crop rotation alone, even when legumes were included, was not able to maintain soil productivity over the long term. Additionally, research at that time showed that even though three times as many nutrients were applied in continuous culture as in rotation farming, continuous culture had not produced an equivalent total return

(Williams, 1926). The need for better physical properties of soil was indicated but not really appreciated at that time.

A major problem facing farmers of the past was soil erosion. Several events in the early 1900s forced changes in farm practices. A prolonged drought in the Great Plains states in the 1930s was a spectacular example. Farmers thereafter were encouraged to reduce tillage and to make use of cover crops and mulches. Research continues today on living mulch/cover crop systems for protection of soil from wind (Lauer and Fornstrom, 1988). A key toward greater acceptance of limited tillage systems was the publication in 1943 of *Plowman's Folly* (Faulkner, 1943). Faulkner maintained that "no one has ever advanced a scientific reason for plowing." The acceptance of his ideas made possible a whole range of limited-tillage or conservation-tillage systems, including living mulch, which are now increasing in importance. Improved soil conditioning results from use of these conservation procedures.

A green manure is a plant material grown and then incorporated into the soil while green, or soon after maturity, for improving the soil (Soil Science Society America, 1973). A cover crop is one grown to protect the soil from weather. Pieters (1927) listed four general types of green manure and cover crops:

1. Main crop (green manure crop grown during the regular growing season)
2. Catch crop (green manure planted after the main crop)
3. Winter cover crop (this planting serves to cover the soil in winter, thus protecting it from erosion; it may be allowed to mature and be harvested)
4. Companion crop (what is called today a "living mulch")

A living mulch is a plant species planted at the same time as the main crop and allowed to grow up during the growing season between crop rows or even after the crop is harvested if desired.

2. Recent Trends with Living Mulches

One idea of a living mulch approach to reducing tillage is sod-planting. When cultivation is reduced, soil aeration and nitrification also decrease, and soil crusting and erosion usually increase. Sod-planting, in which row crops are planted into sod that has been killed or inhibited with herbicides, was developed to address these problems (Shear, 1985). This work led directly to the development of a concept of living mulch. The "sleeping sod" concept of producing crops in a grass sod that has been "put to sleep" with a selective herbicide is said to be a viable procedure (Lilly, 1965). Living mulches (or sleeping sods) were seen as a way to put highly erodible and otherwise untillable hillsides into production or to produce two crops in the same field in the same year: a summer row crop and a spring and fall pasture crop.

Conservation and soil improvement procedures eventually were looked upon as a hindrance to the efficient operation of mechanized farms. This has led to many problems as farms became larger and larger. As a result, in the 1960s, many concerned farmers joined the growing organic movement (Paine and Harrison, 1993). This was a call for returning to traditional farming practices involving reduced inputs with an emphasis on manures, legume crops, living mulches, and green manures. In the 1970s, the concerns of organic farmers really became concerns for all farmers because of the energy crisis thrust upon the world. A different and larger understanding of farm efficiency developed. The organic movement also has done much to focus needs for improvement of soil quality.

In research at Cornell University, Hughes and Sweet (1979) applied the concept of companion crops/living mulches to vegetable production. They really developed the name "living mulch." Many horticulturists believe that it is in vegetable production and other small-scale cropping systems that living mulches can be used most effectively. A living-mulch workshop, Crop Production Using Cover Crops and Sods as Living Mulches, was held in 1982 (Miller and Bell, 1982). A summary of progress was made in 1987 (William, 1987). In spite of much success, research results generally presented a mixed picture of the benefits and risks associated with living mulch systems. Living mulches do help to approach having a natural ecosystem and do provide some soil conditioning values.

Living mulches of course are best in areas having sufficient rainfall. A predictable living mulch is believed to be preferable to unpredictable weeds. Research on water and nutrient relationships with living mulches including use of legumes should continue (Longer and Huneycutt, 1995). A living sod mulch has proved useful in soil management and improving poor soils in pecan orchards of the southwestern USA (Miyamoto and Storey, 1995). Alley cropping is a form of living mulch.

B. Green Manure and Cover Crops

A green manure crop is grown and plowed under primarily for the purpose of improving the soil. A cover crop is grown primarily to prevent or reduce erosion (Rogers and Giddens, 1957). Green manure crops are usually annuals, either legumes or grasses. They add organic forms of nitrogen to the soil. They increase the general level of fertility by mobilizing minerals and building up levels of soil organic matter. They reduce losses from erosion. They improve the physical conditions of the soil. They conserve nutrients by cutting down losses from leaching. A disadvantage is that the cost of growing them may be more than the cost of the commercial nitrogen and soil organic matter they save. An increase of diseases, insects, and nematodes is also possible. Green manure crops may exhaust the supply of soil moisture. They may adversely affect the stand of the next crop. It is often said that green manure is not effective in regions with an annual rainfall of less than 50 cm. Green manure crops may have little influence on the level of soil organic matter if cultivation is continuous. Green, easily decomposable, organic materials added to soils have speeded up decomposition of the soil organic matter already present under some conditions. This seems to be the response of some natural ecosystems to fertilizer nitrogen (Wedin and Tilman, 1996). Green manure crops can be expected to have more effect on soil organic matter when the soil clays are the montmorillonitic type (expanding clay with high cation exchange capacity) than when they are of the kaolinitic type (Paine and Harrison, 1993).

Cover crops prevent leaching from soil of nitrogen and potassium and possibly other elements. Cover crops improve soil aggregation, porosity, bulk density, and permeability (Rogers and Giddens, 1957). A readily available source of food for soil microorganisms will result in an increase in soil aggregation. Soil aggregation is associated with the gums, slimes, and other products of soil microorganisms, while good ground cover reduces soil erosion because of the foliage. Soil loss, however, is greater on land that has an annual winter cover crop than on land on which corn stubble is left on the ground. Seeding without disturbing the surface residues may be the answer. Use of cover crops results in an increase in the rate at which water infiltrates into soil. Leaves and stems catch the rain, and the roots open channels for the water. Deep-rooted crops, such as some of the legumes, help open up soils with restricted subsoils. Bulk density of soils is lowered and porosity is enhanced in turn by the use of green manure crops (Rogers and Giddens, 1957).

In arid lands, cover crops may increase water use even though they increase water penetration (Holtz, 1995). There are benefits, however, that may be obtained by using mixed species as cover crops and rotating the kind of cover crops from year to year. Research on use of cover crops in vineyards is showing improvement in soil physical properties and water relations (Arcamonte, 1995). Benefits from cover crops in arid areas may take years to obtain. Cover crops can lead to soil improvement even though they change water requirements. Long-time studies are showing the values of cover crops on soil characteristics in arid lands (Katz, 1995).

C. Soil Acidity Produced by Leguminous and Other Plants and Fertilizer and Additional Changes in the Rhizosphere and Adjacent Soil

Many aspects of crop production are involved in cycling of protons or hydrogen ions, and on the various reactions that have bearing on soil acidity and the solubility of lime in soil (Wallace and Wallace, 1992). Protons exert various effects on soil quality, and their sources are of concern. Soil acidification can decrease soil quality as calcium is subsequently leached from the soil. In calcareous soils, however, the protons produced can release soluble calcium, which usually results in considerable soil improvement. Some plants whose roots produce and expel protons are very valuable in reclamation of sodic calcareous soils. The additional role of plants in such reclamation is discussed in Chapter 8. Their possible role in bioremediation is discussed in Chapter 21. Plants can be effective soil conditioners.

Greater uptake by plants of mineral anions (NO^-_3, $H_2PO^-_4$, SO^{2-}_4, Cl^-,SiO^{4-}_3) than mineral cations (K^+, Ca^{2+}, Mg^{2+}, Na^+, NH^+_4) results in a decreased acidity in the soil and in the root rhizosphere in particular (Haynes, 1990; Zaharieva and Romheld, 1991). The pH of soil increases. Monocot crops are in this category. A large crop of corn (maize) for this reason can sometimes add the equivalent of over 400 kg/ha base as $CaCO_3$ equivalent to the soil with removal of the grain only because of its nutrient supply. It is assumed the nitrogen was all taken up in the nitrate form and that there was no excess nitrogen fertilizer applied.

Greater uptake of mineral cations than anions by plants results in an increased acidity in the soil (Haynes, 1990; Peirre et al., 1970; Pierre and Banwart, 1973; Ugolini and Sletten, 1991; Wallace et al., 1976). Dicots in general are in this category, but not all are (Pierre and Banwart, 1973; Zaharieva and Romheld, 1991). Dicots with a marked tendency for cation uptake to greatly exceed anion uptake usually have a high oxalate concentration to maintain internal electrical charge balance. Examples are *Galenia* and *Atriplex* (Wallace, 1982). There is a wide range of variation for plant species, some with near zero proton release to soil and some with considerable, such as the equivalent of 500 or more kg/ha sulfuric acid per crop or life cycle of a crop if all the biomass is removed. Usually it is somewhat less.

When nitrate is the source of nitrogen taken up by plants, hydroxide or bicarbonate bases can be released from roots to the soil. In contrast, when legumes and other plants obtain nitrogen by symbiosis with biological fixation of N_2, there is a net result of increased release of protons to the soil (Liu et al., 1989). The net release can be equivalent to 150 or more kg/ha of sulfuric acid for the seed grain removed with soybeans to over 1000 kg/ha for a large crop of alfalfa removed from the field and without return of the manure from livestock (Armstrong et al., 1984). The differential cation–anion uptake, however, does not account for all of the proton release to soil for symbiotic N_2 fixation. That part of the fixed nitrogen that remains in the soil as root residue or straw turned under, which originated from symbiosis and depending on the mineral cation content, is also a potential source of acid when the organic matter is mineralized. This can add approximately another 200 kg/ha sulfuric acid equivalent for soybeans and over 400 more for alfalfa per year. On many soils legume crops do require considerable liming if the soil is not calcareous.

Excess anion over cation uptake, which is very pronounced with rice, involves considerable anionic silicon and results in alkaline reactions in the rhizosphere (Wallace, 1993). Both acidification and alkalization of the rhizosphere are adaptations that enhance the availability of iron and perhaps other micronutrients for different groups of plants but with different mechanisms. Dicot plants in general and with or without symbiotic nitrogen fixation not only have excess cation uptake but also, under iron stress conditions, develop redox mechanisms and may even excrete reductants to the rhizosphere that help plants to absorb iron (Marschner et al., 1986). With rice and other monocots, a different mechanism for alleviating iron stress is used. Such plants produce and excrete to the rhizosphere a type of chelating agent called phytosiderophores (Marschner et al., 1986). The alkaline reaction that is developed helps those agents to chelate iron, and the chelated iron molecules are absorbed by the roots of many monocots.

Plant species at least under some conditions can have profound and differential effects on the chemistry of the rhizosphere in addition to changes in soil pH. The changes induced also have conditioning effects on the localized area. Organic acids can be present as root exudates in the rhizosphere. These can arise through dark fixation or nonphotosynthetic fixation of carbon dioxide via the phosphoenolpyruvate carboxylation reaction. These organic acids can be substrates for soil microorganisms and could have some soil conditioning value. A more important function of some microbes is their effect on solubilization of soil phosphorus to make it more available for plants (Johnson et al., 1996a,b). These workers report that under phosphorus deficiency conditions, the activity of an associated enzymatic reaction in the rhizosphere is increased several fold to make phosphorus more available.

Reductants, organic acids, and phytosiderophores are not the only organic substances that plant roots expel to the soil. A large number of compounds is involved and they have considerable influence on soil quality and stability. Root exudates enhance the growth and production of microorganisms. Root exudates also provide organic matter for soil stabilization in the rhizosphere.

As mentioned above, protons are added to soil when soil organic matter is mineralized. Although not a reaction confined to the rhizosphere, it is important. The nitrogen is released as NH_4OH or equivalent, which means that soil acidification can result. The NH_4OH can be oxidized to HNO_3 (Mengel and Kirkby, 1982). For a soil to lose 1% soil organic matter in the top 30 cm of soil, the equivalent of 9 MG/ha of H_2SO_4 can be released. This may occur gradually over several years but can add to the lime requirement. That part of the NH_4OH which represents net plant uptake or that part of the HNO_3 which represents net plant uptake will decrease the acidity load on the soil resulting from mineralization of soil organic matter if the crop is removed (harvested) and not allowed to decompose (mineralize) in the soil. Any HNO_3 leached to groundwaters also will decrease the acid load on the soil. The denitrification process also decreases the acid load on the soil as HNO_3 is converted to water and gaseous products.

Manure that is returned to the land will have an acidification capacity when mineralized that will be equivalent to its nitrogen content less the differences in plant uptake between mineral cations and mineral anions derived from the manure. This may be in the order of 50 or more kg H_2SO_4 equivalent per MG of manure. Use of poultry manure from laying hens will be different because of large quantities of $CaCO_3$ in the manure. Partially mineralized organic matter may additionally give some acidity due to formation of organic acids.

An increase in level of soil organic matter will consume protons in the form of amino acids or amides synthesized with nitrogen, and the proton potential will remain there stored until the organic matter is mineralized. Any associated bases would be leached from the soil with time. The proton potential on subsequent breakdown of the organic matter is equivalent to the theoretical (3.5 kg H_2SO_4 equivalent for each kg of nitrogen).

Fertilizers also have potential for adding protons and bases to the soil depending upon cation or anion uptake by plant roots or microbes (Stumpe and Vlek, 1991). The potential comes from reactions in soil as well as from plant uptake of cations and anions. Excess of nitrogen fertilizer beyond plant needs can add more protons than the averages used by the fertilizer industry, which is 1.8 kg H_2SO_4 equivalent for each kg of nitrogen (Wallace, 1994c). This average is only half the theoretical. Excess nitrogen can acidify soil at the full theoretical rate. Examples of acids resulting from microbial reactions involving fertilizer in soil are

$$(NH_4)_2SO_4 \xrightarrow{O_2} 2\ HNO_3 + H_2SO_4 + 2\ H_2O$$

$$\text{urea} \xrightarrow{O_2} H_2O + CO_2 + 2\ HNO_3$$

$$NH_3 \xrightarrow{O_2} HNO_3 + H_2O$$

$$(NH_4)_2HPO_4 \xrightarrow{O_2} 2\ HNO_3 + H_3PO_4 + 2\ H_2O$$

The acidifying power of these nitrogen fertilizers is partially neutralized when plants take up the resulting NO_3^- ions or both NO_3^- and H^+. Calcium nitrate and potassium nitrate fertilizers can have a proton-decreasing effect on the soil as a result of anion uptake exceeding cation uptake or with denitrification. These modifying effects of plants on acidity resulting from fertilizers are absent when fertilizers in excess of plant needs are used. The full theoretical acidity results from any excess (Wallace, 1994c).

Lime addition to soil decreases the proton levels. Ash residues from coal or wood burning also do the same. See Section III.

Crop rotations can alternate cycles of proton addition and removal from soil. Alternation of grass crops (for example, corn or maize) with legumes (for example, soybeans) can keep the proton cycle in some balance unless interfered with by excessive amounts of fertilizer or by lime. Continuous corn may have little net effect on the proton balance if the nitrogen source is mineralized soil organic matter, even if the grain is removed, because the acid from the nitrogen in the soil organic matter will balance that lost from greater anion uptake and over cation uptake and crop removal. Nitrogen fertilizer as urea or NH_3 can have the same effect with corn. Weeds with differential cation–anion balance may have a significant effect on the proton relationship for a given crop and soil. Type of weeds and type of crop make differences.

Acid rain also can add significant quantities of protons to soil (Office of Technology Assessment, 1984; Likens et al., 1996) and may require correction with lime. All these effects are important and often require use of soil conditioners for their modification.

D. Root Systems Increase Water Permeability

Many desert soils on which irrigation is used have smectic mineralogy, which makes them subject to swelling especially when irrigated with high-salt water (Mitchell and Van Genuchten, 1993). Problems with desert soils include inadequate soil moisture in the root zone, inefficient water use, plant injury due to water ponding, and salinity (Mitchell et al., 1995). On nonswelling sandy soils, decaying roots can form macropores that result in satisfactory water infiltration (Edwards et al., 1979; Davidson, 1985). In such soils decaying roots may leave long, continuous pores in soil that enhance water and solute movement. In swelling soils, roots do not easily do this.

Roots of various plant species do not all have the same effect on water flow in soil. Alfalfa is exceptionally good in promoting water flow even in swelling clay soils. Mitchell et al. (1995) attribute the effect to the root morphology of alfalfa and to the fact that alfalfa is a permanent crop with 3 to 5 year production cycles. Alfalfa has a large, long, almost straight tap root. At the soil surface alfalfa has a fleshy crown from which many stems originate. When the plant dies and decays, there is a deep extended flow path that connects with the soil surface. The studies of Mitchell et al. (1995) involved dye-stained pores in swelling soil to identify pores. Alfalfa produced stable macropores along both living roots and decaying root channels. Wheat did not do this. Also cracks and earthworm holes did not remain open during irrigation of the clay soil that was used. The authors proposed a requirement for decaying woody matter for the root pores to remain open under irrigation. Mannering and Johnson (1969) earlier had reported greater water infiltration under a soybean crop (a legume) than under a corn crop (a monocot like wheat).

Another explanation may be possible for the presence of more stable macropores with legumes than with nonlegumes like wheat and corn. The dinitrogen fixation associated with legumes results in release of protons (see Section C above). The protons react in most desert soils with soil lime to release calcium ions. The calcium in turn can either replace sodium on soil colloids, and the sodium be leached downward to give a more stable calcium soil, through which water can penetrate, or the calcium can help to stabilize and fix onto soil the organic matter released in the decay process (Wallace and Wallace, 1994a). Both these effects with legumes, which stabilize the soil around holes made by roots, can result in a more stable channel through which water can flow.

E. Trees Help Stabilize Soil

Trees are especially beneficial in stabilizing soil. Oak trees are a good example. "Oaks play a role in stabilizing soils, cycling soil nutrients and enhancing water quality and storage" (California Agriculture, 1995). Living plants of all kinds produce substances that help to stabilize soil properties in addition to the mechanical strength they add to soil through extensive root systems. Trees in windbreaks, along streams, and elsewhere do much for soil improvement (Tilton, 1997). They root deeply. It has been postulated that one reason why small farms produce more per unit of land than large farms is that the small farms average more trees per ha do than do large farms (Wallace, 1994b). In contrast to these effects, some conifers with certain soil types can result in decreased water permeability (Chappell et al., 1996).

II. PEAT AND SPHAGNUM MOSS

Sphagnum peat moss is commonly used as a soil conditioner. Peats generally contain some organic matter that still has some original structure of the plants from which they were derived. Peats are supposed to contain more than 50% organic matter to be legally marketed.

The most important kind of peat moss is sphagnum. More than 350 species of sphagnum exist, and most are in swamps of the North Temperate Zone. Sphagnum grows in dense mats and is soft and spongy. It has no true roots but the walls of stems can store large quantities of water. This makes peat a valuable soil conditioner.

Sphagnum peat moss is low in nitrogen and quite resistant to decomposition. Its use as a soil conditioner is somewhat frowned upon by the environmental community because it represents overexploitation of a natural resource.

Peat moss of the sphagnum type obtained from Canada, Europe, and some parts of the northern United States is useful. California hypnum peat moss has proved to be satisfactory for horticultural use. In general, other types of peat are too uncertain in chemical and physical composition to be included in standard plant growing mixes. For instance, many of the sedge peats of arid areas (for example, the black peat of coastal California) are saline. Some are infested with undesirable organisms.

Peat can be used on sandy arid lands that are not very conducive to cropping without soil conditioners. Peat provides acidity, water retention, biological activity, and some cation exchange (Takamiya, 1994). Peat products can contribute to soil aggregate stability (Almendros, 1994).

Peatlike products can be produced by composting plants and manures (Bakr, 1994). These are acceptable substitutes.

III. WOOD ASHES

Waste products of many kinds are finding uses as soil conditioners (Karlen et al., 1995). Wood ashes are frequently used by homeowners in their gardens, often without regard for what they may be doing to soil pH (Organic Gardening, 1996). Some gardeners have damaged their soil by overapplication. Wood ashes are a good liming material and also a successful source of several different plant nutrients when use is guided by soil analysis.

Production of wood ash from commercial sources in the USA is approximately 1.3 to 2.5 × 10^6 MG per year (Campbell, 1990). Nationally, approximately 90% of wood ash is disposed of in landfills. In the northeastern USA, 80% is land-applied (Campbell, 1990) largely as an agricultural liming and a nutrient source. Wood ash is used to cover landfills, it is added to leachates to retard chemical mobility of potentially toxic substances (Gray and Rock, 1987), and it is used as a bulking and odor control agent in sludge-composting facilities (Logsdon, 1989; Hart, 1986). Wood ash can help stabilize sewage sludge to be used on soil (Logan and Burnham, 1995). Recently wood ash has been tested in gravel road construction.

Most of the previous research in the USA on land application of ash has been done on agricultural soils. Wood ash is a significant source of the nutrients phosphorus, potassium, magnesium, calcium, and lime (Ohno and Erich, 1990; Naylor and Schmidt, 1986; Lerner and Utzinger, 1986). The neutralizing potential of wood ash typically ranges from 50 to 100% of that of $CaCO_3$ on a dry weight basis (Naylor and Schmidt, 1986; Lerner and Utzinger, 1986). Soil pH may be increased for 2 yr or more (Naylor and Schmidt, 1989) after an agricultural application. Rates of application on agricultural soils are as high as 40 MG ha^{-1} $CaCO_3$ equivalent (Naylor and Schmidt, 1989). The primary objection typically cited for repeated applications on agricultural soils is the potential accumulation of heavy metals (Lerner and Utzinger, 1986).

Application of alkaline wood ash to an acidic forest soil resulted in increased exchangeable nutrient cations and higher pH, and decreased available aluminum concentrations within the rooting zone (Kahl et al., 1996). These changes are favorable and can reverse effects of acid rain, which often results in lower amounts of new plant biomass and in erosion of soil.

IV. HUMATES FROM OXIDIZED LIGNITES

Use on farms of humates derived from oxidized lignites has had a history of turmoil. Stevenson (1979) outlined some of the problems and objections. He did recognize, however, that humates could have value on lawns, turf, and with certain problem soils, especially sandy soils.

Commercial humates are defined as oxidized lignites or products derived from them. Oxidized lignite is an earthy, medium-brown, coallike substance associated with lignitic outcrops. Oxidized lignites typically occur at shallow depths, overlying or grading into the harder and more compact lignite, a type of soft coal. Oxidized lignites are undesirable as fuel because of low heating value. While normally discarded during mining, they have been utilized in a limited way in industry. The experimental use of "coal humates" as soil additives dates back at least six decades. A unique feature of oxidized lignites is their unusually high content of humic acids, which is in the order of 30 to 60%.

The term humic acid is used to describe a brown-to-black organic colloid extracted from soils, sediments, and other geological materials with dilute alkalis and precipitated by addition of acid. Humic acids are normally identified as to source (e.g., soil humic acid, peat humic acid, or lignite humic acid). Mined lignites have low water solubilities, the insolubility being due to the tie-up of humic acids with mineral matter. Some oxidized lignites are essentially salts of humic acids mixed with mineral matter. The original mined product is crushed, pulverized, and usually fortified with commercial fertilizer. This makes products easier to register with state agencies. The material itself does not contain significant amounts of nitrogen, phosphorus, or potassium but is assumed to have a favorable effect on soil properties.

Promoters of coal-derived humates often give the impression that humates from oxidized lignites have biological and chemical properties similar to the humus in soil. This is not necessarily the case according to Stevenson (1979). Coal humates can be considered as an advanced stage in the transformation of humus to fossil forms like coal (Fig. 1). The major changes include losses in carbohydrates, proteins, and other biochemical constituents and an increase in the oxidation state of the humic material. Thus, unlike soil humus, coal humates are essentially free of such biologically important compounds as the proteins and polysaccharides; furthermore, they contain little if any fulvic acids. Lignite humic acids are not entirely like soil humic acids; they have higher carbon but lower oxygen contents, and they may have more highly condensed chemical structures. Since coal humates bear only a superficial resemblance to the humus in soil, they cannot be expected to behave in the same way or to perform the same functions (Stevenson, 1979).

The only way commercial humates will permanently increase the cation exchange and buffering capacities of the soil, or, for that matter, materially improve soil physical conditions, is to build up the soil humus. However, at the rates at which commercial humates are normally applied (500 to 1000 kg per

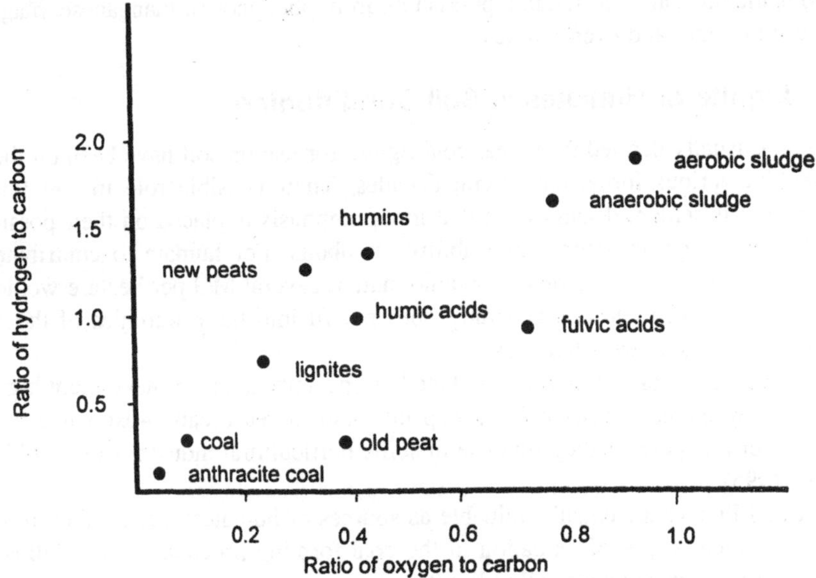

Figure 1 Oxygen, carbon, and hydrogen ratios in various substances used as soil conditioners.

ha), the increases will be small. High application rates of commercial humates would be required to modify significantly the physical conditions of most agricultural soils.

The following terminology is variously used in the humate-lignite industry:

Humus: Unstable, biologically active organic material made up of decayed and decaying plant and animal matter containing most of the biological compounds synthesized by living organisms.

Humate: A geologic term referring to stable, naturally occurring, highly compressed, highly decomposed humus in a matrix of sand, clay, or silt.

Humic acid: The largest and most active fraction of humus and humate. Although it does not actually enter the cells of plant roots, humic acid supposedly will "open" the cell wall, making it more permeable and allowing nutrients to enter more easily.

Humic acids (another definition): The chemical term referring to the collective acid radicals with hydrogen ions attached to the ion exchange sites.

Humic substances: The complete range of organic acids, hormones, organic compounds, and complex molecules found in humus and humate.

Humates: A chemical term referring to the salts of humic acids that hold ions other than hydrogen on the ion-exchange sites. Examples of these held as ions include calcium, sodium, potassium, iron, phosphorus, manganese, magnesium, zinc, and even nitrates.

A. Lignite or Humates in Soil Conditioning

Humates usually derived from near-coal lignite for use on soil have been on the market in various forms for several decades. Their possible role in soil improvement is seldom discussed, in that more emphasis is placed on their possible biological growth stimulating ability on plants. For humate to contribute significantly to the supply of soil organic matter, several MG per hectare would need to be applied and this generally does not fit into the economics of those who prepare and market humates.

Humates are more commonly marketed as specialty items to make them high value. They increase water-holding capacity and increase cation-exchange capacity. For this purpose they find use in some horticultural industries (Reynolds et al., 1995).

Not all lignites are equally valuable as sources of humates because they represent different stages of oxidation in the coal-forming process. Figure 1 illustrates variation in products called lignites.

Howard E. Turrel (personal communication, 1995) studied humate derived from lignite for about 20 years in efforts to commercialize with integrity a large mine of lignite-humate held by his company. His goal included producing materials comparable in cost with nonhumate products and which would be easy to use on farms. Of course performance had to be acceptable. He outlined several benefits obtained. Some of these must relate to effects of physical and biological properties of soils, and certainly more research is needed. The benefits outlined included the following Seed germination time was shortened by 12%; root systems were larger and had greater mass; nutrient uptake was better; micronutrient uptake was greater; moisture requirements were reduced 50% over use of chemical fertilizers; crop yield increase over chemical fertilizer was 8% to 10%; there was no environmental leaching of nitrogen and phosphorus; NPK recommendations were reduced; crop sizes were more uniform; crop sizes were 7 to 40% larger than with equal chemical nutrients; turf was thicker and greener and had stronger root systems; and fruit was larger and a better grade overall. He was comparing humate-nutrient blends with equivalent chemical fertilizer blends.

Viable research with humates can include interactions with other soil conditioners (Wallace and Wallace, 1986). Use with water-soluble polyacrylamide is an example.

B. Some Humate Uses on Turf

Humates are often applied during core aerification on golf courses (Powell, 1995). They are either topdressed before aeration or applied after so that the aerator holes are filled with the humate. Aeration, water permeability, and soil biology are improved.

Humates also improve turf soils that are very sandy. Increased cation-exchange capacity and water-holding capacity result in good root growth and healthier turf (Powell, 1995). In soil preparation for new turf, incorporation of humate in the top few cm resulted in more effective turf establishment (Powell, 1995).

C. Humates as Nutrient Carriers

Humates from lignites are commonly used as carriers of nutrients in fertilizers. A study was conducted some years ago with a 6-3-0 fertilizer formulated with humates from leonardite (Wallace and Khadr, 1966). Relative yields for tobacco plants for a control, the 6-3-0 humate, ammonium nitrate, and half each for the two sources of nitrogen (with equal amounts of nitrogen in all cases except for the control) respectively were 100, 173, 184, 195%. The results among the three treatments receiving nitrogen were not significantly different from each other but may be important.

Some of the benefits of humates may be in their effect on plant roots rather than on the soil. David et al. (1994) reported an effect on increasing the permeability of roots themselves so that nutrient uptake for several different nutrients could be enhanced.

A possibly important effect of humates in the composting process has been reported by Whiteley and Pettit (1994). Lignite treatment of decomposing wheat straw reduced both oxygen consumption and carbon dioxide evolution. The procedure may be used to control rates of composting or the stability point of composting. Compost stability may be reached sooner with less loss of the carbon and/or nitrogen with use of lignite humates. Considerable studies would be needed to establish a workable procedure. Compost-stabilizing agents are being studied elsewhere (Riggle, 1995b).

V. ZEOLITES AND RELATED SUBSTANCES

Zeolites are secondary minerals occurring chiefly in igneous rocks. They occur where igneous rocks were subjected to the action of steam or hot circulating water, which attacked the feldspars and feldspathoids (Pirsson and Knopf, 1947).

Zeolites have high cation-exchange capacity and have internal space or pores (4–6 angstrom units) in crystal lattices that can hold water and nutrient ions. Zeolites have many industrial uses, including building materials, municipal water purification, decreasing ammonia concentrations of effluent water, in detergents, and in feeds for poultry and other animals.

Mumpton (1984) has discussed the structure of zeolites: "Zeolites are crystalline, hydrated aluminosilicates of alkali and alkaline earth metals, having infinite, three-dimensional atomic structures. They are further characterized by the ability to lose and gain water reversibly and to exchange certain constituent atoms, also without major change of atomic structure. Along with quartz and feldspar minerals, zeolites are three-dimensional frameworks of silicate as its central atom, the overall structure is electrically neutral, as is quartz (SiO_2)." In zeolite structures, some of the quadricharged silicon is replaced by triply charged aluminum, which gives rise to a deficiency of positive charge. The charge is balanced by the presence of singly and doubly charged atoms, such as sodium (Na^+), potassium (K^+), calcium (Ca^{2+}), and magnesium (Mg^{2+}), elsewhere in the structure. These ions are exchangeable. The empirical formula of a zeolite is of the type

$$M_{2/n}O \cdot Al_2O_3 \cdot xSiO_2 \cdot yH_2O$$

where M is any alkali or alkaline earth atom, n is the charge on that atom, x is a number from 2 to 10, and y is a number from 2 to 7 (Mumpton, 1984). The chemical formula for clinoptilolite, a common natural zeolite, is

$$(Na_3K_3)(Al_6Si_{40})O_{96} \cdot 24H_2O.$$

The exchange process involves replacing exchangeable atoms in the zeolite by atoms from the solution. The magnitude of such cation exchange in a given zeolite is known as its cation-exchange capability (CEC).

Most of the research and development work on the use of zeolites in soil has been done in nations outside the USA. Postnikov and Illarionova (1990) made a review in Russian with 43 references on zeolite use for soil conditioning, for ammonia absorption from manures, and for regulation of nutrient uptake by plants. In 1991, Postnikov et al. (1991a,b) reported on uses of zeolites for vegetable crops and for use in greenhouses. Yasuda et al. (1995) recently showed that it took from 20 to 30% of zeolite to increase sufficiently the water-holding capacity of a dune sand.

Several different workers have tested zeolites as constituents of golf course greens and tees (Ferguson and Pepper, 1987; Petrovic, 1990, 1993; Watanabe et al., 1994; Ferguson et al., 1986; Barbarick and Pirela, 1985). Drainage and aeration are improved, there is greater compaction resistance, and pesticides and fertilizers less readily leach from soil.

Torri (1978) and Zhang et al. (1994) have reported beneficial effects on yields of field crops with zeolites used as a soil conditioner.

Diatomite is a soft, white, very light rock composed of innumerable microscopic shells of diatoms. Diatoms are exceedingly minute one-celled aquatic organisms (Pirsson and Knopf, 1947). Diatoms proliferate in incredible numbers, as veritable diatom epidemics, and their shells, accumulating at the bottoms of lakes, give rise to extensive deposits. Most varieties of diatomite are white, but some are pale yellow, brown, or gray. Diatomite has a low specific gravity (about 0.4). Loose, scarcely coherent material of this type is called diatomaceous earth; the more coherent material is called diatomite. Beds of fresh-water diatomaceous earth of considerable magnitude occur in many places in the United States and elsewhere.

Diatomaceous earth has many uses as a soil conditioner in landscape and turf programs. Frequently it is calcined (heated), which probably transforms it into a zeolite. Several different companies produce various forms of diatomaceous earth products for soil use. Diatomaceous earth is frequently used as a topping for golf course greens to improve physical properties of soil (Maddox and Goatley, 1995). Pelletized diatomaceous earth has been tested as a means of increasing water retention on golf course greens (Fargerlund et al., 1995). When greens are built with sand they have very little water-holding capacity, poor aeration, too much compaction, and high incidence of disease. Porous soil conditioners plus even with water-soluble polymers (see Chapter 19) could help make better greens on golf courses.

Several different companies are now producing zeolite and diatomaceous earth products for use on turf. Mannion (1996) says that both resist compaction, promote good drainage, and translocate moisture and nutrients deep into the profile. Both result in healthier turf, withstand stress better, and give more uniform turf.

Pozzolan is very much like diatomaceous earth. Pozzolan is formed from volcanic action. Very often pozzolan from volcanic action and diatoms occur together and they become a common product. Pozzolan also has characteristics of zeolites, although the bulk density and specific gravity are much lower. Much is being shipped from the USA to the Middle East, where crops are grown, for use on sandy lands to increase water-holding capacity and to decrease the water infiltration rate, which is much too high in very sandy soils. There is also the potential for pozzolan to increase porosity and aeration in soils high in clay.

VI. SOIL WETTING AGENTS, WATER PENETRANTS

Agents of concern here are those commonly used in soil to enhance water penetration. Several companies make these products. In recent years the scientific literature has said little about them because most of the relevant research was done several years ago. Letey (1975) reviewed the subject, but there has been much progress since then. Gypsum with or without use of water-soluble polyacrylamide (Chapter 15) usually does much better than wetting agents do in improving the physical properties of soil, but gypsum may take longer to give needed effects.

The wetting agents function by decreasing the surface tension of water so that water can penetrate into compacted soil or into soil with very small micropores. This can be beneficial on land where existing plants are growing because water can penetrate deeper into the soil. The wetting agent really does nothing to improve the physical properties of soil. Aeration of soil is not improved with this mechanism. Wetting agents are not able to prevent crusting of soil, but they can increase seed germination and emergence because of better water infiltration and leaching of salts.

We observed some years ago that some wetting agents used on soil can exert a slight toxicity on certain plants (unpublished). A producer of wetting agents from natural products has said that synthetic wetting agents dissolve and encourage loss of soil organic matter with breakdown of soil aggregates. If this is the case, then the improvement from wetting agents must be temporary only, and the need for more effective means of soil improvement is indicated.

One producer of wetting agents for soil says that the poorer the soil, the greater is the percent response after use of the agent. This includes both yields and water savings. Ten percent increases for both have been noted. These results are logical and understandable, but greater results have been reported with the use of gypsum and other materials that actually improve both the soil and the plants in addition to the effects on water movement.

Koski (1996) has recently summarized the information on the use of wetting agents, especially in turf. He says that the time-honored method for dealing with such problems as hydrophobic conditions, stratified layers in root zones, dry spots, poor water penetration, and other problems is the use of wetting agents. He says that when the wetting agent reduces the attraction of water molecules for each other, water is able to spread out more evenly and to move more quickly through small pores and across boundaries.

He emphasizes, as a "point of caution," that wetting agents should not be considered "miracle cures." They do not reduce compaction or affect plant water-use rates. They cannot replace basic cultural practices. Some of the proven advantages include improved water movement in soil, especially in layered soils, and rewetting of hydrophobic root zones. Potential negative effects include phytoxicity when applied to stressed turf or if not properly watered in, root injury, increased thatch accumulation (the healthier the turf, the more thatch it forms), and deflocculation (dispersion) of soil particles. Deflocculation of soil is a potential problem with long-term use of excessive amounts of wetting agents. Regular use of wetting agents is important, however. They do not eliminate the problem but only temporarily modify it. Regular use of wetting agents enhances water infiltration and drainage and results in more efficient water use, in fewer overly wet or dry spots, and in better quality turf.

The active ingredient in at least some soil wetting agents is alkylphenolhydroxypolyoxyethylene, which is nonionic. Another is alkyl naphthalene sodium sulfonate. Some companies blend other products with wetting agents to give claimed multiple benefits. This procedure adds to the cost. Perhaps the bottom line as to the value of wetting agents in farm use is whether or not they are cost effective or are more cost effective than other ways for increasing water penetration. Since gypsum, water-soluble polyacrylamide, and some related products give multiple values to the land, they can be quite cost effective. Wetting agents can improve hydrophobic soils, but gypsum can also.

New wetting agent chemistries have been developed in recent years, and the problems of the past are claimed to be overcome. It is called a technology explosion by some. Manufacturers say that uses on turf can now prevent water-related problems and water repellency before they occur. Basic studies of the water flow through soil have led to new products that are supposed to give uniform water flow through soils. Nutrients as well as water also flow uniformly into the root zone. Wetting agents are now used on overseeding of turf, which helps germination and establishment. Users say they no longer have to hand water localized dry spots. Wetting agents are now available in a variety of forms—liquid, granular, and pellets.

A relatively new use of surfactants or wetting agents as soil conditioners is in bioremediation (Jahan et al., 1995). A biologically derived product is also being used for the same purpose (Robinson et al., 1995). See Chapter 21.

VII. SEAWEED PRODUCTS

Most uses of seaweed products are as biostimulants, but there are some claimed effects as soil conditioners (Smitte, 1991). Composts, which can be effective soil conditioners, are claimed by some also to contain substances that have biostimulating effects, and this is especially claimed for composts containing seaweed products (Smitte, 1991).

Alginic acid, a polyuronide, is a component of seaweed that can increase soil structure by increasing the amount of aggregation (Quastel, 1954; Willis, 1996). A typical analysis of a seaweed showed 56–62% carbohydrates with 26.7% of the carbohydrates as alginic acid (Willis, 1996). Alginate applied at a rate of 0.1% of dry weight of soil markedly improved its structure (Quastel, 1954) but did not appear to be cost effective for field work. Seaweed products contain some compounds identified as growth regulators.

VIII. SOLUTION GYPSUM

Gypsum (see Chapter 7) is traditionally applied to land as powder or granules. Wildman et al. (1988) suggested use of calcium nitrate in irrigation water as a source of calcium, which use was intended for soil improvement to increase water infiltration. Since then several different commercial units have been developed for application of gypsum in irrigation water. This procedure is more economical than using calcium nitrate for this purpose.

At least nine good reasons exist for using gypsum in solution in irrigation water (Traynor, 1986; Wallace and Wallace, 1987; Wallace, 1994d):

1. To increase the solute concentration of irrigation water that is too pure to give good water infiltration into soil (EC of 0.2 dS/m or less)
2. To decrease the sodium adsorption ratio (SAR) of saline irrigation water including reclaimed water so that the water does not result in sodicity of soil by increasing the exchangeable sodium percentage (ESP)
3. To avoid unsightly granules of gypsum lying on turf of golf courses
4. To provide solute calcium to fruit crops in order to avoid low-calcium fruit disorders, such as blossom-end rot of tomatoes and bitter-pit in apples. Over 30 such disorders have been reported (Conway et al., 1994)
5. To preserve effects of soil stabilization created by water-soluble polyacrylamide. See Chapter 15
6. For convenience in use of gypsum
7. To decrease the length of time needed for response to gypsum
8. To increase uniformity in application of the gypsum
9. To decrease existing ESP of the soil surface so that better water infiltration can be achieved (Shainberg et al., 1989)

Not all of these reasons relate to soil conditioning.

Those who have to irrigate with water that is too pure (salt concentration less than around 200 mg/liter or about 0.2–0.3 dS/m EC) are finding that gypsum in the irrigation water gives far better water penetration and less degradation of soil structure (Peacock, 1994; Target Biofiles, 1993). Some workers suggest that gypsum be added to increase the EC to 0.5 or even 1.0 if the SAR of the irrigation water is higher than 7.

When gypsum is used in irrigation water, it is applied at rates of usually between 2 and 10 me/liter. Good grade gypsum that is quite pure and low in $CaCO_3$ is soluble to about 28 me/liter with an EC of about 2.8 (25°C). For gypsum to go readily into solution, it must be finely ground usually to minus 200 to minus 300 mesh size (78 to 118 openings per cm). One producer is grinding to minus 350 mesh to aid solubility and to decrease wear of sprinkler orifices when undissolved gypsum goes through them. The several different machines developed by various companies usually first make a slurry and then inject the slurry into the irrigation stream. Phosphogypsum is generally easier to put into solution than mined gypsum, but phosphogypsum has not been available recently in the USA because of EPA regulations. See Chapter 7. Solution-grade gypsum does cost more than regular-grade gypsum when phosphogypsum is not available.

Fulton (1994) reported that the use of solution gypsum applicators was growing in acceptance and was first introduced commercially about six years previously in California. It was even more popular two years later (Cline,

1996). In 1996 there were at least seven companies marketing both solution-grade gypsum and dispensing machines.

The procedure is driven locally by the fact that over 20% of the irrigated land in California is subject to severe limitations of water penetration, especially as the growing season advances. Losses in 1994 for this factor were estimated at $500–1200 per hectare for trees and vines and $500–750 per hectare for cotton, alfalfa, and other crops (Fulton, 1994).

Water infiltration problems have been increasing in Western states for the past 50 years or so (Fulton, 1994). In some places soil texture and mineralogy were not conducive to good water infiltration, but matters have since become worse with farm traffic, loss of soil organic matter, inefficient irrigation practices, development of subsoil compaction layers, and salinity and lack of it in irrigation water. Use of solution gypsum can help improve these situations.

The EC (electrical conductance) and SAR (sodium adsorption ratio) are good indexes for determining the need for solution gypsum in irrigation water. Usually it is appropriate to add enough calcium to irrigation water to equal the amount of sodium present (Fulton, 1994). It is also appropriate to add enough gypsum so that the EC is increased to at least 0.4 or 0.5 dS/m or more under some conditions of irrigation water (Wildman et al., 1988). If the irrigation water is high in bicarbonate it may be well to consider use of acids in the water (Chapters 8 and 9).

There has been success with the addition of both gypsum and water-soluble polyacrylamide together in the irrigation stream. In a test in Washington state with potatoes, water-soluble polymer was applied to 25 hectares through a center-pivot irrigation system with materials costing about $40 per hectare and involving 3.3 kg of the polymer in a ha 2.5 cm water (Wallace and Wallace, 1990a). See Chapter 17. An equal area was a control. Gypsum was applied simultaneously. The grower obtained a yield increase of 14.5 MG per hectare, a 25% increase. Potato sizes were increased to give better quality. The grower earned about $25 for each $1 for materials. Water-use efficiency was increased in the potato test. This is crucial in drought-prone areas of the arid west. With suitable soil conditions, water-soluble polyacrylamide can, with gypsum also in the solution, increase water-use efficiency by increasing the yield, by increasing the permeability of soils to prevent water runoff, by creating a protective mulch to decrease evaporative water loss from soil, by encouraging deeper rooting for more efficient use of water stored in soil, by enhancing seed emergence so that less water is used to bring on germination, by earlier closing of the canopy between seed rows to decrease evaporation, and by using less water to control dust on roads and fallow ground. See Chapters 15, 17, and 19.

Zhang and Miller (1996) used water-soluble polyacrylamide in solution together with 5 me/liter gypsum to decrease soil surface sealing and crusting on an acid soil in Georgia (USA). They sprayed the solution onto the land at the rate of 15,000 liters per hectare for efficient control with 15 kg/ha of the polyacrylamide.

In November 1991 a windstorm blew dust across the I-5 freeway near Bakersfield, California (Wallace and Wallace, 1994a). A massive car collision resulted in which many people were killed. Dust, however, did not blow off a certain farm that had been treated with water-soluble polyacrylamide and gypsum together in solution, because of improved soil structure. Additionally, seed emergence was aided, growth of the vegetable crop was increased, and clay clods were prevented from adhering to the crop when harvested. The advantages were multiple. In the 9.3 ha test, the grower obtained $8 for every $1 invested in water-soluble polacrylamide and gypsum.

Solution gypsum perhaps with water-soluble polyacrylamide should be used wherever there is a need for wetting agents.

IX. SHREDDED RUBBER TIRES

Between 2 and 3% of the rubber tires being recycled become used as a soil conditioner for turf and other athletic fields (BioCycle, 1996). Metal-free rubber near the surface of the soil is believed or claimed to have a resilient effect that protects athletes from injury (Riggle, 1995a). Incorporation of several MG per hectare of granular rubber is necessary for the desired effect, but since rubber is essentially nondegradable, a treatment will last for many years. Rubber tires contain appreciable amounts of zinc, some of which should be removed from shredded rubber before plants can be safely grown in it. Rubber granules are also being used as a recycling bulking agent in composting (BioCycle, 1996). Although the rubber permits good aeration, it does nothing for the carbon: nitrogen ratio generally expected of a good bulking agent for cocomposting.

X. SEWAGE SLUDGE (BIOSOLIDS) AS A SOIL CONDITIONER FOR LANDSCAPING

About 8 million MG of dry sewage sludge (biosolids) are produced each year in the USA (see Chapter 5). When reasonably free of pollutants, these waste products are useful as a source of plant nutrients and as a soil conditioner (Clapp et al., 1994). Some states are able to use most of their sewage sludge on their own land while other states ship it out of state to be used on land elsewhere (Walker, 1994). Much of the sewage sludge in the USA, however, is still landfilled, incinerated, or put in bodies of water (Oberle and Keeney, 1994). Better land use of sewage sludge needs to be made if full recycling is to be achieved. Those concerned with management of sewage sludge hope that much of it can be used in landscaping especially after cocomposting it with municipal solid waste.

Problems and methodology concerning the use of sewage sludge products or other organics for farms differ from those for home and commercial landscaping. Farm use has concentrated mostly on supplying the nitrogen needs of a crop, and the term "agronomic rate" is used. Landscape use involves much more. The need for improvement in the physical and biological properties of soil generally requires application rates much higher than the agronomic rate. Uncomposted sewage sludge is quite satisfactory to supply the agronomic rate on farms as long as pathogens, heavy metals, and toxic organics are appropriately minimal. The nitrogen concentration is quite high, and much of the organic matter is readily mineralized. But at the relatively low agronomic rate, not much or not enough soil conditioning is achieved.

Composts of sewage sludge, therefore, are more suitable for soil preparation in landscape work than are noncomposted materials. Almost-mature composts are preferred for soil preparation as they are still capable of soil improvement with biological production of compounds that help to flocculate and stabilize soil as the compost further decomposes. Being near a mature stage helps keep the nitrogen from being mineralized too rapidly. Soil conditioning is then possible because higher levels can be used. Composting avoids excesses of available nitrogen even when very high rates of compost are used. With the high rates of application, however, the potential for toxicity from metals can increase.

Three national conferences have been held in a series (Clapp et al., 1994; Page, 1983; National Association of State Universities, 1974) concerning a greatly expanded research program on land application of sewage sludge. The expanded interest includes needs for both protection of the environment and conservation of energy resources. Little was said in the 1993 (Clapp et al., 1994) conference about improvement of the physical properties of soil, however. General consensus in 1993 was that the growing use of land application of sewage sludge is a true success story, but that there are yet many research needs and remaining questions. There was much optimism for the use of sewage sludge on land, which can be both for soil conditioning and as a source of nutrients including in landscaping and land reclamation.

One stated research need, a very important one in the opinion of the author, is long-time, meaning decades-long, investigation of the status of heavy metals with continuous application of sewage sludge, including compost, to soil, especially in landscaping, where large amounts of compost may be used. Present researchers generally have adopted the hypothesis that heavy metal activity is largely controlled by the organic matrix in soil derived from the sludge, especially with a certain mass of additions of the sludge. The metals are mostly immobilized by the organics. Critical research work yet must be done in the field on this hypothesis. More than one plant type must be involved in the research, because there are wide differences between monocots and dicots in responses to levels of heavy metals, and there are even wide differences among species within each of these groups. Woody plants in landscapes generally have low tolerance to heavy metals. Corn (maize) is the major plant used in the studies of the past, but it is quite resistant to heavy metal toxicities. Effects with corn are not easy to generalize to all other plant materials, especially those used in gardens and landscapes, where vigorous, healthy plants are very important.

Even though most land application of sewage sludge is focused on meeting the nitrogen requirement of a crop, more reliable methods than those in use are needed for predicting mineralization rates of nitrogen for different sludges or compost mixes of sludges. Of considerable concern is the potential for nitrate leaching to ground water when mineralization rate of nitrogen exceeds removal by a crop. More study is needed on the fraction of the sewage sludge that becomes somewhat permanent soil organic matter. This could be important, especially if it can be shown that soil structure and stability have been improved with the addition of this form of stable soil organic matter. Since there are several different types of soil organic matter (Paustian et al., 1992), more research needs to be done on the values of the nonmineralizable component of sewage sludge, especially after cocomposting with other materials. Perhaps much of the nitrogen in sewage sludge compost that is applied to land is variously difficult to mineralize and remains in the soil almost permanently. Some of the reported values for potentially mineralizable nitrogen are 40% from a waste-activation process, 30% for primary sludge, 40% for lime-stabilized sewage, 25% for aerobically digested sludge, 15% for anaerobically digested sludge, and from 5 to 10% for composted sludge (Pierzynski, 1994). Much nitrogen and organics can remain for soil improvement, especially when composts are used in landscapes, more so than uncomposted sewage sludge added for agronomic reasons.

Other urgent research needs for both farms and landscapes include risk analyses that involve the soil ecosystems, including the food chain and water supplies for humans, agricultural and wildlife animals, and microorganisms in the soil (Ryan, 1994). Holistic systems research is needed where both on-landscape and off-landscape effects are monitored for different management systems (Fairbrother and Knapp, 1994). The bioavailability of metals in sludge that is directly ingested by humans, livestock, and wild animals including earthworms needs to be better known. Actually, ecological information is also needed for the organic constituents of sludge as they interact with wildlife, water resources, and adjacent plant and animal communities (Walker, 1994).

As previously implied, considerable effort in the 1993 conference was devoted to heavy element questions. It was concluded that past greenhouse studies greatly overestimated the bioavailability and phytotoxicity of metal elements in sludge (Page and Chang, 1994). Results of many field studies, some over several years, showed that plants (unfortunately mostly corn) under field conditions would not accumulate enough heavy metals to be toxic to the plants, although there was a potential for cadmium, molybdenum, and selenium to be taken up in amounts harmful to humans (Chaney, 1994). Except possibly on very acid soils, there were no reported cases of copper, nickel, or zinc toxicities on plants grown on soils amended with sewage sludge. Moreno et al.

(1996) have suggested that it is copper and nickel that are tightly bound to sludge and that zinc and cadmium are much more available to plants. We would like to see more evaluation of these effects over several decades of use. Sludge based products have been used for decades in the Los Angeles area for landscaping, and there are excess-metal problems present for some reason.

The controversy concerning the effects of sewage sludge on soil microorganisms continues. Insam et al. (1996) found a decrease in soil respiration when sewage sludge was applied to land but could not clearly show that it was due to the metals. Loth and Hoefner (1995) found a decreased level of inoculation with VA-mycorrhizae as a result of heavy metals in the sludge. More studies are needed to resolve this aspect of soil conditioning. Holmstrup and Krogh (1996) have suggested that a surfactant in sewage sludge coming from detergents is a factor that can cause toxicity to at least one soil organism.

One possible reason for differences between results in the field versus in greenhouses is the rooting behavior of the plants (Romney et al., 1981; Wallace, 1980). Most of the sludge would be in the top few cm of soil in the field, while the majority of roots may well be below where the sludge is concentrated. Roots then would even avoid the zones of soil with sludge if any toxicity were involved. In landscape work such as tree transplanting, composts are added to deeper soil zones than in field agriculture. Another possible reason for differences is that potted soils usually have higher root temperatures than field soils; more decomposition would occur in the pots. If sludge compost were used as a potting soil in horticulture work, roots could not seek a more heavy metal free zone.

The concentrations in plants of certain heavy metals that cause significant yield reductions or unsightly landscapes do need more study, and the author thinks that the values used for three metals discussed are far too high anyway, especially for some horticultural plants. The values used in mg per kg dry weight of plant leaves were >400 zinc, >40 copper, and >50 nickel (Chaney, 1994). We have seen cases of zinc toxicity at around 100 mg per kg for zinc (Wallace and Wallace, 1994b). Much will depend on plant species, including those used in landscaping, and also on the concentrations present of other elements, especially calcium and phosphorus. High phosphorus levels can decrease the uptake of zinc by plants even with high levels of extractable zinc in soils.

There are many uncertainties in the risk assessment involved in the creation of the 503 sludge rule of the EPA. Methodology used in the analysis is outlined in the 1993 conference proceedings, and it is recognized that continuing efforts for improvement are necessary. Even so, it is declared that "the risk assessment process used for the development of the 503 sewage sludge regulation is without question better than anything (else) done to date and represents the best that could be accomplished within the constraints of time, money and available data" (Ryan, 1994).

The "clean" limits and "ceiling" limits for heavy metals deemed by the USEPA, in 1993, to avoid hazards from sludge use on land are in Table 1. Since

Table 1 Parameters for Use of Sewage Sludge in Terms of Heavy Metals Called Pollutants by the USEPA (1993) When Levels Are Too High[a]

Pollutant	Ceiling concentration in sludge[b] (mg/kg)	Cumulative pollutant loading rate in soil[c] (kg/ha)	"Clean" concentration in sludge[d] (mg/kg)	Annual loading rate in soil[e] (kg/ha)
Arsenic	75	41	41	2.0
Cadmium	85	39	39	1.9
Copper	4300	1500	1500	75
Lead	840	300	300	15
Mercury	57	17	17	0.85
Nickel	420	420	420	21
Selenium	100	100	36	5.0
Zinc	7500	2800	2800	140

[a]Molybdenum and chromium temporarily at least have been withdrawn as of 1996.
[b]Cannot be used on land if any one level is exceeded.
[c]Pollutant concentration in sludge and in soil must be known; when loading rate has been reached for any one metal, sludge use will cease on that parcel of land.
[d]If all metals are below the "clean" levels, there is not limit to the sludge use. The cumulative loading rate should still apply and would require 1000 or more MG/ha use of sludge to reach that level of metals. Supposedly several years would be needed for that much use.
[e]Only if at least one of the metals exceeds its "clean" concentration limit.

originally presented, the values for molybdenum and chromium have been dropped, at least for the time being. One of the suggestions of a participant in the 1993 conference was a plea to help "fuse or defuse the supposition that biosolids use on soils is tantamount to placement of a time bomb that will release tightly bound metals as soon as the soil becomes acid after little or no management, especially in sensitive unmanaged ecosystems" (Walker, 1994).

The landscape industries must seriously evaluate the implications of Table 1. We have suggested that studies should be conducted for up to 40 years or more to ascertain the validity of the point of view involved, especially for landscaping, when effects can be magnified (Wallace and Wallace 1994b). We have observed many cases of zinc and copper toxicities in soils in Southern California, and most often high rates of composted sludge were part of landscaping management systems for many years. Since many homeowners have also been using fertilizers that contain micronutrients, that source could be part of the problem. Zinc based paints around homes could add to the problem. Sludge based products have been used in many landscapes for decades in Southern California and before regulations resulted in more clean products. Rules on the use of sewage sludge really need to consider that other sources of zinc and other metals can be, have been, and will be applied to soil. A safe rule that minimizes induced iron chlorosis is never to apply significant quantities of micronutrients unless they are needed. The 503 rule permits them to be applied in landscaping in large quantities far beyond actual need. Landscaping, therefore, should be considered separately from farm use.

Some other nations have more severe restrictions for sludge use on the land than does the USA (Witter, 1996). Although such restrictions may not be warranted, the point of view is good—that excessive loading rates will not be reached in the foreseeable future.

Many of the studies reported in the 1993 conference were conducted in Minnesota. Mean annual temperatures, including for soil, are higher further south in the northern hemisphere. Consideration in studies even needs to be given areas south of the USA. Will the protection from nonmineralized sludge on heavy metal availability to plants be the same in warmer areas as for that in Minnesota? Perhaps this is a question that needs to be answered in growth-chamber studies where high root temperatures are possible, but over a long period of time. All needed answers will not necessarily come from field studies.

Studies reported in the 1993 conference are among the few that have addressed both organisms and environmental issues, and some have been conducted for 20 years. The focus in the conference on research yet needed is very welcome. Certainly, clean sewage sludge rightfully deserves to be utilized as a valuable resource for land improvement in landscapes and as a source of nutrients on farms. Land reclamation is an appropriate use (Marx et al., 1995). Perhaps, however, the philosophy of its use should be changed or modified from a means of disposal (ways to get rid of it) to one of maximizing its value by spreading it over more land than previously. Soil conditioning value needs greater emphasis to equal, perhaps, the nutrient value, especially on farms. Researchers and operators should not have to direct major attention to the useful life of an application site but rather to the immediate and long-term values for the uses of sewage sludge on land both on farms and in landscaping.

The quantities of sewage sludge applied to land in the USA has steadily increased since 1972. In 1972, 20% of the sewage sludge available was applied to land, while in 1989 it had increased to over 33%. Land application on a weight basis was actually more than doubled in those 17 years (Clapp et al., 1994). The trend is expected to continue for some time to come. More research, public acceptance, and regulations are important for the trend to continue. Most probably it will have to be used on farms, as there are too many competing organics in landscaping. Cocomposting is a procedure that can help landscape use. Currently, about two thirds of biosolids generated by the human populations of the USA are processed in treatment plants. This will increase also.

The cost for sludge management is high. Its use on land involves considerable subsidy by municipalities. More effective use in landscaping would help municipalities. More cost-effective procedures for sludge management are being investigated and are urgently needed.

XI. RECYCLED PLASTERBOARD GYPSUM

With over three million MG of scrap drywall plasterboard being recycled each year in the USA, the pressure for using much of it in agriculture as a source of gypsum (see Chapter 7) will increase. However, much of it will probably be cleaned of paper and used to make more drywall. Of the 30 million MG of gypsum wallboard produced per year in the USA, about 15% is waste. Relatively new drywall is much freer of contaminants than is old drywall, which is probably useless for soil application in agriculture. There is a small amount only of contaminants in the newer drywall. The total composition of drywall gypsum needs to be constantly monitored and possible effects for soil use need to be documented.

One of the questions that frequently arises is whether recycled gypsum from plasterboard (drywall) can be used as a source of gypsum on land. Mined gypsum is officially considered to be "organic" for organic farming, while recycled gypsum is not.

One objection to the use of drywall gypsum in soil in that it usually contains some boron and other items used as fire retardants (Wallace and Stiles, 1995). The danger of boron toxicity depends on the natural boron levels in the soil, the plants grown, the type and quantity of gypsum applied from drywall, the amount of boron leaching from the soil, and other factors. An essential plant element in small quantities, boron is toxic to plants in large amounts.

In a 1993 study, gypsum from drywall was applied to corn at rates of 4 and 8 MG/ha with an increase in yield of grain of 25% for both plots (unpublished study of J. Reindl, Dept. of Public Works, Madison WI). The issue of boron (assumed at 150 parts per million) and other contaminants was studied, and no indications were found of any problems from normal drywall, although further research was recommended regarding moisture-resistant and fire-retardant additives. No one can make a categorical statement that recycled drywall is unsuitable for all agricultural purposes.

The question of the use of plasterboard or drywall gypsum on land certainly is not simple. There are places and conditions where such use is justified. In areas where boron toxicity is a potential problem, however, caution is in order. Drywall gypsum may supply needed calcium and sulfur when low application levels are indicated; but when high levels are needed, such as for reclamation of sodic soils, too much boron may be applied. In areas where boron is in the deficiency range, the amounts supplied with recycled drywall gypsum can be very beneficial.

Harker (1995) has listed some problems of the past concerning the use of old drywall gypsum:

1. Asbestos. Prior to about 1970, asbestos was used in plasterboard at the rate of less than 1%. Since then fiberglass has been used, presumably to increase the insulation value. The asbestos-removal industry has an interest in classifying plasterboard as hazardous, but there is little basis for such a classification.
2. Lead. Lead-based paints have long since been discontinued for use on plasterboard.
3. Mercury. An antifungal chemical Troisan, which contains mercury, was used on plasterboard prior to 1990. It has been officially discontinued since then.
4. Hydrogen sulfide. This gas is formed when plasterboard is added to landfills and then subjected to anaerobic conditions. The sulfate in the gypsum is reduced to sulfide. This, of course, is not a factor when gypsum is recycled and not put in landfills.

Harker's (1995) proposed technique for recycling gypsum is to return it to drywall manufacturers who in turn will use it to make new drywall with 15–20% recycled gypsum. That way, it need not be applied to land.

At least one company on the east coast of the USA uses calcining of the reclaimed plasterboard as a step in preparation for its use on land. This procedure burns out any waste paper residue and all organic fungicides so that only minerals remain. Any boron remains. This creates an interesting and useful product for specialty uses on the land. Its high value is of interest to gardeners and landscapers. Since the east coast of the USA and other areas worldwide are subject to boron deficiencies, the calcined product provides multiple benefits to the land and involves few or none of the objections usually raised against recycling of plasterboard drywall gypsum.

Getting drywall gypsum out of landfills is worthwhile. In some locations, at least, use on land is possible with proper precautions. When anaerobic conditions exist in landfills, microbes produce H_2S from gypsum, which gives problems. The same can happen in anaerobic soils where gypsum is applied. If paper is not removed, the ground gypsum product can be fluffy and hard to apply to land. Residual fungicides need to be monitored. Recycled gypsum usually has higher water concentrations than mined gypsum, so users need to know how much actual gypsum they are receiving. Mined gypsum usually contains a little (sometimes a lot) of lime (calcium carbonate), and this carries over to recycled gypsum. Some lime in gypsum is worthwhile for use on acid soils but it is of no importance when used on calcareous sodic soils. The lime concentration in gypsum from any source should be known.

One general conclusion is that recycled gypsum should be used only with caution on arid, potentially sodic soils. When very high application rates are needed, like one to several MG/ha, recycled gypsum could give problems because of boron. When only a few hundred kg/ha are needed, recycled gypsum should be useful. It would be more useful in areas of boron deficiency than in areas of boron excess.

XII. PLASTIC MILK FROM WASTE PLASTIC

DeBoodt (1992) has mentioned a possible soil conditioner product prepared from waste plastics. See Chapter 12. Recent evolution in waste management promotes better economics for a potential soil conditioner and ease in production. Plastic wrappings offer an opportunity at low cost. The waste is abundant and it is an environmental nuisance. Reasons are that the waste plastics are not biologically degradable, are difficult to burn and pollute the air if they are burned, and are too expensive for regeneration. In agriculture a double advantage could be realized by taking the plastics out of the environment and using them as soil conditioners. Polyethylene and polypropylene are favored, as they can be put easily in an emulsion that is very effective in decreasing water evaporation from soil. Through low heating they melt to a gel. In an ordinary blender a hydrophobic emulsion is obtained by adding water and an emulsifier. This emulsifier is easily available as it can be found in most plants secreting gums or in the waxes produced by bees. The new product, called plastic milk, can be applied with a watercan or sprinkler truck to the top layer of the soil to prevent evaporation and for other advantages. Experience is showing that it is degradable within a season after application to soil (DeBoodt, 1992).

XIII. BITUMEN (ASPHALT) EMULSIONS

Bitumens are insoluble in water and are applied to soil as emulsions. Asphalt bitumen emulsions have been used for soil conditioning for over 50 years (Walker, 1995). Goor et al. (1976) described the chemical and physical properties of such materials used in work with soil. The emulsions contain three basic ingredients: the asphalt, water, and an emulsifying agent. With blending, the emulsifying agent decreases the surface tension of the asphalt sufficiently to obtain a stable suspension that behaves like a liquid. The liquid can then be pumped, stored, and mixed as needed for land application.

These materials are environmentally safe as they result in no leaching of hazardous substances into groundwater (Walker, 1995). They eventually degrade.

Emulsions so prepared are commonly used as tackifyers in hydroseeding operations or as surface sprays to prevent wind and water erosion and to avoid crusting while seeds are emerging. They are frequently used with straw to control slope erosion. The emulsion holds the straw mulch in place until the new plants have grown sufficiently to hold the soil.

Emulsions are used as dust retardants as well as controls for erosion, especially on slopes. Stanley and Kempsey (1990) used various emulsions to reduce erosion on coastal sand dunes and obtained good success. They did say, however, that the use was fairly expensive—$3000 to $4000 per hectare.

Bitumen emulsion has been used to improve water-use efficiency in field crops or very sandy soil (Wahba et al., 1990; Matyn et al., 1990). Relatively low rates of bitumen were used so as to control the costs. Bitumen did give improved water relations but much of that was as water-stress increased. Costwise these materials may be most useful for erosion control of landscapes along highways where costs are less of a problem.

XIV. AMMONIUM LAURETH SULFATE (AGRI-SC) SOIL CONDITIONER

Ammonium laureth sulfate has properties somewhere between those of water-soluble polymer soil conditioners and wetting agents. It is anionic. It is applied to soil surfaces, at relatively low application rates and costs, for the purpose of protecting the soil surface from crusting or sealing when water as rain or irrigation strikes the surface. Depending on the method of application and type of soil, published reports indicate no effect on properties such as aeration, aggregate stability, and hydraulic conductivity (Fitch et al., 1989). It does have the ability to reduce interrill erosion and increase water infiltration (Norton and Nearing, 1994).

The producers of the product report that it can stabilize sandy soils (Northcutt, 1996). On other soils it is claimed to reduce compaction, make tillage more easy to accomplish, make tillage possible in areas otherwise too wet, decrease runoff, and provide other benefits. A minimum till and no-till farmer says that the soil is looser and closes easily over the seeds when the product is used (Northcutt, 1996).

Treatments last around a year according to some users. It functions mostly by breaking the surface tension. Because of its multiple effects, it has some influence on increasing water-use efficiency.

XV. COPOLYMER LATEXES AND RELATED PRODUCTS

Most copolymers used as tackifiers in hydroseeding and as dust and erosion control agents are similar to those used in paints and adhesives. They are long-chained and have some cross-linking. Variously used are vinyl acrylic copolymers, methacrylate copolymers, styrene butadiene acrylates, and others. They are produced as emulsions. They are diluted and sprayed on soil areas to be stabilized. A spray application can penetrate as much as 5 cm of soil. Silicates are sometimes added to aid soil penetration and stabilization. By varying the amount of water used to dilute the concentrates, the strength of these products can be varied to meet the need for any special project.

Major uses of copolymer latexes on soil include controlling dust and blowing sand, controlling erosion on bare slopes, and controlling erosion on roadways and construction sites. They dry and form films around soil particles that hold the soil in place. The membranes so formed are semipermeable and biodegradable. General application rates before dilution are around 200 to 400 liters per hectare depending on purpose. They therefore are moderately expensive to use. Adding a wetting agent can improve penetration into soil, and soluble silicates can help harden treated soil. Some copolymers are used to stabilize almost vertical slopes on construction sites. Several different latex copolymers are available commercially. They are considered preferable to water-soluble polymers for heavy erosion control, but perhaps more research is needed on both for this purpose.

The producers of rubber have been investigating the use of rubber latex as a soil conditioner for about 25 years (Bernas et al., 1995). The active molecule is polyisoprene with molecular weight of the polymer at 1 to 2 million Daltons. This natural rubber is expensive as a soil conditioner and requires couse of an aromatic oil and a stabilizer to maintain a dispersed suspension. Otherwise it is very effective as a soil conditioner and an erosion control agent, especially when mixed with other substances to give more cost effectiveness (Ho and Bachik, 1985). Rubber latex in soil behaves like a coat of paint to cover and coat aggregates, but it has poor penetrating ability.

Bernas et al. (1995) compared rubber latex with poly (diallyl dimethyl ammonium chloride), poly (DADMAC), for soil conditioning ability. Poly (DADMAC) is positively charged and does not require couse of an aromatic oil or stabilizing agent. Poly (DADMAC) was better able to penetrate small pores than latex and was effective in improving soil with poor structure. Its economic viability is unknown as yet, but it appears that higher rates of it will be needed than for water-soluble polyacrylamides. The positive charge, which results in ease in soil clay reactivity, may be a valuable characteristic.

XVI. CEMENT-TYPE FORMING PRODUCTS

Portland cement has been tested for its ability to produce soil stability characteristics (Ahuja and Swartzendruber, 1972). On a silt loam soil, water-stable aggregates were increased threefold, and hydraulic conductivity was much improved over the control. Best results were obtained with about 1% portland cement of weight of soil (1/2% and 2% were also tested) and samples were kept moist for various time periods to test moisture-curing time. Portland cement also has some liming ability, so it could improve that aspect of acidic soils. Treatment costs may not be high if only the top one or two cm of soil are treated.

Recently other cement-type products have been used for protecting soil. Hemihydrate gypsum (plaster of paris) has been used because when it dries it forms a permeable crust over the soil surface. With or without an accompanying fiber, it is used to control dust in arid areas. Ideally it can hold seed, fertilizer, and soil in place until plants are capable of stabilizing a site.

Pozzolan has been used with appropriate lime to create crusts or caps over a site to be protected. The crust comes from the formation of calcium silicate in a cement-type forming process. Appropriate mixtures are sprayed on a site with conventional hydroseeding equipment. The mixture is sometimes fortified with fiber, which can be varied to alter roughness of the crust to slow down the rate of water movement over the crust. A product made by the Chemical Lime Company of Fort Worth, Texas, has been used at a variety of sites including mines, landfills, burned-over wild brush and forest areas, construction sites, and golf courses (Northcutt, 1996).

XVII. ADDITIONAL DUST SUPPRESSANTS

Nearly all soil conditioners help to control dust. Gypsum and water-soluble polyacrylamide can be effective on farms. Those mentioned in this section are additional ones not previously discussed. Among the more effective dust suppressants are lignosulfonate, which is a by-product of paper production, magnesium chloride, and calcium chloride.

Lignosulfonates are water soluble and react in soil much like water-soluble polymers. They react with negatively charged clay to cause flocculation. They are most effective in arid climates. They give good control of dust but are less effective than the chlorides in preventing the formation of potholes (Northcutt, 1996).

The chlorides remove water from the air and generally absorb water. This property helps control dust. They are less effective in dry climates, but magnesium chloride seems to work better in desert areas than calcium chloride.

All these products considerably reduce the cost of maintenance of unpaved roads. Treatment costs vary from \$5000 to \$7000 per km. The US Army Corps of Engineers estimate that nearly two-thirds of 6.5×10^6 km of roads in the continental USA are unpaved (Northcutt, 1996). It is estimated that about one-third of the fine suspended particles (PM10) in the air originate from unpaved roads. Other sources of dust are construction sites, parking lots, racetracks, and airports. All these create health and nuisance problems. Maintenance costs to vehicles and additional travel time also increase when roads are not properly maintained.

XVIII. SOME SPECIALTY USES FOR WATER ABSORBING (GEL) POLYMERS

Hydrophylic gel or cross-linked polymers are synthetic water-absorbing polymers of high molecular weight. See Chapter 12. They have been used as absorbents in the diaper industry for the past 35 years. Polymers differ from each other in the specific monomer building block, amount of water absorbed per gram of material, particle size and distribution, response to salinity, and cost. While there are only four different monomers used in the production of useful gel polymers, there are several dozen gel polymers available commercially under different trade names (Orzolek, 1993).

The chemical components of these hydrophylic polymers include cross-linked acrylamide, sodium polyacrylates, swellable starch, potassium polyacrylates, acrylate copolymers, and acrylonitrile. Cross-linking in polymers contributes to the increased storage of plant-available water in addition to acting as a physical barrier to the outflow of water from the gel (Orzolek, 1993).

Gel polymers have been used as means for increasing water-holding capacity and decreasing water-infiltration rate of very sandy soil (see Chapters 12 and 13). Under high value crop conditions and in very water short countries, this technology has been intriguing. Considerable developmental work has also been done concerning use of gel polymers for special purposes.

A. Greenhouse-Potting Soil Uses

One of the earliest uses for gel polymers was to increase the water-holding capacity of soils and soilless mixes used in production of crops in greenhouses

(Bearse and McCollum, 1977; Gearing and Lewis, 1980; Wang, 1989). A major goal for such usage was to have a means for reducing the number of days needed between irrigations. A second goal was to reduce the total amount of water needed to grow a plant (Orzolek, 1993). The gels were most effective in sand or in media very low in soil organic matter. Some workers have reported poor effects and given reasons (Foster and Keaver, 1990; Letey et al., 1992). Salts, especially divalent cations, do reduce the ability of gel polymers to hydrate (Bowman, 1990).

Use of gel polymers in soil for rooting of cuttings improved rooting of holly and azalea because of more uniform distribution of moisture (Banko, 1984).

B. For Vegetable Crops

Gel polymers have been used in the production of high value vegetable crops especially on an experimental basis during prolonged drought years (Orzolek, 1993). Increase of water-holding capacity is not the only benefit, and depending on soil type and other conditions it may not be a matter of importance. Reducing the total amount of water used and also reducing the amount of fertilizer needed have been major objectives of users. Users have often found these and other benefits associated with drought stress.

Banding of polymers in furrows requires 18 to 45 kg per ha, while broadcasting uses 170 to 900 kg per ha. Both procedures increased yields, but the first procedure was much more economical (Orzolek, 1991). Banding also increased the nutrient-use efficiency (Orzolek, 1991). More needs to be known about these materials under drought conditions.

C. For Fruit and Orchard Crops

Dry gel polymers have been applied by soil application into four or more holes around the drip line of existing trees (Orzolek, 1993). They also have been injected in the hydrated state in various configurations in established orchards. In newly established orchards or as part of the establishment procedure they are reported as useful in sustaining active growth when stress conditions exist. Most important is that their use is associated with lower tree mortality (Orzolek, 1993). Similar results are obtained for water-soluble polyacrylamides, but for different reasons (see Chapter 19). The hydrated gel polymers do add water-holding capacity in addition to improving pore space and aeration with proper application. The water-soluble polyacrylamide in contrast adds to pore space, aeration, and infiltration by reacting with the clays to form more stable aggregates. Application rate of hydrated gels is around 225 grams per 300 liters per tree (Orzolek, 1993).

D. For Turf

Gel polymers have been used in the soil preparation for turf planting and also injected into soil of existing turf (Orzolek, 1993; Piper, 1990). Deeper rooting in contrast to shallow rooting for nontreated turf has been obtained. Two to seven kg per 100 m^2 have been injected into turf and up to 15 kg per 100 m^2 have been incorporated into very sandy soil with good results (Baker, 1991). Gel polymers incorporated into soccer fields have increased water retention, reduced soil hardness in dry weather, and improved health of ground cover (Orzolek, 1993).

Workers have concluded that injection of prehydrated gels into existing turf is more successful and efficient than injection of the dry granules. The remaining pore space when the gels partially dry enables more easy rehydration of the gels and also results in better aeration of the soil.

Piper (1990) recommends lower rates in soil preparation for poorly drained soils than for well drained soils and more for sandy soils than for clay soils. Rates for soil preparation vary from 7 to 22 kg per 100 m^2. All workers say never to apply any kind of polymer to the surface of turf because polymers create slippery conditions.

E. For Establishing New Turf Sod

Piper (1990) has recommended that for a quick start to laying new sod, after the soil preparation is complete and before the sod is placed, 2 kg of gel polymer per 100 m^2 are applied to the soil surface with a drop spreader to obtain uniformity. After the sod is laid, the sod is irrigated as recommended. This results in a layer of water between the soil and the sod to help roots become established and to avoid stress. The sod keeps the polymers in place.

F. Tree Planting with No Irrigation

Gel polymers have been used in the backfill when trees and shrubs are transplanted. Extreme care must be taken when they are used to avoid overly wet conditions. Although water-soluble polyacrylamide (see Chapters 15 and 19) is perhaps safer in the backfill for such planting, the use of gels may be very useful if no irrigation is possible except for the transplant solution. Under these conditions, granules of gel polymers may be blended with the backfill at the rate of 3 kg per m^3 for sandy soil and 1 kg per m^2 for clay soil (Piper, 1990). If the transplant hole is much larger (not deeper) than the plant, the increase in water-holding capacity will protect the tree or shrub against stress. Regular good transplanting procedures do need to be followed otherwise.

Millions of forest trees have been successfully planted with roots saturated with finely ground gel polymers with some additional hydrated polymer in the transplant hole (Piper, 1990).

G. With Water-Soluble Polyacrylamide for Seeding of Arid Wildlands

Seeding for revegetation is different under arid and semiarid compared with mesic conditions because of the uncertainty of the timing and magnitude of rain events. Irrigation is impossible under most arid conditions, and hand application of small amounts of water is not only cumbersome but also ineffective because of rapid dissipation of the water.

A procedure has been developed (Wallace and Wallace, 1990b) in which from 200 to 2000 ml of a gel polymer–water-soluble polyacrylamide mixture in water with a very small amount of gypsum are applied in a depression large enough to hold the mixture. Appropriate seeds are then placed on the gel mixture and covered with a layer of soil suitable to the kind of seed used. The purpose of the water-soluble polyacrylamide is to flocculate the soil above and around the gel so that there is no crust to inhibit seed emergence. Depending on the season, the gel polymer can maintain a moist condition for several weeks, or long enough for the seeds to germinate. Sufficient rain events must follow for the seedlings to become well established.

Concentration of the water-soluble polyacrylamide with the gel mixture can range from 200 to 400 mg per liter. From 2 to 5 grams of dry gel polymer per liter of solution are also used. The concentrations of gypsum need not be more than 2.5 me per liter and may be omitted if the water used has that much or more salt. High salt concentrations can inhibit germination of seeds.

The gels can maintain sufficient moisture for three or more weeks in this procedure, even if daytime temperatures exceed 38°C for part of the time. The tap root of a seedling can penetrate deep enough before summer dormancy occurs so that the plant may resume growth when more favorable conditions return.

H. The Sunbelt System for Growing Crops in Arid Lands

Dan Wofford, Jr., of Western Polyacrylamide, Inc., of Jay, Oklahoma (USA) has helped to develop what he calls a ten-year NO-IRRIGATE/NO-WEED/NO-CULTIVATE dryland vegetable gardening system. It is readily adaptable to small fruit and vegetable production. The basic Hydrosource/DeWitt Sunbelt Dryland Gardening System consists of a long-lasting, synthetic, water absorbing (gel) cross-linked polyacrylamide (CLP) incorporated into the soil and covered with a woven polypropylene weed/evaporation/erosion barrier with a highly effective UV blocker. It was first developed in Colorado in 1992; more than 200 of the Hydrosource/Sunbelt beds have been constructed in the U.S., Russia, Lithuania, and Germany (Wofford, 1992a,b,c; Wofford and Orzolek, 1993). Scientific study of the system with small fruits and some vegetable crops is just beginning. Crop success, weed control, and erosion control are among the benefits.

The essence of the system is that the plastic cover controls weed growth and is effective for several years. It also prevents evaporative water loss. It also allows passage of water from e.g. rains at a very high rate. The gel polymer serves to help retain 12.5 or more cm of water in the effective root zone, which is the minimum needed to grow some crops. Rates of the gel polymer to use are based on retention of about 400 times weight of the gel with water in relationship to what is needed.

I. For Prevention of Soil Crusting

From 5 to 50 kg per ha of gel polymers have been drilled into seed rows for prevention of surface soil crusting or sealing (Pryor, 1988). Success depends on choosing the proper size of the gels; small sizes are the most effective. The procedure is reasonably effective in enhancing seed emergence, enhancing date of maturity, and increasing yields. Where they can be used, however, water-soluble polyacrylamides are more effective and probably less costly than gels for this purpose.

XIX. UREA-SULFURIC ACID ADDUCTS

Commercial forms of urea-sulfuric acid adducts are sold as combination nitrogen fertilizer–soil conditioner products (see Chapter 8). A popular item contains 15% nitrogen and 49% sulfuric acid (Purcell, 1993, 1996). The urea when combined with the sulfuric acid serves to decrease much of the caustic nature of the sulfuric acid. It becomes much easier to use and is essentially nondamaging to application equipment. Each MG of 49% sulfuric acid applied to calcareous soil can result in the production of 1.75 MG of gypsum, or 2 MG of 87% gypsum by reacting with lime in the soil. When the nitrogen in urea becomes acidified, even more gypsum is formed (2.6 MG total per MG of product).

Urea–sulfuric acid adducts are commonly used for anticrusting to enhance seed emergence. The effect results from formation of gypsum from the native lime in the soil. The adduct can also decrease the pH of irrigation water by neutralizing bicarbonates and carbonates, as shown in Chapter 8. In addition to its anticrusting effect, the adduct increases the availability of several nutrients in soil to plants, especially iron. Both chemical and physical forces are involved, as iron chlorosis is preventable with soil conditioners (Wallace et al., 1986b).

XX. AMELIORATION OF SALINE–SODIC SOILS WITH WATER-SOLUBLE POLYACRYLAMIDE

Chapter 15 mentions that water-soluble polyacrylamide (WS-PAM) is perhaps too costly to use for helping to decrease salinity and sodicity in soils. On urban

Table 2 Time in Seconds Needed for Infiltration of Water Through Stabilized and Unstabilized Sodic Soil from Lost Hills, California

Treatment rates WS-PAM kg ha^{-1}	Infiltration time for 25 ml in 25 g soil, s	Strength of aggregates in water
0	270	Feeble
28	95	Marginal
56	70	Very good
112	45	Strong

Source: Wallace et al., 1986a.

and landscape work, WS-PAM can be useful for this purpose and is cost effective. We have many successful projects on which the soil was impermeable to water before treatment. Mechanical cultivation, WS-PAM, and gypsum followed by leaching with water resulted in decreasing both salinity and sodicity and in increasing water permeability to acceptable levels. The WS-PAM was anionic and of high molecular weight.

Successful use of water-soluble polymers to correct salinity–sodicity was reported over 40 years ago (Allison, 1952; Allison and Moore, 1956; Quastel, 1954). See also Chapter 14. In those old studies, the application rate of Krilium (vinylacetate maleic acid copolymer) of 0.1% used as a solution resulted in many-fold increases in the water infiltration. Sweet corn grown in treated soil gave large yield increases. Actually yield on some untreated soils was near zero because of the heavy crusting that inhibited seed emergence. There were strong indications that the water-soluble polymer was able to increase yields because of the improved aeration of the sodic soil even without any sodium replacement by calcium from gypsum.

The present WS-PAM in use is much more effective than was Krilium. The rate of 0.1% is about 2200 kg ha^{-1}. Data in Table 2 for a saline–sodic soil having a pH of 8.6 and an EC of 8.0 dSm^{-1} indicate that much lower rates of WS-PAM can be useful in improving saline–sodic soils. Emergence and growth of tomato plants of soils having various levels of salinity and sodicity have been increased substantially (Wallace et al., 1986a). A rate of 56 kg ha^{-1} of WS-PAM with gypsum very efficiently flocculated the clay in a soil of pH 9.9 and an EC of 25 dS m^{-1} (Wallace et al., 1986a). Trees grew well on the treated soil. Such rates are cost effective under urban landscape conditions. Zahow and Amrhein (1992) were able to increase the saturated hydraulic conductivity of a swelling sodic soil when the exchangeable sodium percentage was less than 15 with 100 kg/ha WS-PAM.

XXI. CONCLUSIONS

This volume cannot identify and describe all the substances that can be used as soil conditioners. The present chapter discusses some products and technologies not fully covered elsewhere in this book, but it does not mention all that are possible. More and more waste products are being used as soil conditioners (Karlen et al., 1995). (Users prefer the term by-products to waste products.) Perhaps the present effort can stimulate further work on many soil conditioners. The soil conditioning values that can result from plants themselves is an exciting field that deserves full exploitation. Actually it is an important part of maximizing soil quality and the development of a more sustainable agriculture. This chapter emphasizes that several innovative procedures are possible for more effective use of various soil conditioners described in earlier chapters.

REFERENCES

Ahuja, L. R., and Swartzendruber, D. (1972). Effect of portland cement on soil aggregation and hydraulic properties. *Soil Sci. 114*:359.

Allison, L. E. (1952). Effect of synthetic polyelectrolytes on the structure of saline and alkali soils. *Soil Sci. 73*:443.

Allison, L. E., and Moore, D. C. (1956). Effect of VAMA and HPAN soil conditioners on aggregation, surface crusting, and moisture retention in alkali soils. *Soil Sci. Soc. Am. Proc. 20*:143.

Almendros, G. (1994). Effects of different chemical modifications on peat humic acid and their bearing on some agrobiological characteristics of soils. *Commun. Soil Sci. Plant Anal. 25*:27.

Arcamonte, M. S. (1995). Wulf vineyard cover crops. *Agribusiness Fieldman*, December 1995, pp. 1–2.

Armstrong, D., Agerton, B., and Martin, S., eds. (1984). The diagnostic approach. *Better Crops/Spring 68*:1–46.

Baker, S. W. (1991). The effect of a polyacrylamide co-polymer on the performance on a *Lolium perenne* L. turf grown on a sand rootzone. *J. Sports Turf Res. Inst. 67*:66.

Bakr, M. A. (1994). Production of peat moss-like substance from local organic wastes. *Egypt. J. Soil Sci. 32*:1.

Banko, T. J. (1984). Medium amendment plus watering system may improve rooting. *Amer. Nurseryman*, May (5):51.

Barbarick, K. A. I., and Pirela, H. J. (1985). Agronomic and horticultural uses of natural zeolites: a review. *Zeo-Agriculture Uses of Natural Zeolite in Agriculture and Aquaculture* (W. G. Pond, and F. A. Mumpton, eds.). Westview Press, Boulder, Colorado.

Bearce, B. C., and McCollum, R. W. (1977). A comparison of peat-lite and noncomposted hardwood-bark mixes for use in pot and bedding plant production and the effects of a new hydrogel soil amendment on their performance. *Flor. Rev. 10*:21.

Bernas, S. M., Oades, J. M., Churchman, G. J., and Grant, C. D. (1995). Comparison of the effects of latex and poly(DADMAC) on structural stability and strength of soil aggregates. *Aust. J. Soil Res. 33*:369.

BioCycle. (1996). Good year for tire recovery. *BioCycle 37*(3):35.

Bowman, D. C. (1990). Fertilizer salts reduce hydration of polyacrylamide gels and affect physical properties of gel-amended container media. *J. Amer. Soc. Hort. Sci. 115*:382.

California Agriculture. (1995). Urbanization crowds out oaks. *Calif. Agr. 49*(5):5.

Campbell, A. G. (1990). Recycling and disposing of wood ash. *Tappi J.* (September):140–145.

Carter, V. G., and Dale, T. (1974). *Topsoil and Civilization.* Univ. of Oklahoma, Press, Norman.

Chaney, R. L. (1994). Trace metal movement: soil-plant systems and bioavailability of biosolids-applied metals. *Sewage Sludge: Land Utilization and the Environment* (C. E. Clapp, W. E. Larson, and R. H. Dowdy, eds.). SSSA Misc. Pub. Madison, WI, pp. 27–31.

Chappell, N., Stobbs, A., Ternan, L., and Williams, A. (1996). Localized impact of sitka spruce (*Picea sitchensis Bong*) on soil permeability. *Plant Soil 182*:157.

Clapp, C. E., Larson, W. E., and Dowdy, R. H., eds. (1994). *Sewage Sludge: Land Utilization and the Environment.* SSSA Misc. Pub., American Society of Agronomy, Inc., Crop Science Society of American, Inc., Soil Science Society of America, Inc., Madison, Wisconsin.

Cline, H. (1996). New gypsum in irrigation highly valued. *California-Arizona Farm Press 18*(10):6–7, 10.

Conway, W. S., Sams, C. E., and Kelman, A. (1994). Enhancing the natural resistance of plant tissues to postharvest diseases through calcium application. *HortSci. 29*:751.

David, P. P., Nelson, P. V., and Sanders, D. C. (1994). A humic acid improves growth of tomato seedlings in solution culture. *J. Plant Nutr. 17*(1):173.

Davidson, M. R. (1985). Numerical calculation of saturated-unsaturated infiltration in a cracked soil. *Water Resour. Res. 21*:709.

DeBoodt, M. (1992). Soil conditioning for better soil management. *J. Korean Soc. Soil Sci. and Fert. 25*:311.

Edwards, W. M., van der Ploeg, R. R., and Ehlers, W. (1979). A numerical study of the effects of non-capillary sized pores in infiltration. *Soil Sci. Soc. Amer. J. 43*:851.

Fairbrother, A., and Knapp, C. M. (1994). Ecological aspects of land spreading sewage sludge. *Sewage Sludge: Land Utilization and the Environment* (C.E. Clapp, W.E. Larson, and R.H. Dowdy, eds.), pp. 75–79.

Fargerlund, J., Jobes, J. A., Marsh, M., and Shouse, P. J. (1995). Soil water retention characteristics of sand-pelletized diatomaceous earth mixtures used for golf greens. *1995 Agron Abs.*, p. 203.

Faulkner, E. H. (1943). *Plowman's Folly*, Univ. of Oklahoma Press, Grosset and Dunlap, New York.

Ferguson, G. A., and Pepper, I. L. (1987). Ammonium retention in sand amended with clinoptilolite. *Soil Sci. Amer. J. 51*:231.

Ferguson, G. A., Pepper, I. L., and Kneebone, W. R. (1986). Growth of creeping bentgrass on a new medium for turfgrass growth: clinoptilolite zeolite-amended sand, *Agron. J. 78*:1095.

Fitch, B. C., Chong, S. K., Arosemena, J., and Theseira, G. W. (1989). Effects of a conditioner on soil physical properties. *Soil Sci. Soc. Am. J. 53*:1536.

Foster, W. J., and Keever, G. J. (1990). Water absorption of hydrophylic polymers (hydrogels) reduced by media amendments. *J. Environ. Hort. 8*(3):113.

Fulton, A. E. (1994). Improving water infiltration with water amendments. *Irrigation News*, July 1, University of California Coop. Ext., Kings County, California.

Fussel, G. E. (1965). *Farming Technique from Prehistoric to Modern Times*. Pergamon Press, Oxford, U.K.

Fussell, G. E. (1971). *Crop Nutrition: Science and Practice Before Liebig*. Coronado Press, Lawrence, Kansas.

Fussell, G. E. (1972). *The Classical Tradition in Western European Farming*, David and Charles, Newton Abbot.

Gearing, J. M., and Lewis, A. J. (1980). Effect of hydrogel on wilting and moisture stress of bedding plants. *J. Amer. Soc. Hort. Sci. 105*:511.

Goor, G., Bernaert, E., Sadones, M., Callebaut, F., and Schamp, N. (1976). The influence of chemical and physical properties of synthetic polymer emulsions on soil conditioning. Proc. Third International Symposium on Soil Conditioning (Ghent, Belgium). *Meded. Fak. Landouww. Rijksuniv. Ghent 41*:187.

Gray, M. N., and Rock, C. R. (1987). *Boiler Ash Addition to Agricultural Soil*. Dept. of Civil Engineering Rep., Univ. of Maine, Orono.

Harker, B. (1995). A technique to recycle gypsum, *C and D Debris Recycling*, Fall, 1995.

Hart, J. R. (1986). Using fly ash as a bulking agent. *BioCycle 27*(1):28–29.

Haskell, S. B., ed. (1926). Symp. on soil deterioration. *J. Amer. Soc. Agron. 18*:89–165.

Haynes, R. J. (1990). Active ion uptake and maintenance of cation-anion balance: a critical examination of their role in regulating rhizosphere pH. *Plant Soil 126*:247.

Ho, C. W., and Bachik, A. T. (1985). Commercial application in the use of natural rubber latex emulsion for hydroseeding. Proc. Int. Rubber Conf. Kuala Lumpur, pp. 710–722.

Holmstrup, M., and Krogh, P. H. (1996). Effects of an anionic surfactant, linear alkylbenzene sulfonate, on survival, reproduction and growth of the soil-living collembolan *Folsomia fimetaria*. *Environ. Toxic. Chem. 15*:1745

Holtz, B. A. (1995). Cover crops can enhance water infiltration. *Nut Grower 15*(10):17, 18, 20.

Hughes, B. J., and Sweet, R. D. (1979). Living mulch: a preliminary report on grassy cover crops interplanted with vegetables. *Proc. Northeast Weed Science Soc.* (R. B. Taylorson, ed.) *33*. Evans, Salisbury, MD.

Insam, H., Hutchinson, T. C., and Reber, H. H. (1996). Effects of heavy metal stress on the metabolic quotient of the soil microflora. *Soil Biol. Biochem. 28*:691.

Jahan, K., Ahmed, T., and Maier, W. J. 1995. Selection of nonionic surfactants in enhancing biodegradation of phenanthrene in soil. *Proc. Water Environ. Fed. Ann. Conf. Expo. 68th* 2:579.

Johnson, J. J., Allan, D. L., Vance, C. P., and Weiblem, G. (1996a). Root carbon dioxide fixation by phosphorus-deficient *Lupinus albus*. *Plant Physiol. 112*:19.

Johnson, J. J., Vance, C. P., and Allan, D. L. (1996b). Phosphorus deficiency in *Lupinus albus*. Altered lateral root development and enhanced expression of phosphoenolpyruvate carboxylase. *Plant Physiol. 112*:31.

Kahl, J. S., Fernandez, I. J., Rustad, L. E., and Peckenham, J. (1996). Threshold application of wood ash to an acidic forest soil. *J. Environ. Qual. 25*:220–227.

Karlen, D. L., Wright, R. J., and Kemper, W. D., eds. (1995). Agriculture utilization of urban and industrial by-products. ASA Special Publication Number 58, Amer. Soc. Agron., Madison, WI.

Katz, M. (1995). The cover crop challenge. *Citograph 81*(1):4–6.

Koski, T. (1996). How do wetting agents do what they do? *California Landscaping*, June–July, p. 30.

Lauer, J. G., and Fornstrom, K. J. (1988). A living mulch system for sugarbeet establishment. Wyoming Agr. Sta., Laramie, Bul. 907:50–53.

Lerner, B. R., and Utzinger, J. D. (1986). Wood ash as soil liming material. *HortSci. 21*:76.

Letey, J. (1975). The use of nonionic surfactants on soils. Chapter 14 in *Soil Conditioners*, SSSA Spec. Pub. #7, Madison Wisconsin pp. 145–154.

Letey, J., Clark, P. R., and Amrhein, C. (1992). Water-sorbing polymers do not conserve water. *Calif. Agriculture. 46*(3):9.

Likens, G. E., Driscoll, C. T., and Buso, D. C. (1996). Long-term effects of acid rain: response and recovery of a forest ecosystem. *Science 272*:244.

Lilly, J. P. (1965). The sleeping sod. *Crops and Soils 18*(6):5–6.

Liu, W., Lund, L. J., and Page, A. L. (1989). Acidity produced by leguminous plants through symbiotic dinitrogen fixation. *J. Environ. Qual. 18*:529.

Logan, T. J., and Burnham, J. C. (1995). The alkaline stabilization with accelerated drying technology to convert sewage sludge into a soil product (D. L. Karlen, R. J. Wright, and W. D. Kemper, eds.). ASA Special Publication No. 58. Madison, Wisconsin pp. 209–223.

Longsdon, G. (1989). Odor control with bioash. *BioCycle 30*(1):30–31.

Longer, D. E., and Huneycutt, H. J. (1995). White clover as a living mulch in row crop systems. *1995 Agron. Abs.*, p. 125

Loth, F. G., and Hoefner, W. (1995). Influence of sewage sludge treated soils on the infectivity of VA-mycorrhizal fungi isolates in different plants. *Agriobiol. Res. 48*:269.

Maddox, V. I., and Goatley, J. M., Jr. (1995). Evaluation of diatomaceous earth topdressing for algae suppression on a bermudagrass green. *1995 Agron Abs.*, p. 153.

Mannering, J. V., and Johnson, C. B. (1969). Effect of crop row spacing on erosion and infiltration. *Agron. J. 61*:902.

Mannion, B. (1996). Topdressing and aeration strategies. *California Fairways 5(4)*:24, 29–30.

Marschner H., Romheld, V., and Kissel, M. (1986). Different strategies in higher plants in mobilization and uptake of iron. *J. Plant Nutr. 9*:695.

Marx, D. H., Berry, C. R., and Kormanik, P. P. (1995). Application of municipal sewage sludge to forest and degraded land (D. L. Karlen, R. J. Wright, and W. O. Kemper, eds.). ASA Special Publication No. 58. Madison, Wisconsin pp. 275–295.

Matyn, M. A., Ghent, M. Y., Tayel, S. A., Wahba, M., and Abd El Hady (1990). Effect of bitumen mulch on the growth of garlic under different field conditions. *Soil Tech. 3*:63.

Mengel, K., and Kirby, E. A. (1982). *Principles of Plant Nutrition*, 3rd ed. International Potash Institute, Bern, Switzerland.

Miller, J. C., and Bell, S. M. (1982). Crop production using cover crops and sods as living mulches. Intl. Plant Protection Center, Oregon State Univ., Corvallis.

Mitchell, A. R., and Van Genuchten, M. Th. (1993). Flood irrigation of a cracked soil. *Soil Sci. Soc. Am. J. 57*:490.

Mitchell, A. R., Ellsworth, T. R., and Meek, B. D. (1995). Effect of root systems on preferential flow in swelling soil. *Commun. Soil. Sci. Plant Anal. 26*:2655.

Miyamoto, S., and Storey, J. B. (1995). Soil management in irrigated pecan orchards in the southwestern United States. *HortTech. 5*:219.

Moreno, J. L., Garcia, C., Hernandez, T., and Pascaul, J. A. (1996). Transference of heavy metals from a calcareous soil amended with sewage-sludge compost to barley plants. *BioResour. Tech. 55*:287.

Mumpton, F. A. (1984). *Natural Zeolites, Zeo-agriculture: Use of natural zeolites in Agriculture and Aquaculture* (W. B. Pond, and F. A. Mumpton, eds.). Westview Press, Boulder Colorado, pp. 33–43.

National Association of State Universities and Land Grant Colleges (1974). Proceedings of the Joint Conference on Recycling Municipal Sludges and Effluents on Land. Champaign, Illinois, 9–13 July 1973. NASULGU, Washington, D.C.

Naylor, L. M., and Schmidt, E. (1986). Agricultural use of wood ash as a fertilizer and liming material. *Tappi J.* (October):199.

Naylor, L. M., and Schmidt, E. (1989). Paper mill wood ash as a fertilizer and liming material: field trials. *Tappi J.* (June):199.

Northcutt, G. (1996). Molecules hold solutions to some tough erosion problems. *Erosion Control 3*(5):54.

Norton, L. D., and Nearing, M. A. (1994). Soil amendments for altering erosion processes to reduce soil loss. *1994 Agron. Abs.*, p. 361.

Oberle, S. L., and Keeney, D. R. (1994). Interactions of sewage sludge with soil-crop water system. *Sewage Sludge: Land Utilization and the Environment* (C. E. Clapp, W. E. Larson, and R. H. Dowdy, eds.). Madison, Wisconsin, pp. 17–20.

Office of Technology Assessment (1984). Acid rain and transported air pollutants: implications for public policy, *OTA-O-204*. U.S. Congress, Washington, D.C.

Ohno, T., and Erich, M. S. (1990). Effect of wood ash application on soil pH and soil test nutrient levels. *Agric. Ecosyst. Environ. 32*:223.

Organic Gardening. (1996). Get your soil tested! *Organic Gardening 43*:30.

Orzolek, M. D. (1991). Reduction of nitrogen requirement for vegetable production with polymers, Proc. 23rd Natl. Agriculture Plastics Congr., pp. 204–210.

Orzolek, M. D. (1993). Use of hydrophylic polymers in horticulture. *HortTech. 3*:41.

Page, A. L., ed. (1983). Utilization of municipal wastewater and sludge on land. Denver, Colorado, 23–25 Feb. 1983. Univ. of California, Riverside.

Page, A. L., and Chang, A. C. (1994). Overview of the past 25 years: technical perspective. *Sewage Sludge: Land Utilization and the Environment* (C. E. Clapp, W. E. Larson, and R. H. Dowdy, eds.). Madison, Wisconsin, pp. 3–6.

Paine, L. K., and Harrison, H. (1993). The historical roots of living mulch and related practices. *HortTech. 3*:137.

Paustian, K., Parton, W. J., and Persson, J. (1992). Modeling soil organic matter in organic-amended and nitrogen-fertilized long-term plots. *Soil Sci. Am. J. 56*:476.

Peacock, B. (1994). Improving water penetration with gypsum. *Grape Grower 26*:20.

Petrovic, A. M. (1990). The potential of natural zeolite as a soil amendment. *Golf Course Management 58*:92–93.

Petrovic, A. M. (1993). *Use of Natural Zeolites in Turfgrass Management*. Zeolite '93. Red Lion Hotel Riverside, Boise, Idaho, pp. 162–164.

Pierre, W. H., and Banwart, W. L. (1973). Excess-base and excess base/nitrogen ratio of various crop species and parts of plants. *Argon J. 65*:91.

Pierre, W. H., Meisinger, J. J., and Birchett, J. R. (1970). Cation-anion balance in crops as a factor in determining the effect of nitrogen fertilizer on soil acidity. *Agron J. 62*:106.

Pierzynski, G. M. (1994). Plant nutrient aspects of sewage sludge. *Sewage Sludge: Land Utilization and the Environment* (C. E. Clapp, W. E. Larson, and R. H. Dowdy, eds.). Madison, Wisconsin, pp. 21–25.

Pieters, A. J. (1927). *Green Manuring, Principles and Practice.* Wiley, New York.

Piper, C. D. (1990). Water absorbing polymers. Published by the author, San Luis Obispo. CA.

Pirsson, L. V., and Knopf, A. (1947). *Rocks and Rock Minerals.* New York: John Wiley; London: Chapman and Hall, p. 81.

Postnikov, A. V., and Illrionova, E. S. (1990). Applications of zeolites in the cultivation of plants. *Agrokhimiya 7*:113.

Postnikov, A. V., Chuprikova, O. A., Baikova, S. N., Yashina, L. P., and Scherbakova, R. V. (1991a). Zeolite for greenhouse soils. *Khim Sel'sk. Khoz. 6*:58.

Postnikov, A. V., Zekunov, A. V., and Eliseeva, N. A. (199lb). Cultivation of vegetable crops on zeolite. *Khim. Sel'sk. Khoz. 11*:22.

Powell, D. (1995). Humates—a plus for turf. *Carolinas Green*, Nov.–Dec.: 38–39.

Pryor, A. (1988). Pretty poly: water-absorbing polymers have been shown to improve yields in processing tomatoes. *California Farmer 269*(10):12.

Purcell, S. (1993). Soil surface crusting, causes and remedies. Unocal Solution Sheet, October 1993, pp. 1–3.

Purcell, S. (1996). Using N-phuric in low volume irrigation systems. Unocal Solution Sheet, July 1996, pp. 1–3.

Quastel, J. H. (1954). Soil conditioners. *Ann. Rev. Plant Physiol. 5*:75.

Reynolds, A. G., Wardle, D. A., Drough, T. B., and Cantwell, R. (1995). Gro-mate soil amendment improves growth of greenhouse-grown chardonnay grapevines. *HortSci. 30*:539.

Riggle, D. (1995a). A finer grind for rubber recyclers. *BioCycle* 36(3):42–43, 55–56.

Riggle, D. (1995b). Soil science with a palette of composts. BioCycle 36(5):74–77.

Robinson, K. G., Ghosh, M. M., Zhou, S., and Hunt, W. P. (1995). The impact of a biologically-derived surfacatant (rhamnolipid R1) on HOC mineralization. Proc. Water Environ. Fed. Annual. Conf. Expo., 68th, 2:569.

Rogers, T. H., and Giddens, J. E. (1957). Green manure and cover crops. *Soil, the Yearbook of Agriculture 1957*. Washington D.C., pp. 252–257.

Romney, E. M., Wallace, A., Cha, J. W., and Mueller, R. T. (1981). Effect of zone placement in soil on trace metal uptake by plants. *J. Plant Nutri. 3*:265.

Ryan, J. A. (1994). Utilization of risk assessment in development of limits for land application of municipal sewage sludge. *Sewage Sludge: Land Utilization and the Environment* (C. E. Clapp, W. E. Larson, and R. H. Dowdy, eds.). madison, Wisconsin, pp. 55–65.

Shainberg, I., Summer, M. E., Miller, W. P., Farina, M. P. W., Pavan, M. A., and Fey, M. V. (1989). Use of gypsum on soils: a review, *Advances in Soil Science 9*:2–111.

Shear, G. M. (1985). Introduction and history of limited tillage. *Weed Control in Limited Tillage Systems* (A. F. Wiese, ed.). Weed Sci. Soc. Amer., Champaign, Illinois.

Smitte, D. (1991). Seaweeds come ashore. *Fine Gardening 22* (Nov./Dec.): 31–33.

Soil Science Society of America (1973). *Glossary of Soil Science Terms.* SSSA, Madison, WI.

Stanley, R. J., and Kempsey, L. A. (1990). Liquid sprays to assist stabilization and revegetation of coastal sand dunes. *Soil Tech. 3*:9.

Stevenson, F. J. (1979). Humates—facts and fantasies on their value as commercial soil amendments. *Crops and Soils 79* (April/May):14–16.

Stumpe, J. W., and Vlek, P. L. G. (1991). Acidification induced by different nitrogen sources in columns of selected tropical soils. *Soil Sci. Soc. Am. J. 55*:145.

Takamiya, N. (1994). *Study of Peat for Desert Greening.* Asahi Garasu Zaidan Josei: Kenkyu Seika Hokoku, pp. 603–621.

Target Biofiles. (1993). Water-injected gypsum improves penetration. *Target Specialty Products Biofiles 1*:4.

Tilton, J. L. (1997). Efforts abound to fight agricultural land erosion. *Erosion Control 4*(1):64.

Torri, K. (1978). Utilization of natural zeolites in Japan. *National Zeolites: Occurrence, Properties, Use* (L. B. Sand, and F. A. Mumpton, eds.). Pergamon Press, Elmsford, New York, pp. 441–450.

Traynor, J. (1986). Coping with low calcium water. *Fruit Grower*, November:16-A.

Tull, J. (1731). *Horse Hoeing Husbandrie: An Essay on the Principles of Vegetation and Tillage.* Published by the author, London.

USEPA (1993). Standards for Disposal of Sewage Sludge; Final Rules: Part II. Federal Register, Feb. 19, 1993, pp. 9247–9415.

Ugolini, F. C., and Sletten, R. S. (1991). The role of proton donors in pedogenesis as revealed by soil solution studies. *Soil Sci. 151*:59.

Wahba, S. A., Abdel Rahman, S. I., Tayel, M. Y., and Matyn, M. A. (1990). Soil moisture, salinity, water use efficiency and sunflower growth as influenced by irrigation, bitumen mulch and plant density. *Soil Tech. 3*:33.

Walker, J. M. (1994). Production, use and creative design of sewage sludge biosolids. *Sewage Sludge: Land Utilization and the Environment* (C. E. Clapp, W. E. Larson, and R. H. Dowdy, eds.). Madison, Wisconsin, pp. 67–74.

Walker, H. F. (1995). The use of emulsified asphalt for mulching. *Land and Water 39* (May/June):43.

Wallace, A. (1980). Trace metal placement in soil on metal uptake and phytotoxicity. *J. Plant Nutr 2*(1&2):35.

Wallace, A. (1982). Cation–anion relationships in a nonfacultative halophye, *Galenia pubescens. Soil Sci. 134*:51.

Wallace, A. (1993). Participation of silicon in cation–anion balance as a possible mechanism for aluminum and iron tolerance in some gramineae. *J. Plant Nutr. 16*:547.

Wallace, A. (1994a). Soil organic matter must be restored to near original levels. *Commun. Soil Sci. Plant Anal. 25*:29.

Wallace, A. (1994b). Small-scale farms as a model for conservation. *Commun. Soil Sci. Plant Anal. 25*:67.

Wallace, A. (1994c). Soil acidification from use of too much fertilizer. *Commun. Soil Sci. Plant Anal. 25*:87.

Wallace, A. (1994d). Use of gypsum on soil where needed can make agriculture more sustainable. *Commun. Soil Sci. Plant Anal. 25*:109–116.

Wallace, A., and Khadr, A. H. (1966). Some plant physiological effects of humic acids derived from leonardite. *Current Topics in Plant Nutrition* (A. Wallace, ed.). Los Angeles, California, pp. 45–52.

Wallace, G. A., and Stiles, S. (1995). Gypsum. *National Gardening 18*(2):84–87.

Wallace, A., and Wallace, G. A. (1986). Effect of polymeric soil conditioners on emergence of tomato seedlings. *Soil Sci. 141*:321–323.

Wallace, A., and Wallace, G. A. (1987). Conditionerigation: new process proves successful. *Irrigation J. 37*(3):12–15.

Wallace, A., and Wallace, G. A. (1990a). Soil and crop improvement with water-soluble polymers. *Soil Tech. 3*:1.

Wallace, A., and Wallace, G. A. (1990b). Using polymers to enhance shrub transplant survival and seed germination for revegetation of desert lands. Proceedings of a Symposium on Cheatgrass Invasion, Shrub Die-Off, and Other Aspects of Shrub Biology and Management (E. D. McArthur, E. M. Romney, S. D. Smith, and P. T. Tueller, Nevada eds.). Intermountain Research Station, USDA Forest Service, Las Vegas, Nevada, April 5–7, 1989.

Wallace, A., and Wallace, G. A. (1992). Some of the problems concerning iron nutrition of plants after four decades of synthetic chelating agents. *J. Plant Nutr. 15*:1487.

Wallace, A., and Wallace, G. A. (1994a). Water-soluble polymers help protect the environment and correct soil problems. *Comm. Soil Sci. Plant Anal. 25*:105.

Wallace, A., and Wallace, G. A. (1994b). A possible flaw in EPA's 1993 new sludge rule due to heavy metal interactions. *Commun. Soil Sci. Plant Anal. 25*:129.

Wallace, A., Wood, R. A., and Soufl, S. M. (1976). Cation–anion balance in lime-induced chlorosis. *Comm. Soil Sci. Plant Anal. 7*:15–26.

Wallace, A., Wallace, G. A., and Abouzamzam, A. M. (1986a). Amelioration of sodic soils with water-soluble polymers. *Soil Sci. 141*:359.

Wallace, A, Wallace, G. A., Abouzamzam, A. M., and Cha, J. W. (1986b). Effects of polyacrylamide soil conditioner on the iron status of soybean plants. *Soil Sci. 141*:368.

Wang, Y. T. (1989). Medium and hydrogel affect production and wilting of tropical ornamental plants. *HortSci. 24*:941.

Watanabe, N., Kawabara, M., Kata, R., Nomura, K., Iwashita, T., and Kammimura, K. (1994). Soil amendment containing zeolites. Jpn. Kokai Tokky Koho JP 06, 49, 447.

Wedin, D. A., and Tilman, D. (1996). Influence of nitrogen loading and species composition on the carbon balance of grasslands. *Science 274*:1720.

Whiteley, C. P., and Pettit, C. (1994). Effect of lignite humic acid treatment on the rate of decomposition of wheat straw. *Biology and Fertility of Soils 17*(1):18.

Wildman, W. E., Peacock, W. L., Wildman, A. M., Goble, G. G., Pehrson, J. E., and O'Connell, N. V. (1988). Soluble calcium compounds may aid low-volume water application. *Calif. Agric. 42*(9):7.

William, R. D. (1987). Living mulch options for precision management of horticultural crops. Oregon State Univ. Extension Service, Corvallis, Ext. Circ. 1258.

Williams, C. G. (1926). The testimony of the field experts of the country. *J. Amer. Soc. Agron. 18*:106.

Willis, H. (1996). Kelp and humic acids for crop production making better soils. *AcresUSA*: 38(Oct.):1, 6–8.

Witter, E. (1996). Towards zero accumulation of heavy metals in soils: an imperative or a fad? *Fert. Res. 43*:225.

Wofford, D. J., Jr. (1992a). Worldwide research suggestions for cross-linked polyacrylamide in agricultural research. The 1992 Intern. Conf. for Agri. Res. Administators, Sept. 13–19, McLean, Virginia.

Wofford, D. J., Jr. (1992b). More on water-absorbing polymers. *Hortideas 12*:133.

Wofford, D. J., Jr. (1992c). Dryland raspberries grown with 14 inches of rain! *Northland Berry News*, Dec. 1992.

Wofford, D. J., Jr., and Orzolek, M. D. (1993). No irrigating or weeding for 10 years. *Amer. Veg. Grower*, November, pp. 30–32

Yasuda, H., Takuma, K., Mizuta, N., and Nishide, H. (1995). Water retention changes in dune sand due to zeolite addition. *Tottori Daigaku Nogakubu Kenkyu Hokoku. 48*:27–34.

Zaharieva, T., and Romheld, V. (1991). Factors affecting cation–anion uptake balance and iron acquisition in peanut plants grown on calcerous soils. *Plant Soil 130*:81.

Zahow, M. F., and Amrhein, C. (1992). Reclamation of a saline sodic soil using synthetic polymers and gypsum. *Soil Sci. Soc. Am. J. 56*:1257.

Zhang, X. C., and Miller, W. P. (1996). Polyacrylamide effect on infiltration and erosion in furrows. *Soil Sci. Soc. Am. J. 60*:866–872

Zhang, J., Yan, L., Guan, L., Changhua, L., Li, H., and Zhang, Z. (1994). Effectiveness of zeolite in increasing crop yield and its effect on the ions of leaching water in soil. *Turan Tongbao 25*:123.

19
Use of Soil Conditioners in Landscape Soil Preparation

Garn A. Wallace *Wallace Laboratories, El Segundo, California*

I. INTRODUCTION

A landscape architect once said, “Good soil and bad plants beat bad soil and good plants any day of the week.” Tilth in soil is critical for many reasons. Drainage and aeration are also extremely important when dealing with high-value plantings. Sensitive plants will have some dieback of roots if they experience as little as 15 minutes deprived of oxygen (Letey et al., 1961). Aesthetics is also all-important in landscaping. This is tied to economics since landscaping around a building is often valued at as much as 10 to 20% of the cost of the building. It is, therefore, extremely urgent that soil preparation for landscaping be as perfect as possible. Unfortunately, this is not a general rule in the landscape industry, at least in the USA. The need for correction of mistakes of failed landscapes is all too common. A large portion of the effort in this laboratory is for correction of failures of previous plantings.

Substandard plants will generally develop into healthy, mature plants when they are grown in properly prepared soil. On the other hand, good plants generally deteriorate, grow slowly, and turn off-color when planted in poor soil. Nearly always no one needs to plant in poor soil since satisfactory soil can usually be prepared from bad soil and can give complete success. Absence of toxicities and adequate drainage are implied when soil is excellent. Poor soil preparation can result in nearly complete failure and gives an impression of poor design work made by the landscape architect. If proper soil preparation is really impossible, raised beds with man-made or imported soil may be used if precautions are taken to prevent soil interfaces.

While soil preparation alone accounts for merely 3 to 4% of the cost of many landscaping budgets, it substantially affects the whole job. It can make the difference between complete success and utter failure. Soil preparation should not be neglected, since poor soil preparation can later cause excessive maintenance expenses. The additional direct costs of replanting can be from two to five times that of the soil preparation expense. Even higher indirect losses are possible from such factors as reduced leasing and sales of new building developments or from longer periods needed to achieve full occupancy of buildings. Well done soil preparation will help guarantee healthy, attractive landscaping; it is like insuring success and is well worth any slight increase in expense.

Poor plant appearance and discoloration, surface root growth in contrast to deep rooting, toppling of trees by strong winds, defoliation, and root rot are examples of results of poorly prepared soils in landscaping. Additionally, slow growth of ground covers, shrubs, and hydroseeded areas can create erosion and weed problems. Waterlogging of soils also occurs in soil with poor tilth or poor drainage.

One of the goals of a good soil preparation program is the creation of proper soil tilth. A soil with appropriate tilth has the clay and silt particles bound together in reasonably large aggregates to improve air and water movement. Humus in soil organic matter can cement the soil particles together, but use of organics for soil improvement is generally a slow process requiring large amounts of plant litter accumulation in the soil in order to upgrade its characteristics. This happened naturally over a long period of time in the prairie lands of the USA with a humid climate and with tall grasses and an abundant supply of available calcium. The result was excellent soil until it had been cultivated continuously for over 100 years when compaction and erosion eventually resulted. Soil conditioners were then needed for creation of soil with properties acceptable for landscaping.

Soils are not all alike; landscape job requirements are not all alike. The different purposes of each landscape site must be considered as treatments with soil conditioners are made (Urban and Craul, 1996). For example, lawn areas need high water infiltration rates while flower and shrub planting areas are more in need of a high water holding capacity. High compaction resistance and higher water infiltration achieved mostly with mineral soil conditioners usually are coupled with low water holding capacity. Organic amendments tend to give high water holding capacity but slower infiltration. High clay soils are improved with addition of organic matter as are sandy soils, but for different reasons. All these factors and more should be considered in landscape soil preparation.

II. COMMON SOIL PROBLEMS THAT NEED CORRECTION OR IMPROVEMENT IN LANDSCAPE WORK

Shallow soil
Crusted soil
Poor water infiltration

Poor drainage
Compacted soil
Highly alkaline saline soil
High exchangeable sodium levels in soil
Calcareous soil
Acidic soil
Dustiness
Erosion
Sticky, muddy soil
Poorly aerated soil
High magnesium containing soil (including Serpentine soil)
Infertile soil
Low water holding capacity of soil
Multiple soil interfaces
Presence of excess available trace or heavy metals
Poor quality of water used for irrigation (including some reclaimed water)
Poor plant growth on the soil
Off-color plants on the soil
Wilting plants on the soil
Defoliated plants on the soil
Diseased plants on the soil
Dying plants on the soil
Iron deficiency symptoms in plants on the soil
Waterlogging of soil
Very low soil organic matter levels
Hydrophobic soil
Interfaces when two or more kinds of soil are present
Very low cation exchange capacity of soil
Problems in soil created by construction
Soil with debris and contaminants from past land use
Various other toxicities in the soil

A. Soil Compaction

Soil compaction is one of the most important problems involved in preparing a soil for landscaping. Research at the University of Florida has demonstrated the need for loose, uncompacted soil for plants to grow well and survive (Gilman et al., 1987). They grew Honey Locust (*Gleditsia triacanthos*) in sandy loam soil for 25 months with and without soil compaction. Soil was removed from a 60 cm diameter, 75 cm deep planting pit, and two-year-old trees were planted. In the uncompacted treatment, the rootball was covered with loose soil. This soil had a dry bulk density of 1.20 g per cm^3 (75 pounds per cubic foot or about 53% compaction). In the other treatment, soil was compacted around the rootball to give a dry bulk density of 1.78 g per cm^3 (111 pounds per cubic foot or about 80% compaction).

Upon the conclusion of the study, the size of the trees was measured and the root systems were carefully excavated and mapped. In the compacted soil, no roots grew deeper than 25 cm. One-fourth of them were in the first 2.5 cm of soil, and three-fourths were in the upper 12.5 cm. In contrast, for the uncompacted soil, 58% of the roots were deeper than 25 cm with no branching roots in the uppermost 2.5 cm of soil.

Measurements of the canopies and root systems showed a direct correlation between the canopy radius and the root radius. Roots spread two to three times beyond the distance from the tree trunk to the drip line in the uncompacted soil. This did not happen in the compacted soil; roots went to the drip line only in the compacted soil. Good development and spreading of the above-ground canopy required good growth of the root system. The data also support the hypothesis that quick establishment of the root system is best for the trees. In the compacted soil, a tap root did not develop. Thus resistance to wind would be less and toppling of trees would be more frequent in compacted soils. Therefore the status of the soil is all-important to the success of any landscape project.

III. USE OF SOIL AND WATER ANALYSES

Successful soil preparation for landscaping must be guided by appropriate laboratory support. Some characteristics of soil can be checked easily by soil analyses. Others can be had from observation. A handful of good soil will compress together as a good potting mix does, but it will not stick together in a tight ball, such as with moist clay. Moderate compression indicates good tilth. Laboratory and field work are both important.

Soil and water analyses with accompanying diagnosis can provide important answers for a soil management program (Wallace and Wallace, 1994). Soil must be managed to give it biological, nutritional, and physical qualities (Youngberg, 1992). This requires careful attention to analyses. Various analyses are important in landscaping. They even help conserve resources (Gardener, 1996). Their use is essential if failures are to be avoided.

A. Soil pH and Whether or Not the Soil is Calcareous

The results of analyses determine whether modifications are necessary and possible. Recommendations for different plant types, including turf, woody plants, are species resistant to iron and other deficiencies, and other nutrient problems can be developed. It is wise to make plant selections that minimize the need for soil manipulation.

B. The Physical Characteristics of the Soil

The analytical results can determine whether conditioners will help. Fortunately, major soil improvement is nearly always possible. Erosion can be minimized for a better environment. The drainage characteristics of all horizons need to be ascertained, and these must be determined on site. This is of urgent importance.

C. The Presence of Salinity and Sodicity

Analysis determines needs for gypsum, leaching, or change of irrigation water or its modification. Salt concentration and exchangeable sodium percentage need to be known for any soil. Analysis also indicates whether soil conditioners other than gypsum will help.

D. The Status of the General Fertility

Analyses should be made for all essential macro- and micronutrients, and fertilizer recommendations can be given for achieving specific goals. Adequate levels and adequate balance are both important. Excesses can be avoided, and this will result in considerable savings and more environmental responsibility. Much research still needs to be done on the interpretation of soil analysis, however, (Dahnke, 1993). Spacial relationships need to be considered, since major differences in soil characteristics can occur in short distances.

E. Evaluation of Possible Toxicity of Nonessential Elements and Also of Some That Are Essential

Past use of soil conditioners and fertilizers frequently has led to the accumulation of trace and heavy metals in toxic concentrations in some soils. Potential problems can be identified by the laboratory staff and procedures developed to overcome any adverse effects. Problems can be avoided if soil conditioners-amendments are also analyzed prior to use.

F. Soil Compaction and Poor Water Penetration

These problems cause surface erosion, water loss, plant stress, root diseases, and unsightly landscapes (Gray and Pope, 1986). Prevention, of course, is much easier than correction; however, correction is usually possible and is guided by laboratory analyses.

G. Reasons for Poor Root Growth

We have more calls for laboratory work to explain dying roots and poor root growth than almost any other problem. Routine testing and use of recommendations will avoid most for root problems, which are caused by soil, water, salt, compaction, and toxic substances.

H. Low Water Quality and Irrigation with Reclaimed Water

Laboratory testing can guide successful use of such waters for irrigation. Both soil and water need testing to guide use of the water so that adjustments can be made before major problems arise.

I. Management for Water Conservation

Use of test results helps to conserve irrigation water. Often, half the water can be saved with use of soil conditioners and conservation irrigation methods.

J. Impact of Cultural Practices

Soil problems frequently have been or are caused by or enhanced by cultural practices. Guidance can come from a laboratory concerning a number of practices that may be harmful. Type of tillage is an example.

K. Biological Status of Soil

Current research in many laboratories has the purpose of finding parameters that identify the biological quality of a good soil (Youngberg, 1992). Healthy soil must be a live functional ecosystem. Content and composition of soil organic matter are both important (Kononova, 1966).

L. Special Problems in Various Urban Soils

Craul (1996) has outlined several problems that are peculiar to urban landscapes. Most of these are related to previous use of the land for buildings and infrastructure of various kinds. The soil often has been cut, filled, layered, and/or mixed with all kinds of contaminants and debris. Soil conditions on land previously used for buildings become very unpredictable, so that detailed site sampling is necessary to identify potential problems. The contaminants include metal, glass, plastic, wood, asphalt, masonry, stones, various organics, oil, pesticides, heavy metals and others. Designers must cope with these materials when they are present. Rainwater coming off sidewalks and buildings can contain substances that increase the soil pH. Soil temperatures can be modified by sidewalks and buildings enough to require special attention.

IV. SOIL CONDITIONERS

A. Some Benefits to Soil from the Use of Soil Conditioners

Soil aeration can be improved.
Drainage can be improved.

Deep rooting of plants can be obtained.
Improved water management is possible and urgent.
Conditioners can be an aid in plant disease management.
Less soil compaction comes from proper use of soil conditioners.
Indirectly, the soil chemistry and fertility can be improved.

B. At Least Seven Types of Soil Conditioners

Except in a very few cases, all soils need conditioning to establish and maintain a superb landscape. The best soils are found in areas of moderate rainfall of around 75 cm per year with temperate conditions and under good management practices that result in a moderately high soil organic matter content. Common problems encountered in native soils are high levels of harmful salts, low levels of soil organic matter, poor water infiltration and drainage, low aeration, compaction, and crusting problems. Soil conditioning involves improvement in the physical properties of soil, almost always with different types of conditioners on a given soil.

The success of soil for landscaping is affected by its structure or the arrangement of soil particles in aggregates or crumbs. Soils lacking those larger composites of sands, silts, and clays are termed structureless or deflocculated. The function of most soil conditioning materials is to cement smaller soil particles into larger water-stable aggregates. Otherwise the aggregates created by tillage break down during irrigation or rain. Soils containing stable aggregates maintain good water infiltration, retain plant-available water, and provide for large amounts of pore space needed for excellent aeration and temperature control of the soil. Stable aggregates favorably influence nutrient availability and prevent accumulation of undesirable salts. Various conditioners are useful for the creation of stable aggregates and otherwise improving the physical properties of soil.

1. Inorganic Minerals

Some useful minerals (see Chapter 7–11) are gypsum, sulfur, sulfuric acid, pyrite, iron sulfate, limestone (calcitic and dolomitic), lime oxides, and calcium silicate. Some of these minerals help to control excess sodium and soil alkalinity and excess lime in soil. Calcium from the gypsum or from acidification of lime in the soil helps to flocculate clay into the more porous block or aggregate form. Sulfur and pyrite produce sulfuric acid with the action of appropriate microorganisms. This oxidation process is slow, especially in poorly aerated and high pH soils. Only a few types of bacteria can oxidize sulfur, and these bacteria are not common in nonacid soils. Some of these minerals can be used to correct soil acidity.

2. Inert Materials

Inert or near inert materials (see Chapter 18) such as sand, gravel, expanded minerals (perlite and vermiculite), zeolite, diatomaceous earth, pozzolan, and nondecomposable wood materials (redwood and cedar sawdust) are used to dilute soil particles. Sometimes clay is used. Often a very large quantity of an inert material is needed, such as 10% to 90% of the volume of soil to be conditioned, as in the case of sand. This is not always cost effective and in most instances is not. Sometimes sand and clay treatments are not effective. For example, it does not take much clay to cement sand into a concrete-like material unless other conditioners are also used.

3. Decomposable Organic Material

These (see Chapters 2–6) have animal and plant origins and are products such as composts, manures, sludges (biosolids), hardwood sawdusts, and dried blood. These products mainly affect the soil indirectly. Most soil microbes use decomposable organic materials as an energy source. In the process, biological gums and mucilages are formed. These gums are well known as natural soil cements that form water-stable soil aggregates (Sarma and Das, 1996). The more decomposable the organic materials added to the soil, the more such gums are formed. However, too much organic matter may cause toxicity such as from salts in manures or heavy metals from use of large amounts of sludges or biosolids. Because the gums and mucilages are also decomposable, the cementing action is often and variously temporary. The warmer the soil, the shorter the period of soil improvement. During the decomposition process, oxygen is consumed and may become depleted.

Here we give a definition of good humus additives for landscaping. Although these specifications are rather strict, their use minimizes future soil problems in landscaping.

1. Humus material most useful in landscaping preferably shall have an ash content of no less than 8% and no more than 50%.
2. The pH of the material shall be between 6 and 7.5.
3. The salt content shall be less than 8 dS per m at 25°C (ECe less than 8) on a saturated paste extract. If the concentrations of both sodium and chloride are 30% or less of all cations and anions, respectively, on an equivalent basis in the saturation extract, the maximum level of acceptable salinity shall be 10 dS per m at 25°C.; and if the sodium and chloride are 10% or less, the maximum acceptable salinity shall be 13 dS per m at 25°C.
4. Boron concentration of the saturated extract shall be less than 1.0 mg per kg for land areas threatened by boron toxicity.
5. Silicon concentration (acid-insoluble ash, soil, and sand) shall be less than 20%.
6. Calcium carbonate shall not be present if the material is to be applied on alkaline calcareous soils.
7. Types of acceptable products are some composts, some manures, mushroom composts, straw, alfalfa, some sludges (biosolids), and peat mosses,

which are low in salts, low in heavy metals, free from weed seeds, and free of pathogens and other deleterious materials.

8. Composted wood products are conditionally acceptable (stable humus must be present). Wood-based products, such as from redwood or cedar, are not acceptable for general soil preparation owing to toxins.
9. Sludge-based materials are not acceptable, especially if the soil already has a high toxic level of zinc, copper, cadmium, or other heavy metal based on soil analysis or if the biosolids have high concentrations of such elements. Conversely, soils low in essential micronutrients can be improved with biosolids.
10. The carbon: nitrogen ratio shall be less than 25:1.
11. The compost shall be aerobic without malodorous presence of decomposition products.
12. Some maximum total permissible pollutant concentrations of heavy metals in organic conditioners used in landscape soil preparation as recommended by the author are in Table 1. *Note*: These are much lower than the "clean" values permitted by the 503 rule of the USEPA (USEPA, 1993). The EPA-503 values really should not be used for landscaping because application rates are usually much higher than agronomic-field farming levels. In some nations, permissible levels may even be much lower than those in Table 1 and lower than they need be (Witter, 1996).

The rationale for the point of view concerning low levels of heavy metals in composts used in landscaping is that a minimum of 2.5 cm deep application is usually made on a project. On a per ha basis, that is about 160 MG. Often two

Table 1 Recommended Maximum Levels of Some Metal Pollutants in Organic Soil Conditioners Used in Landscaping

Element	Level, mg per kg dry weight
Arsenic	40
Cadmium	15
Chromium	500
Copper	400
Cobalt	50
Lead	200
Mercury	10
Molybdenum	60
Nickel	100
Selenium	36
Silver	30
Zinc	400

or three times that much is applied. If the concentration of a given heavy metal in the compost were 300 mg/kg, the loading rate for that metal on the soil would be about 48, 96 or 144 kg per ha for the 160, 320 or 480 MG rates. If after soil application, the extractable (or available) amounts were 10% of the total, then 4.8, 9.6, or 14.4 kg respectively per ha are available. For this one year or one time application, the tolerance level could be approached or even exceeded. If the 2800 mg per kg rate for zinc used by the USEPA for clean sludge were used, the available amount of zinc (10% of that applied) in one year could be 45, 90, or 130 mg per kg for the three use rates, which is too much for most conditions. For multiyear application rates like this, there could be toxicities, and they have been frequently observed in landscapes especially with zinc. Sources of zinc and other metals other than composts find their way into soil, and zinc from paint is an increasing problem. Landscapes are safer at levels shown in Table 1 than at the levels given in the USEPA-503 regulations, which really were meant for farms, where lower rates are used with a focus on nutrient needs rather than on soil conditioning. Also most ornamental plants, especially trees and shrubs, are more sensitive to heavy metals than are field crops like corn on which the USEPA 503 rule is based.

Higher amounts of salinity or boron from decomposable organic matter may be permissible if the soils are to be preleached to reduce the excess or if the plant species used tolerate the salinity and/or boron.

Some organic conditioners should not be used to excess, especially uncomposted or partially composted organic materials. However, uncomposted or partially composted organic minerals will form more gums and mucilages when applied to soil than will highly composted materials. Uncomposted organics at moderate rates must be applied one to three months ahead of planting. If too much decomposition takes place in the soil at time of planting, and if the volatile decomposition products do not readily escape from the soil, toxicity and plant injury can occur.

The bacterial content of soil resulting when partially decomposed organic matter is added can limit plant pathogenic fungi partially due to competition among microbes (Hoitink et al., 1995, Van Bruggen, 1995). Quality and quantity of compost both need consideration with regard for the presence of useful organisms.

4. Synthetic Organic Conditioners

An example of a synthetic soil conditioner (see Chapters 12–17) is anionic water-soluble polyacrylamide (WS-PAM). This polymer functions like the natural gums and mucilages in stabilizing soil aggregates. These conditioners do not have the means to make soil structure. Rather, they stabilize soil aggregates already formed such as from tillage operations. Because synthetic cements or polymers are 100% active, a few kilograms of these are equivalent to several MG of organic materials. In addition, the synthetic conditioners are not readily decomposable. They work for many years. Also, WS-PAM is essentially nontoxic.

For long-term stabilization of soil aggregates, synthetic mucilages such as WS-PAM are highly valuable and essential unless large amounts of decomposable organic materials are applied to the soil several times a year, preferably with a source of calcium (Wallace, 1994). WS-PAM and organics are best used simultaneously (Wallace and Wallace, 1986). Gel polymers are different from water-soluble polymers (Orzolek, 1993).

5. Wetting Agents

These are chemicals (see Chapter 18) used to lower the surface tension of water in order to make water "wetter." They also help to improve hydrophobic soil. They are surfactants, soaps, detergents, and penetrants. They help to move water through the soil in the existing pores. Moisture-holding capacity of soil is thus decreased. Some of these materials are toxic to plants, especially if directions are not followed. Some decrease levels of soil organic matter.

6. Soil Inoculates and Microbial Cultures

These (see Chapter 18) are microbes such as legume bacteria (*Rhizobia*) that have a symbiotic relationship with legumes for making atmospheric nitrogen available to plants (Metting, 1993; Hendrix et al., 1995). Mycorrhizae are beneficial fungi that help plants to assimilate nutrients and to ward off pathogens (Pfleger and Linderman, 1994). Enzymes and vitamins applied to soil have not been proven valuable for enhancing volume of plant growth, although they may have other benefits (Metting, 1993). Many but not all of these have had success as soil additives in landscaping, but only indirectly do they have soil conditioning value.

7. Mechanical Cultivation

Cultivation conditions the soil if soil moisture levels are adequate. Tillage and soil ripping are valuable in improving the physical properties of soil, particularly soil structure and pore space. Unless stabilized, however, soil crumbs formed by tillage do not endure the presence of water. Water slakes the weak binding material, causing the aggregates to disintegrate unless stabilized by organic additions (for several months' duration) or by synthetic conditioners, such as WS-PAM, which can persist for many years. Earthworms naturally cultivate the soil. Their presence can be enhanced with conditioners, especially organics.

The best soil preparation procedures use a combination of techniques. Some soil conditioners cannot be incorporated unless the soils are loosened and tilled. Often, rototillage is ineffective unless the soils are ripped. Even disk plows will not cultivate compacted soil more than a few cm deep unless the soil is ripped. Mechanical tillage is essential for most soil preparation techniques, especially for incorporation of conditioners.

In addition to soil conditioning, proper mineral nutrition in correct proportions of the nutrients is critical. All best management practices should be used simultaneously for their value to be maximum (see Chapter 1).

C. Which Soil Conditioners to Use for Which Soil Problem

1. For Lowering Soil pH

Sulfur, pyrite, sulfuric acid, and acidifying fertilizers (ammonium sulfate, potassium sulfate) are used.

2. For Raising Soil pH

Limestone, lime oxides, dolomite, and alkali-producing fertilizers (calcium nitrate, potassium nitrate) are used.

3. For Correcting Sodicity, Salinity, and Excess Magnesium

Gypsum, sulfuric acid, plus polymer soil conditioners to increase soil porosity are used. Typically, saline and/or alkaline soils need some inorganic minerals or conditioners, especially if the soil is low in soluble calcium. Gypsum or gypsum substitutes are commonly used.

4. For Increasing Water Holding Capacity and Decreasing Bulk Density

Organic matter, superabsorbant polymers, zeolites, pozzolan, and diatomaceous earth are used.

5. For Decreasing Evaporative Water Loss

Mulch, tillage, soil conditioning to increase soil structure, and shading the soil are used.

6. For Increasing Water Infiltration

Soil conditioning with organics and WS-PAM together to increase soil structure and to improve tillage are used. Gypsum also helps.

7. For the Control of Iron Chlorosis

Iron chlorosis in plants can be prevented in many or most cases with creation of adequate physical properties of soil. Conditioners to increase soil aeration, decrease waterlogging conditions, decrease soil pH if it is high, decrease soil and/or irrigation water levels of bicarbonates, decrease activity of heavy metals, provide pockets of available iron, and provide soluble iron all help to decrease the incidence of iron chlorosis.

8. For Soil Preparation (Includes Some of the Above)

Humus-type organic matter, decomposable organic matter, soil aggregating materials such as WS-PAM, and zeolite-pozzolan-type materials are used.

9. For Dilution of Clay Soil

Sand, perlite, vermiculite, slowly decomposable organic minerals (redwood sawdust, fir bark, rice hulls), zeolites, diatomaceous earth, and pozzolan are used.

V. SOME EXPECTATIONS FROM SOIL CONDITIONING

Acceptable plant growth requires control of several soil factors in unison. When a single factor is optimized to the detriment of others, inferior plant growth may result. Sand has been selected in some cases for backfill when shrubs and trees are planted because of its excellent drainage rates and high bulk density. Sometime this can give problems, however. Vine et al. (1981) studied root growth rates of corn and compared them for forest topsoil to fine sand and coarse sand. They discovered that the sands reduced the growth rates of roots by 50%. The depth of root growth was reduced an additional 50% by the presence of solid objects in the soil—in their case gravel.

Sand is a poor substitute for soil not just because sand does not retain plant nutrients and has low water-holding capacity. Sand particles are not compressible and they are restrictive of root growth. The essential properties of soil for proper plant growth were reviewed in *Root Development and Function* (Gregory et al., 1987). Soil conditioners are used to obtain these characteristics. The major essential properties are as follows.

A. A Good Soil Has Low Resistance to Root Penetration

The force required to displace soil by a growing root can easily be measured by inserting a probe called a penetrometer into the soil. For a 0.3 cm diameter penetrometer, the following forces are required to penetrate these soils and sand (Vine et al., 1981): forest topsoil, 6 g (100% growth rate); coarse sand, 300 g (59% growth rate); fine sand, 600 g (45% growth rate); and compacted soil arresting root growth, 1160 g (0% growth rate). For a loamy sand, the point where penetration resistance stops root growth is about 80% compaction (Taylor and Ratliff, 1969). Success in landscaping requires much lower compaction.

B. A Good Soil Is Compressible and Has Low Bulk Density

Because roots have limited abilities to push soil aside, they can more easily penetrate soil of low bulk density and soil whose particles are compressible. In compacted soils, roots will generally grow into worm holes, into old root channels, or along cracks. Roots grow easily in soils having bulk densities of 1000 kg per m^3. Sand hinders root growth because its bulk density is about 1600 kg per m^3 and the particles are noncompressible. Soil conditioning must create proper compressibility.

C. A Good Soil Has Porosity for Aeration, for Water Infiltration, and for Holding an Adequate Supply of Available Water

Root proliferation requires an optimum balance among air, water, and soil. The total available space for water, air, and pores for drainage is the volume that remains after subtracting the space occupied by soil particles from the total soil volume. Thus 80% compacted soil has available 20% space for air and water, and 60% compacted soil has 40% space. An average soil may absorb 20% water. Typical soils before conditioning used in landscaping may be near the zero point for air space, since average soils have 75 to 80% compaction and 20% or more water. There are other factors, such as the average diameter of soil pores, which control air and water movement in soils. See Chapter 4. Pores can be large enough for good drainage but could preclude root penetration in noncompressible soils. Soil with 70% compaction can give seven times better growth than 80% compacted ones, and 60% compacted ones, and 60% compacted ones can give 14 times better growth than 80% compacted ones.

Freshly tilled, unirrigated soils have about 50% compaction, which is an ideal balance of pore space and bulk density. This condition may be only temporary, however. Unfortunately, many soils are adversely affected by water. Tillage-prepared soil crumbs often disintegrate in the presence of water and form mud. Water lubricates the soil particles, allowing dense packing, which decreases pore space. The goal of good soil preparation is the prevention of soil recompaction. Soil conditioners are needed to create such soil.

When soil aeration is sacrificed, roots are restricted to the upper few cm, which is very different from the norm. A rule of thumb for good agricultural crops is that the top 30 cm of soil contains 40% of the roots, the next 30 cm contains 30%, the third 30 cm contains 20%, and the next zone contains 10%. Landscape plants should be the same.

Soils with excellent physical properties exist naturally in the presence of high levels of soil organic matter, such as in the undisturbed (uncultivated) prairie lands. The western USA climate is too arid for much soil organic matter to exist. The organic matter stabilizes the soil crumbs and makes them water-stable (Taylor and Ratliff, 1969). This condition can be accomplished with use of very high rates of *compost* types of organic conditioners. Wood residuals are not compressible and do not make good soil additives. The natural polymers for stabilizing soil crumbs are formed by microbes after the woods or other organic materials decompose (Martens and Frankenberger, 1992). Coarse wood products, if used at high rates, also impede root growth because of penetration resistance. Wood products act as wedges separating soil particles, and they lower the bulk density of the soil. However, because of lack of pores and other additional problems wood products are not very satisfactory.

The best and most economical method to optimize soils for plant growth is to stabilize tilled soil against the compacting action of water. This is done in several ways. Soil can be maintained at low bulk densities filled with voids through the use of a soil aggregating polymer in conjunction with a little bit of a true compost. Water-stable soil crumbs are formed that allow soil to have an ideal balance of particles, air, and water. Compaction is maintained around 50%, resulting in a low bulk density, which in turn gives a soil low in resistance to root penetration. Because of the large quantity of pores, the soil is compressible. Exeptional plant growth is now possible with this practice.

D. A Good Soil Has High Cation-Exchange Capacity

Cation-exchange capacity of the soil can be improved with use of certain soil conditioners such as organics, zeolites, pozzolan, or clay. Better storage of nutrient cations with less leaching of cations results.

E. A Good Soil Is Biologically Alive

Soil quality is now known to be extremely important (see Chapter 1). It is now known that soil will not support plant growth adequately unless the soil has proper microbial health. Adequate soil organic matter with continuous renewal is important to such health. Sometimes it may be necessary to supply certain microorganisms also (Metting, 1993).

F. A Good Soil Has No Differential Soil Interfaces

Interfaces arise if one kind of soil is placed on top of another or if soil amendments-conditioners applied in large quantities are layered to give two different conditions. Water and roots do not easily cross over such interfaces. Drought and wilting can occur when roots are on the wrong side of the interface.

Rather than importing topsoil, the existing soil usually can be improved sufficiently with simultaneous use of multiple soil conditioners including water-soluble polymers. All soil in the root zone can be made to be quite uniform by this procedure. Interfaces then can be avoided.

VI. WATER-SOLUBLE POLYACRYLAMIDE (WS-PAM) AS A SOIL CONDITIONER

A. Differences and Similarities Between Cross-Linked Polymer Gels for Water Absorption and Linear Water-Soluble Polymers (Usually WS-PAM)

The two different types of polymers used in landscaping and horticulture can and do cause confusion. Water-absorbing, gel-forming polymers, commonly known as super slurpers, are aids for increasing the water holding capacity of sandy soils or well-drained synthetic potting media which do not have the ability to hold as much moisture as the loam or clay soils. (see Chapter 12 and 13 in this book). The advantages of the gel polymers are that they reduce the frequency of irrigation and that the soil moisture becomes more constant (Orzolek, 1993). A more constant soil moisture will help encourage proper plant growth.

These super-absorbing polymers are manufactured from similar materials as are the water-soluble polymers used for soil improvement. The soluble polymers bind soil particles together for structural soil enhancement. For gels, the polymers are bridged between adjacent molecules and are thus cross-linked to make them insoluble so that they act like sponges. They absorb 50 to 400 or more times their dry weight in water.

These super slurpers are also used at the rate of 100 million kilograms annually for consumer products such as baby diapers and related products. Because of the large manufacturing capacity and numerous manufacturers seeking to increase their sales, many brands of water-absorbing polymers exist in the landscape trade.

The water-soluble linear polymer is a single chain of a large number of repeating units. These polymers are designed to bind together thousands of soil particles and to form a lattice work in and around soil crumbs (see Chapters 14–17). In time, this polymer also becomes water-insoluble in the soil. The function of the soil-enhancing polymer is not to absorb water but to stabilize cultivated soil by maintaining soil in a loose and friable state. The soil particles then become water-stable.

Cultivated soils most often respond to irrigation and rain by dispersing the soil particles from the crumbs. The result is an unstable amorphous mixture of soil and water. Water lubricates the clay and silt particles, allowing them to pack together to form soil of higher bulk density, which originally existed before tillage. Cultivation is performed for the purpose of lowering bulk densities as well as for weed control. Excellent soils do not increase in bulk density in the presence of water. Their crumbs are water-stable.

Unless large quantities of specific types of organic matter are added to the soil or exist naturally in the soil, the only reliable method of forming water-stable crumbs is through the use of soil binding polymers such as WS-PAM. Wood residues do not bind the soil. When used at very high rates, they lower the bulk density, which results in easier root penetration in the soil, but they do not provide for aeration and drainage. The economics is such that better soil is prepared with the linear polymers and also a small amount of compost usually at a lower net cost than with all-organic conditioners.

Most soils hold adequate moisture for good plant growth. However, with poor soil physical properties, irrigation is wasted as water runs off. Water conservation requires porous soils, which can be developed with linear polymers. Soil moisture can then be maintained longer by decreasing evaporative loss of water. Well-structured soils have less evaporative loss because the top two or more cm of polymer-treated soil acts like a mulch. This is another benefit of linear water-soluble polymers.

The available soil moisture reserve for plants is a function of the depth of the roots. Turf rooted to a 15 cm depth can have six times the moisture reserves as turf rooted to 2.5 cm depth. In lieu of a daily irrigation requirement, deeply rooted plants can be watered weekly or even less frequently. Since roots of most plants do not grow deeper than where oxygen can penetrate, plants will not root deeply in dense soils. Good soil structure concomitant with sufficient aeration, low bulk density and high porosity, need to be maintained. The best technique for creating good soil is through the use of soil enhancing linear polymers together with other soil conditioners.

Both types of polymers have a role in landscaping, but the roles are very different. It is those like WS-PAM that are used in soil preparation.

B. Preliminary Soil Preparation When Using Water-Soluble Polyacrylamide and Other Conditioners

The initial procedure in soil preparation is to cultivate the soil. Only in the cases where the soil is loose and the soil clods easily crumble when worked can this preliminary step be eliminated. No matter which soil conditioner is to be mixed into the soil, the soil must have crumbs no larger than 1 to 4 cm in diameter; most will be smaller. The secret of an excellent soil preparation is to have the conditioner uniformly blended into the soil.

Rototillers work well in some soils. However, on other soils, the soil may be so compacted that it can only be loosened by the rototiller in the top few cm. Good soil preparation normally requires a full 15 cm deep cultivation. When the job is large enough, farm implements would be faster and better. Rippers, chiselers, harrows, or powered tillers can be used. For smaller jobs, backhoes can be used. Even in some cases, jack hammers have been used followed by smashing the clods with picks or other tools. Debris, large rocks and clods can be removed by hand or by power raking.

Another skillful way to loosen soil where irrigation water is available is to irrigate the soil in order to wet it 15 or more cm deep. Some soils may require a week to have the wetting front migrate 15 cm due to very poor water infiltration. The soil then needs to partially dry so that when the soil is cultivated, the clods easily break apart into suitable sized particles. If the clods deform, bend, or ooze under compression, the soil is too wet to work. The required drying period may vary from a day to more than a week. Allowing the soil to dry excessively before tillage results in soil becoming hard once again.

C. Application of Conditioners

After the soil has been cultivated, the conditioners are applied. If the soil is to be conditioned with dry water-soluble materials, such as WS-PAM, the soil surface must also be reasonably dry. These materials become sticky and will not mix into the soil if the soil is not sufficiently dry. If damp conditioners are applied first, they must be allowed to dry before water-soluble conditioners are applied.

For uniform application, use spreaders. Drop spreaders are used for granular, powdery, dusty or light-weight materials. Since WS-PAM is a light-weight material, a drop spreader must be used if it is to be applied dry. Exposure to hands is also minimized; the material causes itchiness in some people when exposed to the skin. Cyclone spreaders can be used only for granular products. If wet materials, such as compost, are to be applied over the WS-PAM, the WS-PAM is first raked into the top 2 cm of the soil to prevent it from becoming sticky.

In the event that low rates of conditioners must be hand applied, they are raked into the soil after hand application in order to have better distribution. One liter of WS-PAM weighs approximately 0.6 kilogram. For hand application of WS-PAM, the following guide can then be used:

Application rate, kg per 100 m^2	Use the following, liters per 10 m^2
3	0.5
5	0.8
8	1.3
10	1.6

The soil conditioner is raked (cross-raking is better) into the soil afterwards.

Any dry WS-PAM spilled on walkways or roadways should be swept up. WS-PAM is very slippery when mixed with water on dry soil or on hard surfaces. Caution must be used to prevent accidents.

Bulky conditioners should be uniformly spread and then leveled with a wide rake. Organic conditioners are much more effective when used with WS-PAM. There is a sparing effect upon the organic matter so that only 1/8 to 1/4 of normal rates of organic matter are needed (1.25 to 2.5 cubic m per 100 m^2).

The next step is to rototill all the conditioners and fertilizers into the soil. All conditioners must be tilled into the soil the same day on which they were applied. If the soil is partially damp, the conditioners need to be tilled into the soil very soon after application. If they become wet they mix poorly in the soil.

Other concerns are to avoid working on very windy or damp days. Also, the soil needs to be very well mixed when shovels are used for tillage.

The dryness of the soil should be tested before applying WS-PAM. A teaspoonful of the soil conditioner is spread over a 15 × 15 cm square. If the material becomes sticky in 10 to 15 minutes, the soil is too damp. If the soil is too damp, the soil is allowed to dry some more or the WS-PAM can be rototilled into the soil immediately after application by working the drop spreader and rototiller in tandem.

D. Curing the Soil

Prior to planting, WS-PAM in the soil needs to be cured with water followed by some drying. WS-PAM will not improve soil until the soil has been wet with water. This mucilaginous soil conditioner is dissolved and then allowed to bind to the soil particles followed by partial or near complete drying. The soil needs to be wetted to 15 cm depth or to the depth of treatment. The soil can be irrigated or rained upon to dissolve the WS-PAM.

Unless the soil is wetted slowly with misting, some compaction will occur. The WS-PAM is unable to condition soil until it dissolves, which takes 5 to 15 minutes depending on particle size of the WS-PAM. Normally, treated soils followed with irrigation and drying need to be recultivated after the soils have dried to recreate aggregates that will then be stable.

1. Water Treatment After Dry Application of WS-PAM

With hoses or with the irrigation system, about 3 to 4 cm of water are applied. Checking below the soil surface can verify that the soil has been wetted to the treated depth. With WS-PAM in the soil, the soils take water slower than normal the *first* time. For small jobs where a hose is being used, the soil can be turned over with a shovel to wet the soil faster and to prevent water runoff.

After the soil has dried to the point at which the soil can be worked (soil will crumble when cultivated), the soil is recultivated. This cultivation is very easy.

2. Test the Effectiveness

One can determine how well the WS-PAM was applied and rotilled into the soil after the curing process has started. Well mixed soil will have stringiness, but poorly mixed soil will have sliminess in spots. After a few days of curing, the stringiness disappears as the soil completely cures. It will never return if mixing has been uniform. Properly prepared soil is illustrated in Fig. 1.

E. Leaching of Soil Salinity

If the soil has a high level of soluble salts, it may be advisable to leach it after it has been conditioned with WS-PAM and prior to planting. In some cases, the salts can be leached with normal irrigation after planting without harmful effects. This depends upon the drainage rates, the sensitivity of the plants to salts, water quality, the amount and frequency of irrigation, the rate of evaporation (daytime temperatures, solar influx, relative humidity), and the addition of nu-

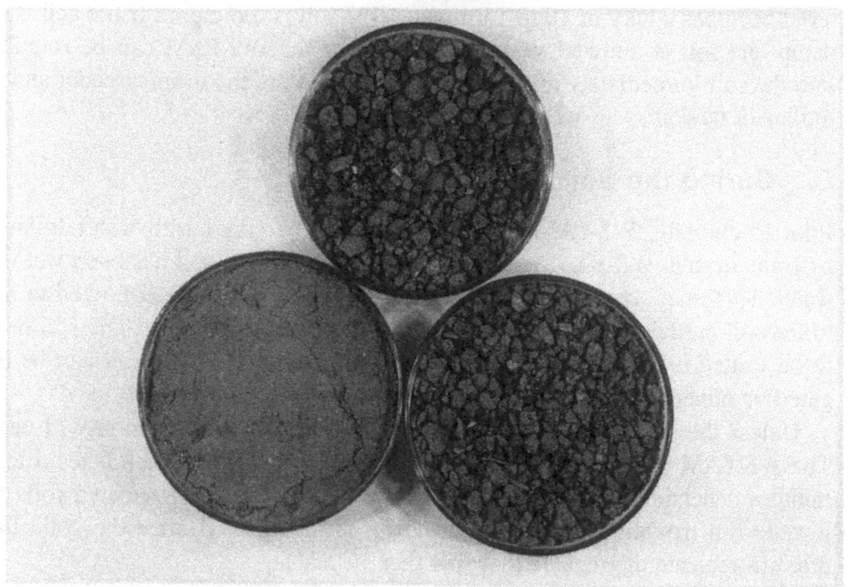

Figure 1 Effect of treatment of cultivated soil aggregates with a solution of WS-PAM followed by drying for the purpose of curing. Top is the soil before wetting with or without WS-PAM. Lower left is the soil without WS-PAM and after wetting with water and drying. Lower right is the dried soil after treatment with WS-PAM in solution. Preservation of existing aggregates is possible for any soil containing some clay.

trients and mineral conditioners to the soil. For typical soils and with good subsoil drainage and good water quality, 7.5 to 15 cm of water for leaching should be sufficient to lower the salinity by 10 dS m^{-1}. If possible, it would be useful to irrigate with water that has 1 or more mg L^{-1} WS-PAM plus 5 to 10 me L^{-1} of gypsum.

F. Alternate Procedure If Desired for Simplification

An alternate procedure to adding dry granules of WS-PAM to soil and one that is useful for large tracts of soil being prepared is to use a water sprinkler truck and apply the WS-PAM, preferably with gypsum, to the land as solution. From 200 to 400 mg L^{-1} of WS-PAM with two to three times as much gypsum are applied in 2.5 to 4 cm of water after the soil has been tilled and any other conditioners added. After curing, the land may not have to be rototilled to recreate desired aggregates. They will be made water-stable. With this procedure, the amount of WS-PAM needed per unit of land can be decreased to 1/4 or less than when applied dry because of better spacial relations (Wallace and Nelson, 1986).

For small land parcels, the soil is wetted with about 40 liters of a 200 to 400 mg L^{-1} solution per square meter. The soil should be wetted to the depth of expected treatment. It may be necessary to use shovels to contain the solution to the area being cured and to avoid runoff.

WS-PAM solutions can be applied with a spray rig (without orifices) or an irrigation system. A siphon/diluter may be used to add a stock solution into the stream of water.

G. Preparation of WS-PAM in Solution

WS-PAM is water soluble but not to a large extent. Depending on particle size, it will completely dissolve in about ten minutes if it is properly added to water. The material needs to be dispersed into the water for it to dissolve quickly, so it is added into a stream of water. If the material is added in a large mass without dispersion, clumps of the material will form that are very slow to dissolve. Rates to use to obtain various concentrations are shown in Table 2.

An easy method to dissolve WS-PAM is to use a 240 ml paper cup and then with it pour the materials slowly so that the particles fall into the stream of water coming out of the hose. The hose should not have a nozzle and the water flow rate should not be too fast (10 to 40 liters per minute for a 1.9 cm hose), otherwise splashing will be a problem. After the soil conditioner has been added, the flow rate can be increased. For small containers, such as less than 50 liters, several agitations over five minutes will keep the particles suspended until they dissolve.

H. Why Organic Conditioners Can Sometimes Be Toxic to Transplants

Inconsistent results in the use of compost to condition soils was the subject of a report from Israel (Avnimelech et al., 1990). In some cases, compost caused definite improvement in soil properties and plant growth, but in other cases, no positive results were found. The goals of the study were to find the causes, to predict the response, and to determine optimal application methods.

Table 2 Some Approximate Amounts to Use When WS-PAM Is Put into Solution for Soil Conditioning

Concentration		Tank size, liters			
percent	mg/liter	20	120	400	4000
0.01%	100	1/2 tsp.[a]	1 tbs.[a]	3 tbs.	0.4 kg
0.02%	200	1 tsp.	2 tbs.	1/2 cup	0.8 kg
0.05%	500	2 tsp.	0.4 cup	1 1/3 cup	2.0 kg
0.10%	1000	4 tsp.	0.8 cup	2 3/4 cups	4.0 kg

[a]Tsp. is a teaspoon (about 5 g). Tbs. is a tablespoon (about 15 g).

They found that low application rates of organic conditioners such as compost were unable to improve soil properties and plant growth. As the application rate was increased, favorable results were observed. However, if the rate exceeded an optimal value, toxic effects were sometimes obtained. The optimal rate in most of their studies was equivalent to between 0.2 and 1 cubic meter of organic conditioner per 100 square meters of land. The range is a low rate to a moderate rate.

Since laboratory experiments gave consistent soil improvement only when the soil was well aerated, the authors concluded that the toxicity with organic matter resulted from insufficient soil aeration. If the applied organic conditioners did not quickly condition the soil and increase the porosity, the level of aeration was worse with the application of organic matter.

Increasingly high level applications of organic matter progressively increase the moisture content of most soils, which decreases the supply of soil oxygen. In addition, the decomposition of applied organic conditioners consumes oxygen, so anaerobiosis could result. The decomposition methodology of organic matter changes under various aeration conditions. Under some conditions, humic compounds are formed that improve soil; while under others, organic acids and ethylene are formed, which are injurious. Excessive amounts of decomposable organic matter with changes in soil aeration cause a change in the method of metabolism. If the soil below the surface smells putrid, there are severe problems.

Also observed by Avnimelech et al. was a loss of nitrate nitrogen by denitrification with high application rates of organic matter, due to the lack of aeration. Another problem was the fact that the positive results decreased with time. They usually lasted for only one growing season.

These studies indicate the need for better soil aeration, especially when organic matter is used. The best technique is to use a small amount of organic matter with a soil conditioner that functions like humus but does not decompose like organic conditioners. Instead of consuming oxygen, much more oxygen is provided because of increased porosity. These effects are achieved with WS-PAM. The soil is immediately conditioned when it is used. In addition, the organic conditioners that are applied are made even more effective because of the improved aeration and the formation of beneficial compounds. Research results from around the world have validated this successful technique.

Landscape professionals have often debated the value of organic products in backfill for transplants and in other soil preparation activities (Morgan, 1986). The information presented here indicates clearly the nature of the so-called dilemma. Organic conditioners can be useful in landscaping only with well-aerated soil. Again, the simultaneous uses of the WS-PAM soil conditioner with organic matter is an excellent means of assuring proper soil aeration. Synergistic responses to both water-soluble polymer and organic amendments can be obtained when used together (Wallace and Wallace, 1986).

Successful landscaping requires the utilization of the proper techniques that give good soil porosity and a high degree of aeration. They are now available for use.

VII. LOSS OF PLANTS TO SOIL-BORNE PATHOGENS CAN BE MINIMIZED

Organic soil conditioners can help manage some plant diseases. Successful cultivation of horticultural plants depends in part upon the microbial relations of the soil (Metting, 1993). The fungal flora of soil contains both beneficial (Hendrix et al., 1995) and parasitic fungi. Beneficial saprophytic fungi and bacteria decompose dead plant residues and form semistable soil organic matter (Kononova, 1996). Other microbes are parasitic and live on hosts. Certain mycorrhizal fungi are symbiotic with plants (Pfleger and Linderman, 1994). These fungi help to solubilize some unavailable nutrients in the soil. In addition, they colonize plant roots. This is also beneficial to the plants because parasitic invasion is reduced. Sometimes the effects are above ground (Rubeiz et al., 1995). Some plant species and cultivars are resistant to various plant diseases, and genetic engineering is even going to improve the number (Abelson, 1995).

Thousands of microbial species exist in the soil. Competition keeps the abundance of any one species low. When disease-causing organisms are not present in the soil in large numbers, the infections are less because of the competition, and the plant is capable of warding off attack (Van Bruggen, 1995).

Root-rot, damping-off, and crown-rot are caused by parasitic fungi. These microbes are opportunists and are nearly always present. With some soil conditions, they are unable to infect plants; but when the conditions are favorable, they readily attack plants. They are suppressed with good soil aeration.

A. Influence of Soil and Soil Conditions Upon Fungal Root Diseases

1. Soil Moisture

Some common pathogenic fungi, such as *Phythium*, *Phytophthora*, and *Fusarium* are phycomycetous organisms that contact the host by means of free-swimming zoospores. When soil moisture content falls below a certain value, infection is usually completely inhibited. Some of these fungi can attack the host at moisture contents of 60% saturation and above, but not at those of 45% saturation or below. The severity of infection increases with rise in moisture content. A few of the *Fusarium* wilts are increased by frequency of irrigation.

In certain situations, high soil moisture content may increase the prevalence of soil-borne diseases in another way, i.e., by flood dispersal. Very high soil moisture contents may also, by reducing soil aeration, injure the roots of the host plants, and thus afford ingress to certain parasites. Soil conditioners can help reduce high water levels in soil by increasing aeration.

2. Soil Types

Sand, silts, and clays can give different responses to soil-borne plant diseases. It is not accurate to conclude, however, that because the activity of a root-infecting fungus is favored by poor soil aeration, the disease will be less severe on sandy soils. Sandy soils can also be poorly aerated. More importantly, it is necessary to remember that the sandy soils are usually naturally poorer in plant nutrients than are clayey soils. Well-nourished plants can usually outgrow many diseases if soil aeration is satisfactory.

3. Soil Organic Matter

Some diseases are reported to be encouraged by the application of organic materials to the soil (Van Bruggen, 1995). The reverse also occurs (Hoitink et al., 1995, Van Bruggen, 1995). Possible stimulating effects on pathogens may be due to increased carbon dioxide levels in the soil atmosphere and trace amounts of volatile substances formed by decomposition of organic matter. Woody materials can serve as food bases from which fungi can infect plant roots.

Hoitink et al. (1995) and others have reported that certain organic matter–microbe combinations directly protect plants from diseases separately from effects on soil aeration. This needs further study.

Attention needs to be paid to the possibility of "biological control" of soil-borne diseases by application of decomposable organic matter (Klinkenborg, 1995). It is assumed that increased activities of saprophytes growing on readily decomposable organic materials must occur at the expense of, and to the detriment of, plant parasites. Parasites must compete with saprophytes. The development of a high population of microorganisms may greatly limit the predominance of any one species. Soil tilth is enhanced by organic matter after its microbial metabolic conversion into mucilages and gums (Muneer and Oades, 1989). The cellulose-utilizing fungi do not generally form mucilages, which are mainly formed by nonwood decomposing fungi and bacteria.

B. Better Plants Through Use of Water-Soluble Polyacrylamide Soil Conditioners

Many plant root diseases can be minimized with soil conditioning polymers due to better soil aeration, to less flooding because of good drainage, and to the need for less frequent irrigation. High moisture content is better for plant growth and not so good for parasitic fungi. Plants that are less stressed but more vigorous are less prone to be infected. The WS-PAM polymer technique of conditioning soils works well with reduced amounts of organic conditioners. Readily decomposable organics are very effective with the polymer and also help to limit the inoculum potential of parasitic fungi. A good method to control most soil-borne diseases is to maintain a good balance of plant nutrients and to have excellent physical and biological soil properties that impart good porosity. WS-PAM will greatly enhance positive soil structure and limit plant root diseases. No one should plant without it when soils contain any clay that is troublesome.

VIII. SOME WAYS TO SAVE IRRIGATION WATER IN A LANDSCAPE SETTING

A major goal in soil preparation is preparing for water management that is to follow. When irrigating or when it is raining, the water either runs in or it runs off. If it runs off, the water is usually wasted and may remove nutrients, soil, and pesticides if they have been used. If it runs into the soil, it either is stored there or it is lost to the ground water via deep percolation. The water that is stored in the soil can remain there or be removed by evapotranspiration. Soil conditioners can modify these effects.

Runoff is the first major source of water loss. If water cannot penetrate into the soil fast enough, it will end up in the gutter and storm drain as runoff. The WS-PAM soil conditioners now available can easily and greatly improve the rate of penetration of water into soils that contain some clay. Water formerly lost can now be mostly saved. This technique, however, will not work for established landscapes but only for new ones at the start. Some gypsum applied to the surface of old lawns may help water penetration.

Excessive evaporation is the second most important reason for water loss. There are several ways to minimize such loss. One way is to irrigate less frequently but deeper. Not only will there be less water on the surface to evaporate but also roots will grow deeper to extract water from a greater volume of soil to use water more efficiently. But roots can grow deeper only if the soil is properly prepared. Again, the water-soluble polymers used at planting time, preferably with some organic matter, can make a large difference in how soil behaves. The polymers can help roots grow deeper. When roots can grow deeper, there can be big savings in the amount of water used, but only if irrigation is wisely done.

Deep percolation is the last of the ways in which water can be lost. The gel polymers swell up in the presence of water, and they can add considerably to the water-holding capacity of soil. Among the advantages of this type of polymer are a decrease in deep water penetration, greater storage of water in the root zone, more time possible between irrigations, and better plants. This type of polymer must be incorporated into the soil, either at planting time or into holes made later in the soil. Success could be better than that generally achieved, however.

Most water can probably be saved when both classes of polymers are used together. These technologies often can result in savings of more than half of the irrigation water used compared with conventional methods of handling soil. They can be cost effective.

IX. PROTECTING AGAINST SLOPE EROSION

The control of soil erosion is difficult and is a major purpose of soil preparation. The maintenance expense of even a modest slope is great, and a catastrophic situation can be a horrendous liability. The problem is formidable. Measured losses of soil from erosion, such as on farmland, where slopes generally are very modest, nationally average approximately 1.2 kg m^{-2} per year. With the proper techniques and knowledge of the causes of erosion, erosion can be properly controlled (Agassi, 1995).

The quantity of soil eroding is a function of the following five simple factors (Lal, 1994; Agassi, 1995). Actually these five factors were researched and outlined by the US Department of Agriculture (1978):

A. Rainfall Erosion Index Factor

The amount of rainfall and the intensity of rainfall influence erosion. Erosion is less when a given amount of rain falls over a longer time period.

B. Soil Erodibility Factor

The ease with which soil particles can be dislodged and subsequently carried away as sediment is a major factor in erosion. This factor includes the amount of water being shed by the slope, which relates to the rate of water penetration into the slope. If water did not run off, soil would not erode from the surface.

C. Slope Length Factor

The amount of water running off a slope with potential for erosion is directly proportional to the length of the slope.

D. Vegetative Ground Cover Factor

The degree to which the slope is protected by vegetation also affects erosion. Not only do the roots anchor the soil in place, but the plant tops break the fall of the raindrops and decrease the velocity of their impact, thus lessening the dislodgment of soil particles.

E. Erosion Control Practice Factor

Control practice can be implemented in various ways. Contouring of slopes is an effective method. Smooth soil sheds water faster than a rough soil surface and usually results in more erosion unless the surface has been hardened.

Since these five factors are multiplied together to calculate the total effect, erosion can be stopped completely if the erosive effect of any one factor is reduced to zero, such as by 100% water penetration, 100% ground cover, or very short slope length, which is accomplished with frequent contours.

The most precarious period for erosion in landscape work exists in the interval between grading and plant establishment. This is the time and indeed the best occasion for proper soil preparation in order to achieve acceptable plant growth. Well-prepared soils have good porosity to allow water infiltration and to reduce the amount of water runoff.

One of the better technologies to control erosion in landscaping is that of WS-PAM soil conditioning together with some organic matter and gypsum. The polymers bind and flocculate the soil particles to achieve open structured soil with excellent aeration conducive to rapid plant growth. Since soils treated with WS-PAM have lower soil compaction, the water infiltration is also increased, resulting in greatly reduced loss of soil.

An additional benefit of the soil polymer technology is the prevention of mud (liquid soil) formation. Since polymer-treated soils are stabilized against the action of water that causes normal soils to loose their "solid" properties and exhibit "liquid" properties, mud will not form. Additionally, compaction does not occur when soil particles cannot reorient, as they do in muddy conditions. This is the crux of the requirement for an ideal and perfect soil that resists erosion.

Care must be exercised especially with steep slopes, since the soil particles are not stabilized below the treatment depth of the soil preparation. Subterranean drain lines should be used to remove drainage water in such cases in order to avoid sudden slope failures. Geological engineering assistance is recommended where property would be damaged by slope failure.

F. Hydroseeding

Hydroseeding is often used as a method for establishing vegetation on slopes. It is used when destroyed or damaged vegetation on hillsides needs to be restored to curtail or decrease soil erosion. Seeds with various additives are applied to the soil with spray rigs. The additives include fertilizers and mulches. Success depends on water relationships and how well the soil is protected until the new plants are able to resist the forces of erosion. Soil conditioner additives include, variously, straw worked into the soil, gypsum, organic glue tackifiers, compost, and paper fiber (Lee, 1996). Water-soluble polymers will help. From 0.02 to 0.04% of WS-PAM in the slurries can be used. One operator has said that he never has been called back for patch work when he has included WS-PAM in the hydroseeding. A procedure for starting seeds under arid conditions using a combination of water-soluble and gel polymers has been suggested (Wallace and Wallace, 1990). Increased available water and soil conditioning are both involved.

X. 100% SUCCESS IN TREE AND SHRUB TRANSPLANTING IS POSSIBLE

The success rate for transplanting nursery seedlings into containers is nearly 100 per cent. The success rate for grafting scions onto decapitated rootstocks is over 95%, although this is a drastic process. But the success rate for establishing transplanted liners, shrubs, and trees into landscapes, which is a less drastic process, is often only about 80%, which results in expensive replacement costs. Except for mechanical injury or inadequate irrigation, the success rate should be 100%. The main cause of this less than 100% success is poor soil management. The nursery grower uses a synthetic potting mix or superbly conditioned soil, which give soils having half pore space (50% soil compaction). The exciting news is that the same can be achieved at almost any landscaping site utilizing native soils with the new polymer soil technology for soil improvement.

The results of properly prepared soils, which are conditioned to achieve 50% compaction with the necessary void volume (empty space) include

1. Vigorous root growth
2. Needed oxygen
3. Porosity for water infiltration without displacing all the air
4. Rapid leaching rates of harmful salts
5. Easy exchange of respiration products

In addition, the physical properties of the potting soil around the transplant rootball and the physical properties of the conditioned backfill soil will be similar, since they both have about 50% void volume. This prevents an interface that would limit plant growth due to failure of water movement. Also, conditioned native soil being taken from site soil for backfill has the same characteristics. This alleviates another potential interface due to dissimilar soil properties caused by high application rates of organics. This problem can cause a "bathtub" that can kill plants, because water does not easily move from the transplant hole through the interface to the surrounding soil.

Creating excellent soil from compacted soil first requires that it be tilled. The final porosity will be no better than which exists after tilling. Planting pits need to be larger than the rootballs. The soils are stabilized to avoid compaction that often occurs with irrigation, because water lubricates soil particles, allowing them to move. The result without soil conditioning is compacted soils upon drying. The water-soluble polymer (not the gel-forming, water-absorbing superabsorbents) binds soil particles together to form water-stable aggregates or crumbs that do not compact with irrigation. This stabilization process is the same as what eventually happens with organic matter, except it is much more complete, far longer lasting, and gives higher chances of having 100% success.

Table 3 Growth Rates of Chinese Elm Trees Planted February 1987 at Retail Center, Tri-City, San Bernardino, California.

	Caliber of tree trunks (in cm)	
Date	Control trees	WS-PAM treated trees
June 26, 1987	3.1 ± 0.3	4.2 ± 0.2
Feb. 11, 1988	3.1 ± 0.5	4.6 ± 0.4
June 12, 1989	3.8 ± 1.0	6.1 ± 0.6

Table 4 Cost of Chinese Elm Trees and Cost of Needed WS-PAM Conditioners

Size of transplant	Purchase cost per tree, $	Backfill per tree, m^3	Cost of soil conditioners per tree, $
1.5 liters	0.82	0.04	0.08
4 liters	3.45	0.07	0.14
Egg can	10.75	0.125	0.25
50 liters	45.00	0.5	1.00
60 cm × 60 cm box	160.00	1	2.00
90 cm × 90 cm box	400.00	2	4.00

An example of the value of soil conditioning with WS-PAM in transplant work is illustrated in Table 3. Table 4 shows the relatively inexpensive cost of such treatments.

A. Treating Transplant Backfill with WS-PAM

When transplanting, backfill soil is best conditioned if done with WS-PAM. The backfill soil is drenched with a WS-PAM solution similar to those used for hydroseeding and before any irrigation or rainfall. The soil is first amended with any needed fertilizers, low amounts of organic matter, and chemical conditioners such as gypsum. Gypsum improves the action of WS-PAM in most soils. The soil should be initially close to air dry for good, very long-term soil improvement. The soil can be drenched after transplanting of liner stock, ground cover, and small container transplants. For large shrubs, trees, and oversized planting pits, the drenching should be done concurrently with backfilling to insure uniform and deep treatment. This can be done by filling the pit part way with WS-PAM solution after the placement of the rootball and then backing the conditioned soil into the solution. Additional solution will need to be added if only part of the soil is drenched. A berm filled with the solution of WS-PAM may be needed to drench the top portion of the backfill soil.

If it is easier, the backfill can be saturated with a spray rig or repetitive filling of a berm. Untreated soil should not be placed on top of treated soil.

The soil needs to dry partially for curing. For curing, leave a day or so before irrigation for drying unless the plants are wilting. If the weather is cool, several days can pass before additional irrigation is needed.

XI. CONCLUSIONS

Soil conditioners are extremely important in landscape work because of the high value of the projects and because soils at construction sites often have harsh characteristics. Nontillage is popular on farms but has little or no place in starting landscaping because of the need for incorporation of soil conditioners and the danger of compaction due to heavy usage of the land. Ideally, soil conditions should be as near perfect as possible so that failure can be avoided. Poor soil preparation too often is the cause of landscape failure, and correction costs usually far exceed those necessary to have solved all problems before planting was done. Soil analysis and site evaluation are essential for making decisions concerning conditioners and other amendments. Conditioners of most value in soil preparation are lime, gypsum, compost, other organics, water-soluble polymers, and zeolites or pozzolans. The conditioners may be of little value unless there is proper drainage. With planning and use of proper soil conditioners, landscapes can be successful without the need for replacement or reworking when landscapes fail. It is better to condition the existing soil as necessary than to import topsoil. Topsoil from elsewhere can result in troublesome interface problems.

REFERENCES

Abelson, P. H. (1995). Plant pathogens in soils. *Science 269*:1027.

Agassi, M., ed. (1995). *Soil Erosion, Conservation and Rehabilitation.* Marcel Dekker, New York, Basel.

Avnimelech, Y., Cohen, A., and Shkedi, D. (1990). The effect of municipal solid waste compost on the fertility of clay soils. *Soil Tech. 3*:275.

Craul, P. J. 1996. Soils of the city. *Turf West.*, June pp. 12–14.

Dahnke, W. C. (1993). Soil test interpretation. *Comm. Soil Sci. Plant Anal. 24*:11.

Gardner, G. (1996). Preserving agricultural resources. *State of the World: A Worldwatch Institute Report on Progress Toward a Sustainable Society* (L. Starke, ed). Norton, New York and London, pp. 78–94.

Gilman, E. F., Leone, I. A., and Flower, F. (1987). Effects of soil compaction and oxygen content on vertical and horizontal root distribution. *J. Environ. Hort. 5*:33.

Gouin, F. (1995). Use of commercial compost in production and maintenance of horticultural plants. Seminar at ProGreen Expo. Denver. CO. February.

Gray, L. E., and Pope, R. A. (1986). Influence of soil compaction on soybean stand, yield, and *Phytophthora* root rot incidence. *Agron. J. 78*:128.

Gregory, P. J., Lake, J. V., and Rose, D. A., eds. (1987). *Root Development and Function*. Cambridge Univ. Press, New York.

Hendrix, J. W., Guo, B. Z., and An, Z.-Q. (1995). Divergence of mycorrhizal fungal communities in crop production systems. *Plant Soil 170*:131.

Hoitink, H. A. J., Grebus, M. E, and Stone, A. G. (1995). Control of diseases caused by soilborne plant pathogens: an update. Ohio Florists Association, Bulletin 790, August.

Klinkenborg, V. (1995). A farming revolution. *Nat'l Geographic 188*(6):60–89.

Kononova, M. M. (1966). *Soil Organic Matter: Its Nature, Its Role in Soil Formation and in Soil Fertility*. 2d ed. Pergamon Press, Oxford.

Lal, R. (1994). *Soil Erosion Research Methods*. 2d ed. St Lucie Press, Delray Beach, Florida, p.352.

Lee, D. 1996. Effective hydroseeding specs. *Erosion Control 3*(4):22, 24, 26–29.

Letey, J., Lunt, O. R., Stolzy, L. H., and Szuszkiewicz, T. E. (1961). Plant growth, water use and nutritional response to rhizosphere differentials of oxygen concentration. *Soil Sci. Soc. Am. Proc. 25*:183.

Martens, D. A., and Frankenberger, W. T., Jr. (1992). Modification of infiltration rates in an organic-amended irrigated soil. *Agron. J. 84*:707.

Metting, F. B., Jr., ed. (1993). *Soil Microbial Ecology*. Marcel Dekker, New York.

Morgan, W. C. (1986). Is organic matter beneficial for planting shrubs and trees? *Landscape and Irrigation 19*(8):22–28, 97–98.

Muneer, W., and Oades, J. M. (1989). The role of calcium–organic interactions in soil aggregate stability. III. Mechanisms and models. *Aust. J. Soil Res. 27*:411.

Orzolek, M. D. (1993). Use of hydrophylic polymers in horticulture. *HortTech. 3*:41–44.

Pfleger, F. L., and Linderman, R. G., eds. (1994). *Mycorrhizae in Reclamation, Mycorrhizae and Plant Health*. American Phytological Soc. Press, St. Paul, Minnesota.

Rubeiz, I. G., Aslam, M., and Chahine, H. (1995). Poultry manures and ammonium sulfate influence on whitefly population in cantaloupe. *1995 Agron. Abs.*, p. 155.

Sarma, P. K., and Das, M. (1996) Effect of aggregating agents on synthesis of microaggregate and physical properties of an Alfisol. *J. Indian Soc. Soil Sci. 44*(1):12.

Taylor, H. M., and Ratliff, L. F. (1969). Root elongation rates of cotton and peanuts as a function of soil strength and soil water content. *Soil Sci. 108*:113.

Urban, J., and P. Craul. 1996. Success with soils in urban landscapes. *Arbor Age 16*(7):18, 20, 22.

US Department of Agriculture. (1978). *Handbook No. 537. Predicting Rainfall Losses—A Guide to Conservation Planning*. Washington, D.C.

USEPA. (1993). Standards for disposal of sewage sludge; final rules: Part II. *Federal Register*, Feb. 19, 1993, pp. 9247–9415.

Van Bruggen, A. H. C. (1995). Plant disease severity in high-input compared to reduced-input and organic farming systems. *Plant Disease 79*:976.

Vine, P. N., Lal, R., and Payne, D. (1981). The influence of sands and gravels on root growth of maize seedlings. *Soil Sci. 131*:124.

Wallace, A. (1994). Soil organic matter must be restored to near original levels. *Com. Soil Sci. Plant Anal. 25*:29.

Wallace, A., and Nelson, S. (1986). Special Issue on Soil Conditioners. *Soil Sci. 141*:311–397.

Wallace, A., and Wallace, G. (1986). Additive and synergistic effects on plant growth from polymers and organic matter applied to soil simultaneously. *Soil Sci. 141*:334.

Wallace, G. A., and Wallace, A. (1990). Using polymers to enhance shrub transplant survival and seed germination for revegetation of desert lands. Proceedings of the Symposium on Cheatgrass Invasion, Shrub Die-Off, and Other Aspects of Shrub Biology and Management, 5–7 April 1989, Las Vegas, Nevada. Gen. Tech. Rept. INT-276, (E.D. McArthur, E.M. Romney, S.D. Smith, and P. T. Tueller, eds.). USDA Forest Service Intermountain Research Station, Ogden, Utah.

Wallace, A., and Wallace, G. (1994). Role of soil and plant analyses in safe, sustainable agriculture. *Commun. Soil Sci. Plant Anal. 25*:55.

Witter (1996). Towards zero accumulation of heavy metals in soils: an imperative or a fad? *Fert. Res. 1995: 43*:225.

Youngberg, G., ed. (1992). Special Issue on Soil Quality. *Amer. J. Altern. Agric. 7*:2–90.

20
Soil Conditioners for Sports Turf Areas

C. Frank Williams *Brigham Young University, Provo, Utah*

Donavon H. Taylor *University of Wisconsin–River Falls, River Falls, Wisconsin*

With the advent of turf for golf courses and sports fields, turfgrass managers have been interested in modifying the soil to enhance soil–plant relationships, to alter playing surface conditions, and to minimize soil compaction and other management problems. Sports turf soils pose difficult problems for turf managers. Frequent foot and vehicle traffic over the turf breaks down natural soil structure, resulting in lowered water infiltration and percolation rates and poor soil aeration. In addition, due to the dense, perennial turf, tillage to alleviate soil compaction is limited to aerification implements that punch holes in the turf, disturbing only a small percentage of the surface. The widespread use of soil conditioners is one method turf managers have used to ameliorate the effects of soil compaction and to manage soil water.

In turf literature the terms soil conditioners and soil amendments have been used interchangeably. We use the term soil conditioner to indicate materials used to improve the physical conditions of soil in an effort to enhance growth of turf in recreational settings. A list of amendments used or proposed for use in turf is given in Table 1 (Carrow, 1993).

I. INORGANIC CONDITIONERS IN THE CONSTRUCTION OF GOLF COURSES AND ATHLETIC FIELDS

Sand is used widely, and in large quantities in the construction of golf greens and athletic fields. It might more appropriately be considered as alternative soil

Table 1 Soil Amendments Used to Modify Soil Physical Properties

Inorganic materials	
Sand	Pumice
Calcined clay	Sintered fly ash
Diatomite	Slag
Expanded shale	Vermiculite
Perlite	Gypsum
Organic materials	
Peat	Composed organic matter
Aged sawdust	Digested sewage sludge

rather than a soil conditioner. In construction or renovation of most golf greens and many athletic fields, sand or sand amended by other materials is brought in to serve as the major soil component making up the root zone soil. In many situations, however, turf managers are using sand as a conditioner to change physical properties of existing soils. Sand characteristics such as average particle size, fineness modulus, and uniformity affect the resultant properties of sand-based soil mixtures, and recommendations for suitable sands have been published extensively (see for example Bingaman and Kohnke, 1970; Adams et al., 1971, Blake, 1980; Hummel, 1993).

Regardless of sand size quality, it must be present in large quantities to increase the general pore size of soil with subsequent increased water infiltration rate and improved soil aeration. If sand particles are simply mixed into a matrix dominated by the existing soil, change in the general pore sizes will be minimal. In order to affect pore size significantly, sand must be made the dominant component of the existing soil matrix. In modified soil mixtures, sand contents of 85 to 90% on a weight basis were required to maintain adequate conditions for turf growth after compaction (Brown and Duble; 1975; Taylor and Blake, 1979). Effective modification of turf soils with sand means applying sufficient amounts that sand becomes the soil rather than the soil conditioner.

Inorganic conditioners other than sand have different purposes in native soil situations and modified high-sand profiles. In native soils, conditioners are added to reduce the impact of compaction on the soil and thus are added to reduce soil bulk density, increase air-filled porosity, and increase water infiltration and percolation rates. In high-sand mixtures used as root zone soils, conditioners are added principally to increase the nutrient and available water holding capacity of mixtures. Some conditioners are added to lend stability to the sand mixtures as well.

Calcined materials have been heated to very high temperatures to form stable aggregates. These are typically crushed and screened to specific size fractions. Clay and diatomaceous earth are two materials commonly calcined and used as turf soil conditioners. Some manufacturers heat these materials to 1500 to 1800° F and then extrude the materials to increase stability and produce a uniform size. These conditioners are marketed as porous ceramics.

Calcined clay has been used as a soil conditioner for many years. In native soils, calcined clay additions reduced soil bulk density, increased air-filled porosity, and increased water infiltration rate (Morgan et al., 1966; Waddington, 1992). Although water holding capacity of calcined clay is high, much of it is unavailable to plants (Beard, 1973; Waddington et al., 1974). The stability of calcined clay amendments varies, depending on clay materials used and the processing method. Particle degradation has been reported on some turf areas (Hummel, 1993). Daniel and Freeborg (1979) found the more stable calcined clay products still retaining a granular structure after 19 years in an experimental putting green.

Diatomaceous earth is a naturally occurring material derived from the remains of single-celled organisms called diatoms. Calcined diatomaceous earth granules have low density and high pore space and are added to sand mixtures to increase available water holding capacity and reduce bulk density.

Although each has unique characteristics, porous ceramics and calcined products such as clay and diatomaceous earth have many properties in common, including low bulk density and high water holding capacity. These products are usually mixed into sand or sand/peat mixtures at 5–20% by volume ratios to increase porosity and available water holding capacity without reducing soil water and percolation rates.

Randomly oriented, interlocking mesh elements have beneficial effects on high-sand root zone mixtures. When mixed at a rate of 2.5 kg m^{-3} into the surface 15 cm of a high-sand root zone, mesh elements resulted in better root zone stabilization, playing surface quality for sports, increased water infiltration and percolation rates, and enhanced moisture retention (Beard and Sifers, 1993).

Polyacrylamides (PAM) have been proposed as conditioners to enhance soil water retention and improve physical properties of turf soils. The hydrogel-type polyacrylamide polymers retain considerable amounts of water (see Chapter 13). In a study by McGuire et al. (1978), PAMs did not affect soil physical properties when used to amend sand or sandy loam; however Baker (1991) found that improved water retention reduced surface hardness and increased ryegrass cover in PAM-amended soil.

Karlik (1995) developed a series of experiments using two turfgrasses to determine the effects of six rates of gel polymer on establishment of two turfgrass species, performance under reduced irrigation, and field hardness. The polymer was incorporated at six rates. The rates used varied from 0 to 90 lbs per 1000

ft^2 (0 to 44 kg per 100 m^2). The polymer was incorporated to a depth of three inches (7.62 cm). After planting the treatment received a light raking and 21 inches (53 cm) of water.

Under the conditions of the research, the polymer did not affect field hardness with only light use, but after heavy use the field hardness of tall fescue was increased by the use of the polymer as measured by both the penetrometer and the accelerometer. There was also no interaction between the polymer and foot traffic. Tall fescue germination and coverage were unaffected by the use of the polymer. The fact that the polymers can affect field hardness, especially under heavy use, may be as useful in the culture of turfgrass as any characteristic related to the conservation of water.

Other inorganic conditioners including vermiculite, perlite, clinoptilolite zeolite, pumice, expanded shale, slag, sintered fly ash, and gypsum are described by Waddington (1992).

The stability of many inorganic conditioners in sports turf soils with normal weathering processes and frequent traffic has been questioned, but few research studies have definitively established which ones are likely to break down or lose their effectiveness in a reasonable time period.

II. ORGANIC CONDITIONERS

Peat is the most popular and widespread organic conditioner used in sports soils. Beneficial effects on soil physical properties ascribed to peat include increased moisture retention in sandy soils, increased water infiltration rates in fine-textured soils, improved aeration, and ease of root penetration (Lucas et al., 1965). Peats vary considerably in organic content, degree of decomposition, source, pH, and moisture retention. Consequently, the degree to which soil properties are changed depends in many respects on the type of peat as well as the amount.

Specifications for peats used in turf soil mixtures have generally been based on organic content of peat as measured by weight loss on ignition. Published recommendations have varied from greater than 80% (McCoy, 1992) to greater than 90% organic content (Beard, 1973; Waddington et al., 1974). The United States Golf Association recommends peats of greater than 85% organic contents when used in soil mixtures for putting green construction (Hummel, 1993).

When used in soil mixtures for golf greens or athletic fields, sand and peats are mixed in volume ratios varying from 80/20 to 95/5 sand to peat. The resulting characteristics of the mixture depend on both sand and peat characteristics. Both McCoy (1992) and Taylor et al. (1997) found that sand/reed sedge peat mixtures had noticeably higher organic content on a mass basis than sand/sphagnum peat mixtures despite identical volume mixing ratios.

Sedimentary peats, muck soils, and "black dirt," though high in organic matter compared to mineral soils, and though widely used as soil conditioners, are almost universally discouraged as suitable organic sources by turf researchers (see for example Lucas et al., 1965; Waddington et al., 1974; McCoy, 1992).

Sawdust and bark products have been used to modify turf soils. Sawdust additions have led to such physical property benefits as increased air-filled porosity, moisture holding capacity, and cation exchange capacity but have also led to decreased germination and seedling growth (Waddington et al., 1974). The potential for nitrogen immobilization when fresh sawdust is used as a soil conditioner has led turf researchers to suggest that sawdusts must be thoroughly composted before use (Hummel, 1993).

Another material that has shown promise for use in sports turf is compost. Landschoot (1995) found that incorporation of several different types of compost in soil before planting increased the rate of turf establishment relative to topsoil and unamended control plots. The treatments that provided the fastest establishment were the ones with the highest rates of biosolids composts and paper mill by-products compost. The effect of compost on soil physical properties can be expected to be beneficial for up to three years. The rate of compost to be added will depend on soil conditions and the type of compost used. In most cases it is appropriate to use between 3 cubic yards per 1000 ft^2 for 1 a inch layer (2.5 m^3 per 100 m^2 for a 2.5 cm layer) and 6 cubic yards per 1000 ft^2 for a 2 inch layer) (5 m^3 per 1000 m^2 for a 5.1 cm layer). Landschoot (1995) also suggests that for soils high in clay content it is best to use compost by weight of the soil with as high as 50% organic matter by weight.

Carrow (1993) also mentions other materials that have been used as soil amendments to improve soil structure for various reasons. Table 2 shows the nontraditional amendments that have been used with their primary claims. Many of these materials have been used for several years with periods of popularity and then periods of less use.

III. CONCLUSION

Soil conditioners have been used in turf for many years and under many different conditions to modify soil physical properties. They have been promoted to improve the physical, chemical, or biological properties or a combination of these factors with the hope that there will be an enhancement of the turf growth. Research carried out has shown this to be true in many situations, yet at other times they have failed to live up to expectations. Many have given success for a short period of time, but research is still needed to demonstrate under what conditions these materials work and what the long term effect on the quality of turf might be.

Table 2 Other Amendments and Their Primary Claims[a]

Soil conditioner	Primary claim
Algae-based soil conditioners	Improve soil structure
Enzymes	Improve soil structure, remove salt or other toxicities
Hormones (biostimulants)	Promote root growth
Humus	Improves soil structure, nutrient retention
Humic acids	Improve soil structure, nutrient retention, growth stimulation
Inoculated composts	Organic composts inoculated with various microorganisms. Some have added nutrients and are also classified as fertilizers. Claims are to improve soil structure and disease suppression, provide better nutrient retention and thatch control, and increase soil organic matter
Polymers (starches, PVA, PAM)	Enhance soil water retention, improve soil physical conditions
Porous ceramics	Enhance water retention, improve soil aeration status, improve wetting
Seaweed extracts	Hormones act as biostimulants
Seaweed plant meal	Increases soil organic matter, increases nutrient retention, acts as biostimulant
Root zone stabilizing agents (mesh elements or mesh blankets)	Stabilize soils, alleviate compaction
Thatch control agents	Some are similar to inoculated composts, while others are organic-rich liquids. These may or may not contain added nutrients to be classified as a fertilizer. Claim is to control thatch
Wetting agents	Wet hydrophobic soils, improve soil structure, improve drainage
Zeolite	Increases CEC and improves soil physical properties

[a]See Table 1 for traditional amendments.

REFERENCES

Adams, W. A., Stewart, V. I., and Thornton, D. J. (1971). The assessment of sands suitable for use in sportsfields. *Sports Turf Research Institute 47*:77–86.

Baker, S. W. (1991). The effect of polyacrylamide co-polymers on the performance of *Lolium perenne* L. turf grown on a sand root zone. *Sports Turf Research Institute 67*:66–82.

Beard, J. B. (1973). *Turfgrass: Science and Culture*. Prentice-Hall, Inc., Englewood Cliffs, N.J.

Beard, J. B., and Sifers, S. I. (1993). Stabilization and enhancement of sand-modified root zones for high traffic sports turfs with mesh elements. *Texas Agricultural Experiment Station Publication* B-1710.

Bingaman, D. E., and Kohnke, H. (1970). Evaluating sands for athletic turf. *Agronomy J. 62*:464–467.

Blake, G. R. (1980). Proposed standards and specifications for quality of sand for sand-soil-peat mixes. *In*: J. B. Beard, ed. *Proceedings of the Third international Turfgrass Research Conference*, Munich, Germany, 11–13 July 1977. International Turfgrass Society and American Society of Agronomy, Madison, Wisconsin, pp. 195–203.

Brown, K. W., and Duble, R. L. (1975). Physical characteristics of soil mixtures used for golf green construction. *Agronomy J. 67*:647–651.

Carrow, R. N. (1993). Eight questions to ask: evaluating soil and turf conditioners. *Golf Course Management*, October:56–70.

Daniel, W. H., and Freeborg, R. P. (1979). *Turf Managers' Handbook.* Harvest, Cleveland, Ohio.

Hummel, N. W., Jr. (1993). *Rationale for the Revisions of the USGA Green Construction Specifications. USGA Green Section Record.* United States Golf association, Far Hills, New Jersey.

Karlik, J. (1995). Effects of pre-plant incorporation of polymers on turf grass. *California Turfgrass Culture* 45, nos. 3 and 4:19–22.

Landschoot, P. (1995). Improving turf soil with compost. *Grounds Maintenance 30*(6):33–39.

Lucas, R. E., Rieke, P. E., and Farnham, R. S. (1965). Peats for soil improvement and soil mixes. *Michigan State University Extension Bulletin* 516.

McCoy, E. L. (1992). Quantitative physical assessment of organic materials used in sports turf root zone mixes. *Agronomy J. 84*:375–381.

McGuire, E., Carrow, R. N., and Troll, J. (1978). Chemical soil conditioner effects on sand soil and turfgrass growth. *Agronomy J. 70*:317–321.

Morgan, W. C., Letey, J., Richards, S. J., and Valoras, N. (1966). Physical soil amendments, soil compaction, irrigation and wetting agents in turfgrass management. Effects of compactability, water infiltration rates, evapotranspiration, and number of irrigations. *Agronomy J. 8*:525–528.

Taylor, D. H., and Blake, G. R. (1979). Sand content of sand-soil-peat mixtures for turfgrass. *Soil Sci. Soc. Am. J. 43*:394–398.

Taylor, D. H., Williams, C. F., and Nelson, S. D. (1997). Water retention in root zone soil mixtures of layered profiles used for sports turf. *Hortsci. 32(1)*:82–85.

Waddington, D. V. (1992). Soils, soil mixtures, and soil amendments. *In*: D. V. Waddington, R. N. Carrow, and R. C. Shearman eds. *Turfgrass*. American Society of Agronomy, Madison, Wisconsin, pp. 331–383.

Waddington, D. V., Zimmerman, T. L., Shoop, G. J., Kardos, L. T., and Duich, J. M. (1974). Soil modification for turfgrass areas. I. Physical properties of physically amended soils. *Penn. Agri. Experiment Station Progress Report* 337.

21
Use of Soil Conditioners to Enhance and Speed Bioremediation of Contaminated Soils

Richard E. Terry *Brigham Young University, Provo, Utah*

I. INTRODUCTION

Society generates an enormous amount of waste, much of which is considered hazardous. In 1993 the United States alone produced approximately 234 million Mg of hazardous waste regulated by the Resource Conservation and Recovery Act (US EPA, 1995). Some of this waste is either intentionally applied or accidentally spilled on land. The release of toxic organic and inorganic compounds into the environment has resulted in widespread pollution of soils, ground water, and marine environments. Methods of cleaning up and removing these contaminants are expensive and difficult to implement. The need to develop and apply innovative technologies to deal with these serious environmental contaminants in a cost-effective manner is urgent (Skladany and Metting, 1993).

It is possible for scientists and engineers to design complex mechanical devices to remove, chemically destroy, or incinerate these pollutants. Such technological wonders are often complex, expensive, and difficult to maintain. Another approach is to take advantage of the degradative capabilities of naturally occurring soil-borne organisms. A group of microorganisms exists in the soil with the genetic blueprints for enzyme systems capable of decomposing or detoxifying many man-made (xenobiotic) and petroleum-based contaminants.

The term bioremediation has been used to describe the use of microorganisms and plants to detoxify an environment, either by complete decomposition of pollutant compounds or by transformation of pollutants to nontoxic, innocuous materials. Madsen (1991) defined bioremediation as "a managed or spontaneous process in which biological, especially microbial, catalysis acts on pollutant compounds, thereby remedying or eliminating environmental contamination." Thus bioremediation is an important and often cost-effective method of detoxifying or removing pollutants from contaminated environments.

The techniques of bioremediation include 1) stimulation of the activity of indigenous microorganisms by addition of nutrients, regulation of redox conditions, and optimization of pH and other chemical and physical soil conditions; 2) inoculation of the sites with microbes with specific metabolic capabilities; and 3) use of plants (phytoremediation) to remove and transform pollutants or to stimulate rhizosphere microbes (Bollag and Bollag, 1995).

In both terrestrial and aquatic environments, the existence of natural analogues to xenobiotic compounds has provided the starting point for the evolution of degradative enzyme systems. The decomposition of complex, high-molecular-weight molecules such as wood lignin by white rot fungus (*Phanerochaete chrysosporium*) is an example of these complex degradative metabolic capabilities. These microorganisms use lignin polymers, containing polyaromatic hydrocarbons (PAHs), as carbon and energy sources. It is not surprising therefore that *P. chrysosporium* and other microbes degrade many complex xenobiotics, including PCBs, wood preservatives, and munititions wastes and other persistent environmental pollutants (Bumpus et al., 1985). Microbial treatment of xenobiotic wastes and other pollutants has been used successfully for decades in municipal waste water treatment and in landfarming of oily sludges produced at well fields and at refineries (Loehr et al., 1992; Zimmerman and Robert, 1991).

The natural conditions at some contaminated sites may provide all the necessary components, including microbes adapted to degradation of the pollutants, adequate levels of nutrients, and oxygen or other electron acceptors, so that bioremediation can occur without human intervention—a process called intrinsic bioremediation. In most cases, however, bioremediation requires the addition of materials such as nutrients and electron acceptors to enhance the growth of native or introduced microbial species capable of degrading or destroying contaminants (Committee on *In Situ* Bioremediation, 1993). It may also be necessary to remove contaminated soil or water to an engineered environment in which these growth promoting materials can be provided.

In this chapter I will report on the current and the potential uses of soil conditioners to enhance and promote the growth of microorganisms and plants involved in bioremediation. This discussion will include the microbial processes leading to the destruction or detoxification of contaminants as well as the soil environmental conditions necessary for optimal growth and reproduction of microbial species capable of degrading contaminants. Examples of the use of various soil conditioners in bioremediation will be presented.

II. SOIL MICROBIAL PROCESSES

Heterotrophic bacteria and fungi in the soil gain life-sustaining energy by the enzymatically controlled, energy producing oxidation–reduction reactions, which break the chemical bonds of reduced forms of organic carbon (the electron donor) and subsequently transfer electrons to electron acceptors such as O_2, NO_3^-, SO_4^{2}, Fe^{3+}. In aerobic respiration microbes use O_2 as electron acceptor, while NO_3^-, SO_4^{2-}, Fe^{3+}, Mn^{4+}, and even CO_2 may serve as electron acceptors under anaerobic respiration (Committee on *In Situ* Bioremediation, 1993). Organic molecules also provide the carbon building blocks of new cell constituents. The goal of bioremediation is to use these metabolic capabilities of microorganisms to decompose, transform, or detoxify organic and inorganic pollutants. The success of bioremediation will depend on a number of factors, including: 1) the concentrations, toxicities, and microbial availability of contaminants; 2) the presence of primary energy substrates and mineral nutrients for microbial growth; 3) the concentrations of appropriate electron acceptors; and 4) physical and chemical conditions of the soil environment conducive to microbial growth (Skladany and Metting, 1993).

Soil conditioners may provide some of the necessary factors for successful bioremediation. The addition of animal wastes, biosolids, composts, crop residues, and cover crops as soil conditioners may reduce the toxicity of contaminants by sorption or dilution. These organic substrates may also provide the primary carbon and energy sources for the microbial population. Addition of these and other conditioners to a bioremediation site may result in an increased population of various microbial species including those involved in degradation and detoxification of contaminants.

A. Cometabolism

Many organic pollutants that are targets of bioremediation serve as sole carbon and energy sources for microbes, while others do not supply the carbon and energy needs of microorganisms but are biodegraded fortuitously. The latter process is called cometabolism (Skladany and Metting, 1993). When the only means of decomposing a contaminant is cometabolism, the presence of a primary substrate is required. As the primary substrate is metabolized, transformation or decomposition of the contaminant occurs incidentally to the reactions catalyzed by enzymes involved in normal cell metabolism or by special detoxification reactions. For example, in the process of oxidizing methane, some bacteria produce enzymes that incidentally destroy chlorinated hydrocarbons, even though the chlorinated compound by itself cannot support microbial growth (Committee on *In Situ* Bioremediation, 1993).

Rotert et al. (1995) reported that the addition of benzoate at 1 mg/L enhanced aerobic degradation of benzene, toluene, and *o*-xylene in laboratory aquifer columns. The number of BTX degraders increased by two orders of magnitude in microcosms when benzoate or phenylalanine (1 mg/L) were added. They concluded that the addition of environmentally benign aromatic substrates, such as benzoate, could stimulate metabolic and population shifts that increase the efficiency of bioremediation.

Cometabolism of a contaminant may be facilitated by the decomposition of a structurally related compound. For example, Saez and Rittmann (1993) reported that the cometabolic degradation of 4-chlorophenol was proportional to the rate of phenol (primary substrate) oxidation. The initial monooxygenation of 4-chlorophenol was rapid only when the primary substrate phenol was being oxidized to create a high intracellular level of the reduced electron carrier NADPH. Manipulations of the cell's primary electron-donor and electron-acceptor substrates greatly accelerate or decelerate degradation or detoxification of soil contaminants (Saez and Rittmann, 1993).

B. Nutrients

The elemental components of microbial cells are relatively constant. The cell is typically 50% C, 14% N, 3% P, 2% K, 1% S, with lesser amounts of micronutrients (Committee on *In situ* Bioremediation, 1993). Microbial populations may be small because of the lack of any one of these nutrients, especially nitrogen, phosphorus, or available carbon. Contaminants may be decomposed by cometabolism, requiring the addition of readily decomposable carbon compounds in plant root exudates, plant residues, or added soil conditioners.

In many cases the growth of microbes on carbon from the targeted soil contaminant is very slow. It may be necessary to supplement the microbes with readily available carbon sources, such as crop residues, animal manures, waste water effluent, or biosolids, or plant root exudates provided by cover crops. Addition of soil conditioners that provide these nutrients or enhance and stabilize the soil aggregates should speed microbial decomposition and transformation of environmental pollutants.

C. Soil Physical Properties and Microbial Growth

The most common method of bioremediation in contaminated soils and aquifers involves the stimulation of the activity of indigenous microorganisms by optimizing the environment in which the organisms must carry out decomposition and detoxification reactions (Glass, 1995; Committee on *In Situ* Bioremediation, 1993). The addition of inorganic nutrients, supplemental carbon sources, regulation of redox conditions, and optimization of pH, moisture, and temperature conditions accelerates detoxification and biodegradation reactions.

Soil physical properties at reclamation sites will usually have some or all of the environmental conditions necessary for successful bioremediation. Hydrocarbon biodegradation rates are influenced by soil texture (Song et al., 1990), moisture content, nutrient addition, temperature (Dibble and Bartha 1979), and salinity (Haines et al., 1994). Soil texture is the proportionate quantity of sand, silt, and clay in the mineral fraction of the soil. Textural properties of the soil influence adsorptive surface area, pore size distribution, water movement and retention, and chemical properties imparted by clay minerals. Soil aggregation results in the arrangement of soil textural components into structural units or peds. Soil structural properties determine the size and distribution of large pores between individual aggregates. These large pores (> 50 μm equivalent pore diameter) are necessary for the infiltration and percolation of water as well as the movement of air into the soil as water drains from the large pores.

Appropriate soil moisture levels are critical to the success of bioremediation. Soil microbes depend on capillary water and water films surrounding soil particles for the solubilization and movement of organic and inorganic nutrients to the cells. Some of the degradative enzymes produced by microbes are extracellular. This means that enzymes are exuded by microbes, which in turn, break down organic polymers and other residue constituents into components small enough to be transported across the plasma membrane into the cell.

Optimal moisture potentials for microbial activity are near –0.01 MPa or –0.1 bar (Paul and Clark, 1996). Activity decreases as the soil becomes saturated and movement of oxygen to microorganisms is limited. Microbial activity again becomes limited as the soil dries and the soil moisture potentials reach greater negative values.

The distribution of various pore size diameters in the soil affects water movement, water storage, and the movement and distribution of microorganisms. Macropores (> 50 μm equivalent pore diameter) are important in the delivery of water by saturated flow. The micropores (< 50 μm equivalent pore diameter) are important in the retention of water and movement of water by capillary action. Once gravitational water has drained from the soil, the interior of the soil aggregate tends to be wetter than the large pores between aggregates, reflecting the ability of the smaller pores within the aggregates to retain water (Turco and Sadowsky, 1995). Microorganisms congregate in pores large enough to contain them comfortably. Ultrastructural studies of the distribution of microorganisms within the soil fabric have shown that bacteria tend to inhabit the micropores inside aggregates while fungal hyphae are restricted to the larger pores between aggregates (Foster, 1988). Small pores within aggregates retain their water longer as the soil dries. Suitable moisture conditions in the micropores may cause bacteria to congregate there. Bacteria located in the smaller pores also escape predation by soil protozoa (Foster, 1988).

The movement of bacteria in soil is also a function of pore size distribution and water potential relationships. Movement of bacteria in soil is tempered by their sorption onto a soil particle or transport into small soil pores (Smith et al.,

1985). Passive movement of bacteria by percolating water was limited to 2.7 cm in soil without plants or earthworms. Percolating water and the presence of earthworms enhanced the vertical transport of *Pseudomonas putida* to a depth of 10 cm (Madsen and Alexander, 1982).

Uncontaminated soil conditioners may be used to dilute the contaminants to tolerable concentrations. Conditioners may also be used to improve aggregation, enhance moisture conditions, and open the soil matrix, allowing the movement of O_2 to microorganisms. Bioremediation is enhanced by preventing saturation with water. Aerobic conditions are also maintained by mechanical mixing or tillage of the soil and by avoiding excessive compaction, thus maintaining aerobic conditions. Improved soil physical properties imparted by conditioners may assist in maintaining aerobic conditions, perhaps reducing the need for tillage.

III. TYPES OF BIOREMEDIATION

Bioremediation is often used as a cost-effective clean up technology for contaminated land and water. Biodegradation and detoxification of contaminants via the various bioremediation methods may cost a fraction of the expense of soil removal and disposal in a toxic waste facility. For instance, bioremediation costs one-third to one-half the cost of incineration (Bollag and Bollag, 1995). In addition, degradation at or near the site lessens the risk of exposure of clean up personnel or members of the public to the contaminants. The chances of further liability or legal action are decreased with bioremediation, which destroys or detoxifies contaminants, in comparison with deposition in a toxic waste facility, where contaminants may remain intact and toxic indefinitely.

Several specific methods of bioremediation have been developed for treatment of contaminated soil and water. Those methods that currently call for the use of soil conditioners or that may benefit from the use of conditioners include landfarming, *in situ* treatment, remove-and-treat, and phytoremediation.

A. Landfarming

For many years the petroleum industry has used landfarming as a method of waste oil disposal (Loehr et al., 1992; Zimmerman and Robert, 1991). The degradation of oily wastes is enhanced by supplementing the soil with the nutrients and oxygen, then tilling and irrigating to create an optimal environment for microbial activity and to increase the chance of contact between contaminants and microorganisms. This method of bioremediation requires the presence (natural or constructed) of a thick layer of clay between the landfarmed surface and the deep water table to prevent ground water contamination (Bollag and Bollag, 1995) The sites are normally surrounded with berms to prevent runoff that could contaminate surface water.

Biodegradation rates of crude oils vary with the composition and class of organic compounds found in the oil. Landfarming techniques have resulted in the removal of approximately 70% of the aromatic and 90% of the aliphatic compounds from oil-contaminated wastes within a 35 d period (Litchfield, 1991). The use of soil conditioners that enhance soil physical properties could greatly enhance bioremediation of landfarmed wastes.

B. Remove-and-Treat

Remove-and-treat techniques may involve the excavation of limited quantities of contaminated soil from a surface spill or from the soil volume surrounding a leaking storage tank. These soils are generally removed to a treatment area that is lined and bermed to prevent further spread of contamination.

Extraction-treatment or pump-and-treat techniques are often used to remove contaminants from ground water. These techniques are designed to shrink or stop the growth of a contaminant plume and include the use of bioreactors that provide nutrient delivery, systems for providing oxygen and mixing, and pumps and delivery systems. These methods produce gaseous products such as carbon dioxide, methane, or nitrogen (Bollag et al., 1994; Committee on *In Situ* Bioremediation, 1993). Litchfield (1991) reported that a number of organic solvents, chlorinated hydrocarbons, pesticides, and aromatic and aliphatic compounds have been successfully removed from ground water by these techniques. The treated soil may be amended with bulking agents, nutrients, and moisture to promote microbial degradation of the spilled contaminants. The addition of conditioners to these soils may promote the movement of oxygen to microbes, enhance water infiltration and retention, and supply microbes with a readily available carbon source promoting cometabolism. The use of cover crops and other conditioners may enhance remediation at these sites.

C. *In Situ* Bioremediation

In situ bioremediation involves the use of indigenous microbes to remove or transform chemical pollutants at the site of contamination (Bollag and Bollag, 1995). Crude oil spills at exploration and production sites are typically less than 0.5 ha in size and are often in very remote locations. Bioremediation is often used as a cost-effective cleanup technology at these sites (McMillen et al., 1995). *In situ* bioremediation is usually enhanced by the addition of nutrients, available carbon substrates, or electron acceptors which speed the growth of microorganisms. In some cases an inoculum of appropriate pure or mixed cultures of degrading microorganisms is added to the site to enhance the removal of the contaminants (Glass, 1995). Numerous examples of successful *in situ* bioremediation of surface soils, beaches, and aquifers have been reported in both research and review articles (Bollag and Bollag, 1995; Madsen, 1991; Pollard et al., 1994; Skladany and Metting, 1993).

IV. CONDITIONERS IN BIOREMEDIATION

In Chapter 1 of this handbook, soil conditioners were described as materials that improve soil physical properties, enhance soil quality, and improve the conditions for the growth of plants and microbes. There is a paucity of information on the use of soil conditioners in bioremediation work. In some cases, cited hereafter, bulking agents have been used in bioremediation research, but either these agents are poorly described or studies have not been performed to demonstrate their efficacy. In the case of phytoremediation, however, there is a growing body of literature on the use and efficacy of plants and root systems in promoting bioremediation.

Bioremediation projects seek to speed and enhance the degradation and detoxification of soil and water contaminants. There are a number of conditioners described in other chapters of this book that may improve the soil environment for bioremediation. Research is needed to demonstrate the effectiveness of conditioners in soil decontamination. Some conditioners may work well to enhance and speed bioremediation, while others may prove to be ineffective.

Conditioner materials that improve soil structure, porosity, aeration, water movement, water holding capacity, pH, and nutrient content need to be considered for use in bioremediation. Several chapters in this book have described the uses of organic waste materials as soil conditioners (see Chapters 2–6). The addition of organic matter improves soil physical properties. Conditioners may be used to ameliorate soil problems at contaminated sites. For instance, sandy soils may require conditioners that improve water retention. Clay soils may require conditioners that improve soil structure and increase the rate of water infiltration and percolation.

Soil erosion prevention is an important consideration at contaminated sites. This concern must perhaps be addressed while the bioremediation project is in the planning stages to prevent the further spread of contaminants. Conditioners that limit the detachment and transport of contaminated soil particles will help to localize the contamination and prevent the further contamination of surface soils and water.

A number of soil conditioners improve the soil fertility and supply soil microbes with both organic and inorganic nutrients. The use of plant cover and the surface application of plant residues and animal wastes to stabilize the soil against erosion are important considerations.

A. Animal Manures

The application of animal manures to agricultural soils has been practiced for thousands of years and has provided benefit both to soil physical properties and to soil fertility. In Chapter 4 of this book Pagliai and Vignozzi have described the effects of pig manure slurry and farmyard manure applications on various soil physical properties. The application of these wastes significantly increased both the microporosity and the macroporosity of a silty clay soil amended with 300 Mg ha^{-1} of liquid manure. Micropores of equivalent pore diameter ranging from 0.5 to 50 μm are the pores that store water for use by plants and microorganisms. Macropores of equivalent pore diameter ranging from 50 to 500 μm are responsible for the transmission of water through the soil.

Khaleel et al. (1981) reviewed the literature on effects of organic waste application on soil physical properties. Most of the waste materials were described as animal manures, biosolids, or composts. In general, the application of wastes, either for plant nutrient supply or for disposal, increased the C content of the soil. The benefits of these organic carbon additions were reported to be in the form of increased aggregation, decreased bulk density, increased water holding capacity, and hydraulic conductivity of soils.

Animal manures, which are high in dissolved organic carbon, significantly increase the size of soil microbial biomass. Manures supply readily mineralizable C and N that directly stimulate microbial activity and growth (Collins et al., 1992). The potential for increasing microbial growth, activity, and biomass by the addition of animal manures should not be overlooked in planning a bioremediation project.

Aprill and Sims (1990) used screened cow, manure to facilitate the incorporation of PAHs (polycyclic aromatic hydrocarbons) into soil samples for phytoremediation studies. The manure was added to all samples, so the effects of this soil conditioner on remediation was not determined. McGinnis et al. (1991) reported that the addition of high levels (2 volumes of manure to 1 volume of soil) of chicken manure to soil contaminated with wood preservatives resulted in 100 and 96% loss of PAHs and PCP (penlachlorophenol), respectively. The loss of PAHs and PCP in the unamended soil was 75 and 46%, respectively. The manure-amended and unamended contaminated soils were incubated outdoors in large boxes for 9 months.

B. Biosolids

Biosolids are the residual solids produced in municipal waste water treatment. These solids may be in the form of activated biosolids, removed from the waste stream following activated digestion, or anaerobically digested biosolids formed as the activated biosolids are subjected to anaerobic digestion for a period of approximately 40 days. Many municipalities compost biosolids with wood chips and green waste gathered from city parks and from the yards of homeowners. The number of coliform bacteria and disease-causing organisms is reduced upon composting, so composted biosolids are potentially safer for those applying the conditioner or working at the bioremediation site. The benefits of biosolids as soil conditioners include the organic matter, which is somewhat resistant to rapid degradation in the soil, improved soil structure and water holding capacity, and the slow release of nitrogen, phosphorus, and other nutrients for microorganisms and plants. In Chapter 5 of this book Olness et al. reported on the beneficial effect of biosolids application on the physical, chemical, and microbiological properties of soils.

Kladivko and Nelson (1979) reported that the application of anaerobically digested biosolids at the rate of 56 Mg ha^{-1} improved the physical properties of two silt loam soils and one sandy loam soil. There were significant increases in the size of water-stable aggregates, large pore space, water holding capacity, and organic carbon and cation exchange capacity as a result of biosolids application. The soil bulk densities were significantly decreased. Khaleel et al. (1981) reviewed several studies of the effects of biosolids and manure applications on soil physical properties. The beneficial effects of organic waste application to soil included decreased bulk density and increased water holding capacity, porosity, and hydraulic conductivity.

Biosolids may be added in dry or wet form, either to the soil surface or injected into the subsurface. EPA regulations under the Clean Water Act, Section 503, establish the levels of nitrogen, phosphorus, and heavy metals that may be added to U.S. soils. Thus it is important that each batch of biosolids be analyzed for these components.

There are few examples of the use of biosolids to improve soil physical properties at bioremediation sites. Knaebel and Vestal (1992) used soil from a cropped field amended with 4600 kg of biosolids ha^{-1} in each of five years in a growth-chamber study of the mineralization of surfactants. The determination of beneficial effects of biosolids on soil physical properties was not among the objectives of that study. Dibble and Bartha (1979) conducted a laboratory study on the effects of various soil environmental parameters of landfarming. One of those parameters was the addition of organic supplements to soil treated with oil sludge. They found that the addition of biosolids to soil at the rate of 1% (wt/wt) increased the CO_2 evolution of soil without oil sludge, but biosolids depressed both CO_2 evolution and hydrocarbon biodegradation in samples treated with oil sludge. They proposed that the presence of biosolids may have selected a microbial population unfavorable for hydrocarbon degradation. The effects of biosolids on soil physical properties were not determined in the laboratory study.

Scientists in South Africa conducted a pilot study of the effects of microbial supplements and anaerobically digested biosolids on landfarming of petroleum hydrocarbons (Pearce et al., 1995). They did not show any data but concluded that the addition of biosolids enhanced the moisture-retaining capacity of the landfarmed soil.

Municipal activated biosolids were used in a two step process to remove phenanthrene from the soil fines left over from a soil washing process (Van Kemenade et al., 1995). The fines were pretreated with chemical oxidants and then were subjected to a 5 day biological oxidation step employing the activated biosolids. The process was found to speed remediation of the contaminated soil.

C. Composts

Composts are prepared by combining organic waste materials (e.g., straw, animal manures, waste water biosolids, wood chips, landscape trimmings, grass clippings, paper, etc.) with water and inorganic nutrients in piles or windrows of sufficient size to retain much of the waste heat of microbial decomposition. Under these conditions the growth of thermophilic microorganisms is encouraged and the temperature of the compost rises to approximately 50°C. The elevated temperature of the compost destroys weed seeds and harmful pathogenic bacteria and promotes the rapid degradation of readily decomposable carbohydrates, fats, and proteins in the waste materials.

Laine and Jorgensen (1996) reported on the use of straw compost as an inoculum for the bioremediation of chlorophenol-contaminated soil. The compost was prepared at a mushroom farm and was made of wheat (*Triticum aestivum*) or rye (*Secale cereale*) straw with horse or chicken litter. The compost was then adapted to pentachlorophenol degradation in a circulating percolator for a period of three months. The PCP-adapted straw compost mineralized 56% of the added ^{14}C-labeled PCP within four weeks. They concluded that the compost could be used as an effective soil inoculant, but they did not add the compost to soil in this study.

Composted yard waste and plants were used at an agrochemical retail facility in Illinois to aid in the remediation of pesticide-contaminated soil (Cole et al., 1994). The compost was used to improve the soil physical properties and to increase the microbial population. Plant growth was significantly greater in contaminated soil amended with compost. Bacterial and fungal populations in compost–soil mixes were several-fold higher than populations in soil without compost.

Hupe et al. (1996) tested the use of compost to enhance the bioremediation of oil-contaminated soils treated in both static and stirred bioreactors. A soil-to-compost ratio of 2 to 1 provided greatest enhancement of bioremediation of the contaminated soil, but a more practical ratio of 8 to 1 also provided some enhancement. The addition of compost to contaminated soil in a paddle stirrer bioreactor provided a benefit for the physical structure of the soil, thus preventing the buildup of large soil pellets. Without compost in the stirrer bioreactor, soil pellets formed, preventing proper operation of the bioreactor. The use of compost in the bioreactor allowed proper operation of the stirring mechanism and likely facilitated better aeration of the contaminated soil. They also reported that the age of the compost did not influence the turnover of oil contaminants in the soil.

D. Other Conditioners and Bulking Agents

Scientists in Finland (Lehtomaki and Niemela, 1975) tested the use of both peat and brewery waste yeast to accelerate oil decomposition in the soil. These materials were added to soil in a greenhouse pot experiment at the rate of 7.5 g kg^{-1}. The addition of peat to the soil did not significantly affect the rate of decomposition of either light fuel oil or refinery waste oil. The addition of waste yeast accelerated oil degradation two-to tenfold, however. The rate of decrease in oil concentration in soil was not affected by the age of the yeast waste (fresh vs. stored 6 months) or death of the cells (70°C heat treatment). It was concluded that the yeast cells were not responsible for the accelerated degradation, rather, the increase in organic and inorganic nutrients introduced to the soil promoted the growth of hydrocarbon-assimilating bacteria. In a subsequent study by Dibble and Bartha (1979) the addition of yeast extract to soils amended with oil sludge failed to stimulate biodegradation.

Elektorowicz (1994) reported on the use of sawdust and nutrients as a treatment of a petroleum-contaminated clay soil. The soil was mixed to 8% with sawdust, and nutrients were added to keep the ratio of added carbon to added nitrogen and phosphorus at 100:10:1, respectively. The added sawdust bulked the soil, lowering the density and improving the movement of water and air within the soil. She reported a loss of 89% of oil and grease in a period of 122 days of incubation. No oil or grease was recorded lost from soil without the sawdust conditioner. It is apparent that bulking the heavy-textured soil greatly improved soil physical conditions for microbial growth. It is likely that degradative enzymes of bacteria and fungi involved in the degradation of sawdust lignin, which contain polyaromatic hydrocarbons (PAHs), were also involved in decomposition of the petroleum wastes.

Sawdust and bermuda grass hay were used as bulking agents in a bioremediation study of field soil contaminated with petroleum hydrocarbons (Rhykerd et al., 1994). These bulking agents enhanced bioremediation compared to nonbulked control soils. The contaminated soils were approximately 10% total petroleum hydrocarbons (TPH) at the beginning of the experiment. After 12 weeks the TPH of the tilled treatment bulked with chopped bermuda grass hay was reduced to 1.4%. The TPH of soil bulked with sawdust was 3.9%, while the nonbulked soil contained 6.1% TPH. The bulking agents were found to enhance aeration and the reduction of TPH concentrations by bioremediation.

E. Plants as Soil Conditioners

1. Phytoremediation

The soil–plant–microbe system can be used to detoxify organic or inorganic pollutants in surface soils. The term phytoremediation has recently been coined to encompass the use of plants in various remediation projects. In a recent review article Cunningham et al. (1996) defined phytoremediation as the use of green plants and their associated microbiota, soil amendments, and agronomic techniques to remove, contain, or render harmless environmental contaminants. Plants can be used to decontaminate the soil matrix by enzymatic degradation, extraction and accumulation of contaminants in the plant tissue, or by volatilization from the leaves. Plants can accumulate various pollutants in vegetative parts, making it possible to harvest the plants to detoxify contaminated soils. The plant–soil system may be used to sequester, solidify, or precipitate contaminants within the soil matrix, thus reducing bioavailability, mobility, and toxicity.

Plants may indirectly aid in the removal of pollutants by increasing the biological activity in the soil through 1) rhizosphere interactions, 2) contribution of dead plant material as readily available C, and 3) improvement of the soil environment for many microorganisms.

Phytoremediation has been successfully used to remove heavy metals from soils (Salt et al., 1995) Certain plants are known to hyperaccumulate metals in their tissues. The plants can be harvested and disposed of in an acceptable manner or the metals can be extracted from the tissues for reuse.

Aprill and Sims (1990) conducted a growth-chamber study on the disappearance of PAHs from nonvegetated soils and soils planted to eight prairie grass species. The disappearance of PAHs was significantly greater in soils containing the fibrous roots of the grass species. They concluded that the roots effectively stimulated the rhizosphere microflora to degrade the contaminants and they hypothesized that a greater portion of the PAHs were incorporated into soil humus in the presence of active root systems.

Schwab and Banks (1994) studied the degradation of PAHs in the rhizospheres of three grass and one legume species grown in previously landfarmed and in previously uncontaminated soils that were then amended with the PAHs, pyrene and anthracene. PAH compounds were reported to dissipate at a faster rate in the planted soils than in unvegetated soil. Growth of the plants in uncontaminated soil without added PAHs, uncontaminated soil amended with PAHs, and previously contaminated soil amended with PAHs was not significantly different, but the trend was toward lower root and shoot biomass in the amended soils.

Bossert and Bartha (1985) conducted a lysimeter study of the germination and growth of corn (*Zea mays*) and soybeans (*Glycine max*) on soils treated for more than 2 years with oily sludge, a petrochemical waste. The treated soil severely inhibited the emergence and growth of corn plants and completely prevented the emergence of soybean seedlings. The use of cover crops as soil conditioners in landfarming may only be possible in moderately contaminated soils or in soils where landfarming has ceased and concentrations of oily wastes have decreased below the level of phytotoxicity. In the case of fuel spills, it may be possible to revegetate the contaminated soil after six weeks (Wang and

Bartha, 1990). Warm summer temperatures and effective bioremediation treatments were found necessary to ameliorate the phytotoxicity during that period.

2. Rhizosphere Bioremediation

The rhizosphere is a zone of enhanced microbial activity in the soil volume immediately surrounding the root surface. This zone is usually within 0 to 3 mm of the root surface and varies from bulk soil, located some distance from the root, in a number of chemical and biological properties. The presence of plant roots changes the conditions of soil pH and availability of phosphate, calcium, and other nutrients. Respiration by roots and microorganisms depresses O_2 levels and increases concentrations of CO_2, changing the redox potential of the soil to more reducing conditions. Transpiration causes changes in the soil moisture levels. The abundance of microorganisms in the rhizosphere is one to two orders of magnitude greater than in bulk soil. Generally microbial populations and biomass in the rhizosphere increase in response to the presence of available organic C in the form of root exudates, sloughed root cells, mucilages, and enzymes (Curl and Truelove, 1986).

The interactions between plant roots and microbial communities in the rhizosphere is complex and has been shown to benefit the growth of both plants and microbes. The plants sustain large microbial populations in the rhizosphere by secreting such substances as carbohydrates and amino acids through root cells and by sloughing root epidermal cells (Anderson et al., 1993). Plants receive benefit from microbial activity in the rhizosphere as nutrient availability increases. An increased population of nonsymbiotic nitrogen fixing bacteria may contribute a portion of the nitrogen needs of the plant (Wullstein, 1980; Curl and Truelove, 1986). Iron and phosphorus may be solubilized by microbes and by the effects of changes in pH and redox potential within the rhizosphere.

A fundamental premise behind the use of plants in bioremediation is that the root environment will stimulate microbial activity and result in increased degradation. In contaminated soils, the presence of the rhizosphere will increase the number of heterotrophic bacteria and fungi that may degrade contaminants. The ability of plants to release photosynthate to soil through exudation and sloughing of dead cells increases the quantity of available organic matter that may be required for cometabolism of organic contaminants. The organic substances in the rhizosphere soil may alter toxicant sorption, bioavailability, and leachability. Rhizosphere microorganisms may also facilitate copolymerization of toxicants with humic acids (Walton et al., 1994a).

Reports by Aprill and Sims (1990), Schwab and Banks (1994), and Bossert and Bartha (1985), cited in the preceding section, provide examples of the benefits of incorporating plants and the rhizosphere environments into bioremediation projects. The reader is referred to a number of articles on the benefits of plants and rhizosphere microbes presented at a recent symposium entitled

Bioremediation Through Rhizosphere Technology (Anderson and Coats, 1994). Reported studies on the role of plants and rhizosphere microorganisms in bioremediation lead to the conclusion that vegetation, in conjunction with associated microbial communities in the root zone, may prove to be a low-cost effective strategy for remediation of chemically contaminated soils.

Barkovskii et al. (1994) studied the growth and nitrogenase activity of *Azospirillum lipoferum* isolated from the rhizosphere of rice (*Oriza sativa*). They report that one strain of this bacterium has the unique ability to use polyphenolics, especially caffeic acid, as alternative terminal respiratory electron acceptors under low oxygen. This type of respiratory transformation of phenolics could contribute to the degradation of xenobiotic and petroleum compounds in the environment.

Walton et al. (1994b) studied the fate of ^{14}C-PAHs in the rhizosphere of white sweet clover (*Melilotus alba*). They found that rhizosphere microbial communities not only enhanced the decomposition of PAHs to CO_2 but also caused the incorporation of ^{14}C-PAHs into the combined humic and fulvic acid fraction of soil. Such an association could reduce the bioavailability and potential phytotoxicity of the PAHs in the soil. They proposed that external protection of the plant from toxic materials by the rhizosphere microbial community may have evolved to the mutual benefit of both microorganisms and plants.

Davis et al. (1994) allowed alfalfa (*Medicago sativa*) plants and soil microbes to adapt to growth in the presence of ground water saturated with both phenol and toluene. The plants were grown in 35 cm deep tanks packed with Kansas river sand and silt. The depth to water table was 25 cm. They reported that only a small amount of toluene evaporated from the soil or exited the ground water and no phenol was detected in either the gas phase or in the exiting ground water. The increased respiratory activity of the soil–plant system with added toluene and phenol in comparison with respiratory activity without the organics suggested that toluene and phenol were being degraded.

3. Mycorrhizae

Most plants will form a symbiotic mycorrhizal relationship with certain soil fungi. This symbiosis between plant and fungus provides benefit to both. The plant provides the bulk of the organic carbon needed by the fungus while the mycorrhizal fungus effectively increases the surface area of the root system, aiding the plant in the uptake of phosphorus, zinc, and other nutrients. While the host plant provides the mycorrhizal fungi with carbohydrates, these fungi retain the ability to utilize other forms of organic residues found in the soil and in the leaf litter layer. Many mycorrhizal fungi are capable of decomposing lignin. Donnelly et al. (1993) and Donnelly and Fletcher (1994) have studied the degradation of chlorinated aromatic compounds, such as 2, 4-dichlorophenoxyacetic acid (2, 4-*D*), atrazine, and PCBs by mycorrhizal fungi. The ability of these fungi to decompose xenobiotics is similar to the metabolic capabilities reported for the white-rot fungus (Bumpus et al., 1985). These studies were conducted *in vitro*, without the host plant, so it is not yet known how well they would perform in the remediation of contaminated soils. The use of plants and mycorrhizal fungi may prove beneficial in bioremediation.

F. Bioaugmentation

One of the critical factors in a successful bioremediation project is the presence of a group of soil microorganisms adapted to decomposition of the targeted contaminants. In most cases indigenous microbes already have the metabolic capabilities or may through adaptation, upon exposure to the contaminants, develop the capabilities to degrade the contaminants. However, sites may be contaminated with recalcitrant xenobiotics, or soil environments may have become hostile to indigenous microbes. These sites may be candidates for inoculation with specially adapted nonnative bacteria. Adapted microbes are sometimes isolated from heavily contaminated sites. In the laboratory they may be produced through classic enrichment techniques (Focht, 1994) or they may be developed with gene transfer technology.

Inoculation of the soil with nonnative microorganisms possessing special metabolic capabilities has been practiced for over a century. A very common and successful example of inoculation is the use of Rhizobia (*Rhizobium* and *Bradyrhizobium*) as seed inoculants for legumes. Inoculation of the legume with an effective strain of the correct *Rhizobium* species enhances symbiotic nitrogen fixation. This is especially important in fields never before cropped to a particular legume. Other examples of environments amended with specialized strains of microorganisms include the use of *Bacillus thuringiensis*, which produces a unique endotoxin protein that kills the larvae of certain insects. Waste water treatment plants are sometimes inoculated with microorganisms selected to optimize the treatment process.

The most common method of bioremediation of contaminated soils relies on the stimulation of indigenous microbes with addition of appropriate organic and inorganic nutrients, electron acceptors, and other factors. In recent years nonnative cultures of microorganisms have been selected from contaminated sites (Glass, 1995) or have been specially adapted to degrade contaminants through laboratory enrichment culture techniques (Focht, 1994; Krockel and Focht, 1987). Glass (1995) has reviewed the literature on the bioaugmentation of contaminated soils with microbial inoculants. He has listed a number of microbial products sold commercially for use in bioremediation of contaminated soils.

There is increased interest in the use of white rot fungus *P. chrysosporium* as a soil inoculum (Bumpus et al., 1985). These fungi are able to degrade lignin, a recalcitrant natural polymer with many aromatic ring structures. These organisms have been shown capable of degrading a wide variety of persistent pollutants in the laboratory, but they have difficulty in the soil with predation and competition with indigenous microbes (Lamar and Dietrich, 1990).

Brodkorb and Legge (1992) studied the decomposition of ^{14}C-labeled phenanthrene in soils with native microflora and in soil bioaugmented with *P. chrysosporium.* Under aerobic laboratory conditions 20% of the labeled phenanthrene was mineralized to CO_2 by the native microbes in 21 days. The addition of *P. chrysosporium* increased mineralization to 38% in the same period. Bioaugmentation of soils with microorganisms specially isolated, adapted, or genetically engineered for the rapid decomposition of contaminants may shorten the time required for effective bioremediation. Bioaugmentation may be necessary in highly contaminated or toxic soils, but in most cases indigenous soil microbes will adapt and mineralize or detoxify the contaminants of concern. In some cases economic considerations may demand that bioaugmentation be used to speed bioremediation.

Bioremediation in cold environments may require the presence of psychrotrophic bacteria capable of growth at low soil temperatures. Isolates from a variety of ecosystems in Canada were examined for their ability to mineralize a number of types of organic molecules (alkanes, aromatics, PAHs, and PCP) found at bioremediation sites (Whyte et al., 1996). All of the 135 isolates were capable of mineralization activity at both 23° and 5°C.

Successful bioaugmentation in the field will require the preparation of microbial supplements in a form that will preserve sufficient numbers of viable organisms and allow distribution to the contaminated soil. Romich et al. (1995) attempted to prepare and preserve cultures of *Alcaligenes eutrophus* H850 by freeze-drying, spray-drying, and freezing. They determined that freezing the cultures produced substantially higher microbial survivals than the two drying techniques. Freezing also preserved the ability of the bioaugmentation supplement to remediate biphenyl contamination.

Microencapsulation of a nonindigenous *Flavobacterium* in agarose was assessed as a microbial supplement for use in a sedimentary aquifer (Petrich et al., 1995). For successful bioaugmentation of aquifer materials, microbeads that encapsulate the cells must be able to move throughout the contaminated portion of the aquifer. Agarose microbeads less than 10 microns in size encapsulating the cells were prepared and injected into the aquifer. In comparison with the movement of bromine and polystyrene latex microspheres (2 to 5 microns), the movement of the agarose microbeads was retarded. Problems with filtration and adsorption of microbeads or of microbial cells must be overcome for successful bioaugmentation of aquifers.

Pearce et al. (1995) reported that the addition of a commercial biosupplement and one prepared with indigenous soil microbes produced the same rate of reduction (94%) in the initial TPH in an outdoor pilot study of landfarming techniques. The preparation of indigenous soil microbes consisted of soil inoculated into a mineral salts broth containing 30 mL petroleum oil L^{-1}.

Microorganisms introduced to the field must compete for resources with native species and therefore appear best suited for extreme conditions where competitors and predators are virtually nonexistent (Thomas and Ward, 1989). Bioaugmentation of field sites has been conducted with mixed success. There is a growing list of reports that inoculation of soils and other environments has enhanced biodegradation (e.g., Chatterjee et al., 1982; Comeau et al., 1993; Edgehill and Finn, 1983; Kilbane et al., 1983; Mayotte et al., 1996). The list of reports of the failure of bioaugmentation to enhance bioremediation is also growing (Lewandowski et al., 1986; Lehtomaki and Niemela, 1975; Pearce et al., 1995). Failures have also been reported where the population density of the inoculant was insufficient to survive the lethal effect of the contaminant (Mileski et al., 1988) or where the concentration of contaminant in nonsterilized soil was insufficient to sustain the acclimated bacteria (Daughton and Hsieh, 1977).

Microorganisms with specialized metabolic capabilities may be isolated from contaminated sites, or microbes may be adapted and selected in the laboratory, but the ability of a microbial strain to metabolize certain contaminants does not assure that the microorganism will survive in the soil environment or that the rate of degradation will be enhanced. The ability to metabolize soil contaminants is important for successful bioaugmentation, but the inoculated microorganisms must also be able to survive and thrive in a contaminated and sometimes toxic soil environment. Goldstein et al. (1985) list a number of problems and pitfalls that limit the success or cause the failure of bioaugmentation: 1) the concentration of contaminant may be insufficient to support growth; 2) the microbial inoculant may not be able to survive the presence of toxins or predators in the soil environment; 3) the microbes may prefer metabolism of other organic substances rather than the targeted contaminant; or 4) the microbes may be unable to move from site to site in the soil.

V. SUMMARY

Bioremediation is a cost-effective method of using plants and microorganisms to detoxify, degrade, or remove pollutants from contaminated environments. Bioremediation techniques are used to optimize the physical and chemical conditions of the soil or other contaminated environments to stimulate the activity of indigenous or augmented microorganisms capable of degrading contaminants. Those involved in planning and conducting bioremediation work on contaminated soils should consider the use of soil conditioners to improve aggregation, moisture conditions, and soil aeration for degradative microbes. In addition to enhanced soil physical properties, some conditioners also provide inorganic and organic nutrients that increase the activity and biomass of microorganisms involved in decontaminating the soil.

There are a number of reports of enhanced bioremediation in planted soils. The presence of green plants and their associated rhizosphere microorganisms has been shown to speed the decomposition of both xenobiotic and petroleum-based contaminants. The presence of roots and associated microbes may also increase the incorporation of contaminants into soil humus.

Bioremediation sites may become too toxic for indigenous microbes or may be contaminated with recalcitrant xenobiotics. Bioaugmentation of these sites with specially isolated or adapted nonnative microorganisms should be considered. Introduced microorganisms must be added in sufficient quantities or otherwise acclimated to the soil environment, so that they can survive extreme conditions, indigenous competitors, and predators. The list of successful uses of bioaugmentation in bioremediation work continues to grow, but so does the list of bioaugmentation failures. Knowledge of the problems and pitfalls associated with bioaugmentation is important in the design of a successful bioremediation project.

Organic waste materials have been used on soils to increase organic matter and to improve soil structure, water infiltration, and water retention. Animal manures, composts, and biosolids have been shown to improve the physical properties of soils in general. There have been few reports of the use of these materials to improve and enhance bioremediation.

The use of many soil conditioners to improve soil properties and quality have been described in this handbook. Many of the conditioners, including the synthetic polymers, have never been tested for use in bioremediation. There is a need for more research in this area to assess the ability of soil conditioners to enhance bioremediation.

REFERENCES

Anderson, T. A., and Coats, J. R. (1994). *Bioremediation Through Rhizosphere Technology*. American Chemical Society, Washington, DC.

Anderson, T. A., Guthrie, E. A., and Walton, B. T. (1993). Bioremediation. *Environ. Sci. Technol. 27*:2630–2636.

Aprill, W., and Sims, R. C. (1990). Evaluation of the use of prairie grasses for stimulating polycyclic aromatic hydrocarbon treatment in soil. *Chemosphere 20*:253–265.

Barkovskii, A. L., Boullant, M.-L., and Balandreau, J. (1994). Polyphenolic compounds respired by bacteria. *Bioremediation Through Rhizosphere Technology* (T. A. Anderson, and J. R. Coats, eds.). American Chemical Society, Washington, DC, pp. 28–42.

Bollag, J. M., and Bollag, W. B. (1995). Soil contamination and the feasibility of biological remediation. *Bioremediation: Science and Applications* (H. D. Skipper, and R. F. Turco, eds.). Soil Science Society of America, Special Publication No. 43, Madison, WI, p. 1.

Bollag, J. M., Mertz, T., and Otjen, L. (1994). Role of microorganisms in soil bioremediation. *Bioremediation Through Rhizosphere Technology* (T. A. Anderson, and J. R. Coats, eds.), American, Chemical Society, Washington, DC, pp. 2–10.

Bossert, I., and Bartha, R. (1985). Plant growth in soils with a history of oily sludge disposal. *Soil Sci. 140*:75–77.

Brodkorb, T. S., and Legge, R. L., (1992). Enhanced biodegradation of phenanthrene in oil tar–contaminated soils supplemented with *Phanerochaete chrysosporium. Appl. Environ. Microbiol. 58*:3117–3121.

Bumpus, J. A., Tien, M., Wright, D., and Aust, S. D. (1985). Oxidation of persistent environmental pollutants by a white rot fungus. *Science 228*:1434–1436.

Chatterjee, D. K., Kilbane, J. J., and Chakrabarty, A. M. (1982). Biodegradation of 2,3,4-trichlorophenoxyacetic acid in soil by a pure culture of *Pseudomonas cepacia. Appl. Environ. Microbiol. 44*:514–516.

Cole, M. A., Liu, X., Zhang, L. (1994). Plant and microbial establishment in pesticide-contaminated soils amended with compost. *Bioremediation Through Rhizosphere Technology* (T. A. Anderson, and J. R. Coats, eds.), American Chemical Society, Washington, DC, pp. 210–222.

Collins, H. P., Rasmussen, P. E., and Douglas, C. L., Jr. (1992). Crop rotation and residue management effects on soil carbon and microbial dynamics. *Soil Sci. Soc. Am. J. 56*:783–788.

Comeau, Y., Greer, C. W., and Samson, R. (1993). Role of inoculum preparation and density on the bioremediation of 2,4-*D*-contaminated soil by bioaugmentation. *Appl. Microbiol. Biotech. 38*:681–687.

Committee on *In Situ* Bioremediation. (1993). *In situ Bioremediation: When Does It Work*? National Academy Press, Washington, DC.

Cunningham, S. D., Anderson, T. A., Schwab, A. P., and Hsu, F. C. (1996). Phytoremediation of soils contaminated with organic pollutants. *Advances in Agronomy 16*:55–114.

Curl, E. A., and Truelove, B. (1986). *The Rhizosphere*. Springer Verlag, Berlin.

Daughton, C. G., and Hsieh, P. H. (1977). Accelerated parathion degradation in soil by inoculation with parathion-utilizing bacteria. *Bull. Environ. Contam. Toxicol. 18*:48–56.

Davis L. C., Muralidharan, N., Visser, V. P., Chaffin, C., Fateley, W. G., Erickson, L. E., and Hammaker, R. M. (1994). Alfalfa plants and associated microorganisms promote biodegradation rather than volatilization of organic substances from ground water. *Bioremediation Through Rhizosphere Technology* (T. A. Anderson, and J. R. Coats, eds.). American Chemical Society, Washington, DC, pp. 112–122.

Dibble, J. T., and Bartha, R. (1979). The effect of environmental parameters on the biodegradation of oil sludge. *Appl. Environ. Microbiol. 37*:729–739.

Donnelly, P. K., and Fletcher, J. S. (1994). Potential use of mycorrhizal fungi as bioremediation agents. *Bioremediation Through Rhizosphere Technology* (T. A. Anderson, and J. R. Coats, eds.). American Chemical Society, Washington, DC, pp. 93–99.

Donnelly, P. K., Entry, J. A., and Crawford, D. L. (1993). Degradation of atrazine and 2,4,-dichlorophenoxyacetic acid by mycorrhizal fungi at three nitrogen concentrations in vitro. *Appl. Environ. Microbiol. 59*:2642–2647.

Edgehill, R. U. and Finn, R. K. (1983). Microbial treatment of soil to remove pentachlorophenol. *Appl. Environ. Microbiol. 45*:1122–1125.

Elektorowicz, M. (1994). Bioremediation of petroleum-contaminated clayey soil with pretreatment. *Environ. Technol. 15*:373–380.

Focht, D. D. (1994). Microbiological procedures for biodegradation research. *Methods of Soil Analysis, Part 2: Microbiological and Biochemical Properties* (R. W. Weaver, ed.). Soil Science Society American, Book Series, No. 5, pp. 407–427.

Foster, R. C. (1988). Microenvironments of soil organisms. *Biology and Fertility of Soils 6*:189–203.

Glass, D. J. (1995). Biotic effects of soil microbial amendments. *Soil Amendments: Impacts on Biotic Systems* (J. R. Rechcigl, ed.). Lewis, Boca Raton, FL, pp. 251–303.

Goldstein, R. M., Mallory, L. M., and Alexander, M. (1985). Reasons for possible failure of inoculation to enhance biodegradation. *Appl. Environ. Microbiol. 50*:997–983.

Haines, J. R., Kadkhodayan, M., Mocsny, D. J., Jone, C. A., Islam, M., and Venosa, A. D. (1994). Effect of salinity, oil type, and incubation temperature on oil degradation. *Applied Biotechnology for Site Remediation* (R. E. Hinchee, D. B. Anderson, F. B. Metting, Jr., and G. D. Sayles, eds.). Lewis, Boca Raton, FL, pp. 75–83.

Hupe, K., Luth, J.-C., Heerenklage, J., and Stegmann, R. (1996). Enhancement of the biological degradation of soils contaminated with oil by the addition of compost. *Acta Biotechnol. 16*:19–30.

Khaleel, R., Reddy, K. R., and Overcash, M. R. (1981). Changes in soil physical properties due to organic waste application: a review. *J. Environ. Qual. 10*:133–141.

Kilbane, J. J., Chatterjee, D. K., and Chakrabarty, A. M. (1983). Detoxification of 2,4,5-trichlorophenoxyacetic acid from contaminated soil by *Pseudomonas cepacia. Appl. Environ. Microbiol. 45*:1697–1700.

Kladivko, E. J., and Nelson, D. W. (1979). Change in soil properties from application of anaerobic sludge. *J. Water Poll. Cont. Fed. 51*:325–332.

Knaebel, D. B., and Vestal, J. R. (1992). Effects of intact rhizosphere microbial communities on the mineralization of surfactants in surface soils. *Can. J. Microbiol. 38*:643–653.

Krockel, L., and Focht, D. D. (1987). Construction of chlorobenzene-utilizing recombinants by progenitive manifestation of a rare event. *Appl. Environ. Microbiol. 53*:2470–2475.

Laine, M. M., and Jorgensen, K. S. (1996). Straw compost and bioremediated soil as inocula for bioremediation of chlorophenol-contaminated soil. *Appl. Environ. Microbiol. 62*:1507–1513.

Lamar, R. T., and Dietrick, D. M. (1990). *In situ* depletion of pentachlorophenol from contaminated soil by *Phanerochaete* spp. *Appl. Environ. Micobiol. 56*:3093–3100.

Lehtomaki, M., and Niemela, S. (1975). Improving microbial degradation of oil in soil. *Ambio 4*:126–129.

Lewandowski, G., Salerno, S., McMullen, N., Gneiding, L., and Adamowitz, D. (1986). Biodegradation of toxic chemicals using commercial preparations. *Environ. Prog. 5*:212–217.

Litchfield, C. D. (1991). Practices, potential, and pitfalls in the application of biotechnology to environmental problems. *Environmental Biotechnology for Waste Treat-*

ment (G. S. Sayler, R. Fox, and J. W. Blackburn, eds.). Plenum Press, New York, pp. 147–157.

Loehr, R. C., Martin, J. H., Jr., and Neuhauser, E. F. (1992). Land treatment of an aged oily sludge—organic loss and change in soil characteristics. *Water Res. 26*:805–815.

Madsen, E. L., and Alexander, M. (1982). Transport of *Rhizobium* and *Pseudomonas* through soil. *Soil Sci. Soc. Am. J. 46*:557–560.

Madsen, E. L. (1991). Determining *in situ* biodegradation. *Environ. Sci. Technol. 25*:1663–1673.

Mayotte, T. J., Dybas, M. J., and Criddle, C. S. (1996). Bench-scale evaluation of bioaugmentation to remediate carbon tetrachloride–contaminated aquifer materials. *Ground Water 34*:358–367.

McGinnis, G. D., Borazjani, H., Hannigan, M., Hendrix, F., McFarland, L., Pope, D., Strobel, D., and Wagner, J. (1991). Bioremediation studies at a northern California Superfund site. *J. Hazard. Materials 28*:145–158.

McMillen, S. J., Requejo, A. G., Young, G. N., Davis, P. S., Cook, P. D., Kerr, J. M., and Gray, N. R. (1995). Bioremediation potential of crude oil spilled on soil. *Microbial Processes for Bioremediation* (R. E. Hinchee, C. M. Vogel, and F. J. Brockman, eds.). Battelle Press, Columbus, OH, pp. 91–100.

Mileski, G.J., Bumpus, J.A., Jurek, M.A., and Aust, S.D. (1988). Biodegradation of pentachlorophenol by the white rot fungus *Phanerochaete chrysosporium*. *Appl. Environ. Microbiol. 54*:2885–2889.

Paul, E. A., and Clark, F. E. (1996). *Soil Microbiology and Biochemistry* 2d ed. Academic Press, San Diego, CA.

Pearce, K., Snyman, H. G., Oellermann, R. A., and Gerber, A. (1995). Bioremediation of petroleum-contaminated soil. *Bioaugmentation for Site Remediation* (R. E. Hichee, J. Fredrickson, and B. C. Alleman, eds.). Battelle Press, Columbus, OH, pp. 71–76.

Petrich, C. R., Stormo, K. E., Knaebel, D. B., Ralston, D. R., and Crawford, R. L. (1995). A preliminary assessment of field transport experiments using encapsulated cells. *Bioaugmentation for Site Remediation*. (R. E. Hinchee, J. Fredrickson, and B. C. Alleman, eds.). Battelle Press, Columbus, OH, pp. 237–244.

Pollard, S. J. T., Hrudey, S. E., and Fedorak, P. M. (1994). Bioremediaton of petroleum- and creosote-contaminated soils: a review of constraints. *Waste Manag. Res. 12*:173–194.

Rhykerd, R. L., Crews, B. Weaver, R. W., and McInnes, K. J. (1994). Bioremediation of crude oil saturated soil: influence of bulking agents, aeration, and tillage. *Agronomy Abstracts*, p. 56.

Romich, M. S., Cameron, D. C., and Etzel, M. R. (1995). Three methods for large-scale preservation of a microbial inoculum for bioremediation. *Bioaugmentation for Site Remediation* (R. E. Hinchee, J. Fredrickson, and B. C. Alleman, eds.). Battelle Press, Columbus, OH, pp. 229–235.

Rotert, K. H., Cronkhite, L. A., and Alvarez, P. J. J. (1995). Enhancement of BTX biodegradation by benzoate. *Microbial Processes for Bioremediation*. (R. E. Hinchee, C. M. Vogel, and F. J. Brockman, eds.). Battelle Press, Columbus, OH, pp. 161–168.

Saez, P. B., and Rittmann, B. E. (1993). Biodegradation kinetics of a mixture containing a primary substrate (phenol) and an inhibitory co-metabolite (4-chlorophenol). *Biodegradation 4*:3–21.

Salt, D. E., Blaylock, M., Kumar, N. P. B. A., Dushenkov, V., Ensley, B. D., Chet, I., and Raskin, I. (1995). Phytoremediation: a novel strategy for removal of toxic metals from the environment using plants. *Biotechnology 13*:468–474.

Schwab, A. P., and Banks, M. K. (1994). Biologically mediated dissipation of polyaromatic hydrocarbons in the root zone. *Bioremediation Through Rhizosphere Technology* (T. A. Anderson, and J. R. Coats, eds.). American Chemical Society, Washington, DC, pp. 132–141.

Skladany, G. J., and Metting, F. B., Jr. (1993). Bioremediation of contaminated soil. *Soil Microbial Ecology: Application in Agricultural and Environmental Management* (F. B. Metting, ed.). Marcel Dekker, New York, pp. 483–513.

Smith, M. S., Thomas, G. W., White, R. E., and Ritonga, D. (1985). Transport of *Escherichia coli* through intact and disturbed soil columns. *J. Environ. Qual. 14*:87–91.

Song, H., Wang, X., and Bartha, R. (1990). Bioremediation potential of terrestrial fuel spills. *Appl. and Environ. Microbiol. 56*:652–656.

Thomas, J. M., and Ward, C. H. (1989). *In situ* biorestoration of organic contaminants in the subsurface. *Environ. Sci. Technol. 23*:760–766.

Turco, R. F., and Sadowsky, M. (1995). The microflora of bioremediation. *Bioremediation: Science and Applications* (H. D. Skipper, and R. F. Turco, eds.). Soil Sci. Soc. Am., Special Publication No. 43, Madison, WI, p. 87.

US EPA (1995). The national biennial RCRA hazardous waste report, 1993 data. WWW.EPA.GOV

Van Kemenade, I., Moo-Young, M., Scharer, J. M., and Anderson, W. A. (1995). Bioremediation enhancement of phenanthrene contaminated soils by chemical preoxidation. *Hazardous Waste Hazardous Materials 12*:345–355.

Walton, B. T., Guthrie, E. A., and Hoylman, A. M. (1994a). Toxicant degradation in the rhizosphere. *Bioremediation Through Rhizosphere Technology* (T. A. Anderson, and J. R. Coats, eds.). American Chemical Society, Washington, DC, pp. 11–26.

Walton, B. T., Holyman, A. M., Perez, M. M., Anderson, T. A., Johnson, T. R., Guthrie, E. A., and Christman, R. F. (1994b). Rhizosphere microbial communities as a plant defense against toxic substances in soils. *Bioremediation Through Rhizosphere Technology* (T. A. Anderson, and J. R. Coats, eds.). American Chemical Society, Washington, DC, pp. 82–92.

Wang, X., and Bartha, R. (1990). Effects of bioremediation on residues, activity and toxicity in soil contaminated by fuel spills. *Soil Biol. Biochem. 22*:501–505.

Whyte, L. G., Greer, W. W., and Inniss, W. E. (1996). Assessment of the biodegradation potential of psychrotrophic microorganisms. *Can. J. Microbiol. 42*:99:106.

Wullstein, L. H. (1980). Nitrogen fixation (acetylene reduction) associated with rhizosheaths of Indian rice-grass used in stabilization of the Slick Rock, Colorado tailings pile. *J. Range Manag. 33*:204–206.

Zimmerman, P. K., and Robert, J. D. (1991). Oil-based drill cuttings treated by landfarming. *Oil and Gas Journal* (April 12), pp. 31–34.

Index

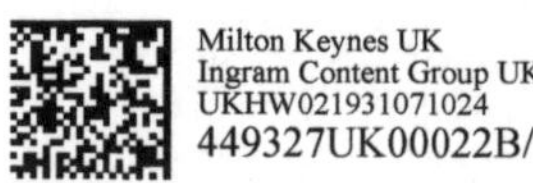

Milton Keynes UK
Ingram Content Group UK Ltd.
UKHW021931071024
449327UK00022B/1770